Oxidative Stress: Biological Aspects

Oxidative Stress: Biological Aspects

Edited by **Nick Gilmour**

R CALLISTO REFERENCE

New York

Published by Callisto Reference,
106 Park Avenue, Suite 200,
New York, NY 10016, USA
www.callistoreference.com

Oxidative Stress: Biological Aspects
Edited by Nick Gilmour

International Standard Book Number: 978-1-63239-505-4 (Hardback)

Printed in the United States of America.

Contents

Preface

This book focuses on the elucidation of the biological aspects of oxidative stress. Researchers have shown great interest in the interaction of free radicals with biological molecules ever since their discovery. As a result of successful ideas and results, oxidative stress is now acknowledged by both general and applied scientists as an advanced steady state level of reactive oxygen species with broad spectrum of biological effects. This book encompasses a broad range of issues and aspects associated with the field of oxidative stress. The relation between elimination and generation of reactive species and impact of oxidative stress have also been discussed, along with analyses of current studies on the signaling role of reactive species in eukaryotic organisms. Contemporary comprehension of homeostasis of reactive species and cellular processes has also been provided in this book for the benefit of the readers, along with valuable information for future references.

This book unites the global concepts and researches in an organized manner for a comprehensive understanding of the subject. It is a ripe text for all researchers, students, scientists or anyone else who is interested in acquiring a better knowledge of this dynamic field.

I extend my sincere thanks to the contributors for such eloquent research chapters. Finally, I thank my family for being a source of support and help.

Editor

Section 1

Introduction

1

Introductory Chapter

Volodymyr I. Lushchak and Halyna M. Semchyshyn
Vassyl Stefanyk Precarpathian National University,
Ukraine

1. Introduction

Under normal conditions in living organisms over 90% of oxygen consumed is used in electron transport chain via four-electron reduction. This is coupled with nutrient oxidation and results in production of energy, carbon dioxide and water. However, less than 5% of oxygen consumed enters partial one-electron reduction via consequent addition of electrons leading to the formation of series of products collectively termed reactive oxygen species (ROS). They comprise both free radical and non-radical species. Figure 1 demonstrates well characterized ways of reduction of molecular oxygen via four- and one-electron ways.

Reactive oxygen species include both free radicals and non-radical molecules. Free radical is any species capable of independent existence that contains one or more unpaired electrons on the outer atomic or molecular orbital. Molecular oxygen possesses at external molecular orbital two unpaired electrons with parallel spins. According to the Pauli exclusion principle, which states that there are no two identical fermions occupying the same quantum state simultaneously, the electrons are located at different molecular shells. Despite O_2 is a biradical, it not easy enters chemical reactions, because it needs the partner reagent possessing at external orbital also two unpaired electrons with parallel spins, what is not common. The addition of one electron to oxygen molecule cancels the Pauli restriction and leads to the formation of more active $O_2^{\bullet-}$. Singlet oxygen belongs to ROS also. It can be formed as a result of change the spin of one of the two electrons at the outer molecular shells of oxygen. The latter cancels the Pauli restriction also, thus singlet oxygen is more reactive than oxygen at its ground state. That is why partially reduced oxygen forms or singlet oxygen have been termed "reactive oxygen species".

In addition to singlet oxygen, H_2O_2, $O_2^{\bullet-}$ and $HO^{\bullet}$, other oxygen-containing reactive species have been described. For example, those can be organic-containing oxyradicals ($RO^{\bullet}$). In combination with nitrogen, oxygen is a component of other reactive species (RS) like nitric oxide ($^{\bullet}NO$), peroxynitrite ($ONOO^-$) and their derivatives, which are collectively named reactive nitrogen species (RNS). Among other RS containing oxygen hypochlorous acid (HOCl), carbonate radical ($CO_2^{\bullet-}$), reactive sulfur-centered radicals ($RSO_2^{\bullet}$) and reactive carbonyl species (α,β-unsaturated aldehydes, dialdehydes, and keto-aldehydes) should be mentioned. All described in this section RS are more active than molecular oxygen.

Reactive oxygen species are extremely unstable and readily enter many reactions. Therefore, it is not correct to tell that "under some conditions ROS are accumulated". They are continuously produced and eliminated due to what it is necessary to say about their steady-state level or concentration, but not about accumulation.

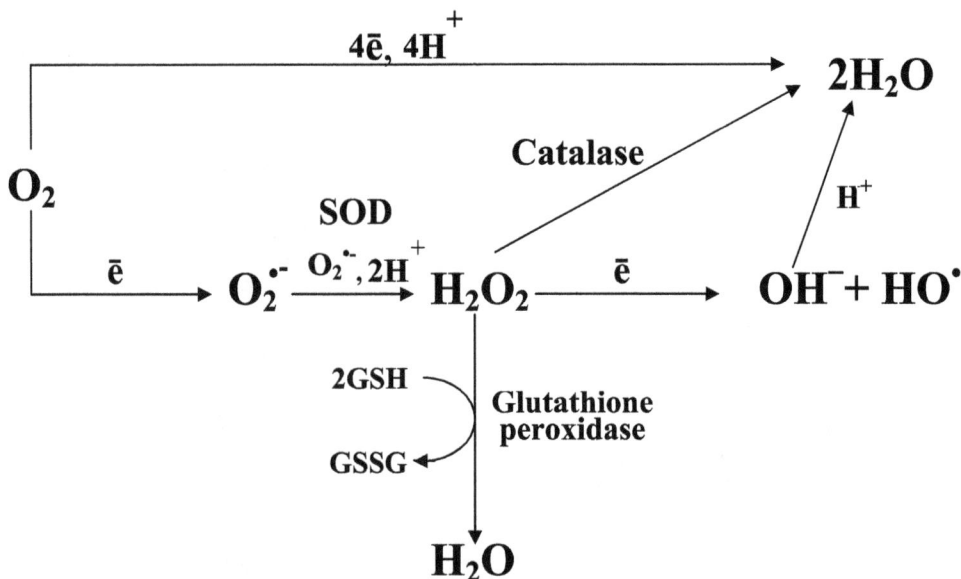

Fig. 1. **Four - and consequent one-electron reduction of molecular oxygen.** The addition of one electron to oxygen molecule results in the formation of superoxide anion radical ($O_2^{\bullet-}$). Being charged $O_2^{\bullet-}$ cannot easily cross biological membranes, but its protonation yields electroneutral $HO_2^{\bullet}$, which readily crosses these barriers. Further addition of one electron to $O_2^{\bullet-}$ leads to the formation of hydrogen peroxide (H_2O_2), which is electroneutral molecule, due to what easily penetrates biological membranes. One-electron reduction of H_2O_2 leads to the formation of hydroxyl radical ($HO^{\bullet}$) and hydroxyl anion (OH^-). The chemical activity of partially reduced oxygen species decreases in the order $HO^{\bullet} > O_2^{\bullet-} > H_2O_2$. It should be noted that two abovementioned partially reduced oxygen species, namely $O_2^{\bullet-}$ and $HO^{\bullet}$, are free radicals, i.e. possess unpaired electron on external molecular orbitals, while H_2O_2 is not a free radical, because all electrons at external molecular orbital are paired. The spontaneous transformation of $O_2^{\bullet-}$, and H_2O_2 is substantially accelerated by certain enzymes, called primary antioxidant enzymes. The conversion of $O_2^{\bullet-}$ to H_2O_2 is catalyzed by superoxide dismutase (SOD), which carries out redox reaction with participation of two molecules of the substrate dismutating them to molecular oxygen and hydrogen peroxide. The next ROS in the chain of one-electron oxygen reduction is H_2O_2 that may be again transformed to less harmful species by several specific enzymes and a big group of unspecific ones. Catalase dismutates H_2O_2 to molecular oxygen and water, while glutathione-dependent peroxidase (GPx) using glutathione as a cofactor reduces it to water. There is no information on specific enzymatic systems dealing with hydroxyl radical. Therefore, it is widely believed that the prevention of $HO^{\bullet}$ production is the best way to avoid its harmful effects.

There are many sources of electrons, which can reduce molecular oxygen, and they will be analyzed within the book. But some of their types should be mentioned here. They are ions of metals with changeable valence, among which iron and copper ions have a great importance in biological systems. Degradation of H_2O_2 resulting in hydroxyl radical formation as well as oxidation of superoxide can occur, for example, in the presence of iron:

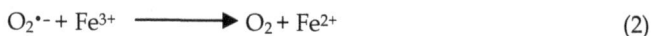

$$H_2O_2 + Fe^{2+} \longrightarrow HO^{\bullet} + OH^- + Fe^{3+} \tag{1}$$

$$O_2^{\bullet-} + Fe^{3+} \longrightarrow O_2 + Fe^{2+} \tag{2}$$

The reaction (1) was firstly described by Fenton and, therefore, called after him as Fenton reaction. The net balance of the reactions (1) and (2) gives Haber-Weiss reaction:

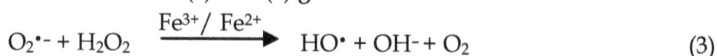

$$O_2^{\bullet-} + H_2O_2 \xrightarrow{Fe^{3+}/ Fe^{2+}} HO^{\bullet} + OH^- + O_2 \tag{3}$$

Reactions 1 and 2 clearly demonstrate that the metal ion (iron in this case) plays a catalytic role and is not consumed during the reactions.

The dismutation of $O_2^{\bullet-}$ to H_2O_2, and H_2O_2 to water and molecular oxygen is substantially facilitated by specific enzymes (Figure. 1). One may note that Figure 1 does not show any enzyme dealing with hydroxyl radical. This is because of its extremely high reactivity, low specificity, and consequently short diffusion distance and life period. Therefore, the best way to avoid injury $HO^{\bullet}$ effects is to prevent its formation. Most cellular mechanisms of antioxidant defense are really designed to avoid $HO^{\bullet}$ production as the most dangerous members of ROS family. However, if produced, it can be neutralized by low molecular mass antioxidants like ascorbic acid, tocopherol, glutathione, uric acid, carotenoids, etc. But certain portion of $HO^{\bullet}$ is not eliminated by the mentioned systems and oxidizes many cellular components.

2. Biological effects of reactive oxygen species

Reactive oxygen species have plural effects in biological systems. These effects may be placed at least in four groups: (*i*) signaling, (*ii*) defense against infections, (*iii*) modification of molecules, and (*iv*) damage to cellular constituents. This division is rather relative and artificial, because in real cell they cannot be separated, i.e. they operate in concert. All these ways are based on ROS capability to interact with certain cellular components. The final effect of the interaction relies on the type of ROS and molecule it interacts with. Generally, at low concentrations ROS are involved in intra- and intercellular communication via specific pathways, while higher concentrations are implicated in more or less specific damage to cellular components. However, one may bear in mind that actually the achieved result depends not on the ROS concentration, but the possibility to interact with certain cellular components. It should be underlined that all biological effects of ROS are based on their interaction with cellular constituents, and the final result depends on the type of cellular component subjected to interaction with specific ROS. Although it is widely believed that the effects of ROS as well as other RS in biological systems are rather unspecific, last years brought understanding that they may have specificity. The latter is provided by the type of RS and target molecules they interact with. Although the issue is under debates, nobody can ignore it now.

Modification of cellular constituents and its evaluation. Above we mentioned that ROS can interact with virtually all cellular components, namely lipids, carbohydrates, proteins, nucleic acids, etc. Damaged molecules of lipids and carbohydrates are further degraded or be important precursors of a variety of adducts and cross-links collectively named advanced glycation and lipoxidation end products (Peng et al., 2011). Similar situation mainly takes place with proteins with several exceptions, where oxidized proteins are reduced by specific systems (Lushchak, 2007). The latter is very true for ROS-based regulatory pathways. Oxidative damage to RNA also leads to followed degradation, but modification of DNA, if not catastrophic, is repaired by complex reparation systems.

Lipid oxidation induced by ROS is well studied. Due to availability of simple and not expensive techniques for evaluation of the products of ROS-promoted lipid oxidation they are frequently used as markers of oxidative stress. Since lipid oxidation in many cases includes the stage of formation of lipid peroxides, ROS-induced oxidation of lipids was termed "lipid peroxidation" (LPO). Several products of LPO are commonly used and probably evaluation of malonic dialdehyde (MDA) levels occupies a chief position. Most frequently it is measured with thiobarbituric acid (TBA). However, this method is rather nonspecific and should be used with many precautions (Lushchak et al., 2011). That is why the measured products, including other compounds besides MDA, are termed thiobarbituric acid-reactive substances (TBARS). Although it is broadly applied to diverse organisms (Semchyshyn et al., 2005; Talas et al., 2008; Falfushynska and Stolyar 2009; Zhang et al., 2008), the abovementioned limitation should be taken into account. Recently HPLC technique was introduced to measure MDA concentration and being more specific may be recommended where it is possible (Fedotcheva et al., 2008). Lipid peroxides may be measured by different techniques and our experience shows that the ferrous oxidation-xylenol orange (FOX) method (Hermes-Lima et al., 1995; Lushchak et al., 2011) may be successfully applied to monitor oxidative damage to lipids in various organisms (Lushchak et al., 2009).

Evaluation of the protein carbonyl levels as an indicator of oxidative modification of proteins is another method very popular among researchers in the field of free radicals. Usually, oxidatively modified proteins are degraded by different proteases. But in some cases they can be accumulated, and like advanced glycation and lipoxidation end products even became the ROS-producers. The level of oxidatively modified proteins is commonly used marker of oxidative stress, and we (Lushchak, 2007) and others (Lamarre et al., 2009) often successfully applied this parameter. It seems that the measurement of protein carbonyls is the most convenient approach and their level can be evaluated with dinitrophenylhydrazine (Lenz et al, 1989; Lushchak et al., 2011).

Oxidation of DNA is one more result of ROS presence in the cell. This type of damages is critically important for cell functions, because it can result in mutations. As abovementioned, this damage is commonly repaired by many specifically designed systems, however some of them can be detected *in vivo*. 8-Oxoguanine is the most frequently evaluated marker of DNA damage, which can be measured by HPLC (Olinski et al., 2006) or immune (Ohno et al., 2009) techniques. So-called Comet assay has been actively applied to monitor extensive damage to DNA in organisms and interested readers may refer to works of Jha and colleagues (Jha, 2008; Vevers and Jha, 2008).

Modification of specific molecules. Reactive species can modify virtually all cellular components. However, this modification not always results in deleterious effects to cellular constituents. In some cases, it regulates their functions. For example, at oxidation of cytosolic form of aconitase, [4Fe-4S] cluster containing enzyme, it may loose one of iron ions. The formed [3Fe-4S]-containing protein cannot catalyze the conversion of citrate to isocitrate, but becomes the protein, regulating iron metabolism. This conversion was described particularly in yeast (Narahari et al., 2000) and mammals (Rouault, 2006).

Defense systems. The respiratory burst, a rapid production of large amounts of ROS during phagocytosis in cells of the human immune system, was discovered in 1933 (Baldridge and Gerard, 1933), but was completely ignored for the next quarter century. Interest in the burst was disclosed around 1960 by work from Karnovsky's and Quastel's laboratories (Sbarra and Karnovsky, 1959; Iyer et al., 1961) indicating that its purpose was not to provide energy for phagocytosis, but to produce lethal oxidants for microbial killing.

The potential applications in biomedicine of the phenomenon discovered and its possible involvement in immune response attracted many researchers that resulted in disclosing of specific system reducing molecular oxygen via one electron scheme. The system was an integral part of leucocyte plasma membrane and needed NADPH for operation. Therefore, it was called "NADPH-oxidase (Noxs)". The latter catalyses one-electron reduction of molecular oxygen yielding superoxide anion, which further either spontaneously or enzymatically can be converted into H_2O_2 and further to $HO^\bullet$. Some of Noxs are called Duoxs ("dual function oxidases") since, in addition to the Nox domain, they have a domain homologous to that of thyroid peroxidase, lacking a peroxidatic activity, but generating H_2O_2 (Bartosz, 2009). These ROS are believed to be responsible for fighting of invaders by immune system cells. Some time later, it was found that leucocytes possess also inducible NO-synthase, which collaborates with NADPH-oxidase. There is a reason in this, because the combination of $^\bullet NO$ with $O_2^{\bullet-}$ gives a very powerful oxidant peroxinitrite. The latter at disproportionation gives $HO^\bullet$.

ROS-based signaling. In early 1990th several groups found that in bacteria some specific systems are involved in ROS-induced up-regulation of antioxidant and some other enzymes (Demple and Amabile-Cuevas, 1991; Storz and Imlay, 1999; Lushchak, 2001, 2011a). A bit later, similar systems were described in yeast (Kuge and Jones, 1994; Godon et al., 1998; Lee et al., 1999; Toone and Jones, 1999; Lushchak, 2010) and higher eukaryotes (Després et al., 2000; Itoh et al., 1999). In most cases, these systems are based on reversible oxidation of cysteine residues of specific proteins (Toledano et al., 2007). However, if in bacteria these proteins may serve both as sensors and regulators of cellular response like transcription regulators such as for example, SoxR and OxyR (Semchyshyn, 2009; Lushchak, 2011a), in eukaryotes the regulatory pathways are much more complicated. That is mainly related with the nucleus presence. Commonly, a sensor molecule is localized in cytoplasm and after signal reception it either directly diffuses into nucleus transducing the signal to transcriptional machinery via special pathway(s) or doing that in collaboration with other components. Although ROS-induced signaling was primary found to regulate cellular ROS-defense systems, now it became clear it coordinates many cellular processes such as development, proliferation, differentiation, metabolism, apoptosis, necrosis, etc. This is a field of interest of many research groups and there is no doubt would gain a great attention in future.

3. Oxidative stress definitions

There are many definitions of oxidative stress, but this term up to now has no rigorous meaning. Of course, there is no "ideal" definition, but it can help in some way to clarify the question someone deals with. Intuitively, it is accepted that oxidative stress is the situation when oxidative damage is increased that, in turn, can be explained as an imbalance between ROS production and elimination in the favor of the first. The term "oxidative stress" was first defined by Helmut Sies (1985) as "Oxidative stress" came to denote a disturbance in the prooxidant-antioxidant balance in favor of the former. Halliwell and Gutteridge (1999) defined oxidative stress as "in essence a serious imbalance between production of ROS/RNS and antioxidant defense". These definitions lack very important element – they ignore the dynamics of ROS production and elimination, i.e., steady-state ROS level should be referred to. The multiple ROS roles must be also mirrored in the definition reflecting also their signaling function. Therefore, we have proposed one more definition such as "Oxidative stress is a situation when steady-state ROS concentration is transiently or chronically enhanced, disturbing cellular metabolism and its regulation, and damaging cellular constituents" (Lushchak, 2011b). However, this definition does not account the ROS effects on cellular signaling, and therefore now it can be formulated as follow "oxidative stress is a transient or chronic increase in steady-state level of ROS, disturbing cellular core and signaling processes, including ROS-provided one, and leading to oxidative modification of cellular constituents up to the final deleterious effects". Not pretending to be ideal or full, it accounts for the information gained in the field of free radical processes in living organisms for the last decades.

Figure 2 may help to understand and systematize modern knowledge on oxidative stress. Under normal conditions steady-state ROS concentration fluctuates in some range, reflecting the balance between ROS generation and elimination. Some circumstances such as oxidative challenges may enhance steady-state ROS concentrations, and the latter may leave the range, leading to oxidative stress when the steady-state ROS concentration is enhanced. If the antioxidant potential is powerful enough, ROS concentration would return into the initial range without any serious consequences for the cell. However, if the antioxidant potential is not sufficient or ROS concentration is too high to cope with enhanced ROS level, the cell may need to increase the antioxidant potential, which finally would result in decreased ROS concentrations. This may have at least two consequences. The first, ROS steady-state concentration would return slowly into initial or close to initial range (so-called chronic oxidative stress) and, the second, it would reach a new steady-state level, so-called "quasi-stationary" one. The latter may not have serious consequences for the cell, but in some cases it can lead to the development of certain pathologies. In other words, the stabilization of increased ROS steady-state levels can be deleterious for the organism. The scheme given in Figure 2 may be of interest to describe the dynamics of ROS level under normal conditions and oxidative insults. Rather similar situation ideologically, but with opposite logic, may be applied to organisms challenged by reductants or under limited oxygen supply. The decrease in ROS steady-state concentration may be called "reductive stress". Despite this term is not commonly used, the situation described can be found in many organisms. For example, Black sea water contains high concentrations of hydrogen sulfide at deep horizons. Although its high concentrations are very toxic for living organisms, they can be exposed to it episodically. The bottom aquatic systems and mud can also be highly reduced and many organisms, particularly worms and mollusks, are very tolerant successfully resisting reductive potential of environment. The reductive stress may be developed in the organisms at oxygen limitations and poisoning of electron-transport chains resulting in increased levels of highly reactive electrons. Although "reductive stress" hypothesis virtually has not been developed, we feel its perspective.

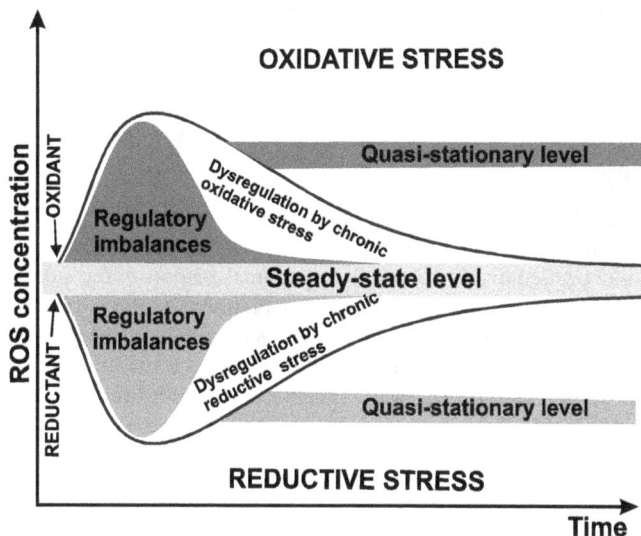

Fig. 2. **Schematic representation of modern ideas on metabolism of reactive oxygen species in biological systems.** The concentration of ROS is maintained at certain range and fluctuates similarly to other parameters in the organism in according to homeostasis theory. However, under some circumstances the concentration may leave this range due to increase/decrease of production or change of efficiency of catabolic system. The state when ROS level is transiently increased is referred to oxidative stress, and when decreased to reductive one. The problem of oxidative stress is investigated rather well, while the reductive stress studies are only at infant state. In the latter even methodological approaches have not been developed. Substantial changes in ROS level, out of certain range "norm" stimulate the systems of feedback relationships. They are abundant and multilevel what provides fine regulation in ROS level in certain range of concentrations. There are two principally different scenarios. In the first case, after induction of oxidative/reductive stress the ROS level returns into initial range. In the second case, the system reaches a new steady-state range and this is a new "normal" range of concentrations. The new steady-state range or quasi-steady state range appears. Both transient and chronic oxidative stresses may have different consequences for the organism and may cause more or less substantial injury to tissues, and if not controlled may culminate in cell death via apoptosis or necrosis mechanisms (modified from Dröge, 2002, and Lushchak, 2011a).

Generally, oxidative stress can be induced in three ways: (*i*) increased ROS production, (*ii*) decreased ROS elimination, and (*iii*) appropriate combination of the two previous ways. Despite it is difficult to demonstrate that oxidative stress can directly lead to pathologies, there are many evidences demonstrating a strong relationship between oxidative stress and many pathologies as well as aging (Valko et al., 2007). In many cases, the application of different antioxidants was shown to be both good prophylactics and cure to certain extent. At least antioxidants were found to be able to reduce some disease symptoms.

In conclusion, it became more and more clear that ROS roles in living organisms are not limited only to damage either in own tissues or invaders. Last two decades, their signaling functions have been disclosed in many organisms to be important not only as adaptive strategies, but also coordinating roles in diverse basic biological processes like differentiation, apoptosis. Knowledge accumulated to date only slightly shed light on the fundamental roles of ROS in biological systems.

4. Acknowledgment

The editors would like to thank all authors who participated in this project for their contributions and hard work to prepare an interesting book on the general aspects of oxidative stress and particular questions of organisms' response and adaptation to it. We also thank to our colleagues from Precarpathian National University who helped us to develop the ideology of this book during many years of collaboration, helpful, creative, and sometimes "hot" discussions, which stimulated us to perfect our knowledge on the role of reactive species in diverse living processes. We are also grateful to the "In-Tech" Publisher personnel, especially to Ms. Sasa Leporic who assisted us in the arrangement of the book and scheduling our activities.

5. References

Baldridge, C. & Gerard R. (1933). The extra respiration of phagocytosis. *American Journal of Physiology*, Vol.103, pp. 235-236.

Bartosz, G. (2009). Reactive oxygen species: destroyers or messengers? *Biochemical Pharmacology*, Vol.77, No.8, pp. 1303-1315.

Demple, B. & Amabile-Cuevas, C. (1991). Redox redux: the control of oxidative stress responses. *Cell*, Vol.67, pp. 837-839.

Després, C.; DeLong, C.; Glaze, S.; Liu, E. & Fobert, P. (2000). The Arabidopsis NPR1/NIM1 protein enhances the DNA binding activity of a subgroup of the TGA family of bZIP transcription factors. *Plant Cell*, Vol.12, pp. 279-290.

Drath, D. & Karnovsky, M. (1975). Superoxide production by phagocytic leukocytes. *The Journal of Experimental Medicine*, Vol.141, pp. 257-262.

Dröge, W. (2002). Free radicals in the physiological control of cell function. *Physiological Reviews*, Vol.82, No.1, pp. 47-95.

Falfushynska, H. & Stolyar, O. (2009). *Responses of biochemical markers in carp Cyprinus carpio from two field sites in Western Ukraine. Ecotoxicology and Environmental Safety*, Vol.72, pp. 729-736.

Fedotcheva, N.; Litvinova, E.; Amerkhanov, Z.; Kamzolova, S.; Morgunov, I. & Kondrashova, M. (2008). Increase in the contribution of transamination to the respiration of mitochondria during arousal. *Cryo Letters*, Vol.29, pp. 35-42.

Godon C.; Lagniel G.; Lee J.; Buhler J.M.; Kieffer S.; Perrot M.; Boucherie H.; Toledano MB. & Labarre J. (1998). The H_2O_2 stimulon in *Saccharomyces cerevisiae*. *Journal of Biological Chemistry*, Vol.273, pp. 22480-22489.

Halliwell, B. & Gutteridge, J. (1999). Free radicals in biology and medicine. Oxford: Clarendon Press.

Hermes-Lima, M.; Willmore, W. & Storey, K. (1995). Quantification of lipid peroxidation in tissue extracts based on Fe(III)xylenol orange complex formation. *Free Radical Biology & Medicine*, Vol.19, pp. 271-280.

Itoh, K.; Ishii, T.; Wakabayashi, N. & Yamamoto, M. (1999). Regulatory mechanisms of cellular response to oxidative stress. *Free Radical Research*, Vol.31, pp. 319-324.

Iyer, G., Islam, M. & Quastel, J. (1961). Biochemical aspects of phagocytosis. *Nature*, Vol.192, pp. 535-541.

Jha, A. (2008) Ecotoxicological applications and significance of the comet assay. *Mutagenesis*, Vol.23, pp. 207-221.

Kuge, S. & Jones, N. (1994). YAP1 dependent activation of TRX2 is essential for the response of *Saccharomyces cerevisiae* to oxidative stress by hydroperoxides. *EMBO Journal*, Vol.13, No.3, pp. 655-664.

Lamarre, S.; Le François, N.; Driedzic, W. & Blier, P. (2009). Protein synthesis is lowered while 20S proteasome activity is maintained following acclimation to low temperature in juvenile spotted wolffish (*Anarhichas minor* Olafsen). *The Journal of Experimental Biology*, Vol.212, pp.1294-1301.

Lee J.; Godon C.; Lagniel G.; Spector D.; Garin J.; Labarre J. & Toledano MB. (1999). Yap1 and Skn7 control two specialized oxidative stress response regulons in yeast, *Journal of Biological Chemistry*, Vol.274, pp. 16040-16046

Lenz, A.; Costabel, U.; Shatiel, S. & Levine, R. (1989). Determination of carbonyl groups in oxidatively modified of proteins by reduction with tritiated sodium borohydride. *Analytical Biochemistry*, Vol.177, pp. 419-425.

Lushchak, V. (2001). Oxidative stress and mechanisms of protection against it in bacteria. *Biochemistry* (Moscow), Vol.66, pp. 476-489.

Lushchak, V. (2007). Free radical oxidation of proteins and its relationship with functional state of organisms. *Biochemistry* (Moscow), Vol.72, No.8, pp. 809-827.

Lushchak, V. (2010). Oxidative stress in yeast. *Biochemistry* (Moscow), Vol.75, pp. 281-296.

Lushchak, V. (2011a). Adaptive response to oxidative stress: Bacteria, fungi, plants and animals. *Comparative Biochemistry and Physiology - Part C Toxicology & Pharmacology*, Vol.153, pp. 175-190.

Lushchak, V. (2011b). Environmentally induced oxidative stress in aquatic animals. *Aquatic Toxicology*, Vol.101, pp. 13-30.

Lushchak, O.; Kubrak, O.; Torous, I.; Nazarchuk, T.; Storey, K. & Lushchak, V. (2009). Trivalent chromium induces oxidative stress in goldfish brain. *Chemosphere*, Vol.75, pp. 56-62.

Muenzer, J.; Weinbach, E. & Wolfe, S. (1975). Oxygen consumption of human blood platelets. II. Effect of inhibitors on thrombin-induced oxygen burst. *Biochimica and Biophysica Acta*, Vol.376, pp. 243-248.

Narahari, J.; Ma, R.; Wang, M. & Walden, W. (2000). The aconitase function of iron regulatory protein 1. Genetic studies in yeast implicate its role in iron-mediated redox regulation. *The Journal of Biological Chemistry*, Vol.275, pp. 16227-16234.

Ohno, M.; Oka, S. & Nakabeppu, Y. (2009). Quantitative analysis of oxidized Guanine, 8-oxoguanine, in mitochondrial DNA by immunofluorescence method. *Methods in Molecular Biology*, Vol.554, pp. 199-212.

Olinski, R.; Rozalski, R.; Gackowski, D.; Foksinski, M.; Siomek, A. & Cooke, M. (2006). Urinary measurement of 8-OxodG, 8-OxoGua, and 5HMUra: a noninvasive assessment of oxidative damage to DNA. *Antioxidants and Redox Signaling*, Vol.8, pp. 1011-1019.

Peng, X.; Ma, J.; Chen, F. & Wang, M. (2011). Naturally occurring inhibitors against the formation of advanced glycation end-products. *Food and Function*, Vol.2, No.6, pp. 289-301.

Rouault, T. (2006). The role of iron regulatory proteins in mammalian iron homeostasis and disease. *Nature Chemical Biology*, Vol.2, pp. 406-414.

Sbarra, A. & Karnovsky, M. (1959). The biochemical basis of phagocytosis. I. Metabolic changes during the ingestion of particles by polymorphonuclear leukocytes. *Journal of Biological Chemistry*, Vol.234, No.6, pp.1355-1362.

Semchyshyn, H. (2009). Hydrogen peroxide-induced response in *E. coli* and *S. cerevisiae*: different stages of the flow of the genetic information. *Central European Journal of Biology*, Vol.4, No.2, pp.142-153.

Semchyshyn, H.; Bagnyukova, T.; Storey, K. & Lushchak V. (2005). Hydrogen peroxide increases the activities of *soxRS* regulon enzymes and the levels of oxidized proteins and lipids in *Escherichia coli*. *Cell Biology International*, Vol.29, pp. 898-902.

Sies, H. (1985). Oxidative stress: Introductory remarks, In: *Oxidative stress*, Sies H, (Ed.), pp. 1-8, Academic Press, London.

Storz, G. & Imlay, J. (1999). Oxidative stress. *Current Opinion in Microbiology*, Vol.2, pp. 188-194.

Talas, Z.; Orun, I.; Ozdemir, I.; Erdogan, K.; Alkan, A. & Yilmaz, I. (2008). Antioxidative role of selenium against the toxic effect of heavy metals (Cd^{+2}, Cr^{+3}) on liver of rainbow trout (*Oncorhynchus mykiss* Walbaum 1792). *Fish Physiology and Biochemistry*, Vol.34, pp. 217-222.

Toledano, M.; Kumar, C.; Le Moan, N.; Spector, D. & Tacnet, F. (2007). The system biology of thiol redox system in *Escherichia coli* and yeast: differential functions in oxidative stress, iron metabolism and DNA synthesis. *FEBS Letters*, Vol.581, pp. 3598-3607.

Toone, W. & Jones, N. (1999). AP-1 transcription factors in yeast. *Current Opinion in Genetics Development*, Vol.9, pp. 55-61.

Valko, M.; Leibfritz, D.; Moncol, J.; Cronin, M.; Mazur, M. & Telser, J. (2007). Free radicals and antioxidants in normal physiological functions and human disease. *The International Journal of Biochemistry and Cell Biology*, Vol.39, pp. 44-84.

Vevers, W. & Jha, A. (2008). Genotoxic and cytotoxic potential of titanium dioxide (TiO_2) nanoparticles on fish cells *in vitro*. *Ecotoxicology*, Vol.17, pp. 410-420.

Zhang, X., Yang, F.; Zhang, X.; Xu, Y.; Liao, T.; Song, S. & Wang, J. (2008). Induction of hepatic enzymes and oxidative stress in Chinese rare minnow (*Gobiocypris rarus*) exposed to waterborne hexabromocyclododecane (HBCDD). *Aquatic Toxicology*, Vol.86, pp. 4-11.

Section 2

General Aspects of Oxidative Stress

2

Interplay Between Oxidative and Carbonyl Stresses: Molecular Mechanisms, Biological Effects and Therapeutic Strategies of Protection

Halyna M. Semchyshyn and Volodymyr I. Lushchak
Vassyl Stefanyk Precarpathian National University,
Ukraine

1. Introduction

Reactive species (RS) are continuously produced and eliminated in variuos groups of organisms: from bacteria to man. Under normal physiological conditions, the steady-state concentrations of RS are maintained at certain range and fluctuate similarly to other parameters in the organism according to homeostasis theory. The persistence of RS in cells demonstrates their evolutionarily selected production in order to perform some useful role in living organisms. The most beneficial among important biological roles of RS is their establishment as important regulators of cell signal transduction and part of immune response controlling cellular defense against various environmental challenges.

However, under some circumstances, RS level may leave the range of normal concentrations due to change of their production or change of efficiency of catabolic system. An increase in steady-state level of reactive oxygen species (ROS) or reactive carbonyl species (RCS) may result in so-called "oxidative stress" or "carbonyl stress", respectively. Generally, ROS and RCS are mainly known for their damaging effects. At molecular level, they are found to disrupt the structure and function of proteins, nucleic acids, lipids, carbohydrates, *etc.* As a consequence of these undesirable effects at cellular and organismal levels, loss of function and even viability can occur.

Recent studies indicate that in many cases increase in RCS steady-state concentrations is a consequence of oxidative stress, whereas increase in ROS steady-state levels is resulted from carbonyl stress. Thus, a vicious cycle can be formed.

Carbonyl/oxidative stress has been found to be implicated in many chronic and degenerative diseases. Different metabolic disorders, diabetes, obesity, kidney and heart diseases, atherosclerosis, and neurodegenerative diseases all have a strong component of carbonyl/oxidative stress. It is unclear however, whether the carbonyl/oxidative stress is causal in disease progression or the result of the cell death associated with cells dying by apoptosis or necrosis.

2. Reactive carbonyl compounds

Reactive carbonyls are commonly generated in vivo as metabolic products (Tessier, 2010; Turk, 2010; Peng et al., 2011; Robert, 2011) or derived from the environment (Uribarri and

Tuttle, 2006; Birlouez-Aragon et al., 2010; Uribarri et al., 2010). Similarly to other RC, RCS can play a dual role in living organisms. For instance, some RCS are implicated as signalling molecules controlling cellular defense against the environmental challenges. On the other hand, due to high reactivity RCS interact with different cellular constituents that may contribute to aging, age-related diseases, and diverse metabolic disorders.

2.1 Structure and reactivity

RCS is a large group of reactive biological molecules mainly with three to nine carbons in length containing one or more carbonyl groups. Most of the biological damages caused by RCS are related to α,β-unsaturated aldehydes, dialdehydes, and keto-aldehydes (Uchida, 2000; Pamplona, 2011). Figure 1 demonstrates the most common RCS found in biological systems. Malondialdehyde (MDA), glyoxal (GO), methylglyoxal (MGO), glucosone, 3-deoxyglucosone (3DG), and ribosone are among highly reactive α- and β-dicarbonyl compounds. Acrolein, crotonaldehyde and 4-hydroxy-trans-2-nonenal (HNE) belong to α,β-unsaturated aldehydes. One of the most biologically important keto-aldehydes is 4-oxo-trans-2-nonenal (ONE). Glycolaldehyde, dehydroascorbate, acetaldehyde, glceraldehyde-3-phosphate and dioxyacetone phosphate are also among the reactive carbonyls ubiquitously generated in biological systems.

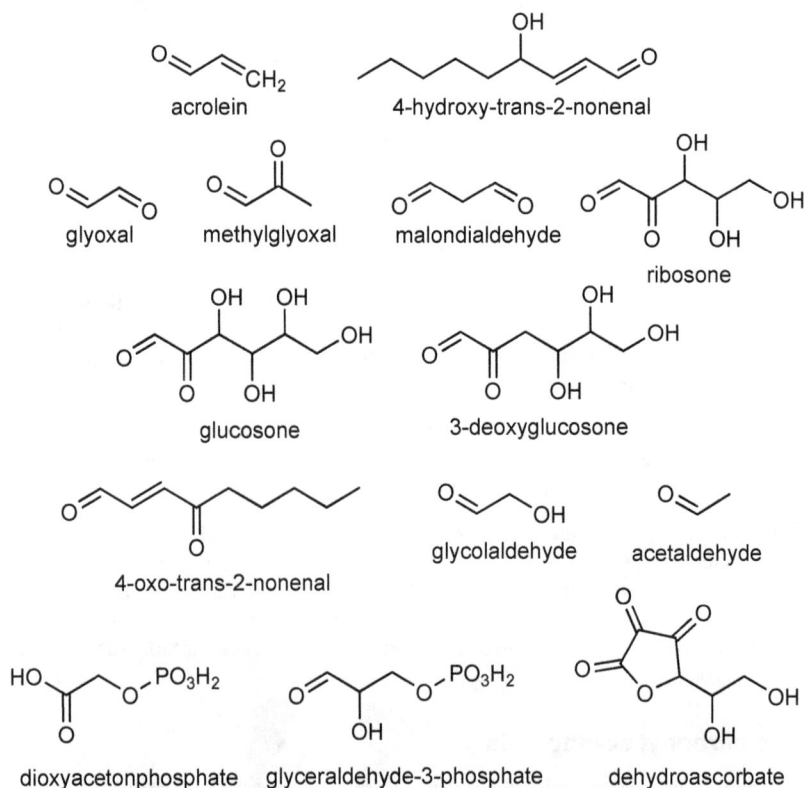

Fig. 1. The structures of the most common biological reactive carbonyl species.

It should be noted that unsaturated RCS are usually an order of magnitude more reactive than their saturated counterparts. α,β-Unsaturated carbonyls are especially reactive, because they have carbonyl group and reactive double bond that makes the C_3 carbon a strong electrophile. Extremely reactive carbonyl compound is HNE which possesses electrophilic double bond, carbonyl and hydroxyl groups. However, such dialdehydes as GO, MGO, and 3-DG are much more active than HNE and MDA (Lankin et al., 2007). In addition, acrolein reacts with thiols 100-fold more rapidly than HNE (Witz, G. 1989, Esterbauer et al., 1991). Like hydroxyl radical is the most powerful oxidant among ROS, acrolein is the most electrophilic, and therefore reactive α,β-unsaturated aldehyde known.

Carbonyl compounds like most other intermediates and by-products of metabolism are electrophilic, and thus are highly reactive with different cellular constituents majority of which are nucleophiles (Zimniak, 2011). Such strong nucleophilic sites as thiol, imidazole, and hydroxyl groups of biomolecules as well as nitrogen and oxygen atoms in purine and pyrimidine bases are the most attractive targets for electrophilic attacks. In general, all mentioned above interactions may lead to chemical modification of proteins, nucleic acids, and aminophospholipids, resulting in cytotoxicity and mutagenicity (Ellis, 2007; Liu et al., 2010). Sience biological effects caused by RCS and ROS are rather similar, chemical properties of both groups seem should be similar as well. However, RCS have a relatively long half life time and therefore higher stability, in contrast to ROS. For instance, reactive carbonyls have average half-life from minutes to hours (Uchida, 2000; Pamplona, 2011). At the same time, half-life of some ROS ranges from 10^{-9} to 10^{-6} s (Halliwell and Gutteridge, 1989; Demple, 1991). It is well known that non-charged ROS such as H_2O_2 and $HO_2^{\bullet}$ are capable to cross biological membranes and diffuse for relatively long distances in the intracellular environment. At the same time, higher stability of non-charged RCS molecules allows them even to escape from the cell and interact with targets far from the site of their generation. That is why, under certain conditions, RCS may have far-reaching damaging effects, and therefore they can be more deleterious than ROS.

2.2 Generation *in vivo*

Carbonyl compounds can be endogenous or exogenously derived. Some RCS (*e.g.* acrolein, crotonaldehyde, acetone and formaldehyde) are ubiquitous industrial pollutants which can readily enter the cell from the environment (Trotter et al., 2006; Liu et al., 2010; Seo and Baek, 2011). Other exogenous sources of reactive carbonyls are products of organic-pharmaceutical chemistry, cigarette smoke, food additives and browned food (Uribarri et al., 2007; Birlouez-Aragon et al., 2010; Colombo et al., 2010; Dini, 2010; Robert et al., 2011). Number of carbonyl compounds is formed under chemical modification of the nutrients during food cooking (browning, Maillard reaction). The browning reaction between amino acids and simple carbohydrates was first observed a century ago by Louis Camille Maillard (Maillard, 1912). About 40 years later Maillard reaction was recognized as one of the main reasons for the occurrence of the non-enzymatic food browning demonstrating an importance in food science (Hodge, 1953; Tessier, 2010). In late 1960s, the products of a non-enzymatic glycosylation similar to the food browning were detected in human organism (Rahbar, 1968; Rahbar et al., 1969). Thus, it took several decades to realize the physiological significance of the reaction discovered by Maillard. In 1980s, the *in vivo* reaction between biomolecule amino groups and monosaccharides, without enzymes, was named "non-enzymatic glycosylation" and several years later renamed "glycation" in order to differentiate it from the enzymatic glycosylation important in the post-translation modification of proteins (Yatscoff et al., 1984). Now it is well documented that glycation is one of the most significant endogenous sources of reactive carbonyls (Tessier, 2010).

More than 20 saturated and unsaturated RCS have been identified in biological samples (Niki, 2009). Table 1 demonstrates that, in general, endogenous RCS can be formed as products of either enzymatic or non-enzymatic processes.

Enzymatic sources			Non-enzymatic sources	
Glycolysis	Polyol pathway	Oxidation of amino acids	Glycation	Peroxidation of lipids
Acetaldehyde Glyceraldehyde-3-phosphate Dioxyacetone phosphate Methylglyoxal	3-Deoxyglucosone 3-Deoxyfructose	Glyoxal Methylglyoxal Acrolein Glycolaldehyde 2-Hydroxypropanal	Glyoxal Methylglyoxal Glucosone 3-Deoxyglucosone Acrolein	Malonic dialdehyde 4-Hydroxy-trans-2-nonenal 4-Oxo-trans-2-nonenal Glyoxal Methylglyoxal Acrolein Crotonaldehyde Hexanal

Table 1. Reactive carbonyl species and sources of their generation *in vivo*

In this section, we will describe common ways of RCS generation *in vivo*, in particular, polyol pathway, amino acid oxidation, lipid peroxidation, and glycation.

2.2.1 Enzymatic reactions

Reactive carbonyls are produced intracellularly through both enzymatic and non-enzymatic pathways. Enzymatically produced RCS, glycolytic intermediates or by-products of metabolic conversion of carbohydrates and amino acids, are presented in Table 1. The effective steady-state concentration of such metabolites as acetaldehyde, glyceraldehyde-3-phosphate and dioxyacetone phosphate is typically low in the cell, because of their rapid utilization by the next step of the pathway (Zimniak, 2011). However, concentration of MGO, a by-product of glycolysis in most living organisms, is not so tightly controlled. Therefore, under certain conditions, biological effects of MGO may be more potent than the effect caused by the glycolytic intermediates.

The elimination of phosphate from glyceraldehyde-3-phosphate and dihydroxyacetone phosphate is the major enzymatic source of MGO *in vivo* (Pompliano et al., 1990; Phillips and Thornalley, 1993; Richard, 1993). In Figure 2 the mechanism of the reaction is given. As seen, enediol phosphate, an intermediate in the above mentioned reactions, may escape from the active site of triosophosphate isomerase and be rapidly decomposed to MGO and inorganic phosphate (Pompliano et al., 1990). MGO can also be formed from hydroxyacetone, an intermediate in the enzymatic oxidation of ketone bodies (Lyles and Chalmers, 1992; Turk, 2010). Oxidation of some amino acids can also lead to MGO formation under physiological conditions. For example, threonine and glycine can be converted to aminoacetone and succinylacetone, MGO precursors (Kalapos, 2008a). It should be noted

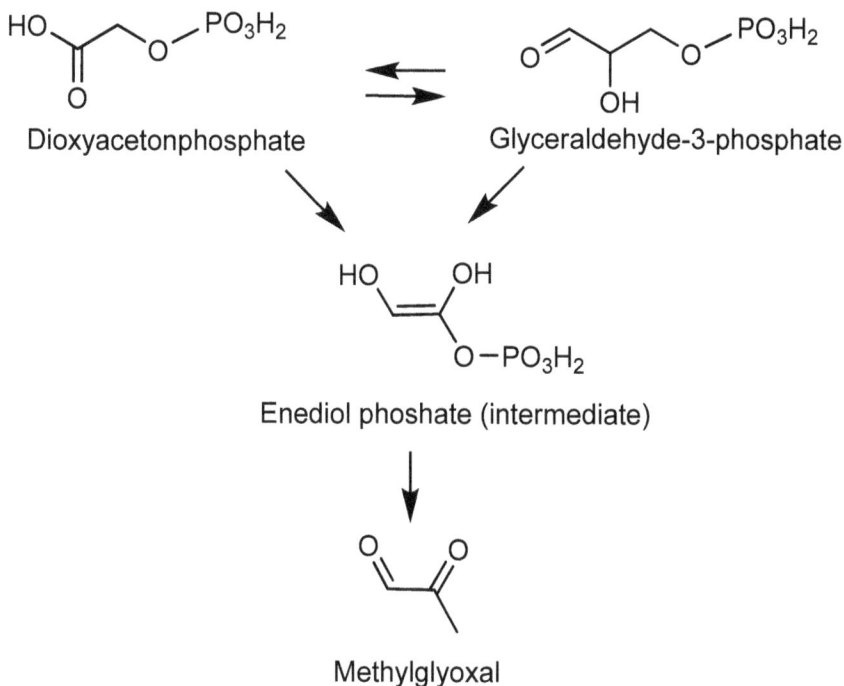

Dioxyacetonphosphate   Glyceraldehyde-3-phosphate

Enediol phoshate (intermediate)

Methylglyoxal

Fig. 2. Formation of methylglyoxal as a by-product of glycolysis.

that MGO can be formed from dihydroxyacetone phosphate in the reaction catalysed by bacterial MGO synthase, however, it is unknown whether this enzyme and this kind of MGO generation also exist in animals (Kalapos, 2008b).

Polyol pathway may be associated with the production of 3-DG one more carbonyl compound ubiquitously generated in biological systems (Niwa, 1999; Chung et al., 2003). The mechanism of its generation is demonstrated in Figure. 3. In one way, 3-DG is formed from fructose, an oxidized product of sorbitol by sorbitol dehydrogenase. In the second one, 3-DG is a hydrolysis product of fructose-3-phosphate, an enzymatic product of fructose phosphorylation. Further enzymatic reduction and oxidation of 3-DG can result in 3-deoxyfructose and 2-keto-3-deoxygluconic acid formation, respectively (Niwa, 1999).

Different RCS can be generated *in vivo* by activated human phagocytes. It has been found that stimulated neutrophils employed the myeloperoxidase-H_2O_2-chloride system to produce α-hydroxy and α,β-unsaturated aldehydes from hydroxy-amino acids in high yield (Anderson et al., 1997). Figure 4 shows possible mechanism of glycolaldehyde formation from L-serine, and acrolein from L-threonine.

In conclusion, *in vivo* detection of enzymatically produced RCS is still quite complicated task, because of their relatively low stability under physiological conditions. It seems, more endogenous sources of RCS would be described with using sophisticated techniques in not far future.

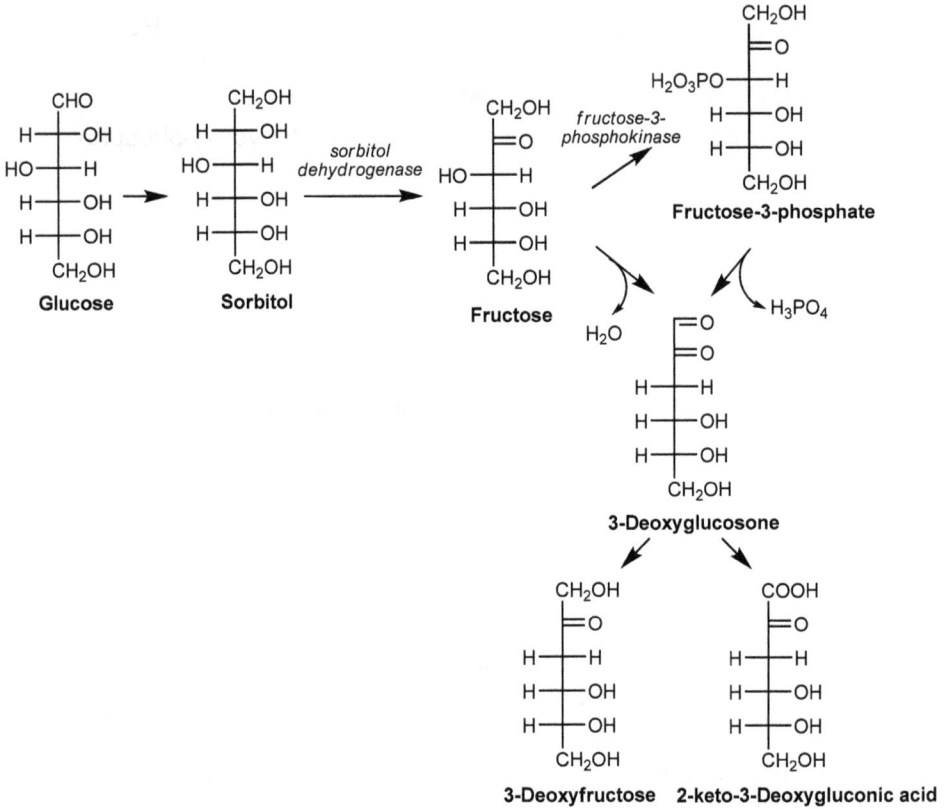

Fig. 3. Polyol pathway as a source of formation of reactive carbonyl species.

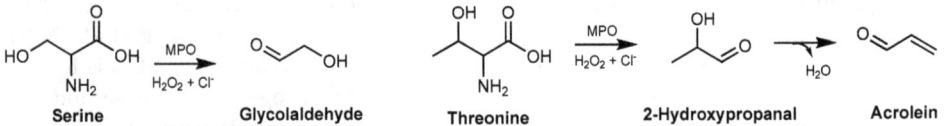

Fig. 4. Possible mechanisms of glycolaldehyde and acrolein generation by activated neutrophils.

2.2.2 Non-enzymatic reactions

Non-enzymatic reactions, bypassing the classic metabolic pathways, play a crucial role in RCS generation *in vivo*. A numerous literature reveals that oxidative degradation of biomolecules is the major way in the non-enzymatic production of RCS. For instance, degradation of nucleic acids and related compounds results in the formation of reactive carbonyls. In model experiments, it was demonstrated that purified RNA, DNA, and their precursors contribute to MGO formation (Chaplen et al., 1996). However, probably because of the higher intracellular steady-state concentrations of lipids, proteins and carbohydrates as compared with nucleic acids, oxidative catabolism of lipids, amino acids and

Interplay Between Oxidative and Carbonyl Stresses: Molecular Mechanisms, Biological
Effects and Therapeutic Strategies of Protection

21

carbohydrates is believed to be the major source of endogenous non-enzymatically produced RCS (Uchida, 2000).

2.2.2.1 Lipid peroxidation

In 1930s, lipid peroxidation (LPO) was first studied in relation to food deterioration (Niki, 2000). Later investigations revealed that LPO products can be formed in living organisms. Similarly to the Maillard reaction, with increased evidences on physiological significance of the process, several decades later LPO received renewed attention in biochemistry, and medicine.

It is well known that different mechanisms underlie LPO process: (*i*) enzymatic oxidation, (ii) ROS-independent nonenzymatic oxidation, and (iii) ROS-mediated nonenzymatic oxidation (Niki 2009). Due to various mechanisms, specific LPO products can be formed. There are many evidences that reactive carbonyls are produced through LPO as a consequence of oxidative stress (Ellis, 2007; Negre-Salvayre et al., 2008; Pamplona, 2008; Zimniak, 2008; Pamplona, 2011; Zimniak, 2011). LPO induced by ROS generates a variety of primary, secondary and end products (Figure 5).

Oxidation of polyunsaturated fatty acids (PUFAs), which are highly susceptible to peroxidation by ROS, involves an allylic hydrogen abstraction to form a tetradienyl radical (L$^{\bullet}$) followed by insertion of molecular oxygen. Addition of oxygen results in peroxyl radical formation (LOO$^{\bullet}$), which is further transformed to hydroperoxide (LOOH) by hydrogen abstraction from another lipid molecule (LH). The latter gives another free radical (L$^{\bullet}$) and propagates oxidation. All radical compounds appeared from oxidation of lipids belong to primary LPO products, and lipid hydroperoxides (LOOH) are named as secondary LPO products. In addition, peroxyl radical (LOO$^{\bullet}$) can undergo further oxidation to form other highly oxidized products such as bicyclic endoperoxides, monocyclic peroxides, serial cyclic peroxides and other complex peroxides (Yin et al., 2002). Most of them are unstable and can be readily decomposed to so-called LPO-derived end products, a wide array of compounds, including RCS (Esterbauer et al. 1991). The most common reactive carbonyls derived from PUFA oxidation are MDA, hexanal and HNE, comprising of 70%, 15%, and 5% of the total produced by lipid peroxidation, respectively (Ellis, 2007). Acrolein was identified as a LPO end product at oxidation of low density lipoproteins (Ellis, 2007).

As mentioned above, reactive carbonyls, end LPO products, can react with nucleophilic groups in biomolecules resulting in their irreversible modifications and formation of a variety of adducts and cross-links collectively named advanced lipoxidation end products (ALEs). In turn, ALEs may lead to ROS formation, and as a consequence, propagation of oxidative modifications.

2.2.2.2 Glycation (Maillard chemistry)

Glycation is a complex series of parallel and sequential reactions, in which reducing free carbonyl groups of carbohydrates react with the nucleophilic amino groups of biomolecules, producing a large number of variuos compounds, including RCS (Finot, 1982; Ellis, 2007; Tessier, 2010; Peng et al., 2011; Robert, 2011). The initial step of glycation, the Maillard reaction, is the covalent interaction between reducing monosaccharide (*e.g.* glucose, fructose, galactose, glucose-6-phosphate) and N-terminal amino acid residues or epsilon amino groups of proteins, lipids, and nucleic acids, which produces an acyclic form of Schiff base rearranging reversibly to cyclic N-substituted glycosylamine (Figure 6).

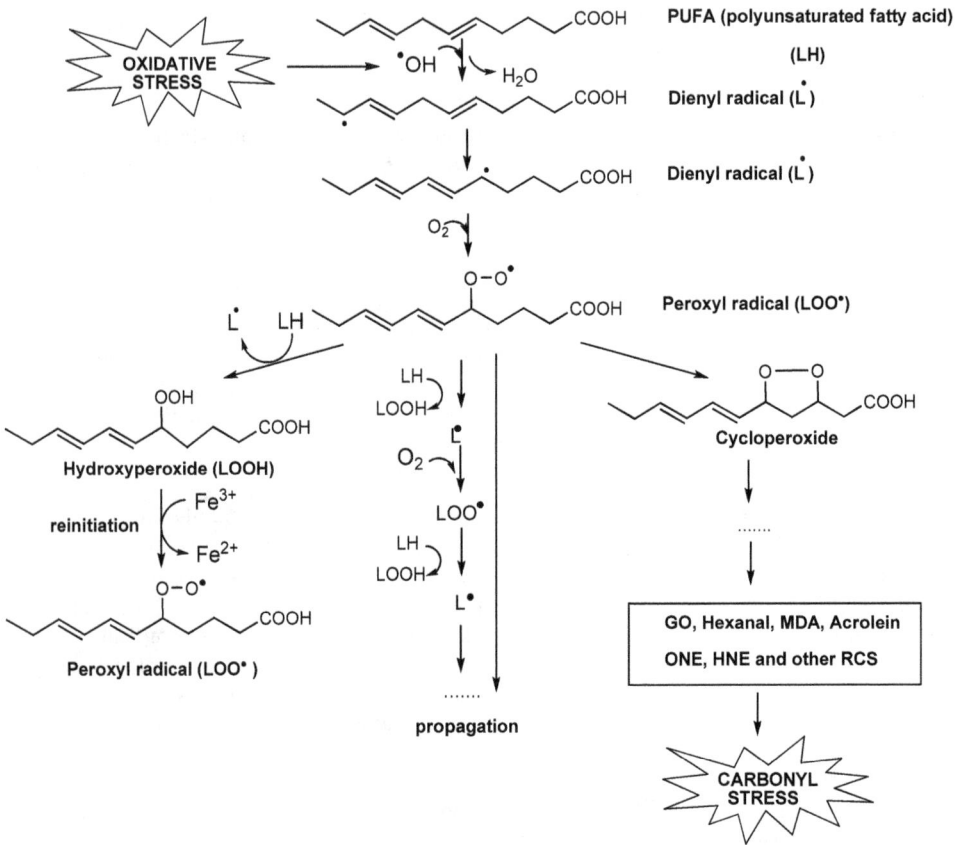

Fig. 5. Suggested pathways of lipid peroxidation and its relation to oxidative and carbonyl stresses (modified from (lushchak et al., 2011c)).

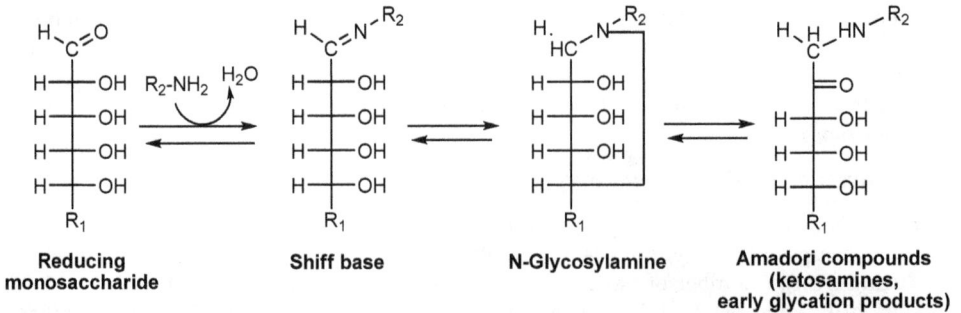

Fig. 6. Formation of Amadori products in the Maillard reaction.

The latter is an unstable compound, which can be subjected to further isomerization called an Amadori rearrangement giving more stable Amadori adducts (early glycation products),

Interplay Between Oxidative and Carbonyl Stresses: Molecular Mechanisms, Biological
Effects and Therapeutic Strategies of Protection

23

namely ketosamines. Amadori products derived from non-enzymatic glycation by hexoses are commonly known as "fructosamine". The carbohydrate moiety of Amadori products can undergo enolization, followed by dehydration, oxidation and/or fragmentation reactions, consequently producing a variety of RCS, including GO, MGO, glucosone, 1-, 2- and 3-deoxyglucosones, 3,4-dideoxyglucosone, erythrosone, ribosone, 3-deoxyerythrosone, and 3-deoxyribosone (Reihl et al., 2004; Thornalley, 2005; Tessier, 2010). Figure 7 shows the mechanism of glucosone formation followed by generation of such ROS as superoxide and hydrogen peroxide.

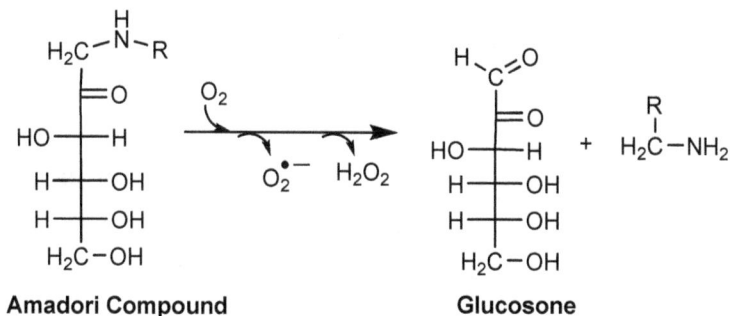

Amadori Compound   **Glucosone**

Fig. 7. Formation of glucosone from the Amadori compound.

In addition, there is an evidence for the fragmentation of the Schiff base, leading to the formation of GO, MGO, and hydrogen peroxide (Figure 8) (Hayashi and Namiki, 1980; Namiki and Hayashi, 1983). The series of reaction pathways in Maillard chemistry established Shiff base fragmentation to α-oxoaldehydes now collectively called the Namiki pathway (Thornalley, 2005; Peng et al., 2011).

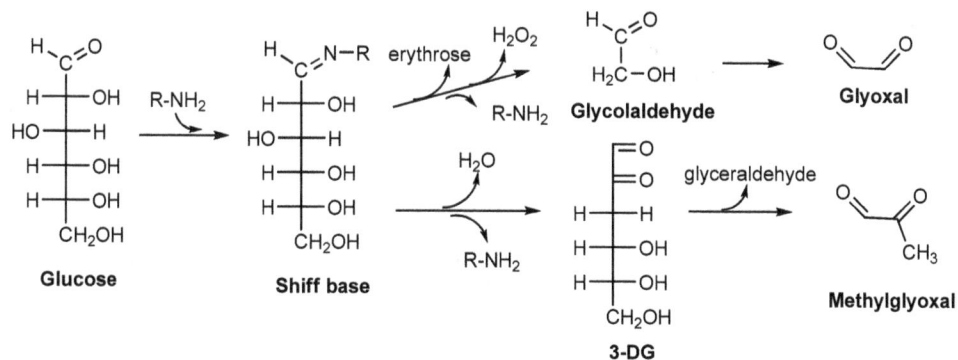

Fig. 8. Namiki pathway.

Slow oxidative degradation of monosaccharides under physiological conditions leads to the formation of α-oxoaldehydes and hydrogen peroxide (Figure 9) (Thornalley et al., 1984; Wolff et al., 1991). This process was called monosaccharide autoxidation or Wolff pathway (Peng et al., 2011). The complicity of glycation with all variety of substrates and products, and almost unpredictable direction of the process is similar to free-radical chain reactions, in particular LPO. That is why the term "Maillard chemistry" is widely used to describe a variety of chemical reactions involved in the glycation processes.

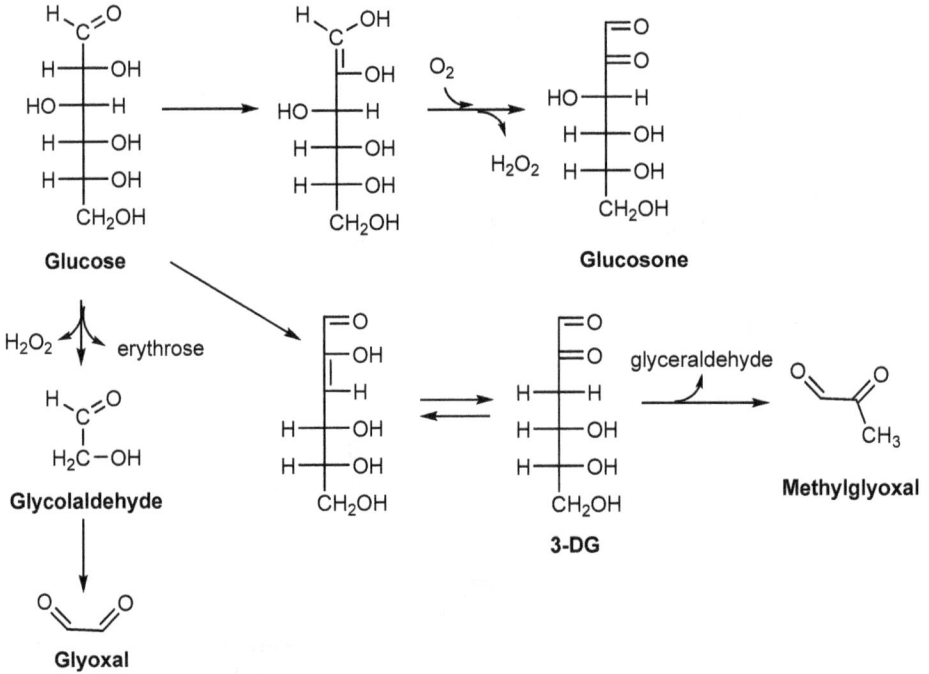

Fig. 9. Wolff pathway.

In the late stage of glycation, these reactive α-oxoaldehydes as well as Amadori compounds again interact with free amino, sulfhydryl and guanidine functional groups of intracellular and extracellular biomolecules leading to crosslinking and formation of advanced glycation end products (AGEs) (Peng et al., 2011; Robert, 2011). Therefore, Amadori products and RCS formed during glycation are believed to be important precursors of glycation adduct formation in biological systems.

2.2.2.3 Advanced lipooxidation and glycation end products

As seen in the above sections, LPO and glycation are complexes of very heterogeneous chemical reactions, leading to the formation of low molecular mass RCS. Further, these RCS, being either LPO end products or glycation intermediates, react with nucleophilic groups of macromolecules like proteins, nucleic acids, and aminophospholipids, resulting in their non-enzymatic, and irreversible modification and formation of a variety of adducts and cross-links collectively named ALEs and AGEs (Figure 10) (Miyata et al., 2000; Ellis, 2007; Tessier, 2010; Pamplona, 2011; Peng et al., 2011).

It is well documented that LPO-derived RCS reacting with proteins produce such ALEs as MDA-Lys, HNE-Lys, propanal-His, propenal-Lys, and S-carboxymethyl–cysteine, as well as such cross-link as MDA-lysine dimmer, among many others (Figure 11) (Uchida et al., 1997; Shao et al., 2005; Pamplona, 2008 and 2011).

Interplay Between Oxidative and Carbonyl Stresses: Molecular Mechanisms, Biological
Effects and Therapeutic Strategies of Protection

25

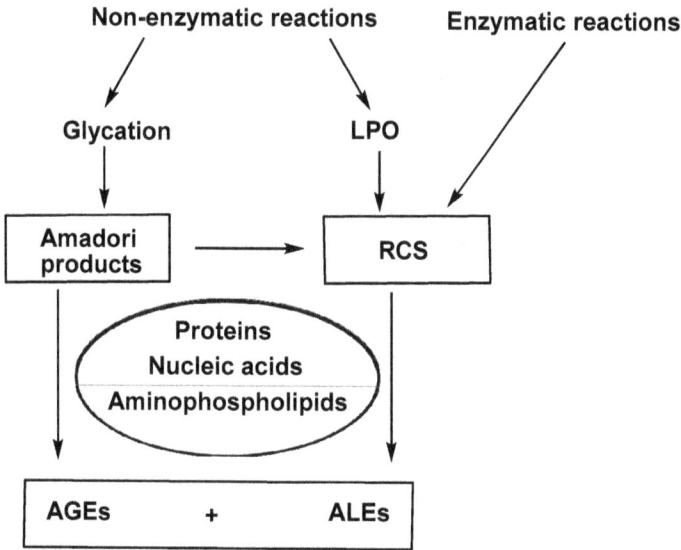

Fig. 10. Formation of reactive carbonyls and advanced glycation and lipoxidation end products in enzymatic and non-enzymatic processes.

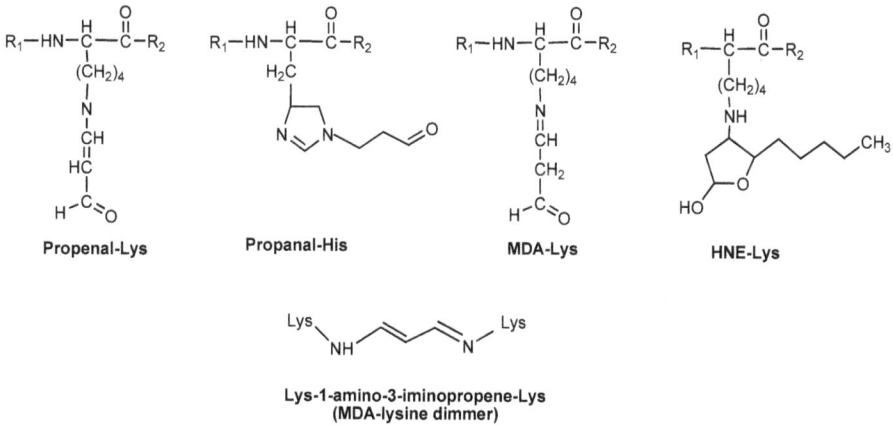

Fig. 11. The structures of the most common biological advanced lipoxidation end products.

LPO end products can also interact with amino groups of deoxyguanosine, deoxycytosine, guanosine to form various alkylated products (Pamplona, 2008). Those are the most common targets for RCS. Interaction between RCS and amino groups of aminophospholipids results in the formation of adducts like MDA-phosphatidylethanolamine, and carboxymethyl-phosphatidylethanolamine (Pamplona, 2008).

Extensive study of AGEs has revealed many stable end-stage adducts derived from the interactions between glycation-derived RCS and biomolecules. For instance, glycation intermediates have been demonstrated to react with guanidine groups of arginine residues, giving arginine-derived advanced glycation adducts: hydroimidazolones, argpyrimidine and Nω-carboxymethylarginine (Thornalley, 2005). Investigation of importance of glycolysis intermediates in the Maillard reaction has shown the formation of lisyl-hydroxy-triosidine and arginyl-hydroxy-triosidine during incubation of glyceraldehyde and glyceraldehyde-3-phosphate with N-alpha-acetyl lysine and N-alpha-acetyl arginine (Tessier et al., 2003). In addition, dihydroxyacetone can also form crosslinking triosidines. Acetaldehyde was shown to react rapidly with proteins, producing deep red macromolecular acetaldehyde-protein condensates (Robert et al., 2010). Among common AGEs found in a biological material are such compounds linking lysine and arginine as fluorescent pentosidine, and non-fluorescent glucosepan (Peyroux and Sternberg, 2006). The structural similarity of glucosepan and pentosidine (Figure 12) makes it obvious some parallelism in the respective pathways of their production.

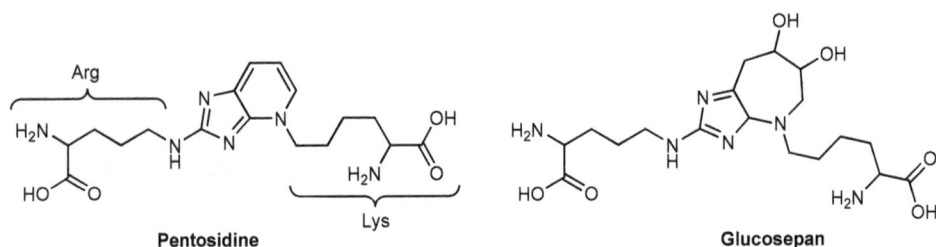

Pentosidine   Glucosepan

Fig. 12. The structures of the most common biological advanced glycation end products.

Physiological processes leading to ALE and AGE formation also involve chemical modifcation of biomolecules by GO and MGO derived from both LPO and glycation processes. Non-fluorescent crosslinks such as GO-lysine dimmer (GOLD) and MGO-lysine dimmer (MOLD), or non-fluorescent, non-crosslinking adducts such as carboxymethyllysine (CML), carboxymethylcysteine (CMC) and argpyrimidine are the most common ALEs/AGEs formed under protein modification (Figure 13). CML was the first AGE isolated from glycated proteins *in vivo* and together with pentosidine and glucosepan was recognized as one of the most important biomarkers of glycation in living organisms (Ahmed et al., 1986; Jadoul et al., 1999; Miyata et al., 1999; Tessier, 2010). Carboxymethyl-phosphatidylethanolamine (CMPE) and carboxymethylguanosine (CMG) represent the ALEs/AGEs derived from GO and MGO interation with nucleic acids and phospholipids, respectively (Figure 13).

In general, ALEs and AGEs are poorly degraded complexes, accumulation of which increases with ageing. The above mentioned ALEs/AGEs were detected in a variety of human tissues and serve as biomarkers of aging and age-related disorders (Tessier, 2010). It should be noted that ALEs/AGEs may continue covalent interations with biomolecules giving more complex cross-links. In addition, ALEs and AGEs are efficient sourses of RCS and ROS *in vivo* (Yim et al., 2001; Takamiya et al., 2003; Thornalley, 2005; Shumaev et al., 2009; Peng et al., 2011). Thus, increase in the concentration of RCS, ALEs and AGEs may result in carbonyl/oxidative stress.

Interplay Between Oxidative and Carbonyl Stresses: Molecular Mechanisms, Biological
Effects and Therapeutic Strategies of Protection

27

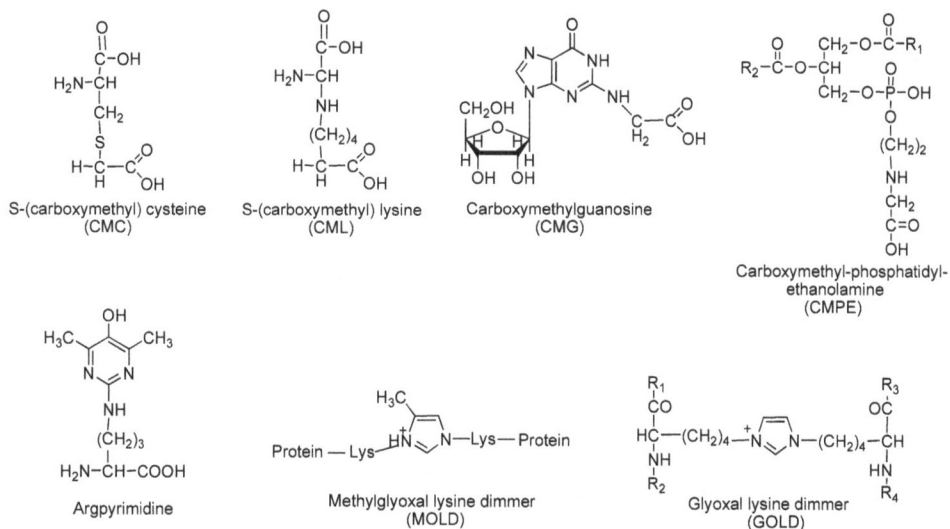

Fig. 13. The structures of the most common advanced end products derived from both glycation and lipoxidation.

3. Steady-state concentration of carbonyl compounds *in vivo* and carbonyl stress

Almost all known RS, in particular ROS and RCS, are continuously produced and eliminated in variuos groups of organisms: bacteria, fungi, plants, and animals (Ponces Freire et al., 2003; Mironova et al., 2005; Yamauchi et al., 2008; Lushchak, 2011b). Since RS are unstable and readily enter many reactions, their concentration is a dynamic parameter and defined as "steady-state". Under normal physiological conditions, the steady-state concentration of RS is maintained at certain range and fluctuates similarly to other parameters in the organism according to homeostasis theory. However, under some circumstances, the parameter may leave this range due to either increase in production or decrease in efficiency of catabolic system. The increase in the steady-state level of ROS or RCS may result in so-called "oxidative stress" or "carbonyl stress", respectively. One of the first definitions of "oxidative stress" as "an imbalance between oxidants and antioxidants in favour of the oxidants, potentially leading to damage" was proposed by Helmut Sies (Sies, 1985). Recently, "oxidative stress" was defined as "an acute or chronic increase in steady-state level of ROS, disturbing cellular metabolism and leading to damage of cellular constituents" (Lushchak, 2011a; b).

The concept of "carbonyl stress" was introduced for the first time by Miyata and colleagues (Miyata et al., 1999). They defined "carbonyl stress" as situation "resulting from either increased oxidation of carbohydrates and lipids (oxidative stress) or inadequate detoxification or inactivation of reactive carbonyl compounds derived from both carbohydrates and lipids by oxidative and nonoxidative chemistry". By analougy with the modern concept of oxidative stress, it can be proposed that "carbonyl stress" is an acute or chronic increase in steady-state level of RCS, ALEs and AGEs disturbing cellular metabolism and leading to damage of cellular constituents.

In some cases, the steady-state ROS/RCS concentration does not return to initial level, but stabilizes at new one called "quasi-stationary level" (Figure 14). This can be found in certain pathologies, for example diabetes mellitus, atherosclerosis, cardiovascular and neurodegenerative diseases.

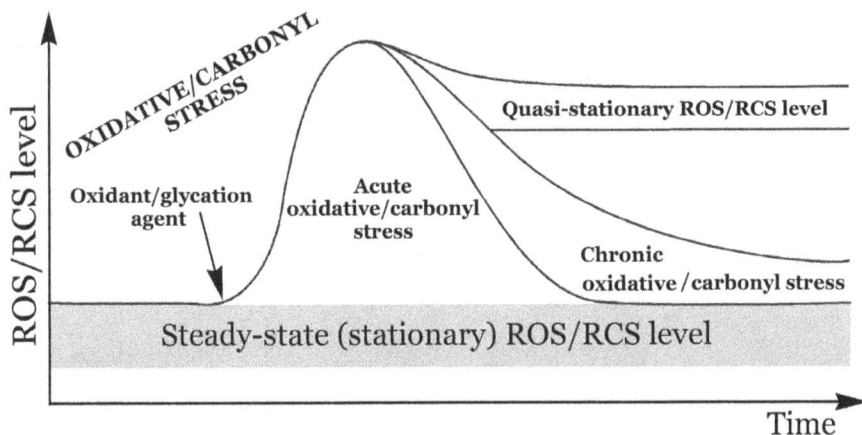

Fig. 14. The dynamics of glycoxidation/lipoxidation-induced perturbations of level of reactive oxygen and carbonyl species in living organisms (modified from (Lushchak, 2011b)).

An obvious question arises: what are the steady-state concentrations of RCS in the cell? Numerous studies demonstrate evaluation of RCS levels in biological systems. For example, it has been reported that overall concentration of LPO end products in plasma of healthy individuals is below 1 µM (Niki, 2009). At the same time, NHE has been found in biomembranes at the concentrations from 5 to 10 mM at oxidative insults (Esterbauer et al., 1991). The concentration of fructosamine in blood plasma of healthy individuals has been found about 140 µM (Kato et al., 1989). Physiological steady-state concentrations of MGO ranged from 120 to 650 nM (Kalapos, 2008a; Talukdar et al., 2009). Since there are no standard methods to evaluate the steady-state levels of RCS, different techniques applying in various laboratories yield different results. It should be noted that sometimes the increase in RCS levels at certain pathologies reported in one study is significantly lower than the normal levels demonstrated in the other. In addition, there are some objective complications in the evaluation of RCS steady-state level: (i) a vast variety of RCS generated by different mechanisms, leading to difficulties in the identification and quantification of all of them; (ii) simultaneous production, degradation and excretion of RCS; (iii) the influence of different factors (intensity of metabolism, oxygen concentration, temperature, etc.) on the rate of the above processes; and (iv) since the cell is not homogenous structure, in different cellular compartments RCS concentrations may differ to large extent.

Therefore, if we operate with some values reflecting RCS concentrations in biological material, one should be kept in mind that they are only approximate values. To know the levels of RCS in biological fluids and tissues is important to evaluate the extensity of carbonyl/oxidative stress, due to which even approximate assessment of RCS steady-state concentrations in biological material is much better, than the absence of any idea on their amounts in living organism.

Interplay Between Oxidative and Carbonyl Stresses: Molecular Mechanisms, Biological
Effects and Therapeutic Strategies of Protection

29

4. Lipoxidation and glycoxidation are processes linking carbonyl and oxidative stresses

Glycation and LPO are closely linked with oxidative stress (Figure 15). Oxidation reactions and ROS have been shown to be involved and frequently accelerate the advanced glycation process (Tessier, 2010). That is why "glycoxidation" term has been introduced (Dyer et a., 1991).

In 1980s, oxidative steps in Maillard reaction pathway were described for the first time (Hayashi and Namiki, 1980; Namiki and Hayashi, 1983). The authors found that the unstable Shiff bases could lead to the formation of reactive α-oxoaldehydes and hydrogen peroxide (Figure 8). Later other ROS were identified as products of Shiff base decomposition (Mullarkey et al., 1990). Wolff pathway (Figure 9), another oxidative pathway, was described several years later (Thornalley et al., 1984; Wolff et al., 1991). It has been suggested that metal-catalyzed autooxidation of reducing carbohydrates could be involved in the formation of AGEs and ROS. Amadori products were also shown to be capable to autoxidation (Figure 7) leading to the formation of reactive dicarbonyl compounds, superoxide anion and hydroxyl radical (Hunt et al., 1988; Mullarkey et al., 1990). Thus, all stages of glycoxidation generate ROS, some of them are common with LPO reactions (Figure 15). Chemical modification of amino groups in biomolecules during LPO is called lipoxidation. Products of lipoxidation were found to increase the concentration of some AGEs (Requena et al., 1996; Miyata et al., 2001). The level of proteins modified during LPO may serve as indicators of oxidative and carbonyl stresses *in vivo*. An example of a non-enzymatically modified amino acid is CML. Being either AGE or ALE, it is a good marker of both glycoxidation and lipoxidation reactions (Requena et al., 1996).

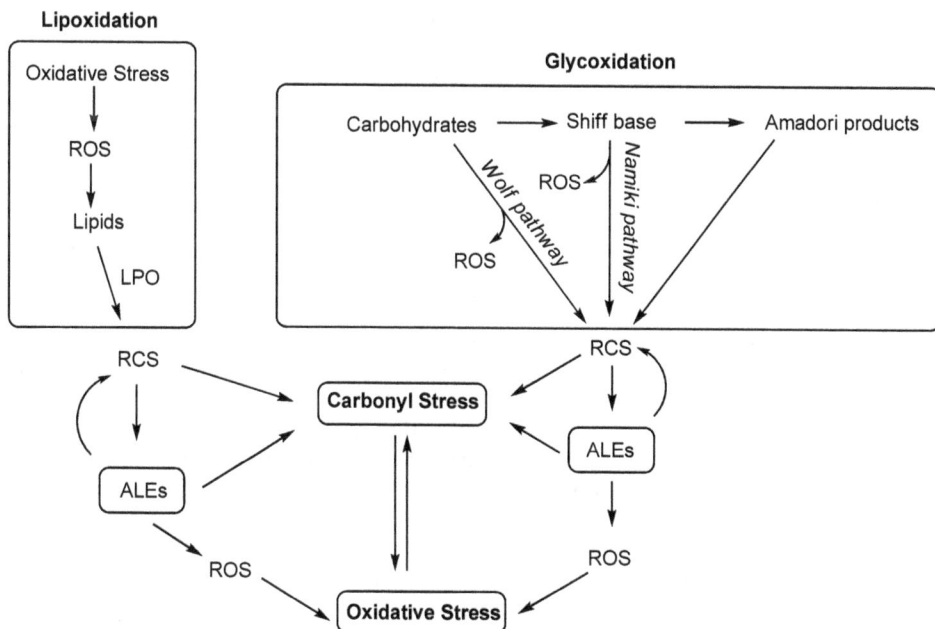

Fig. 15. Relation between lipoxidation, glycoxidation and carbonyl/oxidative stress.

The existence of a strong relation between carbonyl and oxidative stresses can be proved by ROS formation in the presence of low molecular mass reactive carbonyls. In 1993, MGO-induced ROS production was observed (Kalapos et al., 1993). A dose-dependent effect of MGO on ROS formation, mainly hydrogen peroxide, was detected in cultured rat hepatocytes. Enhanced superoxide anion generation was registered in Jurkat cells incubated with MGO also (Du et al., 2001). It is interesting that the inhibition of superoxide dismutase (SOD) by diethyldithiocarbamate doubled the rate of superoxide formation in the stressed cells.

Glycation is suggested to amplify oxidative stress in several ways. First of all, glycation of antioxidant enzymes can lead to increased ROS level. For instance, Cu, Zn-SOD was found to be highly susceptible to glycation (Arai et al., 1987; Takamiya, et al., 2003). Arai and colleagues (1987) demonstrated that glycation lead to gradual inactivation of the enzyme. The inactivation, in turn, can increase the ROS steady-state level and amplify the oxidative stress. Takamiya with co-authors (2003) showed that the mutated Cu,Zn-SOD was highly susceptible to glucation and fructation. Both glucated and fructated mutant also produced higher levels of hydrogen peroxide than the wild type. The authors suggested that high susceptibility of mutated Cu,Zn-SOD to glycation could be the origin of the oxidative stress associated with neuronal dysfunction in familial amyotrophic lateral sclerosis (Takamiya, et al., 2003). Low molecular mass reactive carbonyls (e.g. MGO) were found to decrease the level of reduced gluthatione (GSH) in different cells in vivo (Kalapos, 2008a). The situation seems to be more dramatic if one would take into account that GSH is a cofactor for certain antioxidant and antiglycated enzymes, thus an effective cellular protector against both oxidative and carbonyl stresses. Additionally, MGO was found to inhibit several other defensive enzymes (Kalapos, 1994), that makes the cell more succeptible to oxidative/carbonyl stress.

One more additional way to amplify the oxidative stress by glycation is the activation of membrane receptors by AGEs. For instance, AGEs were found to bind to the cell surface receptor (RAGE). This binding, in turn, triggers cellular events through p38 MAP Kinase, NF-kB, P21 Ras and Jak/STAT pathways (Uchida, 2000; Edeas et al., 2010). It is well known that NF-kB is a transcription factor that regulates different cellular functions. In particular, it activates TNF-α production that increases ROS generation. In conclusion, the increase in RCS steady-state concentrations can be also a consequence of oxidative stress. Thus, a vicious cycle can be formed.

5. Reducing carbohydrates as a factor of carbonyl/oxidative stress

From the above sections it may be concluded that, beside highly reactive low molecular mass RCS, carbohydrates such as glucose are quite important glycating agents. Despite some RCS demonstrate 20,000-fold higher reactivity than glucose (Turk, 2010), the latter is much more abundant intra- and extracellular glycation agent. Although only 0.001% of the total glucose in vivo present in relative unstable acyclic isomer form (capable of initiating glycation), it should be kept in mind that some highly reactive RCS are glycolytic intermediates or by-products of glucose metabolism. Thus it is widely believed that glucose is among the most important contributors to the glycation process. That is why potential role of reducing carbohydrates in the Maillard chemistry is extensively studied with different model systems. It was reported that some carbohydrates react with amino groups of biomolecules much faster than glucose (Sakai et al., 2002; Robert et al., 2010). However, since glucose is the most abundant intra- and extracellular monosaccharide in living

Interplay Between Oxidative and Carbonyl Stresses: Molecular Mechanisms, Biological
Effects and Therapeutic Strategies of Protection

31

organisms, most studies in the field of glycation are focused on the glucation. Despite fructose having a stronger reducing capacity and is a faster glycating agent than glucose (Sakai et al., 2002; Robert et al., 2010), fructation has attracted only a minor attention. At the same time, fructose is commonly used as an industrial sweetener and excessively consumed in human diets and as a glucose substitute by diabetes mellitus patients (Gaby, 2005; Tappy et al., 2010; Tappy and Lê, 2010). It is well documented that long-term consumption of excessive fructose is causative in the development of metabolic disorders (Levi and Werman, 1998; Johnson et al., 2007; Moheimani et al., 2010; Yang et al., 2011). But the mechanisms underlying fructose-induced metabolic disturbances are under debates. Recently, using baker's yeast as a model system we demonstrated that cells growing on fructose had higher levels of carbonyl groups in proteins, α-dicarbonyl compounds and ROS as compared with yeast cells growing on glucose (Semchyshyn et al., 2011). Possible mechanism for the generation of α-dicarbonyls and ROS by glucation and fructation is presented in Figure 16.

According to the Maillard reaction, the labile Schiff bases are formed as result of interaction between amino groups of biomolecules and carbonyl groups of reducing monosaccharides. The latter rearrange to form Amadori or Heyns products. Both compounds can be transformed to enediol proteins and then to alkoxyl radicals. Alkoxyl radicals readily react with molecular oxygen generating superoxide anion radical and α-dicarbonyl compounds. Then superoxide can be converted to hydrogen peroxide which produces highly reactive hydroxyl radical. In turn, α-dicarbonyl compounds and ROS are the major factors associated with oxidative/carbonyl stress.

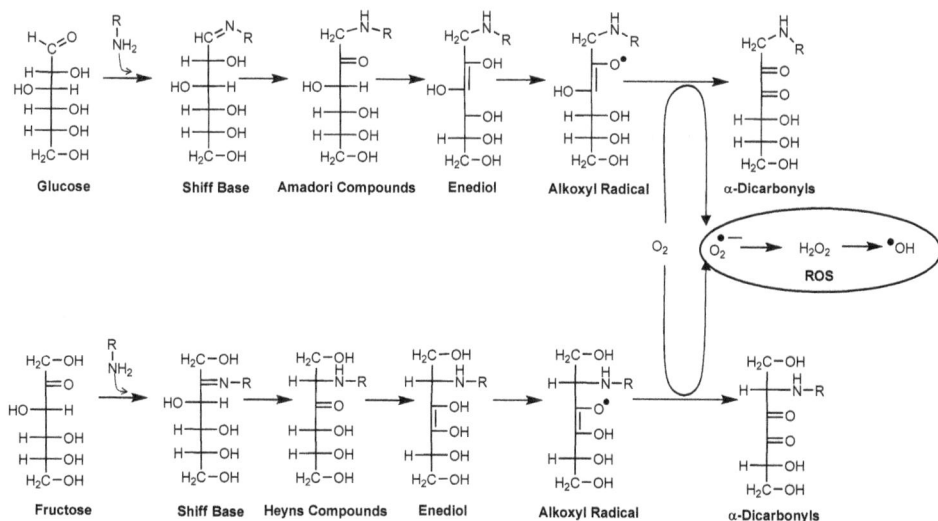

Fig. 16. Formation of α-dicarbonyl compounds and ROS by the glucation and fructation (modified from (Semchyshyn et al., 2011))

In general, our data explain the observation that fructose-supplemented growth as compared with growth on glucose resulted in more pronounced age-related decline in yeast reproductive ability and higher cell mortality (Semchyshyn et al., 2011; Lozinska and

Semchyshyn, 2011). We suggest that fructose rather than glucose is more extensively involved in glycation and ROS generation *in vivo*, yeast aging and development of carbonyl/oxidative stress.

6. Dual biological role of reactive carbonyl and oxygen species

Generally, RS are better known for their cytotoxic effects. As discussed in the preceding sections, glycation of amino groups of various biomolecules leads to diverse types of modifications. At the molecular level, modifications caused by RCS can: (*i*) disrupt the structure and function of proteins and enzymes, (*ii*) lead to formation of nucleic acid adducts, and (*iii*) damage lipids. Generated during the glycoxidation process ROS cause additional harmful impact. As a consequence of these undesirable effects at the cellular and organismal level, the loss of function and even viability can occur.

However, the persistence of RS in cells indicates their evolutionarily selected production in order to perform some useful role in cellular metabolism. The involvement of RCS and ROS in cell signal transduction and immune response is a prominent demonstration of their beneficial role. Understanding of the role of RS and the oxidation/glycation products in signalling has evolved rapidly during the last decades.

6.1 Cytotoxic, genotoxic and mutagenic effects

Proteins can be modified by a large number of reactions involving RCS, ROS as well as lipoxidation and glycoxidation advanced products (Uchida, 2000; Ellis, 2007; Lushchak 2007; Pamplona, 2011). The abovementioned compounds were found to interact directly with proteins causing alterations in physico-chemical properties such as conformation, charge, hydrophobicity, elasticity, solubility, electrophoretic mobility, and many others. Among proteins that can be potentially damaged are enzymes, membrane cytosolic and extracellular transporters, signalling components, transcription and growth factors, microtubules. Modification of proteins is often associated with the appearance of additional carbonyl groups, which leads to: (*i*) decrease in enzyme activity; (*ii*) inactivation or modification of regulatory properties; (*iii*) formation of intra- and intermolecular protein cross-links and adducts; (*iv*) alteration of protein degradation and formation of toxic aggregates which can inhibit proteasomes; (*v*) altered folding, processing, and trafficking of proteins; (*vi*) modification of extracellular matrix properties and cell-matrix interactions; and (*vii*) stimulation of autoimmune response (Ellis, 2007; Lushchak 2007; Thornalley, 2008; Lesgards et al., 2011; Pamplona, 2011).

Nucleic acids are the most favoured targetes for interaction with RS. It is well known that single stranded nucleic acids are particularly susceptible to modification by RS. The mutagenicity of such RCS as MGO and involvement of ROS in this effect is known for a long time (Kalapos, 2008a). Reactive carbonyl and oxygen species induce mutations frequently detected in oncogenes or tumor suppressor genes from human tumors, and correlate to alterations in cell cycle control and gene expression. Nucleic acid lipoxidative and glycoxidative damages were found in DNA of healthy humans and different animal species at biologically significant levels (Pamplona, 2011). In general, interaction between RS and nucleic acids was suggested to be implicated in carcinogenic, mutagenic and teratogenic actions of RCS and ROS, and products of these interations have powerful

Interplay Between Oxidative and Carbonyl Stresses: Molecular Mechanisms, Biological
Effects and Therapeutic Strategies of Protection

33

effects on signal transduction pathways (Ellis, 2007; Kalapos, 2008a; Thornalley, 2008; Niki, 2009; Pamplona, 2011).

Lipid reactions with RCS and ROS can contribute to: (*i*) modification of mitochondrial membranes disturbing cellular energetics, (*ii*) Ca^{2+} release and uncontrolled activation of Ca^{2+}-dependent pathways due to disruption of endoplasmic reticulum and mitochondrial membranes by RCS and ROS, and (*iii*) permeabilization of biological membranes and cell lysis (Negre-Salvayre et al., 2008; Pamplona, 2011). As noted above, increase in the levels of lipoxidation and glycoxidation products may lead to propagation of oxidative modifications.

In conclusion, the moleclar modifications caused by RCS, ROS, ALEs and AGEs are nonenzymatic, spontaneous, and random chemical reactions, therefore, the detailed mechanisms of their harmful effects are mostly unknown.

6.2 Reactive carbonyl and oxygen species are potential aging and age-related pathogenetic factors

Carbonyl/oxidative stress induces biomolecule dysfunctions and damages in different tissues, which are related to aging and age-related pathologies. In 1950s, Gercshman and colleagues proposed the free radical theory of diseases where suggested ROS involvement in damage of living cells and tissues (Gercshman et al., 1954). This was the background for free radical theory of aging developed by Harman somewhat later (1956). The free radical theory of aging states that ROS continuously damage proteins, lipids, and nucleic acids and thus cause the accumulation of molecular and cellular damages that are responsible for aging and age-related diseases. In 1980s, Monnier and Cerami postulated that the Maillard reaction has a causative role in aging and age-related pathologies (Monnier and Cerami, 1981). This theory called the "glycation hypothesis of aging" was at the origin of the growing interest in the field of the Maillard chemistry *in vivo* (Tessier, 2010). Nowadays it is obviously that there is a considerable overlap between carbonyl and oxidative stresses in relation to aging and age-related disturbances. As could be seen from the above sections, RCS can be produced as a consequence of oxidative stress, and ROS can be generated due to carbonyl stress.

The formation and accumulation of AGEs and ALEs have been known to progress in a normal aging process, and at an accelerated rate under age-related disorders. A survey of literature reveals the increase in the steady-state concentrations of ROS, RCS, AGEs and ALEs is related to hypertension, kidney and heart diseases, cancer growth and metastasis, obesity, metabolic syndrome, diabetes, degenerative bone disease, *etc.* (Uchida, 2000; Atanasiu et al., 2006; Ellis, 2007; Lushchak, 2007; Negre-Salvayre et al., 2008; Kalapos, 2008a; Yamagish, 2011; Yang et al., 2011). Chronic hyperglycemia is a major inducer of vascular complications of diabetes, which are responsible for disabilities and high mortality rates in patients with diabetes. Among the various biochemical pathways which are supposed to be involved in vascular complications in diabetes (*e.g.* heart disease, stroke, blindness and end-stage renal failure), the enhanced production of ROS, RCS, AGEs and ALEs and their action are most preferable.

Interestingly, the significant increase in dietary fructose over the past 30 years (Gaby, 2005; Tappy et al., 2010; Tappy and Lê, 2010) has recently been associated with the development and progression of various age-related disorders such as obesity, glucose intolerance, vascular complications of diabetes, cardiovascular and neurodegenerative diseases, fatty liver and non-alcoholic steatohepatitis (Levi and Werman, 1998; Johnson et al., 2007; Moheimani et al., 2010; Yamagish, 2011; Yang et al., 2011). However, short-term application of fructose demonstrated better protective action against oxidative stress than glucose (Spasojević et al., 2009a; 2009b). It was suggested that any excess of fructose should be eliminated from normal diets. On the other hand, short-term acute application of fructose has protective effects under pathophysiological conditions related to oxidative stress (Spasojević et al., 2009a; 2009b).

Carbonyl and oxidative stresses have been found to be causative in the activation of cell death pathways such as apoptosis and necrosis (Ellis, 2007; Negre-Salvayre et al., 2008; Uchida, 2010; Lushchak, 2011b; Yamagish, 2011). For instance, HNE alters mitochondrial calcium uptake and cytosolic calcium homoeostasis, which results in necrosis or apoptosis. Low molecular mass reactive carbonyls, MGO and GO, and AGEs are found to be pro-apoptotic through mechanisms involving calcium deregulation, GSH depletion, and activation of stress kinases (Bohlender et al., 2005; Negre-Salvayre et al., 2008; Yamagish, 2011).

6.3 AGE-Receptors, signal transduction and oxidative stress

In the middle of the 1980s, it was demonstrated that macrophages could specifically recognize, uptake and degrade AGE-modified proteins *in vitro* (Vlassara et al., 1985). This observation led to an active search for high affinity AGE receptors on various cells. Now it is well known that the influence of RCS, ROS, ALEs and AGEs on cell is mostly mediated by receptors (Robert 2010; Tessier, 2010). The first discovered cellular surface multiligand receptor capable to bind AGE-modified proteins with high affinity was RAGE (the receptor for AGE) (Schmidt et al., 1992). Extensive search for new AGE receptors resulted in identification of macrophage scavenger receptors (MSR) type A and B1 (CD36), oligosaccharyl transferase-48 termed AGE receptor 1 (AGE-R1), 80K-H phosphoprotein (AGE-R2), and galectin-3 (AGE-R3), but the best studied is the RAGE receptor (Bohlender et al., 2005; Peyroux and Sternberg, 2006). The latter is a 35-kDa protein belonging to the immunoglobulin superfamily. RAGE is a transmembrane receptor consisting of 394 amino acid residues with a single hydrophobic transmembrane domain of 19 amino acids and a COOH-terminal cytosolic tail of 43 amino acids (Bohlender et al., 2005).

In the presence of extracellular AGE, susceptible cells can rapidly upregulate expression of RAGE. Intracellular oxidative stress is among multiple effects caused by interaction of AGEs with RAGE on the membrane of different cells. RAGE activation induces intracellular generation of hydrogen peroxide dependent on NADPH oxidase. The identified signalling cascade involved p21ras, p38, protein kinase C, and MAP kinases (Bohlender et al., 2005; Ucida, 2000). The activation of nuclear factor NF-kB due to AGE and RAGE interaction was also shown to be involved in the regulation of the gene transcription for various factors: endothelin-1, vascular endothelial growth factor (VEGF), transforming growth factor β (TGF-β), and tumor necrosis factor α (TNF-α) (Peyroux and Sternberg, 2006). Also, NF-kB controls the expression of almost 100 proinflammatory genes encoding cytokines, adhesion molecules, and ROS/RCS generating enzymes such as NADPH-oxidase, superoxide dismutase, inducible nitric oxide synthase and myeloperoxidase (Bohlender et al., 2005; Anatasiu et al., 2006; Peyroux and Sternberg, 2006; Yamagishi, 2011).

Interplay Between Oxidative and Carbonyl Stresses: Molecular Mechanisms, Biological
Effects and Therapeutic Strategies of Protection

35

Therefore, with identification of AGE-recognizing receptors it was realized that glycoxidation plays an important role in age-dependent pathologies and chronic diseases such as diabetes and its complications, inflammation, neurodegeneration, atherosclerosis, amyloidoses, and tumors. It was also found that glycoxidation influenced cell death and proliferation. It cannot be excluded that cross-talk between different receptors activated by AGEs may also be involved in the observed effects. Obviously, the complex RS metabolism is complicated by no less complex molecular mechanisms of their biological effects.

6.4 Potential beneficial impacts of RCS

Although RCS are better known for their harmful effects, their persistence in living organisms indicates that RCS production was evolutionarily selected in order to perform some useful role in cellular metabolism. For instance, phagocytic white blood cells are of central importance in host defense mechanisms implicate RS against invading pathogens. It is demonstrated that, besides ROS, myeloperoxidase generates such RCS as glycolaldehyde, 2-hydroxypropanal, and acrolein (Anderson et al., 1997). Synthesis of RCS by myeloperoxidase required a free hydroxy-amino acid. Since the total concentration of free amino acids in plasma is about 4 mM (Anderson et al., 1997), it seems RCS derived from amino acids are the major products of phagocyte activation *in vivo*, being important biological weapon fighting infections. Mavric and colleagues (2008) found 3-DG, GO, and MGO in New Zealand Manuka honey possessing very high antibacterial activity. Interestingly, MGO was found to be present at concentrations from 38 to 761 mg/kg, which is up to 100-fold higher compared to conventional honey. Minimum concentrations needed for inhibition of bacterial growth for GO was 6.9 mM (*Escherichia coli*) or 4.3 mM (*Staphylococcus aureus*). At the same time, MGO was found to inhibit growth of both bacteria at concentration 1.1 mM, and 3-DG showed no inhibition in concentrations up to 60 mM. The results clearly demonstrated that the pronounced antibacterial activity of New Zealand Manuka honey directly originated from MGO. Besides its antibacterial activity, MGO demonstrates antiviral effect against New-Castle disease, influenza *etc.* (Talukdar et al., 2009). Antimalarial and anticancer activities are also among potential beneficial impacts of MGO (Pavlovic-Djuranovic et al., 2006; Talukdar et al., 2009).

7. Cellular defense against glycation

The use of a cytotoxic chemical as a signalling molecule, immunological weapon or therapeutic agent obviously has potential risks, so it is no surprise that RCS elimination is tightly regulated in the cell. Similarly to the enzymatic defence against oxidative stress, cell possesses the enzymatic protection system against glycation and carbonyl stress. The antiglycation system includes the enzymes that suppress the formation of RCS and glycoxidative products operating at the respective sites of the process.

Many RCS derived from lipoxidation or glycoxidation can be efficiently metabolized in the enzymatic reactions. They are oxidation, reduction and conjugation, which involve aldehyde and alcohol dehydrogenases, aldo-keto reductases, carbonyl reductases, cytochromes P450, and glutathione-S-transferases (Atanasiu et al., 2006; Ellis, 2007). In these reactions, reactive carbonyls and their advanced products are mainly converted to less toxic compounds, and excreted from cell or organism.

One of the main pathways known to be involved in the catabolism of α-oxoaldehydes to α-hydroxy acids is catalyzed by the glyoxalase system possessed by animals, plants, protoctista, fungi, and bacteria cells (Atanasiu et al., 2006; Ellis, 2007; Xue et al., 2011). The glyoxalase system comprises two cytosolic enzymes: glyoxalase I (Glo I) and glyoxalase II (Glo II), and catalytic amount of GSH. For example, it catalyzes MGO conversion to D-lactate through a specific 2-step pathway (Figure 17). Glyoxalase I catalyzes isomerization of the hemithioacetal, formed spontaneously from α-oxoaldehydes and GSH, into S-2-hydroxyacylglutathione derivates. Glyoxalase II catalyzes the conversion of S-2-hydroxyacylglutathione derivates into α-hydroxyacids with subsequent regeneration of GSH consumed in the reaction catalyzed by glyoxalase I. The glyoxalase system was discovered in 1913 (Dakin and Dudley, 1913; Neuberg, 1913; Xue et al., 2011). However, for a long period the biological sence of the pathway catalyzed by the system was under debates. In 1980-1990s, Thornalley suggested the fundamental function of glyoxalases in the metabolism of reactive dicarbonyl metabolites to less reactive products (Thornalley, 1990). Nowadays, numeruos experimental data confirm the relation of decrease in the glyoxalase activities to aging and age-related pathologies (Xue et al., 2011).

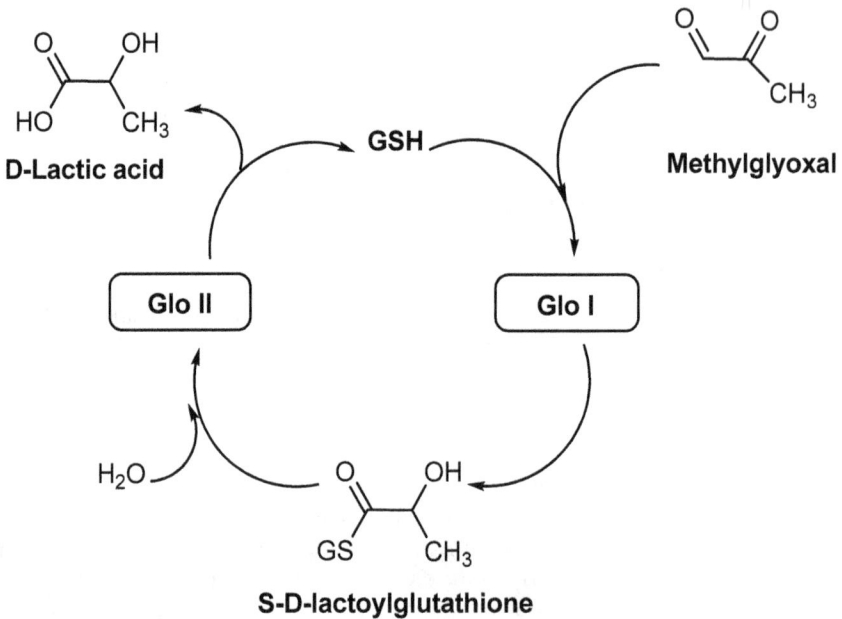

S-D-lactoylglutathione

Fig. 17. Methylglyoxal degradation by the glyoxalase system.

The antiglycation system also includes highly specific and efficient enzymes recognizing and decomposing Amadori products. Commonly they are called "amadoriases". The latters are divided into two groups, operating via different deglycating mechanisms (Wu and Monnier, 2003): fructosylamine oxidases (Figure 18) and fructosylamine kinases (Figure 19). As seen from the figures, a significant action of the amadoriases is connected to the transformation of reactive carbonyls to less reactive products.

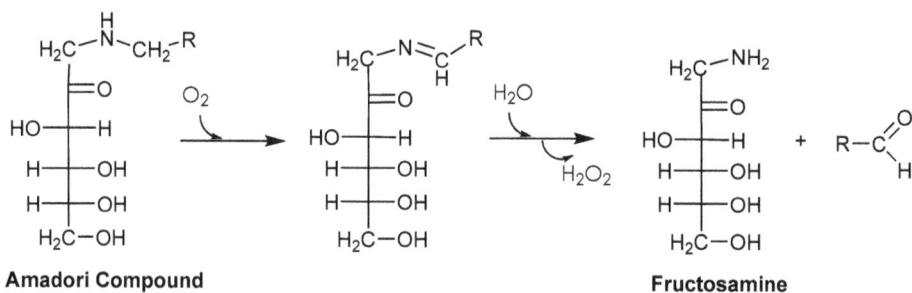

Amadori Compound   **Fructosamine**

Fig. 18. Degradation of Amadori compound by fructosylamine oxidase.

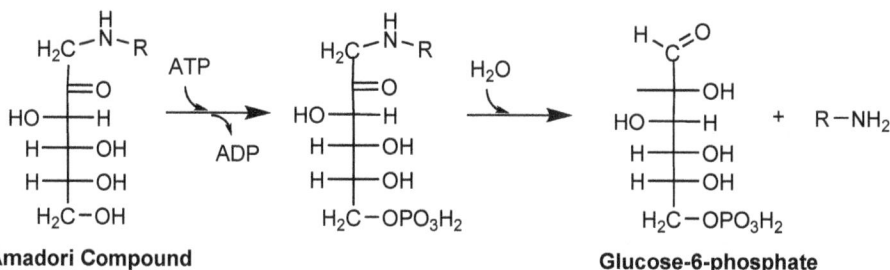

Amadori Compound   **Glucose-6-phosphate**

Fig. 19. Degradation of Amadori compound by fructosylamine kinase.

The enzymatic mechanisms described in this section let us conclude that reactive carbonyls, and related carbonyl stress are not only natural attributes of life, but they have also been a determinant factor, which demanded functional adaptations living organisms that, in turn, determined their viability and longevities. Moreover, in the past years glyoxalase as well as amadoriases have been targeted for the development of novel antitumor, antiprotozoal, antifungal and antibacterial agents.

8. Therapeutic strategies to fight against carbonyl/oxidative stress

Considering the emerging deleterious role of reactive carbonyls and their advanced products in different human diseases, various potential therapeutic strategies have been developed last years. Different stages of RCS, ALE and AGE formation and AGE-mediated damage are suggested as therapeutic intervention strategies (Peyroux and Sternberg, 2006; Aldini et al., 2007; Peng et al., 2011). Peyroux and Sternberg (2006) classified these strategies as follows: (*i*) trapping of reactive dicarbonyl species, (*ii*) AGE cross-link cleavage, (*iii*) AGE receptor blocking, (*iv*) AGE receptor signalling blocking, (*v*) glycemia reduction by anti-diabetic therapy, (*vi*) aldose reductase inhibition; (*vii*) shunting of trioses-phosphate toward the pentose-phosphate pathway by transketolase activation; (*viii*) antioxidant therapy by transition metal chelation and free radical scavenging. As a consequence, many compounds that prevent the formation of AGEs or degrade the existing AGEs have been produced in recent years. Among others are amadoriases, aminoguanidine, pyridoxamine, drugs used in the treatment of type 2 diabetes such as metformin and pioglitazone, angiotensin receptor blockers and inhibitors of angiotensin converting enzyme, amino group capping agents such as aspirin, compounds that mostly break alpha-dicarbonyl cross-links such as phenacylthiazolium bromide.

Since oxidative stress accompanies and accelerates carbonyl stress, antioxidant compounds appear to be promising agents for the prevention of AGE and ALE formation. The antiglycation activities of medical plant materials and naturally occurring phenolic compounds with antioxidant properties are of particular interest (Peyroux and Sternberg, 2006; Edeas, 2010; Peng et al., 2011).

9. Concluding remarks and perspectives

The continuous production and catabolism of RCS result in maintaining of certain steady-state level in living organisms. Being intermediates of certain metabolic pathways, their level to some extent may be regulated. But substantial RCS portion is produced in non-enzymatic chemical processes and, therefore, is difficult to be controlled by living organisms. The same can be said on the elimination of RCS due to concerted activity of enzymatic and non-enzymatic mechanisms. Therefore, the steady-state RCS level is controlled by the cell to large extent, but some room is left out of the cellular control.

High chemical reactivity of RCS determines their biological activity mainly deleterious for living organisms. If at normal physiological state the production and catabolism are well balanced, at certain conditions RCS steady-state level may be transiently increased leading to development of so-called "acute carbonyl stress". Since the cell possesses multilevel defense system, it can combat RCS and return their level into initial steady-state range. The protective potential can be enhanced by up-regulation of defense mechanisms such as glutathione production, specific enzymes like glutathione-S-transferases or glyoxalases. In this case, short-term RCS-induced stress may have no series consequences for living organisms, and even can be beneficial because of increased capability to eliminate RCS. Moreover, it can also increase cell capability to combat other RS like ROS. However, under some circumstances the defense systems of organisms may be overwhelmed and the RCS steady-state level can be maintained increased for a long time. The situation may be called "chronic carbonyl stress". This one can result in the development of different pathological states alone or in combination with other mechanisms leading to diseases and aging. To date, general aspects of RCS homeostasis have been delineated, but a lot of aspects have not been clarified.

Some data shed light on the interplay between the stresses induced by RCS and ROS and in some cases the both may lead to the formation of a vicious cycle. On the other hand, interplay between these stresses, especially under well controlled conditions may be used as preventive approach. For example, the activity of glutathione-S-transferase is up-regulated by both ROS and RCS. It can be used to enhance a potential of RCS defense mechanisms by the induction of mild oxidative stress. Since the molecular details of up-regulation of antioxidant mechanisms is much better studied, the lessons from it can be transferred to RCS field. In some cases, the antioxidant potential can be enhanced ever without induction of oxidative stress and this provides some clues to protect organisms against deleterious RCS effects. Since RCS have been shown to be related to many diseases, the described above aspects can be used to identify appropriate therapeutic targets.

Interplay Between Oxidative and Carbonyl Stresses: Molecular Mechanisms, Biological
Effects and Therapeutic Strategies of Protection

39

Although there are many identified RCS, there is no doubt that in future new species will be identified. Especially it is true for big biologically important molecules like proteins and nucleic acids. However, the development of standardized conventional methods for RCS identification would provide better conditions for inter-laboratory comparison of data received. These new techniques are more than welcomed. Due to important RCS role in the development of various pathologies, the reliable measurement methods would provide a progress not only in basic investigations, but also in applied ones.

The role of RCS in regulation of cellular processes is another topic waiting for investigation. In addition to described to date influence on NF-kB, MAP kinases, TNF one can expect at least their modulatory effects on diverse regulatory cascades via direct interaction with components of regulatory cascades, or indirect via change in level or properties of connected partners.

Currently, we are only at the beginning of therapeutic targeting disorders connected with deleterious RCS effects. Probably, glutathione and glutathione-dependent enzymes along with other components of antioxidant system are the best studied targets from medical point of view. Much less is known on the use of specific compounds which may help to combat RCS-induced or supported pathologies. Although the detail molecular mechanisms are not known, dipeptide carnosine (β-alanyl-L-histidine) and related compounds was proposed to be used as potential beneficial agents to cure diseases related with the stress induced by reactive carbonyls. Since carnosine is found in animals, it is used as an argument for carnivorous against vegetarian diets. There is no doubt that new discoveries in RCS homeostasis will open new avenues for modulation of processes caused by them.

10. Acknowledgment

We would like to express our sincere gratitude and appreciation to Professor M. Carini and Professor M. P. Kalapos for sending a valuable information. Our special thank to PhD students Liudmyla Lozinska and Olexandr Lozinsky for excellent technical assistance in the illustration preparation.

11. Abbreviations

3DG, 3-deoxyglucosone; AGEs, advanced glycation end products; ALEs, advanced lipoxidation end products; CMC, carboxymethylcysteine; CMG, carboxymethylguanosine; CML, carboxymethyllysine; CMPE, carboxymethyl-phosphatidylethanolamine; Glo I, glyoxalase I; Glo II, glyoxalase II; GO, glyoxal; GOLD, glyoxal-lysine dimer; GSH, reduced glutathione; HNE, 4-hydroxy-trans-2-nonenal; LPO, lipid peroxidation; MDA, malondialdehyde; MGO, methylglyoxal; MOLD, methylglyoxal-lysine dimer; ONE, 4-oxo-trans-2-nonenal; PUFAs, polyunsaturated fatty acids; RAGE, the receptor for AGE; RCS, reactive carbonyl species; ROS, reactive oxygen species; RS, reactive species; SOD, superoxide dismutase;TNF-α, tumor necrosis factor α; TGF-β, transforming growth factor β; VEGF, vascular endothelial growth factor.

12. References

Ahmed, MU., Thorpe, SR. & Baynes, JW. (1986). Identification of N epsilon-carboxymethyllysine as a degradation product of fructoselysine in glycated protein, *J. Biol. Chem.* 261(11):4889-4894.

Aldini, G., Dalle-Donne, I., Facino, RM., Milzani, A. & Carini, M. (2007). Intervention strategies to inhibit protein carbonylation by lipoxidation-derived reactive carbonyls, *Med. Res. Rev.* 27(6):817-868.

Anderson, MM., Hazen, SL., Hsu, FF. & Heinecke, JW. (1997). Human neutrophils employ the myeloperoxidase-hydrogen peroxide-chloride system to convert hydroxy-amino acids into glycolaldehyde, 2-hydroxypropanal, and acrolein. A mechanism for the generation of highly reactive alpha-hydroxy and alpha, beta-unsaturated aldehydes by phagocytes at sites of inflammation, *J. Clin. Invest.* 99(3):424-432.

Atanasiu, V., Stoian, I., Manolescu, B. & Lupescu, O. (2006). The glyoxalase system – a link between carbonilic stress and human therapy, *Rev. Roum. Chim.* 51(9):861–869.

Birlouez-Aragon, I., Morales, F., Fogliano, V. & Pain, JP. (2010). The health and technological implications of a better control of neoformed contaminants by the food industry, *Pathol. Biol. (Paris).* 58(3):232-238.

Bohlender, JM., Franke, S., Stein, G. & Wolf, G. (2005). Advanced glycation end products and the kidney, *Am. J. Physiol. Renal. Physiol.* 289: F645 – F659.

Chaplen, FWR., Fahl, WE. & Cameron DC. (1996). Detection of methylglyoxal as a degradation product of DNA and nucleic acid components treated with strong acid, *Anal. Biochem.* 236:262–269.

Chung, SS., Ho, EC., Lam, KS. & Chung SK. (2003). Contribution of polyol pathway to diabetes-induced oxidative stress, *J. Am. Soc. Nephrol..* 14:S233-236.

Colombo, G., Aldini, G., Orioli, M., Giustarini, D., Gornati, R., Rossi, R., Colombo, R., Carini, M., Milzani, A. & Dalle-Donne I. (2010). Water-soluble α,β-unsaturated aldehydes of cigarette smoke induce carbonylation of human serum albumin, *Antioxid. Redox Signal.* 12(3):349-364.

Dakin, HD., Dudley, HW. (1913). An enzyme concerned with the formation of hydroxyl acids from ketonic aldehydes, *J. Biol. Chem.* 14:155–157.

Demple, B. (1991). Regulation of bacterial oxidative stress genes, *Annu. Rev. Genet.* 25:315-337.

Dini, L. (2010). Phagocytosis of dying cells: influence of smoking and static magnetic fields, *Apoptosis.* 15(9):1147-1164.

Du, J., Suzuki, H., Nagase, F., Akhand, AA., Ma, XY., Yokoyama, T., Miyata, T. & Nakashima, I. (2001). Superoxide-mediated early oxidation and activation of ASK1 are important for initiating methylglyoxal-induced apoptosis process, *Free Rad. Biol. Med.* 31:469–478.

Dyer, DG., Blackledge, JA., Katz, BM., Hull, CJ., Adkisson, HD., Thorpe, SR., Lyons, TJ. & Baynes, JW. (1991). The Maillard reaction *in vivo*, *Z. Ernahrungswiss.* 30(1):29-45.

Edeas, M., Attaf, D., Mailfert, A.-S., Nasu, M. & Joubet, R. (2010). Maillard Reaction, mitochondria and oxidative stress: Potential role of antioxidants, *Pathol. Biol. (Paris).* 58(3):220-225.

Ellis, EM. (2007). Reactive carbonyls and oxidative stress: potential for therapeutic intervention, *Pharmacol. Ther.* 115(1):13-24.

Esterbauer, H., Schaur, RJ. & Zollner, H. (1991). Chemistry and biochemistry of 4-hydroxynonenal, malondialdehyde and related aldehydes, *Free Radical .Biol. Med.* 11:81–128.

Esterbauer, H., Schaur, RJ. & Zollner, H. (1991). Chemistry and biochemistry of 4-hydroxynonenal, malondialdehyde and related aldehydes, *Free Radic. Biol. Med.* 11:81–128.

Finot, PA. (1982). Nonenzymatic browning products: physiologic effects and metabolic transit in relation to chemical structure. *Diabetes*. 31(3):22-28.

Gaby, AR. (2005). Adverse effects of dietary fructose, *Altern. Med. Rev.* 10(4):294-306.

Gerschman, R., Gilbert, DL., Nye, SW., Dwyer, P. & Fenn, WO. (1954). Oxygen poisoning and x-irradiation: a mechanism in common, *Science*. 119(3097):623-626.

Halliwell, B., Gutteridge, J.M.C. (1989). Free Radicals in Biology and Medicine. Clarendon Press, Oxford.

Harman, D. (1956). Aging: a theory based on free radical and radiation chemistry, *J. Gerontol*. 11(3):298-300.

Hayashi, T. & Namiki, M. (1980). Formation of two-carbon sugar fragments at an early stage of the browning reaction of sugar and amine, *Agric.Biol.Chem.* 44: 2575-2580.

Hodge, JE. (1953). Chemistry of browning reactions in model systems, *J. Agric. Food Chem.* 1:928-943.

Hunt, JV., Dean, RT. & Wolff, SP. (1988). Hydroxyl radical production and autoxidative glycosylation, *Biochem. J.* 256: 205-212.

Jadoul, M., Ueda, Y., Yasuda, Y., Saito, A., Robert, A., Ishida, N., Kurokawa, K., Van Ypersele De Strihou, C. & Miyata, T. (1999). Influence of hemodialysis membrane type on pentosidine plasma level, a marker of "carbonyl stress", *Kidney Int.* 55(6):2487-2492.

Johnson, RJ., Segal, MS., Sautin, Y.,Nakagawa, T., Feig, DI.,Kang, DH., Gersch, MS.,Benner, S. & Sanchez-Lozada, LG. (2007). Potential role of sugar (fructose) in the epidemic of hypertension, obesity and the metabolic syndrome, diabetes, kidney disease, and cardiovascular disease, *Am. J. Clin. Nutr.* 86:899–906.

Kalapos, MP. (2008a). The tandem of free radicals and methylglyoxal, *Chem. Biol. Interact.* 171(3):251-271.

Kalapos, MP. (2008b). Methylglyoxal and glucose metabolism: a historical perspective and future avenues for research, *Drug. Metabol. Drug. Interact.* 23(1-2):69-91.

Kalapos, MP., Littauer, A. & de Groot, H. (1993). Has reactive oxygen a role in methylglyoxal toxicity? A study on cultured rat hepatocytes, *Arc. Toxicol.* 67:369–372.

Kato, M., Nakayama, H., Makita, Z., Aoki, S., Kuroda, Y., Yanagisawa, K. & Nakagawa, S. (1989). Radioimmunoassay for non-enzymatically glycated serum proteins, *Horm. Metab. Res.* 21:245-248.

Lankin, VZ., Tikhaze, AK., Kapelko, VI., Shepelkova, GS., Shumaev, KB., Panasenko, OM., Konovalova, GG. & Belenkov, YN. (2007). Mechanisms of oxidative modification of low density lipoproteins under conditions of oxidative and carbonyl stress, *Biochemistry (Mosc)*. 72(10):1081-1090.

Lesgards, J.-F., Gauthier, C., Iovanna, J., Vidal, N., Dolla, A. & Stocker, P. (2011). Effect of reactive oxygen and carbonyl species on crucial cellular antioxidant enzymes, *Chem. Biol. Interact.* 190:28–34.

Levi, B. & Werman, MJ. (1998). Long-term fructose consumption accelerates glycation and several age-related variables in male rats, *J. Nutr.* 128(9): 1442-1449.

Liu, X.-Y., Zhu, M.-X. & Xie, J.-P. (2010). Mutagenicity of acrolein and acrolein-induced DNA adducts, *Toxicol. Mech. Meth.* 20(1):36-44.

Lozinska LM., Semchyshyn HM. (2011). Fructose as a factor of carbonyl/oxidative stress development and accelerated aging in the yeast *Saccharomyces cerevisiae*, *Ukrainian Biochem. J.* 83(4):62-71.

Lushchak, VI. (2007). Free radical oxidation of proteins and its relationship with functional state of organisms, *Biochemistry (Mosc)*. 72(8):809-827.

Lushchak, VI. (2011a). Environmentally induced oxidative stress in aquatic animals, *Aquat. Toxicol.* 101(1):13-30.

Lushchak, VI. (2011b). Adaptive response to oxidative stress: Bacteria, fungi, plants and animals, *Comp. Biochem. Physiol. C Toxicol. Pharmacol.* 153(2):175-190.

Lushchak, VI., Semchyshyn, HM., Lushchak OV. (2011c). "Classic" methods for measuring of oxidative damage: TBARS, xylenol orange, and protein carbonyls, in textbook: Oxidative Stress in Aquatic Ecosystems, editors D. Abele, T. Zenteno-Savin, J. Vazquez-Medina, Blackwell Publishing Ltd., 420-431.

Lyles, GA. & Chalmers, J. (1992) The metabolism of aminoacetone to methylglyoxal by semicarbazide-sensitive amine oxidase in human umbilical artery, *Biochem. Pharmacol.* 43:1409–1414.

Maillard, LC. (1912). Action des acides aminés sur les sucres: formation des mélanoïdines par voie méthodique. *C. R. Acad. Sci.* 154:66–68.

Mavric, E., Wittmann, S., Barth, G. & Henle, T. (2008). Identification and quantification of methylglyoxal as the dominant antibacterial constituent of Manuka (Leptospermum scoparium) honeys from New Zealand, *Mol. Nutr. Food Res.* 52(4):483-489.

Mironova, R., Niwa, T., Handzhiyski, Y., Sredovska, A. & Ivanov, I. (2005). Evidence for non-enzymatic glycosylation of *Escherichia coli* chromosomal DNA, *Mol. Microbiol.* 55(6):1801–1811.

Miyata ,T., Saito, A., Kurokawa, K. & van Ypersele de Strihou C. (2001). Advanced glycation and lipoxidation end products: reactive carbonyl compounds-related uraemic toxicity, *Nephrol. Dial. Transplant.* 16(4):8-11.

Miyata, T., van Ypersele de Strihou, C., Kurokawa, K. & Baynes, JW. (1999). Alterations in nonenzymatic biochemistry in uremia: origin and significance of "carbonyl stress" in long-term uremic complications, *Kidney Int.* 55(2):389-399.

Moheimani, F., Morgan, PE., van Reyk, DM. & Davies, MJ. (2010). Deleterious effects of reactive aldehydes and glycated proteins on macrophage proteasomal function: possible links between diabetes and atherosclerosis, *Biochim. Biophys. Acta.* 1802(6):561-571.

Monnier, VM. & Cerami, A. (1981). Nonenzymatic browning *in vivo*: possible process for aging of long-lived proteins, *Science.* 211:491 – 493.

Mullarkey, CJ., Edelstein, D. & Brownlee, M. (1990). Free radical generation by early glycation products: a mechanism for accelerated atherogenesis in diabetes, *Biochem. Biophys. Res. Commun.* 173(3):932-939.

Namiki, M. & Hayashi, T. (1983). A new mechanism of the Maillard reaction involving sugar fragmentation and free radical formation. In: Waller, GR. & Feather, MS. (Eds). The Maillard reaction in foods and nutrition. ACS Symposium, Series 215. Washington DC: American Chemical Society.

Negre-Salvayre, A., Coatrieux, C., Ingueneau, C. & Salvayre, R. (2008). Advanced lipid peroxidation end products in oxidative damage to proteins. Potential role in diseases and therapeutic prospects for the inhibitors. *Br. J. Pharmacol.* 153(1):6-20.

Negre-Salvayre, A., Coatrieux, C., Ingueneau, C. & Salvayre, R. (2008). Advanced lipid peroxidation end products in oxidative damage to proteins. Potential role in diseases and therapeutic prospects for the inhibitors, *Br. J. Pharmacol.* 153(1):6-20.

Neuberg, C. (1913). The destruction of lactic aldehyde and methylglyoxal by animal organs, *Biochem. Z.* 49:502–506.

Niki, E. (2000). Free radicals in the 1900's: from *in vitro* to *in vivo. Free Radic. Res.* 33(6):693-704.

Niki, E. (2009). Lipid peroxidation: Physiological levels and dual biological effects. *Free Radic. Biol. Med.* 47:469–484.

Niwa, T. (1999). 3-Deoxyglucosone: metabolism, analysis, biological activity, and clinical implication, *J. Chromatogr. B. Biomed. Sci. Appl.* 731:23–36.

Pamplona, R. (2008). Membrane phospholipids, lipoxidative damage and molecular integrity: A causal role in aging and longevity. *Biochim. Biophys. Acta.* 1777:1249–1262.

Pamplona, R. (2011). Advanced lipoxidation end-products, *Chem. Biol. Interact.* 192(1-2):14-20.

Pavlovic-Djuranovic, S., Kun, JF., Schultz, JE. & Beitz, E. (2006). Dihydroxyacetone and methylglyoxal as permeants of the Plasmodium aquaglyceroporin inhibit parasite proliferation, *Biochim. Biophys. Acta.* 1758(8):1012-1017.

Peng, X., Ma, J., Chen, F. & Wang, M. (2011). Naturally occurring inhibitors against the formation of advanced glycation end-products, *Food Funct.* 2(6):289-301.

Peyroux, J. & Sternberg, M. (2006). Advanced glycation endproducts (AGEs): Pharmacological inhibition in diabetes, *Pathol. Biol. (Paris).* 54(7):405-419.

Phillips, SA. & Thornalley, PJ. (1993). The formation of methylglyoxal from triose phosphates. Investigation using a specific assay for methylglyoxal, *Eur. J. Biochem.* 212(1):101-105.

Pompliano, D.L, Peyman, A. & Knowles, JR. (1990). Stabilization of a reaction intermediate as a catalytic device: definition of the functional role of the flexible loop in triosephosphate isomerase, *Biochemistry.* 29(13):3186-3194.

Ponces Freire, A., Ferreira, A., Gomes, R. & Cordeiro, C. (2003). Anti-glycation defences in yeast, *Biochem. Soc. Trans.* 31(6):1409-1412.

Rahbar, S. (1968). Hemoglobin H disease in two Iranian families, *Clin. Chim. Acta.,* 20(30):381-385.

Rahbar, S., Blumenfeld, O. & Ranney, HM. (1969). Studies of an unusual hemoglobin in patients with diabetes mellitus, *Biochem. Biophys. Res. Commun.* 36(5):838-843.

Reihl, O., Rothenbacher, TM., Lederer, MO. & Schwack, W. (2004). Carbohydrate carbonyl mobility--the key process in the formation of alpha-dicarbonyl intermediates, *Carbohydr. Res.* 339(9):1609-1618.

Requena, JR., Fu, MX., Ahmed, MU., Jenkins, AJ., Lyons, TJ. & Thorpe, SR. (1996). Lipoxidation products as biomarkers of oxidative damage to proteins during lipid peroxidation reactions, *Nephrol. Dial. Transplant.* 11(5):48-53.

Richard, JP. (1993). Mechanism for the formation of methylglyoxal from triosephosphates, *Biochem. Soc. Trans.* 21(2):549-553.

Robert, L., Labat-Robert, J. & Robert, AM. (2010). The Maillard reaction. From nutritional problems to preventive medicine, *Pathol. Biol. (Paris).* 58(3):200-206.

Robert, L., Robert, A.-M. & Labat-Robert, J. (2011). The Maillard reaction – Illicite (bio)chemistry in tissues and food, *Pathol. Biol. (Paris).* 59(6):321-328.

Sakai, M., Oimomi, M. & Kasuga, M. (2002). Experimental studies on the role of fructose in the development of diabetic complications, *Kobe J. Med. Sci.* 48(5-6):125-136.

Schmidt, AM., Vianna, M., Gerlach, M., Brett, J., Ryan ,J., Kao, J., Esposito, C., Hegarty, H., Hurley, W., Clauss, M., Feng, W., Pan, YCE., Tsang, TC. & Stern, D. (1992). Isolation and characterization of two binding proteins for advanced glycosylation end products from bovine lung which are present on the endothelial cell surface, *J. Biol. Chem.* 267(21):14987-1497.

Semchyshyn, HM., Lozinska, LM., Miedzobrodzki, J., Lushchak, VI. (2011). Fructose and glucose differentially affect aging and carbonyl/oxidative stress parameters in *Saccharomyces cerevisiae* cells, *Carbohydr. Res.* 346(7):933-938.

Seo, Y.-K. & Baek, S.-O. (2011). Characterization of carbonyl compounds in the ambient air of an industrial city in Korea, *Sensors.* 11(1):949-963.

Shao, B., Fu, X., McDonald, TO., Green, PS., Uchida, K., O'Brien, KD., Oram, JF. & Heinecke, JW. (2005). Acrolein impairs ATP binding cassette transporter A1-dependent cholesterol export from cells through site-specific modification of apolipoprotein A-I, *J. Biol. Chem.* 280(43):36386-36396.

Shumaev, KB., Gubkina, SA., Kumskova, EM., Shepelkova, GS., Ruuge, EK. & Lankin VZ. (2009). Superoxide formation as a result of interaction of L-lysine with dicarbonyl compounds and its possible mechanism, *Biochemistry (Mosc).* 74(4):461-466.

Sies, H. (1985). Oxidative stress: introductory remarks. In Sies, H. (Ed.). Oxidative Stress. London: Academic Press. pp. 1–8.

Sies, H. (1986). Biochemistry of oxidative stress. Angewaudte Chemie, International Edition in English 25:1058-1071.

Sies, H. (1991). Oxidative stress: introduction. In Oxidative Stress: Oxidants and Antioxidants. Sies, H. (Ed.). pp. xv-xxii. Academic Press, London.

Spasojević, I., Bajić, A., Jovanović, K., Spasić, M. & Andjus, P. (2009b). Protective role of fructose in the metabolism of astroglial C6 cells exposed to hydrogen peroxide, *Carbohydr. Res.* 344(13):1676-1681.

Spasojević, I., Mojović, M., Blagojević, D., Spasić, SD., Jones, DR., Nikolić-Kokić, A. & Spasić, MB. (2009a). Relevance of the capacity of phosphorylated fructose to scavenge the hydroxyl radical, *Carbohydr. Res.* 344(1):80-84.

Takamiya, R., Takahashi, M., Myint, T., Park, YS., Miyazawa, N., Endo, T., Fujiwara, N., Sakiyama, H., Misonou, Y., Miyamoto, Y., Fujii, J. & Taniguchi, N. (2003). Glycation proceeds faster in mutated Cu, Zn-superoxide dismutases related to familial amyotrophic lateral sclerosis, *FASEB J.* 17(8):938-940.

Talukdar, D., Chaudhuri, BS., Ray, M. & Ray, S. (2009). Critical evaluation of toxic versus beneficial effects of methylglyoxal, *Biochemistry (Mosc).* 74(10):1059-1069.

Tappy, L. & Lê, KA. (2010). Metabolic effects of fructose and the worldwide increase in obesity, *Physiol. Rev.* 90(1):23-46.

Tappy, L., Lê, KA., Tran, C. & Paquot, N. (2010). Fructose and metabolic diseases: new findings, new questions, *Nutrition.* 26(11-12):1044-1049.

Tessier, FJ. (2010). The Maillard reaction in the human body. The main discoveries and factors that affect glycation, *Pathol. Biol. (Paris).* 58(3):214-219.

Tessier, FJ., Monnier, VM., Sayre, LM. & Kornfield, JA. (2003). Triosidines: novel Maillard reaction products and cross-links from the reaction of triose sugars with lysine and arginine residues, *Biochem. J.* 369(3):705-719.

Thornalley PJ. (2005). Dicarbonyl intermediates in the Maillard reaction, *Ann. N Y Acad. Sci.* 1043:111-117.

Interplay Between Oxidative and Carbonyl Stresses: Molecular Mechanisms, Biological
Effects and Therapeutic Strategies of Protection

45

Thornalley, PJ. (1990). The glyoxalase system: new developments towards functional characterization of a metabolic pathway fundamental to biological life, *Biochem. J.* 269(1):1-11.

Thornalley, PJ. (2008). Protein and nucleotide damage by glyoxal and methylglyoxal in physiological systems--role in ageing and disease, *Drug Metabol. Drug Interact.* 23(1-2):125-150.

Thornalley, PJ., Wolff, SP., Crabbe, J. & Stern, A. (1984). The autoxidation of glyceraldehyde and other simple monosaccharides under physiological conditions catalysed by buffer ions, *Biochim. Biophys. Acta.* 797:276-287.

Trotter, EW., Collinson, E J. Dawes, IW. & Grant CM. (2006). Old yellow enzymes Protept against acrolein toxicity in the yeast *Saccharomyces cerevisiae* , *Appl. Environ. Microbiol.* 72(7):4885–4892.

Turk, Z. (2010). Glycotoxines, carbonyl stress and relevance to diabetes and its complications, *Physiol Res.* 59(2):147-156.

Uchida, K. (2000). Role of reactive aldehyde in cardiovascular diseases, *Free Radic. Biol. Med.* 28(12):1685-1696.

Uchida, K., Sakai, K., Itakura, K., Osawa, T. & Toyokuni, S. (1997). Protein modification by lipid peroxidation products: formation of malondialdehyde-derivedNε-(2-Propenal)lysine in proteins, *Arch. Biochem. Biophys.*346(1):45-52.

Uribarri, J. & Tuttle, KR. (2006). Advanced glycation end products and nephrotoxicity of high-protein diets, *Clin. J. Am. Soc. Nephrol.* 1(6):1293-1299.

Uribarri, J., Cai, W., Peppa, M., Goodman, S., Ferrucci, L., Striker, G. & Vlassara H. (2007). Circulating glycotoxins and dietary advanced glycation endproducts: two links to inflammatory response, oxidative stress, and aging. *J. Gerontol. A. Biol. Sci. Med. Sci.* 62(4):427–433.

Uribarri, J., Woodruff, S., Goodman, S., Cai, W., Chen, X., Pyzik, R., Yong, A., Striker, GE. & Vlassara H. (2010). Advanced glycation end products in foods and a practical guide to their reduction in the diet. *J. Am. Diet. Assoc.* 110(6):911-916.

Vlassara, H., Brownlee, M. & Cerami, A. (1985). High-affinity-receptor-mediated uptake and degradation of glucose-modified proteins: A potential mechanism for the removal of senescent macromolecules, *Proc. Natl. Acad. Sci. USA.* 82: 5588-5592.

Witz, G. (1989). Biological interactions of α,β-unsaturated aldehydes, *Free Radical. Biol. Med.* 7:333–349.

Wolff, SP., Jiang, ZY. & Hunt, JV. (1991). Protein glycation and oxidative stress in diabetes mellitus and ageing, *Free Radic. Biol. Med.* 1(5):339-352.

Wu, X. & Monnier, V.M. (2003). Enzymatic deglycation of proteins. *Arch. Biochem. Biophys.* 419:16–24.

Xue, M., Rabbani, N. & Thornalley, PJ. (2011). Glyoxalase in ageing, *Semin. Cell Dev. Biol.* 22(3):293-301.

Yamagishi, S. (2011). Role of advanced glycation end products (AGEs) and receptor for AGEs (RAGE) in vascular damage in diabetes, *Exp. Gerontol.* 46(4):217-224.

Yamauchi, Y., Furutera, A., Seki, K., Toyoda, Y., Tanaka, K. & Sugimoto, Y. (2008). Malondialdehyde generated from peroxidized linolenic acid causes protein modification in heat-stressed plants, *Plant. Physiol. Biochem.* 46(8-9):786-93.

Yang, K., Feng, C., Lip, H., Bruce, W.R. & O'Brien PJ. (2011). Cytotoxic molecular mechanisms and cytoprotection by enzymic metabolism or autoxidation for glryceraldehyde, hydroxypyruvate and glycolaldehyde, *Chem. Biol. Interact.* 191(1-3):315-21.

Yatscoff, RW., Tevaarwerk, GJM. & MacDonald, JC. (1984). Quantification of nonenzymically glycated albumin and total serum protein by affinity chromatography, *Clin. Chem.* 30:446–449.

Yim, MB., Yim, HS., Lee, C., Kang, SO. & Chock, PB. (2001). Protein glycation: creation of catalytic sites for free radical generation, *Ann. N Y Acad. Sci.* 928:48-53.

Havrilla, CM., Morrow, JD. & Porter, NA. (2002). Formation of isoprostane bicyclic endoperoxides from the autoxidation of cholesteryl arachidonate, *J. Am. Chem. Soc.* 124:7745–7754.

Zimniak, P. (2008). Detoxification reactions: relevance to aging, *Ageing Res. Rev.* 7(4):281-300.

Zimniak, P. (2011). Relationship of electrophilic stress to aging. *Free Radic. Biol. Med.* 51(6):1087-10105.

3

Iron, Oxidative Stress and Health

Shobha Udipi[1], Padmini Ghugre[1] and Chanda Gokhale[2]
[1]Department of Food Science & Nutrition,
[2]S.P.N. Doshi Women's College & Dr. Nanavati BM College of Home Science,
S.N.D.T. Women's University,
India

1. Introduction

Iron is an element of crucial importance to living cells. It has incompletely filled 'd' orbitals and exists in a range of oxidation states, the most common being ferrous [Fe II (d^6)] and ferric [Fe III(d^5)] forms. By virtue of this unique electrochemical property, iron is an ideal redox active cofactor for many biological processes and fundamental biochemical activities in all cells. It can associate with proteins; bind to oxygen (O_2), transfer electrons and mediate catalytic reactions. Enzymes of the citric acid cycle – succinate dehydrogenase and aconitase are iron-dependent. Iron is a critical component of heme in hemoglobin (Hb), myoglobin, cytochromes as well as iron–sulfur complexes of the electron transport chain. Iron is also required for activity of ribonucleoside reductase, the rate-limiting enzyme of the first metabolic reaction committed to DNA synthesis. Therefore, iron plays an important role in metabolic processes including O_2 transport, electron transport, oxidative phosphorylation and energy production, xenobiotic metabolism, DNA synthesis, cell growth, apoptosis, gene regulation and inflammation (Zhang and Enns, 2006; Cairo and Recalcati, 2007; He et al., 2007; Outten and Theil, 2009; Wang and Pantopoulos, 2011). In the central nervous system, iron is required for myelogenesis and myelin maintenance by oligodendrocytes. It is also a necessary cofactor for the synthesis of neurotransmitters, dopamine, norepinephrine and serotonin (He et al., 2007). Therefore deficiency of iron can result in myriad disorders. Even mild iron deficiency can adversely affect cognitive performance, behavior, physical growth of infants, preschool and school age children and physical work capacity and work performance of adults (Brabin et al, 2001; Haas and Brownlie, 2001).

2. Iron homeostasis

The redox reactivity of iron makes it extremely useful but the same property makes it a toxic entity, because of its propensity to generate free radicals if it is not tightly bound and/or it is present in excess. Due to this dual nature, the human body possesses elegant and elaborate control mechanisms to maintain iron homeostasis by coordinately regulating iron absorption, iron recycling and mobilization of stored iron. Disruption of these processes can result in deviation from normal iron levels both systemically and at cellular level, which can lead to deleterious consequences.

Homeostasis of all essential metal ions share common features. However, compared to other metals, the body contains much higher levels of iron (3.5 - 4 g of iron versus 100 mg of copper), aqueous ferric ions exhibit low solubility and unlike other metals, excess iron is not actively excreted via the kidney. Also free iron can catalyze formation of reactive oxygen species (ROS) and must be safely bound by specialized proteins. Therefore additional iron-specific homeostatic features are required (Theil and Goss, 2009; Wessling-Resnick, 2010). Iron homeostasis must be tightly controlled at both systemic and cellular levels to provide optimum amounts of iron at all times and yet maintain the delicate balance between iron nutrition and toxicity, especially because the human body does not possess mechanisms for getting rid of the excess metal.

(a)Systemic homeostasis: Maintenance of stable extracellular iron concentration requires the coordinated regulation of iron transport into plasma from (i) dietary iron absorbed in the intestines (ii) recycled senescent red blood cells (RBC) and (iii) storage in hepatocytes (Figure 1). A major indicator and determinant of systemic iron homeostasis is the saturation of plasma transferrin (Tf). Saturation levels are predominantly determined by the amount of iron entering from the above three sources and the amount utilized for erythropoiesis (Mucken et al., 2008).

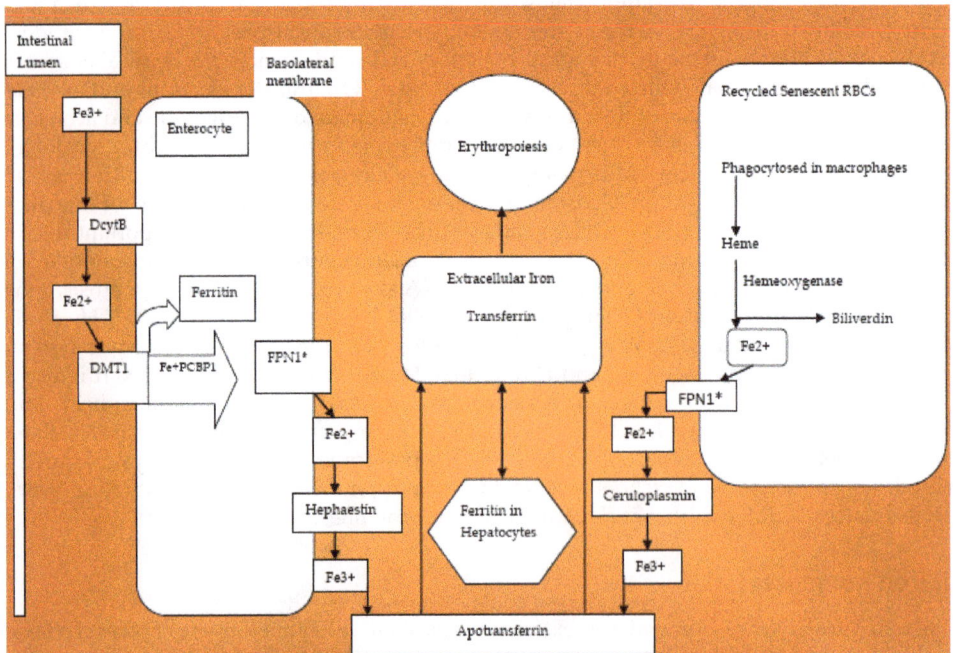

DcytB- Ferric Reductase Enzyme- Duodenal cytochrome B. DMT- Divalent Metal Transporter. PCBP- Poly(rc)-binding proteine. FPN- Ferroportin
*regulated by Hepicidin

Fig. 1. Systemic Iron Homeostasis

Intestinal absorption of dietary iron involves transport across the apical membrane followed by translocation through the cytoplasm and across the basolateral membrane into portal circulation. Non-heme iron which comprises a major component of dietary iron is taken up from the lumen via the transmembrane protein divalent metal transporter (DMT1) located at the enterocyte apical membrane and actively transported as Fe (II) ion. Prior to transport, dietary Fe (III) is reduced by the ferric reductase enzyme – duodenal cytochrome B (DCytb) (Knutson, 2010). The percentage of iron absorbed from the amount ingested is generally higher in case of heme than non-heme iron. As the iron intake increases, the percent of ingested dose that is absorbed reduces; however, the absolute amount absorbed and likely to enter systemic circulation is more when intakes are high as encountered in supplementation and perhaps with fortified foods. Within the enterocyte, due to its reactive nature, iron is bound to carrier proteins- poly(r c) -binding protein (PCBP-1) which acts as a cytoplasmic chaperon in delivering excess iron to ferritin (Shi et al., 2010). Apparently PCBP1 translocates iron to the basolateral membrane where iron is transported out of the enterocytes by the iron export protein ferroportin (FPN1). Hephaestin is thought to oxidize Fe (II) to Fe (III) which is rapidly sequestered by apotransferrin (Knutson, 2010; Muchenthaler, Galy and Hentz, 2008).

Intestinal sources provide only 1-2 mg of iron per day. The daily demand for 20 mg of iron for erythropoiesis is largely met by recycled iron from senescent erythrocytes processed by the reticuloendothelial system (RES) consisting of specialized macrophages present mainly in the liver, spleen and bone marrow. Within macrophages, heme derived from phagocytized RBCs is catalyzed by hemeoxygenase (HO). The liberated iron is released via FPN1, donated to Tf and reutilized (Kohgo et al., 2008; Tanna and Miller, 2010). The iron that is not used for metabolic purposes is stored in ferritin. The RES represents a major iron storage compartment. Ferritin messenger RNAs (mRNAs) are among the most abundant mRNAs in monocyte-derived macrophages. Liver has a high iron storage capacity and accumulates iron mostly in the periportal regions with a decreasing gradient towards the centrilobular areas. Iron stores are depleted in case of increased requirements or excessive iron loss (Muckenthaler, Galy and Hentz, 2008).

Iron export from duodenal enterocytes, macrophages and hepatocytes through FPNI appears to be a limiting step, modulated by the peptide hormone, hepcidin. Binding of hepcidin to FPN1 triggers the internalization and degradation of the receptor-ligand complex. Once internalized, the hepcidin-FPN1 complex is degraded in lysosomes and cellular iron export ceases (Nemeth and Ganz, 2009; Knutson, 2010). Hepcidin expression is regulated by a number of proteins in response to both extracellular and intracellular iron (Figure 2). Two main iron-related signaling pathways, one involving bone morphogenetic proteins (BMPs – these comprise a group of at least 20 soluble molecules belonging to the transforming growth factor β superfamily) and the other dependent on transferrin receptor II (TfRII) regulate hepcidin regulation. Iron sensing by BMP pathway is mediated through BMP-6, as their levels in mice increased with dietary iron overloading and decreased with deficiency (Kantz et al., 2008). BMP-6 binds to BMP receptor (BMPR) complex comprised of BMPRI and BMPRII. After binding of BMP-6, BMPRII phosphorylates BMPRI which then propagates the signal by catalyzing phosphorylation of cytoplasmic Smad1/Smad5/Smad8. Its association with the common mediator Smad4 is followed by translocation to the nucleus where they act as transcription factors. BMP signaling is modulated by coreceptor hemojuvelin (HJV) a cell surface protein (Knutson, 2010). All these proteins are important for hepcidin expression. Disruption of Smad 4 and HJV in mice have been reported to result in iron overload (Nemeth and Ganz, 2009).

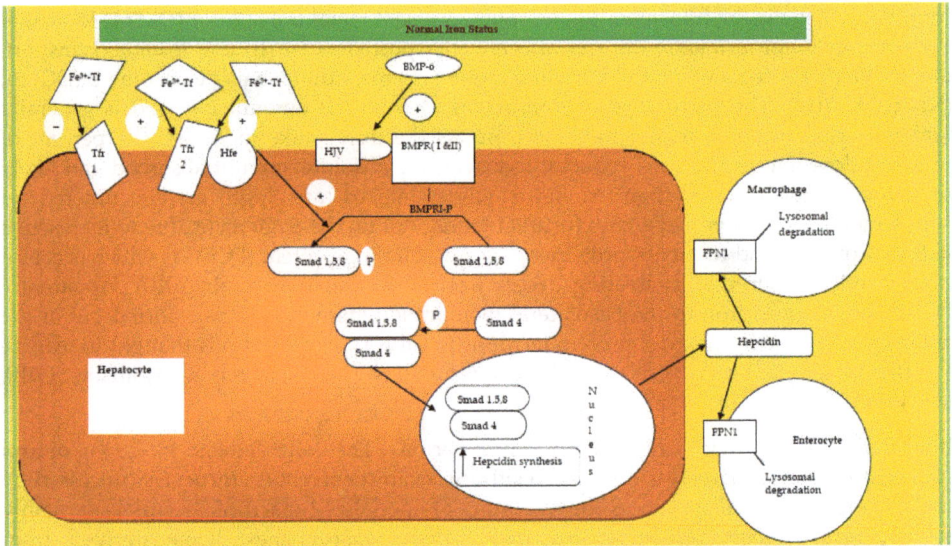

HJV- Hemojuvenil, BMP Bone Morphogenic Protein, BMPR- Bone Morphogenic receptor, FPN-ferroportin

Fig. 2. Regulation of Hepidin Expression

Hepcidin regulation by the TfRII dependent pathway requires holotransferrin and high-Fe gene (HFE) which shuttles between TfRI and TfRII depending on the holotransferrin concentration. In high iron conditions, holotransferrin binds to TfRI, displaces HFE and allows it to interact with TfRII. Gao et al., (2000) suggested that interaction between TfRII and HFE is required for hepcidin induction in response to iron. There appears to be crosstalk between the BMP and TfRII dependent signaling converging into a common pathway for hepcidin expression in response to iron (Nemeth and Ganz, 2009; Knutson, 2010). Hepcidin expression is apparently regulated by iron stores, although the mechanism of intracellular sensing is unclear. Hepcidin is down regulated in response to iron deprivation but how this occurs remains to be fully elucidated. Cell surface HJV appears to be cleared by Matriptase-2, thereby reducing BMP signaling. Matripatase deficiency leads to iron overload phenotypes (Knutson, 2010).

Hepcidin expression is suppressed in anemias with ineffective erythropoiesis, mediated by proteins produced by erythroid precursors. Hypoxia also decreases hepcidin, being modulated by hypoxia-inducible factor. However hepcidin synthesis rapidly increases during infection and inflammation (Nemeth and Ganz, 2009). Inappropriate products of hepcidin have been implicated in pathogenesis of various iron disorders.

(b) Cellular homeostasis: The cytosolic iron pool is highly regulated because it is an important source of iron for numerous cytosolic and nuclear iron proteins, as well as being a source from which mitochondria and other organelles derive iron (Rouault and Cooperman, 2006). Cellular iron levels are balanced by divergent yet coordinated regulation of proteins involved in iron uptake (TfR), export, storage (ferritin) and distribution. Within cells, iron is stored in ferritin. Also, labile iron pool (LIP) is present within cells. Most LIP is free iron bound to small organic chelators - citrate or adenosine phosphate, carboxylate, polypeptides, (Kohgr et al., 2008; Jiang et al., 2009). LIP is biologically active in intracellular metabolism but toxic if present in excess. Maintenance of appropriate LIP levels is critical for homeostasis (Mackenzie, Iwasaki and Tsuji, 2008).

Iron regulatory proteins (IRP) are critical components of a sensory and regulatory network controlling iron homeostasis by exerting genetic control at multiple steps (Goforth et al.,2010). The expression of TfR and ferritin are coordinately regulated upon binding of IRP1 and IRP2 to iron regulatory elements (IRE) in the untranslated regions (UTRs) of their respective genes. IRPs are activated by iron deficient conditions, bind to their target IREs and modulate translation of mRNA (Figure 3). The location of IREs in target mRNA determines whether regulation is positive or negative. In case of TfR1 mRNA, five IREs are located in 3' UTR whereas the mRNA coding for ferritin contains a single IRE in the 5' UTR. When iron is depleted, IRE-IRP interaction at the 3' UTR of TfR1 stabilizes mRNA, while imposing a steric blockade to ferritin mRNA translation. As a result, TfR1 is upregulated and stimulates acquisition of iron from plasma Tf to counteract low iron levels. Simultaneously ferritin synthesis is inhibited as storage of iron is not needed.

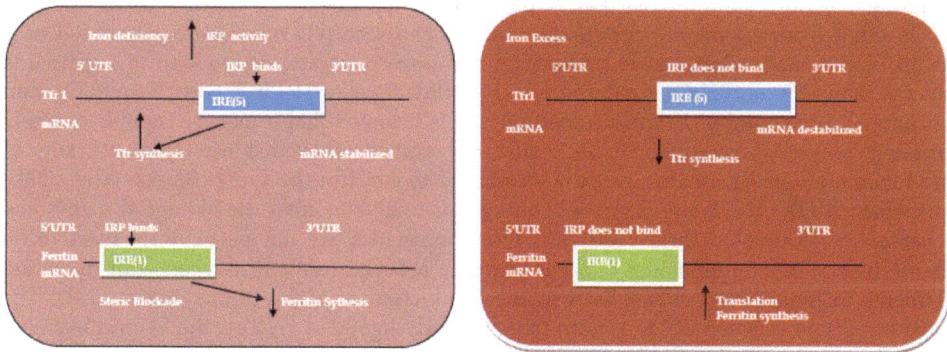

Fig. 3. Iron Status and IRP Regulation of Transferrin Receptors and Ferritin Synthesis

Conversely, when iron concentration is high, IRPs are inactivated resulting in degradation of TfR1 mRNA while translation of ferritin mRNA occurs. Thus when iron levels exceed the cellular needs, the IRE/IRP switch minimizes iron uptake and promotes iron storage (Mackenzie, Iwasaki and Tsuji, 2008; Wang and Pantopoulos, 2011). However, in erythroid progenitor cells which require high amounts of iron for heme synthesis, TfR1 stability is uncoupled from iron supply and TfR1 level is transcriptionally regulated by erythroid active element in its promoter region (Wang and Pantopoulos, 2011). Also TfR1 is regulated in hypoxic conditions, wherein dimerized hypoxia-induced factors (HIF-1α and HIF 1β) bind to hypoxia response element located upstream of the transcription start site (Mackenzie,Iwasaki and Tsuji, 2008).

IRPs, in addition to modulating ferritin and TfR1 levels can regulate the mRNAs or other proteins required for iron utilization (erythroid 5-aminolevulanata synthase), iron uptake (DMT1) and export (FPN1). DMT1 mRNA has IRE in its 3' UTR and is upregulated by iron deficiency while the other two proteins have IREs in their 5' UTR and are therefore upregulated under high iron conditions (Cairo and Recalcati, 2007; Wang and Pantopoulos, 2011). It appears that the functional significance of IRE/IRP system is beyond cellular iron uptake and storage. The regulation of FPN1 at the basolateral surface of enterocytes by both IRE/IRP and hepcidin indicates that the two regulatory systems are interconnected (Muckenthaler et al., 2008). Similar mechanisms may apply to other cells and work is warranted to understand the functional interconnection between hepcidin and IRE/IRP systems.

3. Iron as a pro-oxidant

A variety of mechanisms tightly regulate iron homeostasis in the human body. However, iron can be toxic if its level and/or distribution are not carefully regulated. Iron toxicity is developed through the production of (ROS). ROS are formed during normal cellular metabolism. Superoxide anion ($O_2 \bullet^-$) is generated continuously by the mitochondrial electron transport system, myeloid cells as well as during several cellular oxidase catalyzed reactions. Mitochondrial electron transport system is not the only electron transport system producing $O_2 \bullet^-$. Non mitochondrial electron transport chains such as mono oxygenase system (microsomal or P_{450}), photosynthetic electron transport chain and others are critically important. Besides xanthine oxidase, NADPH oxidase is one of the major sources of $O_2 \bullet^-$, while other oxidases mainly produce hydrogen peroxide (H_2O_2). H_2O_2 is generated as a result of enzymatic (superoxide dismutase) and non-enzymatic destruction of superoxide anions. Both $O_2 \bullet^-$ and H_2O_2 are found in extracellular spaces and in blood. Since H_2O_2 is an uncharged compound of low molecular mass, it is believed that it may be capable of freely crossing biological membranes and diffusing considerable distance from its site of production. However, further work is warranted to confirm the same (Branco et al., 2004; Lushchak, 2011). $O_2 \bullet^-$ being negatively charged is unable to cross the biological membrane whereas its protonated form $HO\bullet$ can. These molecules are not particularly reactive by themselves. It is the interaction of these partially reduced forms of oxygen with transition elements including iron, which lead to the production of highly damaging radicals. The most important reaction of H_2O_2 with free or poorly liganded $Fe(II)$ is the Fenton reaction, generating the highly reactive hydroxyl ions.

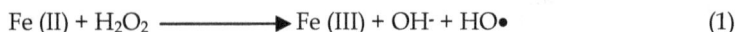

$$Fe\,(II) + H_2O_2 \longrightarrow Fe\,(III) + OH^- + HO\bullet \qquad (1)$$

Superoxide can also react with ferric iron in the Haber – Weiss reaction to produce Fe (II) again thereby effecting redox cycling.

$$O_2\bullet^- + Fe\,(III) \longrightarrow O_2 + Fe\,(II) \qquad (2)$$

Further, ascorbic acid has a number of known interactions with metal ions. Ascorbate can replace $O_2\bullet^-$ within the cell for reducing the Fe (III) to Fe (II), thereby facilitating the Fenton reaction. Also, metal ion catalysis of oxidation of ascorbic acid leads to formation of H_2O_2, ultimately generating hydroxyl ions. This generation of hydroxyl ions from the simple system containing metal ions, ascorbic acid and oxygen has potentially deleterious consequences.

Hydroxyl radical ($HO\dot{}$) is highly toxic among the partially reduced oxygen species. It reacts with all kinds of biological macromolecules including cellular DNA, protein, carbohydrates and lipids. In the nucleus, hydroxyl radicals cause DNA damage, especially double strand breaks as well as chemical changes in deoxyribose, purines and pyrimidines. Hydroxyl radical can be added onto C-8 of guanine leading to guanine modification. 8-hydroxy-deoxyguanosine is an important indicator of DNA damage caused by iron-mediated Fenton reaction. Many of the damaged proteins are enzymes, hence critical cellular functions including ATP generation are adversely affected. Fe (II) reacting with unsaturated fatty acids in the presence of H_2O_2 can initiate lipid peroxidation cascade in biological membranes and lipoproteins by production of HO•. Lipid peroxidation increases membrane fragility of cell organelles such as mitochondria, lysosomes and endoplasmic reticulum leading to impaired cell function (Fardy and Silverman, 1995; Casanueva and Viteri,2003; Kell, 2009 and de La Rosa et al., 2008). Thus HO˙ mediate many higher level manifestations of tissue damage, disease and organ failure and ultimately death.

Normally, the potential toxicity of iron is minimal because it is complexed with Tf in plasma and with ferritin which "minimizes opportunities for uncontrolled iron dioxygen chemistry and oxidative stress", within the cells (Kapralov et al, 2009). In addition, there are several chaperones and iron-binding proteins, reducing the chances of free or unligated iron becoming available to participate in the Fenton reaction. However, any small amount of 'free iron' due to elevated iron levels or released due to specific mechanisms that operate during iron deficiency or dysregulation of iron homeostasis, can increase risk of iron-induced oxidative stress.

3.1 Iron deficiency and oxidative stress

In iron-deficiency anemia, the lifespan of the RBCs is reduced. There is increased membrane stiffness and a decrease in deformability, thus making it relatively difficult for the RBCs to pass through the spleen. Thus in iron deficiency there is suicidal death of RBCs. This will liberate iron which increases the potential for oxidative stress. In turn, the presence of oxidative stress increases membrane stiffness and decreases deformability, setting up a vicious cycle. Further, there is increased lipid peroxidation, combined with depletion of antioxidant enzymes such as glutathione peroxidase, exacerbating the production of ROS. With less number of RBCs and reduced Hb, there is lower partial pressure of oxygen. The partially oxygenated Hb increases autooxidation of Hb which is converted to methemoglobin and superoxide is generated. Heme degradation has been reported to correlate with immunoglobulin G binding to the RBC membrane, which in turn is associated with denaturation of Hb. Denatured Hb and heme degradation products as well as free iron, promote removal of RBCs from circulation. Since the ROS concentration can be increased near the membrane, not only are the RBCs themselves damaged, ROS can be released to the vasculature and neighboring tissues (Nagababu et al., 2008).

3.2 Elevated body iron and oxidative stress

Iron stores tend to increase with age (Kell, 2009). Hofera et al., (2008) studied the iron content and RNA oxidation in skeletal muscle of 6m and 32m old Fischer 344/Brown Norway rats. They observed that with age, non-heme iron levels increased by 233% and RNA oxidation by 85%. Simultaneously, HJV and TfR1 decreased by 37% and 87%, respectively. There appears to be dysregulation of iron homeostasis with age, favoring increased intracellular free iron which in turn can induce oxidative stress. It has been shown that lysosomes are a source of rapidly mobilized, chelatable iron, with accumulation of iron being higher with ageing. Lipofuscin, an age-related iron rich pigment is non-degradable and not hydrolyzed by lysosomal enzymes. As a result, lipofuscin accumulates in the lysosomes. Lysosomes do not have the capacity to remove intralysosomal iron bound to lipofuscin, leading to excessive iron accumulation. This accumulation is evident in the RES, hepatocytes and post-mitotic cells such as neurons. Low internal pH i.e. 4 to 5 , presence of reducing equivalents, such as cysteine, ascorbic acid and glutathione favour intralysosomal Fenton reaction (Kurz et al., 2008). Further, iron is rapidly taken up into mitochondria by calcium uniporters where it catalyzes the ROS cascade (Uchiyama et al., 2008). ROS formation initiates mitochondrial permeability, depolarization, uncoupling of oxidative phosphorylation, mitochondrial swelling followed by membrane rupture, ultimately leading to necrosis or apoptosis(de La Rosa et al., 2008; Uchiyama et al., 2008). Thus the process of ageing and the associated morbidity depends on the oxidative stress-related alterations occurring in long-lived post-mitotic cells.

In women, besides the normal process of ageing, menopause also contributes to iron accumulation. Three-fold higher ferritin values have been reported in post-menopausal women compared to pre-menopausal women (Crist et al., 2009). Ferritin sequesters potentially reactive iron and is cytoprotective. Ferritin is autophagocytosed and degraded through lysosomal pathway to liberate iron which is then transported to the cytosol (Kurtz et al., 2004). Aberrant release of ferritin iron may lead to production of ROS (Mackenzie et al., 2008). Jiang, Pelle and Huang (2009) hypothesized that this increase in body iron may have a role to play in postmenopausal hot flashes since they found that serum ferritin levels were higher among African-American, white, and Hispanic women who reported more hot flashes. Iron is involved in heat production. Poor temperature regulation and thereby intolerance to cold has been observed in iron deficient persons, due to impaired heat production.

Iron status may be an important determinant for occurrence of breast cancer. In young women, deficiency may be linked to breast cancer recurrence whereas in postmenopausal women, higher iron may contribute to breast cancer (Jiang, Pelle and Huang, 2009). Hemochromatosis has been linked with high risk of osteoporosis. Osteopenia has been observed in patients with iron overload although vitamin D status and parathyroid hormone levels were normal. Similarly in patients with sickle cell disease, osteoporosis and osteopenia have been reported and the low bone mineral density was found to be associated with lower Hb and higher levels of ferritin. Iron status may also affect bone health. In animal models, both deficiency and excess of iron have been found to affect bone. D'Amelio et al., (2008) observed that patients with osteoporosis were iron deficient, had lower serum iron and higher Tf levels. A case report suggests that frequent phlebotomies normalized serum while the bone mineral density of the lumbar spine increased.

Iron and oxidative stress may be involved in the dryness, thinning and wrinkling of skin during menopause. It has been suggested that iron accumulation in the skin increases oxidative stress when the skin is exposed to UV rays. This in turn affects downstream genes and increased body iron stores may cause damage and photoaging of skin as well as make the skin more susceptible to damage by UV rays. Prolonged exposure to sun and UVrays was found to be associated with accumulation of non heme iron while treatment with iron chelators reduced the amount of skin damage.

Besides ageing, iron supplementation can expose the body, especially mucosal cells to very high levels. In this situation, large proportion of iron is retained in the mucosal cell as ferritin. However there may be accumulation of small amounts of iron, increasing the risk of oxidative stress. Iron supplementation has been shown to generate free radicals in a murine model, increasing intestinal susceptibility to oxidative stress (Ianotti et al., 2006). Iron accumulation and mucosal abnormalities such as reduction of microvillus height, necrosis have been reported in rats having high iron dose, especially those who were previously anemic (Casnueva and Viteri, 2003). In addition to local effects, iron supplementation can also affect other organs. Studies conducted in rats have indicated that there is continuous absorption of small proportion of supplemental iron at a rate higher than the normal, probably by passive diffusion. Under such circumstances, iron levels may exceed the binding capacity of Tf, giving rise to non-transferrin bound iron (NTB1). NTB1 is taken up by liver and has been shown to contribute to iron loading and hemochromatosis (Mackenzie et al., 2008). Unlike Tf–bound iron, cellular uptake of NTBI is not dependent on TfR and therefore the resulting iron is diffusely distributed throughout the organ (Kohgo et al., 2008). Whether long-term iron supplementation will have adverse effects in humans is also determined by one's antioxidant status.

Conversely, oxidative stress itself influences iron metabolism and iron proteins. Intracellular iron levels are increased in response to oxidative stress. There is reduction in ferritin synthesis, increased expression and uptake of TfR and upregulation of HO-1(Deb et al., 2009). Hb can act as a peroxidase in the presence of H_2O_2. Normally, tight regulation prevents this from happening, except in case of severe inflammation or generation of superoxide radicals by immune cells. Peroxidase activity of Hb may induce self-oxidation of proteins, with covalent cross-linking and aggregation of Hb being consequences which can induce oxidative stress in plasma and macrophages (Kapralov et al., 2009).

Oxidative stress and iron have been linked to several pathological states. Herein we review the role of ROS and iron in reproduction, the central nervous system and liver.

4. Reproduction, oxidative stress and iron

Human reproduction is a complex but not very efficient process, as less than one –third of fertilized human embryos have a chance of surviving upto a term delivery (Ruder et al., 2008). The reproductive process in males and females is regulated by myriad factors-nutritional and non-nutritional, with ROS having an important role. Excess generation of ROS and oxidative stress can adversely affect normal reproductive processes and supplementation with antioxidants may improve menstrual cycle regularity, prevent ovulatory disorders as well as enhance fertility (Ruder et al., 2009).

4.1 Sperm and fertility

Spermatogenesis is a very active process and about 1000 sperm are generated per second (Aitken and Roman, 2008). Corresponding to this high rate of cell division, the germinal epithelium mitochondria utilize oxygen. However, poor vascularization of the testes and the associated low oxygen tension may provide protection to spermatogenesis and Leydig cell steroidogenesis which are susceptible to ROS. Iron-induced lipid peroxidation, protein carbonyl expression and depletion of lipid - soluble antioxidants in testicular tissue results in disruption of spermatogenesis.

Human spermatozoa produce ROS which enhance tyrosine phosphorylation through the suppression of tyrosine phosphorylase activity by H_2O_2 and stimulation of cAMP production, that play a role in regulating signal transduction pathways that control sperm capacitation. A low, steady state of ROS production appears to be important for gamete cell stability (Hsieh et al., 2006) and capacitation (Baker and Aitken, 2005). Seminal fluid is an important source of antioxidants (the enzymes superoxide dismutase, catalase, glutathione peroxidase and vitamin C, vitamin E, hypotaurine, taurine, L-carnitine and lycopene) which are important for protecting spermatozoa from oxidative injury, especially because sperm do not possess much cytoplasmic fluid which generally contains the antioxidant enzymes. Hence the antioxidant capacity of sperm is relatively poor (Zini et al., 2009). In the event of excessive ROS production, a state of oxidative stress ensues. Since spermatozoa contain relatively high amounts of unsaturated fatty acids vis-a-vis the low antioxidant capacity, they become susceptible to oxidative stress. Persistent infertility has been linked to lipid peroxidation of unsaturated fatty acids in the sperm membrane (Hsieh et al., 2006). ROS have been associated with impaired metabolism, morphology, motility and male infertility (Baker and Aitken, 2005; Hsieh et al., 2006), retention of excess residual amorphous cytoplasm that human spermatozoa are not capable of remodeling or discharging. Also, oxidative stress results in DNA damage i.e. strand breaks, formation of DNA base adducts such as 8-hydroxyl-2'-deoxyguanosine O 6-methylguanine that have been reported to inhibit methylation of adjacent cytosine residues which in turn results in DNA hypomethylation (Tunc et al., 2009). This adversely affects the sperm's ability to fertilize.

ROS may be produced by sperm that are immobile, morphologically abnormal, or morphologically normal but functionally abnormal. Hsieh et al.,(2006) reported that sperm motility or concentration were negatively correlated with malondialdehyde concentrations, an indicator of lipid peroxidation. Zribi et al., (2011) recently reported a positive correlation between oxidation and DNA fragmentation in a study wherein 21 nonasthenozoospermic (sperm motility $\geq 50\%$) were compared with 34 asthenozoospermic (sperm motility $< 50\%$). DNA fragmentation was found to be negatively correlated with several parameters of sperm motility and vitality.

Administration of high dose of iron to male mice was reported to induce a dominant lethal effect of male mice characterized by high embryonic loss in females that were mated with male mice exposed to iron. The authors postulated that this embryonic loss was due to non-viable embryos. Lucesoli et al., (1999) exposed male rats to varying doses of iron dextran (250 to 1000 mg per kilogram of body weight) administered intraperitoneally. Iron was found to accumulate in the sperm and other testes cells. In the group administered the highest dose, spermatogenesis was markedly lower and concentrations of antioxidants (alpha-tocopherol, ubiquinol-9, and ubiquinol-10) in the testes were inversely correlated with oxidation. Oxidative protein products viz protein thiols and protein-associated carbonyls were higher in rats exposed to the high iron dose (Doreswamy and Muralidhara, 2005). Supplementation with alpha–tocopherol partially mitigated the oxidative damage caused by chronic iron intoxication.

Wise and coworkers (2003) observed in boars that concentrations of iron and ferritin were higher in small testes, with a decrease as testes weight increased. With increasing testicular iron concentrations, sperm production declined. These authors cited reports in the literature that human males with β-thalassaemia major have decreased pituitary function from Fe overload, and serum ferritin is highly correlated with the presence of hypogonadism. Doreswamy and Muralidhara (2005) administered acute sub-lethal doses of iron dextran to adult mice (50, 100 and 200 mg per kilogram of body weight). Multiple doses induced increase in lipid peroxidation in mitochondrial and microsomal fractions. There was significantly higher DNA damage in testes indicated by increased single strand breaks. Although sperm count remained unaffected, there was a 3- to 7-fold increase in abnormal sperm. The authors attributed the genotoxic consequences to early oxidative damage in testis due to iron overload. Thus, iron overload leads to oxidative stress that damage sperm DNA, making the embryos non-viable(Aitken and Roman, 2008). Therefore, several investigators have examined the effect of antioxidant supplements. Some investigators have observed that short term supplementation with antioxidants to men with high levels of sperm damage, resulted in better fertility potential and improved pregnancy rates (Zini et al., 2009). Tunc et al., (2009) compared 45 men with known male factor infertility with 12 fertile controls who were sperm donors. They were given a capsule containing 500 µg of folate 100 mg of Vitamin C, Vitamin E - 400 IU, Lycopene - 6 mg, zinc -25 mg, selenium - 26 µg, and garlic oil -33 µg for three months in order to cover one full cycle of spermatogenesis. The authors reported that the supplement significantly improved sperm DNA integrity and methylation. Recently, 690 infertile men with primary or secondary male factor infertility for at least one year, were given a daily supplement of 200 µg with vitamin E(400 units) for more than 3 months by Moslemi et al.,(2011). The investigators observed that in about half the subjects, there were improvements in sperm motility, and/ or morphology. In one-tenth of the cases, spontaneous pregnancy occurred compared to none in the control group.

As in case of males, ROS play a role in the entire lifecycle of females, affecting various physiological functions from oocyte maturation, modulation of ovarian germ cell, stromal cell physiology, transcription factors, ovarian steroidogenesis, corpus luteal function and luteolysis, implantation, formation of blastocyst, pregnancy, parturition to perhaps even menopause. Iron nutriture during pregnancy is of concern since iron requirements during pregnancy are significantly higher than in the non- pregnant state and iron deficiency during pregnancy has been linked to maternal morbidity and mortality, low birth weight and infant mortality (Ramakrishna, 2002). Aune et al., (2004) studied 19 healthy pregnant women given 26 mg of iron (as ferrous aspartate) per day. After 4 weeks, total antioxidant activity and conjugated diene levels (indicative of lipid peroxidation) were raised. Casanueva and Viteri(2003) reported that supplementation of pregnant women with 100 mg iron and 500 mg vitamin C in the third trimester of pregnancy was associated with higher levels of serum iron and thiobarbituric acid-reactive substances than in the control group. Vyas (2005) observed an increase in lipid peroxide levels with an increase in serum iron of 85 pregnant Indian women from lower socio-economic strata. Women who delivered at term tended to have slightly higher levels of lipid peroxides than those who delivered preterm. However, there are few studies in the literature on iron deficiency and oxidative stress, *per se* particularly related to pregnancy. Many investigators have reported on iron proteins and their roles in oxidative stress, which is reviewed in this section.

4.2 ROS and pregnancy

Oxidant status may influence early embryonic development. Pregnancy is characterized by increased basal oxygen consumption and susceptibility to oxidative stress, although little data is available on oxidative stress in the preconception and early conception period. Fetal development and survival depend on the placenta as it plays a pivotal role by anchoring the conceptus, providing an interface between the mother and fetus, transfers nutrients, is important for exchange of gases, synthesis of a number of hormones and provides an immune barrier. Any problems associated with placental growth and differentiation ultimately affect its function and therefore fetal growth and development. The placenta itself produces ROS since it is rich in mitochondria, is hemomonochorial and allows electrons to pass that are converted into O2•- radicals (Casanueva and Viteri, 2003; Fuji, et al., 2005; Agarwal et al., 2005). The half lives of these ROS differ which influences their diffusion across the placenta. Superoxide is unable to diffuse across beyond the cell of its origin/synthesis because of short half life whereas the nitric oxide has a higher diffusion distance and can be a paracrine mediator in adjacent cells and be potentially more damaging (Myatt, 2010).

Proliferation and differentiation of trophoblasts and invasion by them is essential for placentation. It is important that trophoblasts colonize the endometrium and sequester the blood supply that is essential for the developing placenta. There is also considerable vascular remodeling in order to provide for the requirements of the developing fetus. In the early stages of pregnancy, a low oxygen milieu prevails. This placental hypoxia in the first trimester is physiological, regulates both morphogenesis and functioning of the placenta and is important for normal embryonic development. Maternal blood flow to the intervillous spaces commences only around 11 weeks of gestation (Pringle et al, 2010). This increase in oxygen concentration is vital for active placental transport and synthesis of proteins required by the rapidly developing fetus in the second and third trimesters (Burton, 2009). Jauniaux and coworkers (2000) studied 30 women at 7-16 weeks of gestation and observed a rapid increase in oxygen tension within the placenta at the end of the first trimester. Maternal blood flow starts at the periphery of the placenta and then gradually extends towards the centre, accompanied by an increase in the oxygen tension. This is associated with increased oxidative stress in the placenta and especially the synctiotrophoblast. Physiologically this is necessary for trophoblast differentiation. A deviation from this normal phenomenon such as early onset of blood flow would result in premature rise in oxidative stress since the placenta at early stages of pregnancy (8-10 weeks) does not have adequate antioxidant enzymes.

Further, the low levels of oxygen in the uterus are advantageous in terms of having relatively lower metabolic rates and thereby minimizing the production of ROS. However, impairment of these processes or hypoxia at later stages in pregnancy is pathological. Insufficient trophoblast invasion, defective vascular remodeling in the first trimester have been found to increase blood flow, damage villous architecture, increase the risk of spontaneous vasoconstriction and ischemia - reperfusion which can result in oxidative stress (Burton et al., 2009). Higher intensity of oxidative stress can have adverse effects on the developing embryo, and can lead to embryo fragmentation (Jauniaux et al., 2000). In studies on *in vitro* fertilization, fluid from follicles whose oocytes did not fertilize contained higher activity of superoxide dismutase than those that fertilized. Also, higher levels of 8-hydroxy-2-deoxyguanosine, an indicator of DNA damage caused by oxidative stress, have been observed in granulosa cells of patients with endometriosis and were associated with lower fertilization rates and poor embryo quality (Agarwal et al., 2005).

Most studies reported in the literature relate oxidative stress to pregnancy complications and some to the outcome of pregnancy. The focus in most investigations has been on the role of oxidative stress in preeclampsia.

4.2.1 Pregnancy complications – Preeclampsia

During early gestation, the syncytiotrophoblast has low concentrations of antioxidant enzymes compared to other villous tissues, making it susceptible to oxidative stress (Jauniaux et al., 2000). Hence, disturbances in oxygen tension and ROS can result in pregnancy complications. Preeclampsia, miscarriages, preterm labour, birth defects, endometriosis and tubal factor infertility, intrauterine growth retardation and stillbirth have been linked to insufficient trophoblast invasion, development, vascular remodeling and impaired placental development and function (Jauniaux et al., 2003, Szczepanska et al., 2003, Pringle et al, 2010) In the event of placental hypoxia, due to defective placentation, ischemic-reperfusion injury occurs and the ensuing oxidative stress stimulates the release of cytokines and prostaglandins. This has been implicated in endothelial cell dysfunction and possible development of preeclampsia (Agarwal et al., 2005). In patients with preeclampsia, antioxidant nutrient levels as well as antioxidant response was lower while lipid peroxidation was higher (Takagi et al, 2004; Aydin et al, 2004 and Mikhail et al, 1994). Also, free 9-isoprostane levels that are produced by non-enzymatic random oxidation of tissue lipids by ROS were found to be significantly higher in women with preeclampsia as compared to healthy pregnant women (Zhang et al., 2008).

Placentas obtained from women with preeclampsia had higher lipid peroxides and xanthine oxidase activity as well as nitrotyrosine residues. Burton (2003, 2009) proposed that complications such as preeclampsia and intrauterine growth restriction may be related to fluctuations in oxygenation and not hypoxia alone due to inadequate trophoblastic invasion which in turn result in spontaneous constriction of the spiral arteries. Cindrova-Davies et al., (2007a) observed in an *in vitro* study that placenta explants exposed to hypoxia and reoxygenation, exhibited increased oxidative stress while addition of vitamin E and C mitigated it.

Several events during pregnancy from placentation, maintenance of uterine quiescence, regulation of hemodynamic control within the uterus and placenta, protection against preeclampsia, regulation of apoptosis and inflammatory cascades in trophoblast cells among others are influenced by HO and its degradative products viz biliverdin, carbon monoxide and ferrous ions. HO-1 has antioxidative, anti-inflammatory and anti-apoptotic effects. Centlow et al., (2008) observed that placental tissue from preeclamptic women exhibited increased expression of $Hb\alpha$ and $Hb\gamma$ mRNA than placentas from normal pregnant women. Increased number of cells expressed Hb in the preeclamptic patients' placentas suggesting that either the cells were migrating into or out of the vessels or there may be binding sites on the vessel walls. Also, the placental vessels in preeclamptic patients had high numbers of Hb+ nucleated cells, which were probably fetal erythroblasts migrating in response to the poor perfusion and low oxygen levels in the pre eclamptic placenta. The authors postulated that if the turnover of Hb producing cells is high, these cells could well release heme into the placental blood vessels and extra villous space as evidenced by their observations of high Hb levels in the blood vessel lumen of pre eclamptic placentas. In contrast, this phenomenon did not occur in the normal control placentas.

Free heme generates ROS and thus can oxidize lipids including low density lipoproteins into cytotoxic peroxides that will damage the endothelium, oxidize membrane proteins resulting in increased membrane permeability and lysis of cells. Normally, HO degrades free heme but in preeclampsia, it has been observed that there is decreased expression of this enzyme and the activity is also less. This may result in accumulation of heme in the placenta. To further exacerbate the situation, carbon monoxide production from heme by HO may be less, thus contributing to reduced placental blood flow, inducing inflammatory response, activation of nuclear factor (NF)-κB transcription factors by the heme-generated ROS. This in turn may stimulate production of adhesion molecules and cytokines, recruitment of leukocytes, migration of neutrophils as well as their activation and induce further increasing risk of damage. Heme can also induce secretion of tumor necrosis factor-α and activate toll like receptor 4 leading to an immune response.

Gandley and coworkers(2008) reported that levels of placental myeloperoxidase, a hemoprotein produced and released by activated monocytes and neutrophils, were elevated in 27 placentas obtained from women with preeclampsia compared to normal gestationally - matched placentas. Myeloperoxidase may be involved in the generation of potent ROS and has been associated with mediation of oxidation of lipoproteins, catalyzing nitration of tyrosine residues, depleting endothelially derived nitric oxide, vascular inflammatory diseases and utilization of antioxidants.

4.2.2 Pregnancy outcome

Zhaoa et al., (2009) observed that early embryonic death in mice occurred when the enzyme HO-1 was deficient, indicating a possible role in early placentation or embryonic development. The authors suggested that deficiency of this enzyme may also affect complement activity and play a role in fetal resorption. Deficiency of this enzyme was associated with rise in diastolic blood pressure, and less dilated spiral arteries in the placenta.

Exposure to ROS during organogenesis can increase risk of birth defects. Burton (2009) reported that in animals with genetically impaired antioxidant enzyme activity and in diabetic rats in whom ROS levels were increased, the offspring had congenital abnormalities (Hagay et al. 1995; Eriksson, 1999; Nicol et al. 2000). Slonima and coworkers (2009) compared differences in gene expression between euploid fetuses and second trimester Down's syndrome fetuses, in uncultured amniotic fluid supernatant samples and suggested that oxidative stress may have a significant role in Down's syndrome. Prater and colleagues (2008) used methylnitrosourea to induce oxidative stress in C57BL/6 mice. They observed that several important placental proteins such as Hgf, Kitl, IFNα4, Ifrd, and interleukin (IL)-1β which are important for development of the placenta and fetal skeleton were altered. Also there was damage to placental endothelial cells and trophoblasts. In the group given methylnitrosourea with quercetin, an antioxidant, the levels of these proteins were normalized. Also, in the group treated with methylnitrosourea, the pups had disproportionately short limbs as well as distal limb malformations which were not observed in the quercetin supplemented group. The authors suggested that oxidative stress could alter normal placental osteogenic signaling and skeletal formation in the fetus.

Guller (2009) reported that in preeclamptic condition, there may be reduced nutrient supply to the fetus, resulting in intrauterine growth retardation. Levels of malondialdehyde were higher in small for gestational age newborns compared to adequate for gestational age controls. Markers of oxidative stress were significantly correlated with maternal Hb. Further, SGA newborns had higher number of "unprotected' erythrocytes i.e. content of antioxidant protectants were lower. This would make the erythrocytes more susceptible to lysis which would in turn release heme and increase risk of ROS generation

During pregnancy, iron supplements are routinely recommended, but there is little direct evidence to determine whether iron supplementation would contribute to oxidative stress in either direction i.e. mitigate the stress by reducing anemia which leads to oxidative stress or produce iron overload which may increase ROS generation (Casaneuva and Viteri, 2003).

5. Iron and central nervous system

Neurons are long-lived postmitotic cells which are rarely replaced through division and differentiation of stem cells. This permits biological waste materials to accumulate in these cells with ageing, possibly because lysosomal degradation is not adequate or unsuccessful, resulting in functional decay and ultimately cell death. Among the waste materials that can be accumulated are lipofuscin, irreversibly damaged mitochondria and aberrant or abnormal proteins with consequent selective loss of neurons which may be age-dependent. Abnormal proteins that accumulate are Ab-amyloid neuritic plaques and neurofibrillary tangles in Alzheimer's disease (AD), α-synuclein, and Lewy bodies in Parkinson's disease (PD) (Gaeta and Hilder, 2005). Ageing of such long lived post mitotic cells has been attributed ROS formed mostly within the mitochondria. Due to its high content of poly unsaturated fatty acids, the brain is very sensitive to attack by ROS. Lipid peroxidation and its aldedydic products, 4-hydroxynonenal (4 HNE) and acrolein have been implicated in neurological ailments. Due to oxidative stress, there may be cytoskeletal damage, mitochondrial dysfunction and altered signal transduction. Neurons stressed and injured by free radicals are rendered non-functional, function and transport of mitochondrial to synaptic regions is impaired and synaptic function decreases resulting in neurodegeneration. The superoxide radical is the most abundant ROS and has been reported to play a role in brain edema and hippocampal neuronal death (Ansari et al, 2008). Diseased neurons can remain viable for 10 years or longer and as such must have sufficient protective mechanisms to maintain normal homeostasis. However, at some point in a neurodegenerative disorder, the oxidative insults may overwhelm cellular antioxidant defense systems leading to cellular dysfunction and death (Siedlak, 2009). In fact oxidative stress is regarded as one of the earliest changes in the pathogenesis of AD that may be present several years before the disease overtly manifests itself (Smith et al., 2010).

5.1 Iron, ROS and neurodegeneration

The brain is vulnerable to oxidative stress because, not only does it have a high content of polyunsaturated fatty acids (George et al., 2009) but also contains substantial amounts of iron (Gaeta and Hilder, 2005). The blood brain barrier serves to limit entry of iron into the brain, thus protecting it from overload (Wang and Michaelis, 2010). Within the brain, there are mechanisms to regulate the cellular iron level ensuring adequate supply to allow normal function of the cells and nervous system while protecting them from toxicity. Iron deficiency affects cell division of neuronal precursor cells, astrocytes, and oligodendrocytes. However, when the amount of iron exceeds what can be safely sequestered by ferritin, heme proteins and iron-sulfur clusters; oxidative stress results especially in regions containing higher amounts of iron – globus pallidus, substantia nigra, red nucleus, putamen, caudate nucleus. It has been suggested that excess iron or "iron invasion" may be the primary event and cause degeneration of axons as well as neuronal cell death. Evidence from studies on inherited disorders of iron metabolism indicates that neurodegeneration occurs with iron accumulation (Mills et al., 2010).

Olivares et al., (2009) reported that iron accumulation is probably a primary event and not a consequence of degenerative diseases. In a rat model, when $FeCl_3$ was injected into the *substantia nigra*, there was selective decrease in striatal dopamine suggesting that iron is responsible for dopaminergic neurodegeneration. The phenomenon was prevented by infusion of desferrioxiamine, an iron chelator. Mutations in the gene that codes for a ferritin subunit result in a disease similar to PD. Catecholamine neurons are selectively degenerated by 6-hydroxydopamine which is oxidized to cytotoxic chatecholaminergic semiquinones and quinines along with hydrogen peroxide and hydroxyl radicals. Iron deficient rats were resistant to toxicity induced by 6-hydroxydopamine. Further, its toxicity was reversed by an iron chelator, desferal. Similarly in iron models of stroke, use of iron chelating agents could reduce the infarct size and brain injury as well as improve neurologic outcome (Stankewicz et al, 2007). In mice, deletion of the gene encoding for IRP2 which interferes with regulation of iron metabolism has been found to result in abnormal iron deposition in the brain, as well as ataxia, tremors and neurodegeneration. Similarly mutation in the gene coding for a ferritin subunit and polymorphisms in genes related to iron homeostasis have been associated with iron deposits, neuronal degradation and sometimes PD (Olivares et al, 2009).

Another possible mechanism is the ready binding of iron to advanced glycation end-products that accumulate with age in the endothelial internal elastic lamina or basement membrane. Iron and copper can scavenge nitric oxide and thus prevent its action on the smooth muscle resulting in prolonged vasoconstriction (Eaton and Qian, 2002). If this occurs in the blood vessels of the brain, it could contribute to neurodegeneration and may therefore contribute to the efficacy of the iron chelator desferroxiamine in AD (Stankiewicz et al., 2007).

Lysosomal (autophagic) degradation is rapid and effective but is not completely successful. Even under normal conditions, some iron-catalyzed peroxidation occurs intralysosomally (lysosomes are rich in redox-active iron), resulting in oxidative modification of autophagocytosed material and yielding lipofuscin, a polymeric compound slowly accumulating within ageing postmitotic cells at a rate that is inversely correlated to the longevity across species (Kurz et al, 2008). When iron-rich, non-degradable substances like lipofuscin and hemosiderin (which is not degradable and composed of polymerized ferritin residues) accumulate in some lysosomes, there is an increase in iron content. This increases the lysosomes' sensitivity to ROS.

The central nervous system contains considerable amounts of iron although the regional distribution of the metal differs, thus conferring varying sensitivity to oxidative stress (Zecca et al., 2004; Hall et al., 2010). Normally, the cytosolic iron is regulated and maintained at a low level adequate to ensure synthesis of essential iron containing molecules and yet to prevent free or of redox-active iron at low levels to prevent damage. Although both ferritin and Tf have high affinity for iron at neutral pH, the metal dissociates easily from the two proteins at pH below 6.0, which may occur in injured areas.

Whenever there is a requirement for iron, it is taken up by the cell from its surroundings. However, in non-dividing cells, iron that is available is determined primarily by turnover and reutilization. Cells are unable to get rid of the intralysosomal iron bound to lipofuscin. Thus, over a prolonged period, iron accumulates despite regulation of iron uptake, which is especially seen in long-lived postmitotic cells like neurons. Histologic and magnetic resonance imaging studies indicate that there is increased iron level in the gray matter in PD and AD as well as other neurological disorders. Abnormal iron deposits have been observed in neurons and oligodendrocytes of patients with multiple sclerosis, with increased in the ferric iron content in the caudate and putamen. Magnetic resonance imaging studies on humans indicate a possible relationship between damage to gray matter and iron deposition in patients with multiple sclerosis (Stankewicz et al., 2007). Thus, as these cells age and especially as the end of their lifespan approaches, they become sensitive to oxidative stress. When there is enhanced autophagy, such as when there is repair following injury coexisting with oxidative stress, there is more rapid formation of lipofuscin.

Hemoglobin is another source of catalytically active iron, a situation that may be encountered in hemorrhage. The iron released from the Hb is likely responsible for the Hb-mediated oxidative stress. In conditions of hypoxia or ischemia, due to brain injury for example in children with anoxia or in animal models of stroke, iron deposition has been reported in basal ganglia, thalamus, periventricular and subcortical white matter. Insufficient oxygen levels increase lactic acid locally which reduces the ability of Tf to bind iron and simultaneously result in release of the iron bound to ferritin (Stankewicz et al, 2007).

In addition to directly affecting the brain adversely, oxidative stress may cause damage through overexpression of inducible nitric oxide synthase (iNOS) that in turn would augment the production of nitric oxide. Nitric oxide reacts with superoxide anion and produces the highly reactive peroxynitrite, resulting in oxidation and aggregation of proteins. Consequently there can be conformational changes in the proteins exposing hydrophobic residues which may result in loss of structural or functional activity, accumulation of oxidized proteins in the cytoplasm as aggregates e.g tau and amyloid-β aggregates as seen in AD. Mitochondrial dysfunction and damage from intracellularly produced ROS may lead to age-related neurodegenerative disease. In mice, mitochondrial deficiency of superoxide dismutase was associated with hyperphosphorylation of tau and exacerbation of amyloid burden (Melov et al., 2007).

Increased intensity of oxidative stress results in peroxidation of the lysosomal contents and the membrane, rendering the lysosome leaky. This phenomenon may occur even after only brief exposure to oxidative stress. Lysosomes contain a fairly high amount of redox active iron originating from degraded iron-containing proteins. The low pH in the lysosomes favours iron catalysed oxidative reactions. Kurz et al., (2004) stated that ferritin needs to undergo autophagocytosis and degradation before the iron the released and incorporated into ferritin for storage or in other iron-containing proteins. Each molecule of ferritin can store about 4500 atoms of iron. As a result, even if there is a relatively small number of autophagocytosed ferritin molecules, there can be substantial reduction of lysosomal iron. Autophagy occurs continually, with the degraded lysosomal ferritin being replaced (Kurz et al., 2008). Kurz et al., (2004) stated that when the lysosome ruptures, it releases hydrolytic enzymes, The rupture also induces coordinated aopoptotic cell death. Besides release of lytic enzymes, iron is also released into the cytosol. This iron may reach the nucleus where hydroxyl radicals may be generated if the oxidative stress continues.

It is not only excess iron and consequent oxidative stress that can induce lysosomal damage. Iron depletion has been found to damage lysosomes although it has been suggested that the mechanism may be a consequence of other apopotogenic factors (Kurz et al., 2008). Iron has been implicated in dysfunction of the central nervous system. Neurodegenerative diseases such as PD, AD, amyotrophic lateral sclerosis, Huntington's disease, Friederich's ataxia and aceruloplasminemia have been linked to iron (Achcar et al., 2011) Sites of neuronal death in the brain have been observed to be sites where iron accumulates. In cases with multiple sclerosis, redox active iron but not total iron content of cerebrospinal fluid was significantly higher.

Besides this, monoamine oxidase (MAO) has also been linked to oxidative stress (Zecca et al., 2004; Mandel et al., 2005). Activity of MAO in both humans and animals is influenced by iron levels (Symes et al., 1969; Youdim et al., 1975). MAO, a flavo-protein located on the outer mitochondrial membrane, can exist in two forms A and B-the B form activity is greater in basal ganglia. In rodents, MAO-A is present in the extraneuronal compartment and within the dopaminergic terminals, it plays a role in the metabolism of intraneuronal and released dopamine. In the process of its role in oxidative deamination of primary, secondary and tertiary amines to the corresponding aldehyde and free amine, H_2O_2 is generated (Moussa, Youdhim and Bakhle, 2006). H_2O_2 is inactivated in the brain by glutathione peroxidase. However, if brain glutathione oxidase levels are low, H_2O_2 can accumulate and participate in the Fenton reaction whereby iron as the ferrous ion, generates hydroxyl radicals that are highly active free radicals.

Both monoamine oxidase activity and brain iron concentration increase with age, thus increasing the potential for generation of hydroxyl radicals and oxidative stress with age. Mandel et al., (2005) reported increased activity of MAO in patients with PD and AD. In PD, samples of substantia nigra from PD patients were found to be deficient in aldehyde dehydrogenase . This could lead to accumulation of neural and toxic aldehydes derived from dopamine by MAO (Moussa, Youdhim and Bakhle, 2006. Ansari et al (2008) reported that oxidative stress results in oxidation of synaptic proteins and has adverse effects on synaptic plasticity, connection. Impaired mitochondrial transport to synapses contributes to neuronal degeneration and death. Loss of synapsin-I and PSD-95 which influence neuronal function and survival was caused by oxidative stress.

5.2 Parkinson's disease

PD is a progressive neurodegenerative disorder. There is degeneration of nigrostratial dopaminergic neurons in the substantia nigra, resulting in depletion of dopamine. Consequently there is loss of motor functions, formation of intracytoplasmic inclusions (Lewy bodies) and Lewy neuritic inclusions (LNs) in the surviving dopaminergic neurons. Dopamine is oxidatively deamined to its aldehyde-3,4-dihydroxyphenylacetaldehyde by MAO and then to 3,4-dihydroxyphenylacetic acid by aldehyde dehydrogenase. 4-HNE and malondialdehyde, products of lipid peroxidation are potent inhibitors of mitochondrial dehydrogenase but do not inhibit MAO. Even low levels of oxidative stress yield the aldehyde. The aldehyde has been shown to be toxic towards dopaminergic cells via oxidative stress and other mechanisms. Thus, cells located close to dopamine neurons such as astrocytes and microglia may well be exposed to high concentrations of the aldehyde during or following oxidative stress. The aldehyde may target and modify proteins important for dopamine neuron homeostasis including α-synuclein (α-syn) (Jinsmaa et al, 2009).

LNs are derived from fibrillar aggregates of α-syn belonging to the synuclein family. α-syn is largely expressed at presynaptic terminals specially in some parts of the brain like neocortex, hippocampus, striatum, thalamus and cerebellum. The central hydrophobic peptide in α-syn has been implicated in Alzheimer's disease and is an important component of amyloid plaques. Oligomers of α- syn through crosslinking of di-tyrosine may occur through nitration, an oxidative mechanism. Nitrated α-syn is not processed well by proteases and has been detected in brains of persons with PD. If α-syn aggregates, there is decrease in the soluble α-syn and it ultimately permits large amounts of dopamine to enter the cell, consequently increasing the potential for generating ROS.

Further iron associated proteins and receptors have been reported to be upregulated in the striatum and substantia nigra of patients with PD. Regan et al. (2008) reported that ferritin levels in the central nervous system increased with normal aging. However, in patients with PD, there was little increase in the substantia nigra, although there was iron accumulation, suggesting that less sequestration into ferritin may make the brain vulnerable to toxic effects of iron and oxidative stress. These authors observed an inverse relationship between cell death in mixed cortical cultures and ferritin expression.

Neuromelanin is a pigment that is protective against oxidative stress because it can bind and concentrate transition metals like iron. In normal brain, only 50% of this pigment is saturated with iron, such that it can still chelate iron in the neurons if there is an onslaught of the mineral. However, in PD, when brain iron concentration is increased, the pigment may become saturated and iron would be bound to low affinity sites, so that some amount of the bound(iron is redox-active and potentially toxic(Gaeta and Hider, 2005).

Brain tissues of patients with Parkinson's have been found to contain substantial amounts of iron in the substantia nigra (a 25% to 100% increase) compared to healthy age-matched controls. α-syn can interact with cations and this binding in turn catalyzes protein aggregation (Stankewicz et al., 2008). Metal-induced aggregation is attributable to the oxidation by redox metals. Increase in the redox active metal, like iron results in oxidation and therefore in the degeneration of dopaminergic neurons in the substantia nigra. Besides aggregation/oligomerization of α-syn, other effects of oxidation can be mitochondrial dysfunction, cytotoxicity and increase in cytosolic free Ca^{2+} resulting in cell death. Redox elements like iron also accelerate the enzymatic and non-enzymatic catabolism of dopamine. High iron concentrations in the substantia nigra can catalyze conversion of hydrogen peroxide during catabolism of dopamine with the production of highly reactive hydroxyl radicals.

Oxidative stress can further exacerbate the situation by removal of iron from ferritin, Hb and cytochrome c peroxides, iron sulfur protein by the action of ONOO-, which in turn will increase the iron levels. Iron also catalyses conversion of excess dopamine to neuromelanin. This is an insoluble pigment that has been found to accumulate in aged dopaminergi neurons. Neuromelanin sequesters redox ions that have affinity for ferric ions. If it is bound to an excess of ferric ions, neuromelanin functions like a prooxidant, reduces the ferric to ferrous and increases the amount of iron available to react with H_2O_2 (Olivares et al., 2009). Iron not only induces oxidative stress but damages proteins, membranes, and nucleic acids and eventually leads dopaminergic neuron death in the substantia nigra (Logroscino et al., 2008).

Powers et al (2009) studied dietary intakes of 266 men and 154 women who were newly diagnosed patients with idiopathic PD. They were compared with a control group consisting of 351 men and 209 women. In men, the highest level of iron intake was associated with increased risk of PD. A combination of low saturated fat intake with high iron was linked to moderate risk in both sexes, with the association being stronger for men. The authors suggested that cholesterol is a protective factor for PD, possibly through the relationship between cholesterol and coenzyme Q10 which is an antioxidant and may reduce oxidative stress generated by dopamine metabolism.

Brains of patients with PD have been observed to have increased amounts of nonheme iron compared to controls but ferritin and Tf content were low. Logroscino and coworkers (2008) stated that this may indicate that there is altered regulation of synthesis of proteins including ferritin that are involved in iron metabolism. The low ferritin levels in the tissues may reflect inadequate neuronal response to the excess amount of nonheme iron entering the brain. Alternatively in PD there may be disturbance in the mechanism for iron transport through the neuronal membrane. Lactoferrin activity was found to be significantly higher in patients' mesencephalon where loss of dopamine is more severe. Gorell et al., (1999) reported that in autosomal recessive juvenile Parkinsonism associated with mutations in the Parkin gene, iron levels were increased in the substantia nigra(Moussa, Youdhim and Bakhle, 2006).

5.3 Alzheimer's disease

In selected regions of the brain, there are deposits of the amyloid precursor protein, amyloid-β (Aβ) deposits in plaques and vessel walls and neuritic plaques (Liu et al., 2010). Amyloid precursor protein (APP) is the precursor of the neurotoxic Ab-amyloid. It contains an IRE in its mRNA which allows cellular levels of iron to control translation (Moussa, Youdhim and Bakhle, 2006). Gaeta and Hider (2005) reported that when the Aβ deposits accumulate sufficiently for oxidative stress to occur; it induces its own production, starting a vicious cycle. Also, there are structural lesions due to collapse of the neuronal cytoskeleton and accumulation of hyperphosphorylated and polyubiquinaed microtubule –associated proteins with formation of neurofibrillary tangles. Cerebral atrophy progressively becomes worse with loss of nerve fibres and cells and there is disconnection of the synaptic circuitry. There is activation of prodeath genes and signaling pathways;. Mitochondrial dysfunction occurs, energy metabolism is impaired and chronic oxidative stress is typical along with DNA damage (De la Monte and Wands 2008).

Aβ has multiple binding sites for iron and has greater affinity for Fe(II) than does transferrin. When Aβ binds iron, ferric becomes more soluble and thus the Fe(III) present in complexes now becomes available to cellular reductants. This can disrupt iron homeostasis and perhaps is why iron is present along with Aβ aggregates in senile plaques. Further, Aβ may accumulate iron from the LIP. The implications of this in generating oxidative stress need further study (Jiang et al., 2009). In animal models increased iron worsened the course of increased risk of developing the disease. Data from postmortem pathology studies indicate iron deposition in neurons, neurofibrillary tangles and plaques of patients with AD. *In vivo* and *in vitro* studies show that amyloid protein aggregation was more and amyloid-induced neuronal injury in human neuroblastoma cell line M1783 was worsened by iron (Stankewicz, 2007). Clinical trials with iron chelators showed favourable results in terms of decreased beta amyloid levels, improved cognition and living skills of patients lend support to results of reduced toxicity in *in vitro* studies on neurons treated with iron chelator and then to beta amyloid (Stankewicz et al., 2007).

6. Iron and liver

Liver is the site of important metabolic pathways including detoxification of ammonia, alcohol and xenobiotics. It serves as a storage organ and acts as a filter by removing bacteria and debris through phagocytic action of Kupffer cells. Liver faces the onslaught of iron and is vulnerable to iron overload. Iron absorbed from food is first transported to the liver. Hepatocytes express TfR as well as its homolog, TfRII, both mediating the uptake of Tf-bound iron. Unlike TfRI, TfRII lacks IRE and is not regulated reciprocally in response to iron levels, thereby allowing iron uptake by hepatocytes. Further, there are reports that hepatocytes can import iron by involving routes not involving TfRs. NTBI is absorbed primarily by liver via SLC39A zinc transporter–ZIP14 (Takami and Sakaida, 2011). Import of iron by more than one route, can contribute to iron loading in hepatocytes, thereby increasing their potential for iron-induced oxidative stress. Iron or iron-induced oxidative stress have also been found to activate cell signaling cascades triggering apoptosis and necrosis pathway via NF-κB and AP-1 pathways respectively. NF-κB promotes the synthesis and release of cytotoxic, proinflammatory and fibrogenic factors such as tumor necrosis factor (TNFα), IL-6 and MIP-1 that alter Kuppfer cells and hepatocyte function, and trigger hepatic stellar cell activation. In hepatic stellar cells, AP-1 transcription factors are involved in the regulation of procollagen (I). In addition, AP-1 and NF-κB-dependent gene products modulate hepatocyte death induced by oxidative stress (Zuwała-Jagiełło et al., 2011).

Oxidative stress leads to formation of glycoxidation products, including advanced glycation end products (AGEs) – among them Nε-(carboxymethyl) lysine and advanced oxidation protein products (AOPPs). Both AGEs and AOPPs trigger the inflammatory response via interaction with receptors for advanced glycation end products (RAGE). RAGE is a signal transduction receptor that binds both AGEs and AOPPs and by causing activation of NF-κB. *In vitro* experiments have shown that AGEs enhance transcription of genes for pro-inflammatory cytokines such as IL-6 and TNF-α. They may increase C-reactive protein (CRP) production. CRP is primarily produced byhepatocytes, and its chief inductor is the pro-inflammatory cytokine IL6. Glycosylated and oxidized proteins indirectly up-regulate CRP expression in hepatocytes by stimulating monocytes to produce IL-6 (Harrison-Findik, 2007). Further iron catalyzes the formation of NO from peroxynitrite and can modulate gene expression in cells leading to altered cell functions (Zuwała-Jagiełło et al., 2011).

Lipid peroxidation products such as 4-HNE and malondialdehyde can react with DNA bases and the ε-NH2 group of lysine and histidine residues. Acetaldehyde which is a product of ethanol oxidation, increases the binding of MDA and its own binding to proteins in a synergistic manner, resulting in the generation of new hybrid adducts i.e. MDA-acetaldehyde adducts. These adducts may play a role in the development and progression of liver fibrosis as they have been shown to stimulate the secretion of several cytokines and chemokines by liver endothelial cells (Zuwała-Jagiełło et al., 2011).

Oxidative stress is implicated in a number of pathological disorders. It is of importance that oxidative stress and associated liver disorders can influence iron metabolism and thereby have tremendous implications for further generation of ROS and exacerbation of disease.

Hepatic iron overload is common in many liver diseases where iron is a risk factor in disease progression (Hou et al., 2009). Hemochromatosis is a well-defined syndrome characterized by normal iron-driven erythropoiesis and toxic accumulation of iron in parenchymal cells of vital organs that can be caused by mutations in any gene that limits iron entry into the blood.

6.1 Hepatitis C virus infection

Hepatitis C virus (HCV) infection affects nearly 2% of the human population and is a major cause of liver disease worldwide. In majority of persons who suffered HCV infection, chronic state can be established which is often accompanied by alterations in liver function culminating in cirrhosis or hepatocellular carcinoma in 20% of infected individuals (Price and Kowdley, 2009). Total iron stores were reported to be elevated in patients with chronic hepatitis C compared to those with chronic hepatitis B, although in both sets of patients, the degree of necroinflammation did not differ (Muller et al., 2009). HCV infection more than hepatitis B infection has been reported to be associated with iron deposition, increased hepatic iron concentration, iron overload, increased serum Tf saturation, and serum ferritin. Franchini et al.,(2008) reported that chronic HCV infection is the main cause of hyperferritinemia.

Elevation in iron indices was reported by Girelli et al., (2009) to be correlated with progression of liver disease. Iron is mostly deposited in Kupffer cells and portal macrophages and interferon therapy reduced hepatic iron. The authors stated that hepatic iron overload is a consequence of hepatocyte necrosis which releases ferritin from hepatocytes to be subsequently taken up by macrophages. Alternatively, these authors suggested that increased iron absorption may contribute to increased iron stores. Further, hepcidin levels are decreased which in turn increase the expression of FPN, thereby increasing iron transport in the duodenum, release by macrophages and accumulation of iron in the liver(Muller et al., 2009). Price and Kowdley (2009) reported that increased levels of TfRII on the hepatocyte membrane may increase iron uptake in chronic HCV infection.

Excess iron which causes ROS generation, leads to lipid peroxidation, protein and DNA damage. The excess iron and lipid peroxidation products, such as 4-HNE can activate hepatic stellate cells as well as cause their proliferation and increase the synthesis of smooth muscle actin and collagen, all of which can contribute to hepatic fibrogenesis (Tanaka et al., 2008). Increased 4-HNE levels upregulate procollagen and inhibits the expression of metalloproteinase – 1 gene. The latter plays an important role in degrading collagen. Thus, oxidative stress contributes to hepatic fibrosis (Gomez et al., 2010)

Results of in vitro studies indicate that iron deposition in hepatocytes enhances HCV replication thereby aggravating the viral infection. Further, hepatocytic DNA damage results in accumulation of 8-hydroxy-2'-deoxyguanosine, which has been implicated in hepatocellular carcinoma. Muller et al., (2009) suggested that reducing body iron by phlebotomy or by consumption of a low iron diet may help to retard disease progression and risk of hepatoceullar carcinoma (Tanaka et al., 2008).

6.2 Alcoholic liver disease (ALD)

Alcohol related liver disease contribute substantially to disability and mortality, globally. Chronic alcohol consumption results in fatty liver, alcoholic hepatitis and cirrhosis and manifests itself in ALD (Phillipe et al., 2007). Ethanol oxidation increases the NADH/ NAD⁺ ratio, which leads to increased fatty acid synthesis, inhibits β-oxidation and the tricarboxylic acid cycle and increases the accumulation of triacylglycerols. Chronic alcohol consumption inhibits AMP kinase and activates sterol regulatory element binding protein-1, which increase lipogenesis resulting in hepatic steatosis(Mantena et al., 2008). Harrison-Findik(2009) suggested that mitochondrial DNA damage and ribosomal defects may inhibit mitochondrial protein synthesis and compromise oxidative phosphorylation in chronic alcohol consumption. Mitochondrial dysfunction has been reported to possibly play an important role in development of non-alcoholic fatty liver disease (NAFLD) and non alcoholic steatohepatitis (NASH).

Hou et al., (2009) reported that ingestion of even mild to moderate amounts of alcohol can increase iron overload, with the iron being localized in Kupffer cells and hepatocytes. On the other hand, alcohol metabolism to form acetaldehyde generates ROS. Thus iron and alcohol together contribute to and aggravate liver disease (Harrison-Findik, 2007). Although the mechanisms underlying iron accumulation are not well understood, it has been reported that alcohol-induced oxidative stress in hepatocytes downregulates hepcidin expression in liver, by altering transcription factor C/EBPα activity, thus dysregulating iron homeostasis (Hou et al., 2009; Harrison-Findik, 2009). To make matters worse, hepcidin expression is not responsive to body iron levels in alcohol consumption (Hou et al., 2009). H_2O_2 increases the expression of TfRI leading to iron accumulation and thus may support ALD progression (Girelli et al., 2009).

6.3 Nonalcoholic fatty liver disease

NAFLD is an important cause of liver damage ranging from fatty liver to nonalcoholic steatohepatitis (NASH). Simple steatosis is usually considered benign, but the development of NASH is recognized as a precursor of more severe liver disease, and in some cases, cirrhosis and hepatocellular carcinoma (Machado et al., 2009). Machado et al., (2009) reported that pathogenesis of NASH can be explained by the "two-hit" hypothesis, (a) insulin resistance and the resultant steatosis being the "first hit". Hyperinsulinemia, increased peripheral lipolysis and reduced β- oxidation lead to fat accumulation (b) the "second hit"- increased ROS/RNS. Once steatosis sets in, the liver progresses to more severe pathologic states when it is exposed to the second hit or other environmental stressors. Iron can increase the potential magnitude of the "second hit". Genetic risk factors also influence the susceptibility and severity of fatty liver disease.

Further, Fujita et al.,(2009) reported that NASH patients had elevated levels of hepatic 8-oxodG which is is a DNA base-modified product generated by hydroxyl radicals. The level of this product was reduced by iron reduction therapy. These authors also reported that in steatosis, saturation of the β-oxidation pathway with excess free fatty acids leads to further generation of H_2O_2, and other ROS. Also, serum iron, transferring saturation and ferritin levels were higher in NASH patients, which could be attributed to iron overload.

The first hit i.e. insulin resistance has also been reported to be linked to iron overload, through mobilization of peripheral fat to the liver and development of hyperinsulinemia. Hyperinsulinemia and peripheral lipolysis promote hepatic uptake of fatty acids as well as fatty acid synthesis, resulting in accumulation of fat and fatty liver (Mantena et al., 2008). A syndrome of "insulin resistance-associated iron overload" has proposed where in the presence of unexplained hepatic iron overload at least one component of the insulin resistance may be present. Insulin resistance also seemed to be closely linked to total body iron stores in the general population. Circulating prohepcidin was reported to be associated with parameters of glucose and iron metabolism, in subjects with altered glucose tolerance (Fujita et al., 2009). The researchers suggested that failure to increase synthesis of prohepcidin which can decrease iron absorption, could contribute to iron-induced disorders of glucose metabolism. Further support comes from reports that iron depletion can improve insulin sensitivity. ROS (induced by iron overload) can interfere with insulin signaling by inhibiting translocation of glucose transporter 4 to the plasma membrane, thus impairing insulin uptake. In insulin resistance, glycation of proteins including Tf occurs, which decreases its ability to bind ferrous iron, and increasing the free iron pool and amplifying oxidative stress (Fernandez-Real et al., 2009). Oxidative stress itself induces insulin resistance, thus creating a vicious cycle.

7. Pros and cons of high iron intakes

Normally in a well-nourished individual, the antioxidant defense mechanisms should protect against ROS injury, however in case of undernutrition, including micronutrient deficits, the antioxidant system would be compromised. In developing countries, anemic children, adolescents and women who receive iron supplements are also malnourished and chronically consume poor quality diets. During the life cycle, the individual is more vulnerable at certain life stages either because of physiological immaturity or senescence or due to alterations in physiology and metabolism as encountered in pregnancy. Thus infants, pregnant women and the elderly may be at higher risk, especially if their nutritional status is compromised by either deficiency or excess of any one nutrient.

Iron supplementation is widely adopted since iron deficiency anemia is a global health problem. However, in public health systems, the issue of iron overload and its possible adverse consequences need to be paid attention in terms of well-designed studies and may need to be monitored closely, if deleterious effects are observed. There are reports in the literature on the effect of iron on the gastrointestinal tract, risk of cancer etc but these are not taken into account in public health programs. In animal models, iron-induced increase in oxidative damage was associated with gastric inflammation, induction of gastric ulcers and increase in colorectal cancer (Seril et al., 2002; Naito et al., 1995). In patients with Crohn's disease treatment with 120 mg ferrous sulphate per day for 7 days increased clinical symptoms of disease activity (Erichsen et al., 2003). There is a dire need to examine the effects of iron deficiency, supplementation and their interaction on several parameters of oxidative stress, besides Hb which is routinely used as an indicator of iron deficiency anemia. In pregnancy, iron supplements are routinely recommended and there needs to be a consideration of the possible effects on ROS generation and pregnancy complications. Per se, the role of ROS in pregnancy in well-nourished and iron deficient women warrants attention. Fortification of foods with iron is adopted as an important public health measure to combat iron deficiency. When persons consume iron fortified foods, along with supplements, intakes could well be in excess of the recommended dietary allowances (RDA), which may pose risk for iron overload. Fisher and Naughton (2004) reported that daily intake of iron was 1874% of the Korean RDA for those who used supplements as compared to 62% of the RDA for non-users. The potential for oxidative stress could be high, if vitamin C is co-ingested, especially through supplements and fortified foods. High concentration of iron taken as bolus dose along with reducing substances like vitamin C can have deleterious effects. Multivitamin/ multimineral complexes are most common supplements administered. Unlike minerals in food where they are part of bioorganic substances, minerals in supplements and in fortified foods are in inorganic forms. In this form, minerals are capable of catalyzing free radicals generation (Rabovsky et al., 2010). Hence there is a need to reformulate nutritional supplements to obtain the benefits while simultaneously minimizing the risks.

A number of diseases have been shown to be associated with the intracellular accumulation of iron especially within mitochondria and lysosomes making them more susceptible to oxidative stress. Even a modest increase in the iron concentration in the substantia nigra may increase oxidative stress and neurodegeneration (Logroscino et al., 2008). Logroscino et al., (2008) stated that brain iron is responsive to peripheral iron status, hence brain iron status can be modulated by the iron content of the diet, both restriction and supplementation. The iron that accumulates in the brain is mostly nonheme iron. Case control studies and a prospective study in the US on suggest that dietary iron intake may be associated with higher risk of Parkinson's disease. The prospective study on a cohort of 47,406 men and 76,947 women indicated that risk was greater for persons whose vitamin C intakes were lower. The risk for Parkinson's disease was higher among those whose diets contained large amounts of non-heme iron primarily derived from fortified grains and cereals. Men were found to have some risk with the use of supplemental iron. The investigators did not find an association with heme iron (Logroscino et al., 1998; Gorell et al., 1999; Powers et al., 2003).

Besides this, with the growth of nanotechnology industry metallic nanoparticles are being increasingly used for variety of purposes including food products and medicines. Nanoparticles are able to cross semipermeable membrane via transcellular pathway. Since these particles are small in size and have a large surface area they will be highly reactive which in turn will increase their ability to produce ROS. Pujalté et al., (2011) have reported that metallic nanoparticles, such as, zinc oxide and cadmium sulphite exerted cytotoxic effects on glomerular and tubular renal cells. These effects were correlated with metal size and metal solubility. In an *in vitro* study, Apopa et al., (2009) have demonstrated that iron-nanoparticles induced an increase in endothelial cell permeability through the production of ROS.

In summary, both iron deficiency and excess are risk factors for oxidative stress. The emphasis should be on well-balanced diets to meet the RDA of all essential nutrients as well as inclusion of foods rich in phytochemicals that provide protection against oxidative stress.

8. Acknowledgement

The authors are extremely grateful to Ms. Mitravinda Savanur and Ms. Neha Lohia for technical support.

9. References

Achcar, F.; Camadro, JM. & Mestivier, D. (2011) A Boolean Probabilistic Model of Metabolic Adaptation to xygen in Relation to Iron Homeostasis and Oxidative Stress. *BMC Systems Biology*, 2011, Vol.5, pp.51−69.

Agarwal, A.; Gupta, S. & Sharma, RK. (2005) Role of Oxidative Stress in Female Reproduction. *Reprod Biol Endocrinol*, Vol.3, pp.28-48.

Aitken, RJ. & Roman, SD. (2008) Antioxidant Systems and Oxidative Stress in the Testes. *Oxid Med Cell Longev*, Vol.1, No.1, (Oct-Dec) pp.15–24.

Ansari, MA.; Roberts, KN. & Scheff, SW. (2008) Oxidative Stress and Modification of Synaptic Proteins in Hippocampus After Traumatic Brain Injury. *Free Radic Biol Med*, Vol.45, No.4, pp.443–452.

Apopa, PL.; Qian, Y.; Shao, R.; Guo, NL.; Schwegler-Berry, D.; Pacurari, M.; Porter, D.; Shi, X.; Vallyathan, V.; Castranova V. & Flynn, DC. (2009) Iron oxide nanoparticles induce human microvascular endothelial cell permeability through reactive oxygen species production and microtubule remodeling. *Particle and Fibre Toxicology*, Vol.6, pp.1-14.

Aune, R.; Kersti, Z.; Klaar, U.; Karro, H, Tiiu K & Zillmer M (2004) Ferrous Iron Administration During Pregnancy and Adaptational Oxidative Stress (Pilot Study). *Medicina (Kaunas)*, Vol.40, No.6, pp.547-552.

Aydin, S.; Benian, A.; Madazli, R.; Uludag, S.; Uzun, H. & Kaya, S. (2004) Plasma Malondialdehyde, Superoxide Dismutase, sE-Selectin, Fibronectin, Endothelin-1 and Nitric Oxide Levels in Women with Preeclampsia. *Eur J Obstet Gynecol Reprod Biol*. Vol.113, pp.21-25.

Baker, MA. & Aitken, RJ. (2005) Reactive oxygen species in spermatozoa: methods for monitoring and significance for the origins of genetic disease and infertility. *Reprod Biol Endocrinol*, Vol.3, pp.67-75.

Barbeito, AG.; Garringer, HJ.; Baraibar, MA.; Gao, X.; Arredondo, M.; Marco Núñez, T.; Smith, MA.; Ghetti, B. & Vidal, R. (2009) Abnormal Iron Metabolism and Oxidative Stress in Mice Expressing a Mutant Form of the Ferritin Light Polypeptide Gene. *J Neurochem*, Vol.109, No.4, pp.1067–1078.

Barbeito, AG.; Levade, T.; Delisle, MB.; Ghetti, B. & Vidal, R. (2010) Abnormal Iron Metabolism in Fibroblasts From A Patient with the Neurodegenerative Disease Hereditary Ferritinopathy. *Mol Neurodegeneration*, Vol.5, pp.50-59.

Brabin, BJ.; Premji, Z. & Verhoeff, F. (2001) An Analysis of Anemia and Child Mortality. *J Nutr*, Vol.131, pp.636S-648S.

Branco, MR.; Marinho, HS.; Cyrne L. & Antunes F. (2004) Decrease of H_2O_2 Plasma Membrane Permeability during Adaptation to H_2O_2 in *Sacchromyces cerevisiae*. *J Biol Chem*, Vol.279, No.8, pp.6501-6506.

Briley-Saebo, KC.; Cho, YS. & Tsimikas, S. (2011) Imaging of Oxidation-Specific Epitopes in Atherosclerosis and Macrophage-Rich Vulnerable Plaques. *Curr Cardiovasc Imaging Rep*, Vol.4, pp.4–16.

Burton, GJ. (2009) Oxygen, The Janus Gas; Its Effects on Human Placental Development and Function. *J Anat*, Vol.215, pp.27–35.

Burton, GJ. & Hung, TH. (2003) Hypoxia-Reoxygenation; A Potential Source of Placental Oxidative Stress in Normal Pregnancy and Preeclampsia. *Fetal Mat Med Rev*, Vol.14, pp.97–117.

Cairo, G. & Recalcati, S. (2007) Iron-Regulatory Proteins: Molecular Biology and Pathophysiological Implications. *Expert Rev Mol Med*, Vol.9, No.33, pp.1-13.

Callens, C.; Coulons, S.; Naudin, J.; Radford-Weiss, I. et al. (2007) Targeting Iron Homeostasis Induces Cellular Differentiation and Synergizes With Differentiating Agents in Acute Myeloid Leukemia. *J Exp Med*, Vol.207, No.4, pp.731-750.

Casanueva, E. & Viteri, FE. (2003) Iron and Oxidative Stress in Pregnancy. *J Nutr*, Vol.133, pp.1700S-1708S.

Centlow, M.; Carninci, P.; Nemeth, K.; Mezey, E.; Brownstein, M. & Hansson, SR. (2008) Placental Expression Profiling in Preeclampsia: Local Overproduction of Hemoglobin May Drive Pathological Changes. *Fertil Steril*, Vol.90, No.5, pp.1834-1843.

Cindrova-Davies, T.; Spasic-Boskovic, O.; Jauniaux, E.; Charnock-Jones, DS. & Burton, GJ. (2007a) Nuclear Factor-κB, p38, and Stress-Activated Protein Kinase Mitogen-Activated Protein Kinase Signaling Pathways Regulate Proinflammatory Cytokines and Apoptosis in Human Placental Explants in Response to Oxidative Stress Effects of Antioxidant Vitamins. *Am J Path*, Vol.170, No.5, pp.1511-1519.

Cindrova-Davies, T.; Yung, HW.; Johns, J.; Spasic-Boskovic, O.; Korolchuk, S.; Jauniaux, E.; Burton, GJ. & Charnock-Jones, DS. (2007b) Oxidative Stress, Gene Expression, and Protein Changes Induced in The Human Placenta During Labor. *Am J Path*, Vol.171, No.4, pp.1168-1179.

Clark, JF.; Loftspring, M.; Wurster, WL.; Beiler, S.; Beiler, C.; Wagner, KR. & Pyne-Geithman, GJ. (2008) Bilirubin Oxidation Products, Oxidative Stress, and Intracerebral Hemorrhage. *Acta Neurochir Suppl*, Vol.105, pp.7–12.

Crampton, N.; Kodiha, M.; Shrivastava, S.; Umar, R.; & Stochaj, U. (2009) Oxidative Stress Inhibits Nuclear Protein Export by Multiple Mechanisms That Target FG Nucleoporins and Crm1. *Mol Biol Cell*, Vol.20, pp.5106-5116.

Crist, BL.; Alekel, DL.; Ritland, LM.; Hanson, LN.; Genschel, V. & Reddy, MB. (2009) Association of Oxidative Stress, Iron and Centralized Fat Mass in Healthy Postmenopausal Women. *J Women's Hlth*, Vol.18, No.6, pp.795-801.

Cutler, RG.; Kelly, J.; Storie, K.; Pedersen, WA.; Tammara, A.; Hatanpaa, K.; Troncoso, J. & Mattson, MP. (2004) Involvement of Oxidative Stress-Induced Abnormalities in Ceramide and Cholesterol Metabolism in Brain Aging and Alzheimer's Disease. *Proc Natl Acad Sci*, Vol.101, No.7, pp.2070-2075.

D'Amelio, P.; Cristofaro, MA.; Tamone, C.; Morra, E.; Di Bella, S.; Isaia, G.; Grimaldi, A.; Gennero, L.; Gariboldi, A.; Ponzetto, A.; Pescarmona, GP. & Isaia, GC. (2008) Role of Iron Metabolism and Oxidative Damage in Postmenopausal Bone Loss. *Bone*, Vol.43, pp.1010–1015.

De la Monte, S. & Wands, JR. (2008) Alzheimer's Disease is Type 3 Diabetes—Evidence Reviewed. *J Diabetes Sci Technol*, Vol.2, No.6, pp.1106-1113.

de la Rosa, LC.; Moshage, H. & Nieto, N. (2008) Hepatocyte Oxidant Stress and Alcoholic Liver Disease. *Rev ESP Enferm Dig*, Vol.100, No.3, pp.156-163.

Deb, S.; Johnson, EE.; Robalinho-Teixeira, RL. & Wessling-Resnick, M. (2009) Modulation of Intracellular Iron Levels by Oxidative Stress Implicates A Novel Role For Iron in Signal Transduction. *Biometals*, Vol.22, No.5, pp.855–862.

Deck, KM.; Vasanthakumar, A.; Anderson, SA.; Goforth, JB.; Kennedy, MC.; Anatholine, WE. & Eisentein, RS. (2009) Evidence that Phosphorylation of Iron Regulatory Protein 1 at Serine 138 Destabilizes the [4fe-4s] Cluster in Cytosolic Aconitase by Enhancing 4fe-3fe Cycling. *J Biol Chem*, Vol.284, No.19, pp.12701–12709.

Doreswamy, K. & Muralidhara. (2005) Genotoxic Consequences Associated with Oxidative Damage in Testis of Mice Subjected to Iron Intoxication. *Toxicology, Vol.206*, No.1, pp.169-78.

Eaton, JW. & Qian, M. (2002) Molecular Bases of Cellular Iron Toxicity. *Free Radic Biol Med*, Vol.32, pp.833-840.

Erichsen, K.; Hausken, T.; Ulvik, RJ.; Svardal, RJ.; Berstad, A. & Berge, RK. (2003) Ferrous Fumarate Deteriorated Plasma Antioxidant Status in Patients with Crohn Disease. *Scand J Gastroenterol, Vol.38*, pp.543-548.

Fardy, CA. & Silverman, M. (1995) Antioxidants in Neonatal Lung Disease. *Arch Dis Child Fetal Neonatal*, Vol.73, No.2, pp.F112–F117.

Fernández-Real, JM.; Equitani, F.; Moreno, JM.; Manco, M.; Ortega, F. & Ricart, W. (2009) Study of circulating prohepcidin in association with insulin sensitivity and changing iron stores. *J Clin Endocrinol Metab*, Vol.94, No.3, pp.982- 988.

Fisher, AE. & Naughton, DP. (2004) Iron Supplements: The Quick Fix with Long Term Consequences. *Nutr J*, Vol.3, No.2, pp.1-5.

Franchini, M.; Targher, G.; Capra, F.; Montagnana, M. & Lippi, G. (2008) The Effect of Iron Depletion on Chronic Hepatitis C Virus Infection. *Hepatol Int*, Vol.2, No.3, pp.335-340.

Fujii, J.; Iuchi, Y.; Okada, F.; Watson, AL.; Skepper, JN.; Jauniaux, E. & Burton, GJ. (1998) Susceptibility of Human Placental Syncytiotrophoblastic Mitochondria to Oxygen-Mediated Damage in Relation to Gestational Age. *J Clin Endocrinol Metab*, Vol.83, pp.1697–1705.

Fujita, N.; Miyachi, H.; Tanaka, H.; Takeo, M.; Nakagawa, N.; Kobayashi, Y.; Iwasa, M.; Watanabe, S. & Takei, Y. (2009) Iron Overload is Associated with Hepatic Oxidative Damage to DNA in Nonalcoholic Steatohepatitis. *Cancer Epidemiol Biomarkers Prev*, Vol.18, No.2, pp.424-32.

Gaeta, A. & Hilder, RC. (2005) The Crucial Role of Metal Ions in Neurodegeneration: The Basis for a Promising Therapeutic Strategy. *Br J Pharmacol*, Vol.146, pp.1041–1059.

Gandley, RE.; Rohland, J.; Zhou, Y.; Shibata, E.; Harger, GF.; Rajakumar, A.; Kagan, VE.; Markovic, N. & Hubel, CA. (2008) Increased Myeloperoxidase in the Placenta and Circulation of Women with preeclampsia. *Hypertension*, Vol.52, No.2, pp.387–393.

Gao, J.; Ehen, J.; Kramer, N.; Tsukamoto, H.; Zhang, A. & Enns, C. (2009) Interaction of Hereditary Hemochromatosis Protein HFE with Transferrin Receptor 2 is Required for Transferrrin –Induced Hepcidin Expression. *Cell Metab*, Vol.9, pp.217-227.

Gao, X.; Hu, X.; Qian, L.; Yang, S.; Zhang, W.; Zhang, D.; Wu, X.; Fraser, A.; Wilson, B.; Flood, PM.; Block, M. & Hong, JS. (2008). Formyl-Methionyl-Leucyl-Phenylalanine-Induced Dopaminergic Neurotoxicity via Microglial Activation: A Mediator between Peripheral Infection And Neurodegeneration? *Environ Hlth Pers*, Vol.116, No.5, pp.593-598.

Gautier, CA.; Kitada, T. & Shen, J. (2008). Loss of PINK1 Causes Mitochondrial Functional Defects And Increased Senitivity to Oxidative Stress. *Proc Natl Acad Sci*, Vol.15 No.32, pp.11364-11369.

George, JL.; Mok, S.; Moses, D.; Wilkins, S.; Bush, AI.; Cherny, RA. & Finkelstein, DI. (2009). Targeting The Progression Of Parkinson's Disease. *Curr Neuropharmacol*, Vol.27, pp.9-36.

Girelli, D.; Pasino, M.; Goodnough, JB.; Nemeth, E.; Guido, M.; Castagna, A.; Busti, F.; Campostrini, N.; Martinelli, N.; Vantini, I.; Corrocher, R.; Ganz, & Fattovich, G. (2009). Reduced Serum Hepcidin Levels In Patients With Chronic Hepatitis C. *J Hepatol*, Vol.51, No.5, pp.845–852.

Goforth, JB.; Anderson, SA.; Nizzi, CP. & Eisenstein, RS. (2010). Multiple Determinants within Iron-Responsive Elements Dictate Iron Regulatory Protein Binding and Regulatory Hierarchy. *RNA*, Vol.16 pp. 154-169.

Gomez, EV.; Perez, YM.; Sanchez, H.; Forment, GR.; Soler, EA.; Bertot, LC.; Garcia, AY.; Vazquez, MRA.; Fabian, LG. (2010). Antioxidant and Immunomodulatory Effects of Viusid in Patients with Chronic Hepatitis C. *World J Gastroenterol*, Vol.16 No.21, pp.2638-2647.

Gorell, JM.; Johnson, CC.; Rybicki, BA.; Peterson, EL.; Kortsha, GX.; Borwn, GG. & Richardson, RJ. (1999). Occupational Exposure to Manganese, Copper, Lead, Iron, Mercury and Zinc and the Risk of Parkinson's Disease. *Neurotoxicology*, Vol.20, No.2-3, pp.239–247.

Guller, S. (2009). Role of The Syncytium in Placenta-Mediated Complications of Preeclampsia. *Thromb Res*, Vol.124, No.4, pp.389-392.

Gupta, P.; Narang, M.; Banerjee, BD. & Basu, M. (2004). Oxidative Stress In Term Small For Gestational Age Neonates Born To Undernourished Mothers: A Case Control Study. *BMC Pediatrics*, Vol.4, No14, doi:10.1186/1471-2431-4-14

Haas, JD. & Brownlie, T. (2001). Iron Deficiency and Reduced Work Capacity. A Critical Review of Research to Determine a Causal Relationship. *J Nutr*, Vol.131, pp.676S-688S.

Hagay, ZJ.; Weiss, Y.; Zusman, I.; Peled-Kamar, M.; Reece, EA.; Eriksson, UJ. & Groner, Y. (1995). Prevention of Diabetes Associated Embryopathy by Overexpression of the Free Radical Scavenger Copper Zinc Superoxide Dismutase in Transgenic Mouse Embryos. *Am J Obstet Gynecol*, Vol.173, pp.1036–1041.

Hall, ED.; Vaishnav, RA. & Mustafa, AG. (2010). Antioxidant Therapies for Traumatic Brain Injury. *Neurotherapeutics*, Vol.7, No.1, pp.51-67.

Harrison-Findik, DD. (2007). Role of Alcohol in the Regulation of Iron Metabolism. *World J Gastroenterol*, Vol.13, No.37, pp.4925-4930.

Harrison-Findik, DD. (2009). Is The Iron Regulatory Hormone Hepcidin A Risk Factor For Alcoholic Liver Disease? *World J Gastroenterol*, Vol.15, No.10, pp.1186-1193.

He, X.; Hahn, P.; Lacovelli, J.; Wong, R.; King, C.; Bhisitkul, R.; Massaro-Giordana, M. & Duraief, J. (2007). Iron Homeostasis and Toxicity in Retinal Degeneration . *Prog Retin Eye Res*, Vol.26, No.6, pp.649-673.

Hintze, KJ. & Theil, EC. (2005). DNA and mRNA Elements with Complementary Responses to Hemin, Antioxidant Inducers, and Iron Control Ferritin-L Expression. *Proc Natl Acad Sci*, Vol.102, No.42, pp.15048-15052.

Hofe, T.; Marzetti, E.; Xu, J.; Seo, AY.; Gulec, S.; Knutson, MD.; Leeuwenburgh, & Dupont-Versteegden, EE. (2008). Increased RNA Oxidative Damage And Iron Content In Skeletal Muscle With Aging And Disuse Atrophy. *Exp Gerontol*, 43(6):563-570.

Holgado-Madruga, M. & Wong, AJ. (2003). Gab1 is an Integrator of Cell Death Versus Cell Survival Signals in Oxidative Stress. *Mol Cell Biol*, 23(12): 4471-4484

Hou, WH.; Rossi, L.; Shan, Y.; Zheng, JY.; Lambrecht, RW.; Bonkovsky, HL. (2009). Iron Increases HMOX1 and Decreases Hepatitis C Viral Expression in HCV-Expressing Cells. *World J Gastroenterol*, Vol.15, No.36, pp.4499-510.

Hsieh, Y.; Chang, CC. & Lin, CS. (2006). Seminal Malondialdehyde Concentration But Not Glutathione Peroxidase Activity Is Negatively Correlated With Seminal Concentration and Motility. *Int J Biol Sci*, Vol.2, No.1, pp.23-29.

Huang X (2008) Does Iron Have A Role in Breast Cancer? *Lancet Oncol*, Vol.9, No.8, pp.803–807.

Ianotti, L.; Tielsch, J.; Black, MM. & Black, R. (2006). Iron Supplement in Early Childhood: Health Benefits and Risks. *Am J Clin Nutr*, Vol.84, pp.1261-1276.

Jaillard, T .; Roger, M .; Galinier, A .; Guillou, P .; Benani, A.; Leloup, C.; Casteilla, L .; Penicaud, L. & Lorsignol, A. (2009) Hypothalamic Reactive Oxygen Species are Required for Insulin-Induced Food Intake Inhibition an NADPH Oxidase-Dependent Mechanism. *Diabetes*, Vol.58, pp.1544-1549.

Jana, A. & Pahan, K. (2007). Oxidative Stress Kills Human Primary Oligodendrocytes via Neutral Sphingomyelinase: Implications for Multiple Sclerosis. *J Neuroimmune Pharmacol*, Vol.2, No.2, pp.184–193.

Jauniaux, E.; Watson, AL.; Hempstock, J.; Bao, YP.; Skepper, JN. & Burton, GJ. (2000). Onset of Maternal Arterial Blood Flow and Placental Oxidative Stress: A Possible Factor In Human Early Pregnancy Failure. *Am J Pathol*, Vol.157, No.6, pp.2111-2121.

Jauniaux, E.; Hempstock, J.; Greenwold, N. & Burton, GJ. (2003). Trophoblastic Oxidative Stress in Relation to Temporal and Regional Differences in Maternal Placental Blood Flow in Normal And Abnormal Early Pregnancies. *Am J Pathol*, Vol.162, No. 1, pp.115-125.

Jian, J.; Pelle, E. & Huang, X. (2009). Iron and Menopause: Does Increased Iron Affect the Health of Postmenopausal Women? *Antioxid Redox Signal*, Vol.11, No.12, pp.2939–2943.

Jiang, D.; Li, X.; Williams, R.; Patel, S.; Men, L.; Wang, Y. & Zhou, F. (2009). Ternary Complexes Of Iron, Amyloid-B And Nitrilotriacetic Acid: Binding Affinities, Redox Properties, And Relevance To Iron-Induced Oxidative Stress In Alzheimer's Disease. *Biochemistry*, Vol.48, No.33, pp. 7939–7947.

Jinsmaa, Y.; Florang, VR.; Rees, JN.; Anderson, DG.; Strack, S. & Doorn, JA. (2009). Products of Oxidative Stress Inhibit Aldehyde Oxidation and Reduction Pathways in Dopamine Catabolism Yielding Elevated Levels of a Reactive Intermediate . *Chem Res Toxicol*, Vol.22, No.5, pp.835–841.

Jones, DP. (2006). Extracellular Redox State: Refining the Definition of Oxidative Stress in Aging. Rejuvenation Res Vol.9, No.2, pp.169-181.(Abstr)

Manusco, C. (2004). Heme Oxygenase and Its Products in the Nervous System. *Antioxd Redox Signal*, Vol.6, No.5, pp.878-887.(Abstr)

Kaczmarek, M.; Cachau, RE.; Topol, IA.; Kasprzak, KS.; Ghio, A.; Salnikow, K. (2009). Metal Ions-Stimulated Iron Oxidation in Hydroxylases Facilitates Stabilization of HIF-1a Protein. *Toxicol Sci*, Vol.107, No.2, pp.394-403.

Kapralov, A.; Vlasova, II.; Feng, W.; Maeda, A.; Walson, K.; Tyurin, VA.; Huang, Z.; Aneja, RK.; Carcillo, J.; Bayır, H. & Kagan, VE. (2009). Peroxidase Activity of Hemoglobin Haptoglobin Complexes.Covalent Aggregation and Oxidative Stress in Plasma and Macrophages *J Biol Chem*, Vol.284, No.44, pp.30395-30407.

Kastman, EK.; Willette, AA.; Coe, CL.; Bendlin, BB.; Kosmatka, KJ.; McLaren, DG.; Xu, G.; Canu, E.; Field, AS., Alexander, AL.; Voytko, ML.; Beasley, TM.; Colman, RJ.; Weindruch, R. & Johnson, SC. (2010). A Calorie-Restricted Diet Decreases Brain Iron Accumulation and Preserves Motor Performance in Old Rhesus Monkeys. *J Neurosci*, Vol.30, No.23, pp.7940–7947.

Kautz, L., Meynard, D.; Monnier, A.; Darnaud, V. & Baunet, R. (2008). Iron Regulates Phosphorylation of Smad1/5/8 and Gene Expression of BMP6, Smad 7, Id1 And Atoh 8 in Mouse Liver. *Blood*, Vol.112, pp.1503-1509.

Kell, DB. (2009). Iron Behaving Badly: Inappropriate Iron Chelation as a Major Contributor to the Etiology of Vascular and Other Progressive Inflammatory and Degenerative Diseases. *BMC Med Genomics*, Vol.2, No.2, pp.1-79.

Knutson, MD. (2010). Iron-Sensing Proteins that Regulate Hepcidin and Enteric Iron Absorption. *Ann Rev Nutr*, 30:149-171.

Kohgo, Y.; Ikuta, K.; Ohtake, T.; Torimoto, Y. & Kalo, J. (2008). Body Iron Metabolism and Pathophysiology Of Iron Overload. *Int J Hematol*, Vol.88, No.1, pp.7-15.

Kurz, T.; Leake, A.; von Zglinick, T.; & Brunk, UT. (2004). Relocalized Redox-Active Lysosomal Iron Is An Important Mediator Of Oxidative-Stress-Induced DNA Damage. *Biochem J*, Vol.378, pp.1039-1045

Kurz, T.; Terman, A.; Gustafsson, B. & Brunk, UT. (2008). Lysosomes in Iron Metabolism, Ageing and Apoptosis. *Histochem Cell Biol*, Vol.129, pp.389-406.

Liu, G.; Mena, P.; Kudob, W.; Perry, G. & Smith, MA. (2009). Nanoparticle -Chelator Conjugates as Inhibitors of Amyloid-B Aggregation and Neurotoxicity: A Novel Therapeutic Approach for Alzheimer Disease . *Neurosci Lett*, Vol.455, No.3, pp.187–190.

Logroscino, G.; Gao, X.; Chen, H.; Wing, A. & Ascherio, A. (2008). Dietary Iron Intake and Risk Of Parkinson's Disease. *Am J Epidemiol*, Vol.168, pp.1381-1388.

Logroscino, G.; Marder, K.; Graziano, J.; Freyer, G.; Slakovich, V.; Lojacono, N.; Cote, L. & Mayeux, R. (1998) Dietary Iron, Animal Fats, And Risk Of Parkinson's Disease. *Mov Disord*, 13(suppl 1):13–16.

Lucesoli, F.; Caligiuri, M.; Roberti, MF.; Perazzo, J.C. & Fraga, C.G. (1999) Dose-Dependent Increase Of Oxidative Damage in the Testes of Rats Subjected to Acute Iron Overload. *Toxicology*, Vol.132, No. 2-3, pp.179-186.

Lushchak, VI. (2011) Adaptive Response to Oxidative Stress: Bacteria, Fungi, Plants & Animals. *Comp Biochem Physiol C Toxicol Pharmacol*, Vol.153, No.2, pp.175-190.

References and further reading may be available for this article. To view references and further reading you must purchase this article.

Machado, MV.; Ravasco, P.; Martins, A.; Almeida, MR.; Camilo, ME. & Cortez-Pinto, H. (2009). Iron Homeostasis And H63D Mutations in Alcoholics with and without Liver Disease. *World J Gastroenterol*, Vol.15, No.1, pp.106–111.

Mackenzie, EL.; Iwasaki, K. & Tsuji, Y. (2008). Intracellular Iron Transport and Storage: From Molecular Mechanisms To Health Implications. *Antioxid Redox Signal*, Vol.10, No.6, pp.998-1016.

Mandel, S.; Weinreb, O.; Amit, T. & Youdim, MB. (2005). Mechanism Of Neuroprotective Action of the Anti-Parkinson Drug Rasagiline and its Derivatives. *Brain Res Brain Res Rev*, Vol.48, pp.379–387.

Mantena, SK.; King, AL.; Andringa, KK.; Eccleston, HB. & Bailey, SM. (2008). Mitochondrial Dysfunction And Oxidative Stress In The Pathogenesis of Alcohol and Obesity-Induced Fatty Liver Diseases. *Free Radic Biol Med*, Vol.44, No.7, pp.1259-1272.

Martinez-Outschoorn, UE.; Balliet, RM.; Rivadeneira, DB.; Chiavarina, B.; Pavlides, S.; Wang, C.; Whitaker-Menezes, D.; Daumer, KM.; Lin, Z.; Witkiewicz, AK.; Flomenberg, N.; Howell, A.; Pestell, RG.; Knudsen, ES.; Sotgia, F. & Lisanti, MP. (2010). Oxidative Stress in Cancer Associated Fibroblasts Drives Tumor-Stroma Co-Evolution: A New Paradigm For Understanding Tumor Metabolism, the Field Effect and Genomic Instability in Cancer Cells. *Cell Cycle*, Vol.9, No.16, pp.3256-3276.

Melov, S.; Adlard, PA.; Morten, K.; Johnson, F.; Golden, TR.; Hinerfeld, D.; Schilling, B.; Mavros, C.; Masters, CL.; Volitakis, I.; Li, QX.; Laughton, K.; Hubbard, A.; Cherny, RA.; Gibson, B. & Bush, AI. (2007). Mitochondrial Oxidative Stress Causes Hyperphosphorylation of Tau. PLoS ONE 2(6): e536. doi:10.1371/ journal.pone.0000536   mice subjected to iron intoxication. *Toxicology*, Vol.206, pp.169-78.

Mikhail, MS.; Anyaegbunam, A.; Garfinkel, D.; Palan, PR.; Basu, J.; Romney, SL. (1994). Preeclampsia And Antioxidant Nutrients: Decreased Plasma Levels Of Reduced Ascorbic Acid, Alpha-Tocopherol, And Beta-Carotene In Women With Preeclampsia. *Am J Obstet Gynecol*, Vol.171, pp.150-157.

Mills, E.; Dong, XP.; Wang, F.; Xu, H. (2010) Mechanisms Of Brain Iron Transport: Insight Into Neurodegeneration And CNS Disorders. *Future Med Chem*, Vol.2, No.1, pp.51-64.

Moslemi, MK. & Tavanbakhsh, S. (2011). Selenium–Vitamin E Supplementation In Infertile Men: Effects On Semen Parametersand Pregnancy Rate. *Int J Gen Med* Vol.4, pp.99-104.

Moussa, B.H.; Youdim. & Bakhle, YS. (2006). Monoamine oxidase: Isoforms and inhibitors in Parkinson's disease and depressive illness. *Br J Pharmacol* Vol.147, pp.S287-S 296.

Muckenthaler, M.; Galy, B. & Hentze, M. (2008). Systemic Homeostasis and IRE/IRP Regulatory Network. *Ann Rev Nutr*, Vol.28, pp.197-213.

Mueller, S.; Millonig, G.; & Seitz, HK. (2009). Alcoholic Liver Disease and Hepatitis C: A Frequently Underestimated Combination. *World J Gastroenterol*. Vol.15, No.28, pp.3462-3471.

Mulder, DJ.; van Haelst, PL.; Wobbes, MH.; Gans, RO.; Ziljstra, F.; May, JF.; Amit, AJ.; Teravaert, AWC. & van Doormal, JJ. (2007) The Effect of Aggressive versus Conventional Lipid-Lowering Therapy on Markers of Inflammatory and Oxidative Stress. *Cardiovasc Drugs Ther*, Vol.21, pp.91-97.

Myatt, L. (2010). Reactive Oxygen and Nitrogen Species and Functional Adaptation of the Placenta. *Placenta*, Vol.31(Suppl), pp.S66–S69.

Nagababu, E.; Gulyani, S.; Earley, C J.; Cutler, RG.; Mattson, MP. & Rifkind, JM. (2008). Iron-Deficiency Anemia Enhances Red Blood Cell Oxidative Stress. *Free Radic Res*, Vol.42, No.9, pp.824–829.

Naito, Y.; Yoshikawa, T.; Yoneta, T.; Yagi, N.; Matsuyama, K.; Arai, M.; Tanigawa, T.; Kondo, M. (2009) A New Gastric-Ulcer Model in Rats Produced by Ferrous Iron and Ascorbic-Acid Injection. *Digestion*, Vol. 56, pp.472-478.

Nemeth, E. & Ganz, T. (2009) The Role of Hepcidin in Iron Metabolism *Acta Hematol* Vol.12, pp.78-86.

Nicol, CJ.; Zielenski, J.; Tsui, LC. & Wells, PG. (2000). An Embryoprotective Role for Glucose-6-Phosphate Dehydrogenase in Developmental Oxidative Stress and Chemical Teratogenesis. *FASEB J*, Vol.14, pp.111– 127.

Nuhn, P.; Kunzlim, BM.; Hennig, R.; Mitkus, T.; Ramanauskas, T.; Nobiling, R.; Meuer, SC.; Friess, H. & Berberat PO. (2009). Heme Oxygenase-1 and Its Metabolites Affect Pancreatic Tumor Growth in Vivo. *Mol Cancer*, 8:37

Olivares, D.; Huang, X.; Branden, L.; Grieg, NH. & Rogers, JT. (2009). Physiological and pathological role of alpha-synuclein in parkinson's disease through iron mediated oxidative stress:; the role of a putative iron-responsive element. *Int.J. Mol. Sci.* 10: 1226-1260.

Outten, FW. & Theil, EC. (2009). Iron-Based Redox Switches in Biology. *Antioxid Redox Signal* Vol.11, No.5, pp. 1029-1046.

Perez, LR. & Franz, KJ. (2010). Minding Metals: Tailoring Multifunctional Chelating Agents for Neurodegenerative Disease. *Dalton Trans*, Vol.39, No.9, pp.2177–2187.

Philippe, MA.; Ruddell, RG.; Ramm, GA. (2007). Role of Iron in Hepatic Fibrosis: One Piece in The Puzzle. *World J Gastroenterol*, Vol.13, No.35, pp.4746-4754.

Picard, E.; Fontaine, I.; Jonet, L.; Guillou, F.; Behar-Cohen, F.; Courtois, Y. & Jeanny, JC. (2008). The Protective Role of Transferrin in Müller Glial Cells after Iron Induced Toxicity. *Mol Vis*, Vol.14, pp.928-941.

Pivotraiko, VN.; Stone, SL.; Roth, KA. & Shacka, JJ. (2009). Oxidative Stress and Autophagy in the Regulation of Lysosome-Dependent Neuron Death. *Antioxid Redox Signal*, Vol.11, No.3, pp.481-496.

Powers , KM.; Smith – Weller, T.; Franklin, G.; Longstreth, WT.; Swanson, PD. & Checkoway, H. (2009). Dietary Fats, Cholesterol and Iron as Risk Factors for Parkinson's Disease. *Parkinsonism Relat Disord*, Vol.15, No.1, pp.47–52.

Powers, KM.; Smith-Weller, T.; Franklin, GM.; Longstreth, WT.; Swanson, PD. & Checkoway, H.(2003). Parkinson's Disease Risks Associated with Dietary Iron, Manganese, and Other Nutrient Intakes. *Neurology*, Vol.60, No.11, pp.1761–1766.

Pujalté, I.; Passagne, I.; Brouillaud, B.; Tréguer, M.; Durand, E.; Ohayon-Courtès, C. & L'Azou, B. (2011) Cytotoxicity and Oxidative Stress Induced by Different Metallic Nanoparticles on Human Kidney Cells. *Particle and Fibre Toxicology, Vol.*8, pp.10-26.

Prater, MR.; Laudermilch, CL.; Liang, C. & Holladay, SD. (2008). Placental Oxidative Stress Alters Expression of Murine Osteogenic Genes and Impairs Fetal Skeletal Formation. *Placenta* Vol.29, No.9, pp.802–808.

Price, L. & Kowdley, KV. (2009). The Role of Iron in the Pathophysiology and Treatment of Chronic Hepatitis C. *Can J Gastroenterol*, 23(12): 822–82

Pringle, KG.; Kind, KL.; Ferruzzi-Perri, AN.; Thompson, JG. & Roberts, CT. (2010). Beyond Oxygen: Complex Regulation and Activity of Hypoxia Inducible Factors in Pregnancy. *Hum Reprod Update*, Vol.16, No.4, pp.415–431.

Rabovsky, AB.; Komarov, AM.; Ivie, JS. & Buettner, GR. (2010) Minimization of Free Radical Damage by Metal Catalysis of Multivitamin/Multimineral Supplements. *Nutrition Journal*, Vol.9, pp.61-68.

Ramakrishnan, U. (2002). Prevalence of Micronutrient Malnutrition Worldwide. *Nutr Rev*, Vol.60, pp.546-62.

Regan, R.; Li, Z.; Chen, ZM.; Zhang, X. & Chen-Roetling, J.(2008). Iron Regulatory Proteins Increase Neuronal Vulnerability to Hydrogen Peroxide. *Biochem Biophys Res Commun*, Vol.375, No.1, pp.6–10.

Roualt, T. & Cooperman, S. (2006). Brain Iron Metabolism. *Semin Pediatr Neurol*, Vol.13, pp.142-148.

Ruder, EH.; Terry, J.; Hartman, J.; Blumberg, J.; & Goldman, MB. (2008). Oxidative Stress and Antioxidants: Exposure and Impact on Female Fertility. *Hum Reprod Update*, Vol.14, No.4, pp.345–357.

Ruder, E H.; Hartman, TJ. & Goldman, MB. (2009). Impact of Oxidative Stress on Female Fertility. *Curr Opin Obstet Gynecol*, Vol.21, No.3, pp.219–222.

Rutherford, JC. & Bird, AJ. (2004). Metal-Responsive Transcription Factors that Regulate Iron, Zinc and Copper Homeostasis in Eukaryotic Cells. *Eukaryotic Cell*, Vol.3, No.1, pp.1-13.

Ryter, SW. & Choi, AMK. (2009). Heme Oxygenase-1/Carbon Monoxide from Metabolism to Molecular Therapy. *Am J Respir Cell Mol Biol*, Vol.41, pp.251–260

Sanchez-Ortiz, E .; Hahm , BK.; Armstrong, DL. & Rossie, S. (2009). Protein Phosphatase 5 Protects Neurons Against Amyloid B Toxicity. *J Neurochem*, Vol.111, No.2, pp.391–402.

Seril, DN.; Liao, J.; Ho, KL.; Yang, CS. & Yang, GY. (2002) Dietary Iron Supplementation Enhances Dss-Induced Colitis and Associated Carcinoma Development in Mice. *Dig. Dis Sci, Vol.*47, pp.1266-1278.

Shi, H.; Bencze, KZ.; Stemmlet, TL. & Philpott, S. (2008). A Cytosolic Iron Chaperone that Delivers Iron to Ferritin. *Science*, Vol.30,320 No.5880, pp.1207-1210.

Siedlak, SL.; Casadesus, G.; Webber, KM.; Pappolla, MA.; Atwood, CS.; Smith, MA. & Perry, G. (2009). Chronic Antioxidant Therapy Reduces Oxidative Stress in a Mouse Model of Alzheimer's Disease. *Free Radic Res*, Vol.43, No.2, pp.156–164.

Slonima, DK.; Koidec, K.; Johnson, KL.; Tantravahid, U.; Cowanc, JM.; Jarrah, Z. & Bianchi, DW. (2009). Functional Genomic Analysis of Amniotic Fluid Cell-Free Mrna Suggests that Oxidative Stress is Significant in Down Syndrome Fetuses . *Proc Natl Acad Sci*, Vol.106, No.23, pp. 9425–9429.

Smith, MA.; Zhua, X.; Tabatonb, M.; Liuc, G.; McKeel, DW.; Cohen, ML.; Wang, X.; Siedlak, SL.; Hayashi, T.; Nakamura, M.; Nunomura, A. & Perry, G. (2010). Increased Iron and Free Radical Generation in Preclinical Alzheimer Disease and Mild Cognitive Impairment. *J Alzheimers Dis*, Vol.19, No.1, pp.363–372.

Sompol, P.; Ittarat, W.; Tangpong, J.; Chen, Y.; Doubinskaia, I.; Batinic-Haberle, I.; Abdul, HM.; Butterfield, DA. & St. Clair, DK. (2008). A Neuronal Model of Alzheimer's Disease: An Insight into the Mechanisms of Oxidative Stress-Mediated Mitochondrial Injury. *Neuroscience*, Vol.153, No.1, pp.120-30.

Stankiewicz, J.; Panter, SS.; Neema, M.; Arora, A.; Batt, C. & Bakshi, R. (2007) Iron in Chronic Brain Disorders: Imaging and Neurotherapeutic Implications. *Neurotherapeutics*, Vol.4, No.3, pp.371–386.

Szczepanska, M.; Kozlik, J.; Skrzypczak, J. & Mikolajczyk, M. (2003). Oxidative Stress may be a Piece in the Endometriosis Puzzle. *Fertil Steril*, Vol.79, pp.1288-1293.

Takagi, Y.; Nikaido, T.; Toki, T.; Kita, N.; Kanai, M.; Ashida, T.; Ohira, S. & Konishi, I. (2004). Levels of Oxidative Stress and Redox-Related Molecules in the Placenta in Preeclampsia and Fetal Growth Restriction. *Virchows Arch*, Vol.444, pp.49-55.

Takami, T. & Sakaida, I. (2011). Iron Regulation by Hepatocytes and Free Radicals. *J Clin Biochem Nutr*, Vol.48, No.2, pp.103- 106.

Tanaka, H.; Fujita, N.; Sugimoto, R.; Urawa, N.; Horiike, S.; Kobayashi, Y.; Iwasa, M.; Ma, N.; Kawanishi, S.; Watanabe, S.; Kaito, M. &Takei, Y. (2008). Hepatic Oxidative DNA Damage is Associated with Increased Risk for Hepatocellular Carcinoma in Chronic Hepatitis C. *Br J Cancer*, Vol.98, No.3, pp.580–586.

Thiel, C. & Goss, DJ. (2009) Living with iron (and oxygen): questions and answers about iron homeostasis. *Chem Rev*, Vol.109, No.10, pp.4568-4579.

Tunc, O. & Tremellen, K J. (2009). Oxidative DNA Damage Impairs Global Sperm DNA Methylation in Infertile Men. *Assist Reprod Genet*, Vol.26, pp.537–544.

Uchiyama, A.;, Kim, JS.; Kon, K.; Jaeschke, H.; Ikejima, K.; Watanabe, S. & Lemasters, JJ. (2008). Translocation of Iron from Lysosomes into Mitochondrria is a Key Event during Oxidative Stress-Induced Hepatocellular Injury. *Hepatology*, Vol.48, No.5, pp.1644-1654.

Vegeto, E.; Belcredito, S.; Etteri, S.; Ghisletti, S.; Brusadellli, A.; Meda, C.; Krust, A.; Dupont, S.; Clana, P.; Chambon, P. & Maggi, A. (2003). Estrogen Receptor-_ Mediates the Brain Antiinflammatory Activity of Estradiol. *PNAS*, Vol.100, No.16, pp.9614-9619.

Vyas, K. (2005). Iron And Zinc Status, Oxidative Stress and Pregnancy Outcome. Unpublished M.Sc. dissertation submitted to S.N.D.T. Women's University, Mumbai, India.

Wang, J. & Pantopoulos, K. (2011). Regulation of Cellular Iron Metabolism, *Biochem J*, Vol.434, pp.365-381.

Wang, X. & Michaelis, EK. (2010). Selective Neuronal Vulnerability to Oxidative Stress in the Brain. *Frontiers in Aging NeuroSci*, doi: 10.3389/fnagi.2010.00012.

Wessling-Resnick, M. (2010). Iron Homeostasis and Inflammatory Response. *Ann Rev Nutr*, Vol.30, pp.105-122.

White-Gilbertson, S.; Kasman, L.; McKillop, J.; Tirodkar, T.; Lu, P. & Voelkel-Johnson, C. (2009). Oxidative Stress Sensitizes Bladder Cancer Cells to Trailmediated Apoptosis by Downregulating Anti-Apoptotic Proteins. *J Urol*, Vol.182, No.3, pp.1178–1185.

Wise, T.; Lunstra, DD.; Rohrer, A. & Ford, JJ. (2003) Relationships of Testicular Iron and Ferritin Concentrations with Testicular Weight and Sperm Production in Boars. J Animal Sci, Vol.81, No. 2, pp.503-511.

Wu, M.; Bian, Q.; Liu, Y.; Fernandes, AF.; Taylor, A.; Pereira, P. & Shang, F. (2009). Sustained Oxidative Stress Inhibits Nf-Kb Activation Partially via Inactivating the Proteasome. *Free Radic Biol Med*, Vol.46, No.1, pp.62–69.

Zecca, L.; Stroppolo, A.; Gatti, A.; Tampellini, D.; Toscani, M.; Gallorini, MMH.; Giaveri, G.; Arosio, P.; Santambrogio, P.; Fariello, RG.; Karatekin, E.; Kleinman, MH.; Turro, N.; Hornykiewicz, O. & Zucca, FA. (2004). The Role of Iron and Copper Molecules in the Neuronal Vulnerability of Locus Coeruleus and Substantia Nigra during Aging. *Proc Natl Acad Sci*, Vol.101, No.26, pp.9843–9848.

Zecca, L.; Youdim, MB.; Riederer, P.; Connor, JR. & Crichton, RR. (2004). Iron, Brain Ageing and Neurodegenerative Disorders. *Nat. Rev. Neurosci*, Vol.5 , pp.863–873.

Zhang, A. & Enns, C. (2009). Iron Homeostasis: Recently Identified Proteins Provide Insight Into Novel Control Mechanisms. *J Biol Chem*, Vol.284, No.2, pp.711-715.

Zhang, J.; Masciocchi, M.; Lewis, D.; Sun, W.; Liu, A. & Wang, Y. (2008). Placental Anti-Oxidant Gene Polymorphisms, Enzyme Activity, and Oxidative Stress In Preeclampsia. *Placenta*, Vol.29, No.5, pp.439–443.

Zhaoa, H.; Wonga, RJ.; Kalish, FS.; Nayak, NR. & Stevenson DK. (2009). Effect of Heme Oxygenase-1 Deficiency on Placental Development. *Placenta*, Vol.30, No.10, pp.861–868.

Zini, A.; San Gabriel, M. & Baazeem, AA. (2009). Antioxidants and Sperm DNA Damage: A Clinical Perspective. *J Assist Reprod Genet*, Vol.26, pp 427–432.

Zribi, N.; Chakroun, NF.; Elleuch, H.; Ben Abdallah, F.; Ben Hamida, AS.; Gargouri, J.; Fakhfakh, F. & Keskes, LA. (2011). Sperm DNA Fragmentation and Oxidation are Independent of Malondialdheyde. *Reprod Biol Endocrinol*, Vol.9. pp 47-54.

Zuwała-Jagiełło, J.; Pazgan-Simon, M.; Krzysztof, S. & Warwas, M. (2011). Advanced Oxidation Protein Products and Inflammatory Markers in Liver Cirrhosis: A Comparison between Alcohol-Related and HCV-Related Cirrhosis, *Acta Biochim Pol*. Vol.58 No.1, pp. 59-65.

4

Oxidative and Nitrosative Stresses: Their Role in Health and Disease in Man and Birds

Hillar Klandorf and Knox Van Dyke
West Virginia University,
USA

1. Introduction

The concept of oxidative stress (OS) was originally used by Professor Helmut Sies who described it as "an imbalance between oxidants and antioxidants in favor of the oxidants, potentially leading to damage" with the ratio of oxidants to antioxidants as >1 (Sies, 1997). Oxidative/nitrosative stress represents the bodies' imbalance in the production and the elimination of reactive oxygen and nitrogen species and various reducing or antioxidant chemical systems of the body which destroy reactive intermediates and prevent/or repair the resultant damage. This is a particularly useful concept to establish a common basis for the longevity of a particular species as well as the many different disease states. However, it has also been found to be involved in a variety of other processes including immune protection of the body, gene control, growth as well as cell death. A parallel process is nitrosative stress (NS) which is defined as the ratio of nitrosants to antioxidants as >1 similarly to oxidative stress, but with involvement of reactive nitrogen species. This is a similar process that is involved with a variety of oxygen-nitrogen species causing excessive oxidation and/or nitrosylation compared to antioxidation or reduction. The question thus arises "What endogenous and exogenous strategies are available to cope with the damage associated with chronic oxidative and nitrosative stress?" It is the recombination of short - lived radicals forming new strong oxidants which mainly cause the majority of the driving force for damage in oxidative and nitrosative stresses. Consequences of these particular inflammatory stresses are discussed as are endogenous and exogenous strategies to cope with this burden. Future research in antioxidant/repair therapy will result in the more effective treatment of diseases but it is better used as a preventative strategy. Limitations to the effective management of inflammatory stress are numerous and include the nature of the oxidants generated in association with the form and half-life of the particular antioxidant administered.

2. Localized oxidative and nitrosative stresses

These oxidative and nitrosative stresses (Zorov et al, 2004) could have a more narrowing or selective definition because often these O/N stresses occur locally in a single or selected areas whereas the body as a whole is experiencing control over oxidative and nitrosative stresses and no general damage occurs. So it is easy to conceive of the idea that in a disease state for e.g., damage due to "reactive oxygen" species that cause O/N stresses might occur in a very localized area e.g. a single area of infection.

2.1 What are reactive oxygen and nitrogen species?

This is a term which encompasses molecules composed of oxygen, oxygen and hydrogen as well as oxygen and nitrogen etc. Even oxygen and carbon structures are known to be reactive species and many molecules that are as yet unknown are likely involved as well. Examples of some oxidants should include the following:

Reactive Oxygen Species (Halliwell and Gutteridge, Table 1)

Oxidant	Description
$O_2^{-\cdot}$ (superoxide anion)	Oxygen with an extra electron formed in mitochondrial electron transport with NADPH oxidases and , xanthine oxidase and other reactions. Superoxide forms hydrogen peroxide upon reduction.
H_2O_2 (hydrogen peroxide)	Hydrogen peroxide is formed via dismutation of superoxide (above). Lipid solubility of H_2O_2 aids diffusion across membranes; destroyed by peroxidases.
$\cdot$OH (hydroxyl radical)	Short lived highly energetic radical formed via Fenton chemistry. It attacks almost all chemicals. It lives and decomposes very quickly essentially where formed.
ROOH (organic hydroperoxide)	Formed by free radical attack on lipids and nucleobases.
RO$\cdot$ (alkoxy) and ROO$\cdot$ (peroxy radicals)	Oxygen centered radicals. Participates in lipid peroxide chain reaction. Production occurs via attack of oxygen radicals on unsaturated lipids.
OCl$^-$ (hypochlorite anion)	Formed from the reaction of hydrogen peroxide and chloride anion with myeloperoxidase. Lipid soluble and highly oxidizing (bleaching effect).
OONO$^-$ peroxynitrite	Peroxynitrite formed via the reaction of superoxide anion ($O_2^{-\cdot}$) and nitric oxide (NO$^+$). Lipid soluble degrades to ($\cdot$OH) and N_2O_2.
N_2O_3 (dinitrogen trioxide)	Dinitrogen trioxide is formed via the reaction of $\cdot$NO and NO_2. N_2O_3 is strongly oxidizing and causes nitrosylation of phenols. It is the anhydride of HNO_2 (nitrous acid).

Table 1. REACTIVE OXYGEN/NITROGEN SPECIES (Causing Oxidative and Nitrosative Stress)

1. superoxide (oxygen with an unpaired electron) $O_2^{-\cdot}$
2. hydrogen peroxide (H_2O_2)

3. hydroxyl radical ($^\bullet OH$)
4. organic hydroperoxides (ROOH)

5. alkoxy ($RO^\bullet$.) and peroxy ($ROO^\bullet$) radicals
6. hypochlorite (OCl^-)
7. peroxynitrite ($OONO^-$)

Reactive Nitrogen Species (Droge, 2001)

1. nitric oxide ($\cdot NO\cdot$)
2. peroxynitrite ($OONO^-$)
3. nitrosoperoxycarbonate ($ONOOCO_2^-$)
4. nitrogen dioxide ($NO_2\cdot$)
5. dinitrogen trioxide (N_2O_3)
6. dinitrogen tetraoxide (N_2O_4)

2.2 Antioxidants which oppose ROS and RNS

There are two major types of antioxidants systems which act to control damage caused by reactive oxygen and reactive nitrogen species (Sies, 1997). They are chemicals low molecular mass compounds and enzymes - which produce products that eliminate ROS and RNS. The major antioxidants in the bodies of man and various species of animals are the following:

1. vitamin C (ascorbic acid)
2. tocopherols and tocotrienols (forms – alpha, beta, gamma, and delta of both types)
3. glutathione is a tripeptide (reduced form-glycine-cysteine-glutamic acid)
4. urate - salt forms of uric acid
5. vitamin A
6. various thiols eg., N-acetyl cysteine
7. bilirubin
8. ubiquinol-10 (reduced form of coenzyme Q-10)
9. flavonoids
10. estrogens
11. salicylates (aspirin compounds) or non-steroidal antiinflammatories
12. lazaroids-21, aminosteroids
13. mannitol, dimethyl sulfoxide, dimethyl thiourea, hydroxyl radical scavengers
14. various drugs – captoprll, beta blockers, calcium antagonists, amiodarone, carvedilol, epinephrine, norepinephrine, dopamine, etc.
15. allopurinol and oxypurinol – xanthine oxidase inhibitors
16. fish oil with omega 3 fatty acids
17. polyphenols from fruits and vegetables, tea etc.
18. melatonin

Endogenous Enzymes or Proteins

1. superoxide dismutatase (converts (dismutatses) superoxide into hydrogen peroxide)
2. catalases and other peroxidases (degrades organic peroxides e.g. hydrogen peroxide and other peroxidases into water and oxygen)
3. glutathione peroxidase (uses reduced glutathione and hydrogen peroxide to produce oxidized glutathione and water)
4. peroxiredoxins reacts with inflammatory- cytokine- producinged peroxides to produce water
5. sulfiredoxin (reduces sulfinic acid to an antioxidant thiolsulfoxide)
6. transferrin and lactoferrin
7. ceruloplasmin,
8. albumen

2.3 Effects of oxidative and nitrosative stresses

When there is a large increase in oxidants or nitrosating compounds, damage to normal cells and tissues can and does occur when the various antioxidant defense mechanisms become depleted. Extracellularly, it is depletion of urate and ascorbate, - while intracellularly, it is the depletion of the reduced form of glutathione that are the most important water-soluble antioxidants. The extent of the antioxidant depletion is the key factor -; small antioxidant perturbations are readily corrected by the cell; , but a large depletion of antioxidants can cause cell death - either apoptosis or even necrosis in cells and tissues. The latter can cause inflammation - either acute or chronic depending on how long the imbalance continues - in oxidants compared to antioxidants.

The induction of oxidative stress via enhanced generation of reactive oxygen species on a continuous basis generally consists of free radicals and peroxides but lesser reactive species like superoxide can be transformed to more reactive species by the reaction with metal ions or other redox-cycling compounds, which can cause damage to cells or tissues. (Figure 1). DNA is a key target of this damage and this could be either somatic or mitochondrial in origin. In particular, mitochondrial DNA repair is less complete than chromosomal DNA repair. Therefore damage to mitochondrial DNA, which controls sugar carbohydrate and lipid metabolism and is key to producing energy in ATP form, can limit energy production. When ATP is depleted, the cell dies via apoptosis or necrosis.

2.4 Oxidant production or depletion

Normally mitochondrial leakage of electrons from electron transport chain to molecular oxygen forming superoxide (S) is about 1- 2% of the total electron flux produced during oxidative phosphorylation during the at production of ATP (Ames et al, 1993). Flavoproteins may contribute some of the total production of (S). Xanthine oxidase generates superoxide when substrates xanthine and hypoxanthine are oxidized. NADPH oxidases and the different cytochrome P-450s also contribute to total production of superoxide.

A variety of different oxidases produce hydrogen peroxide with both along with superoxide and hydrogen peroxide and other peroxides contributing to the total of reactive oxygen species.

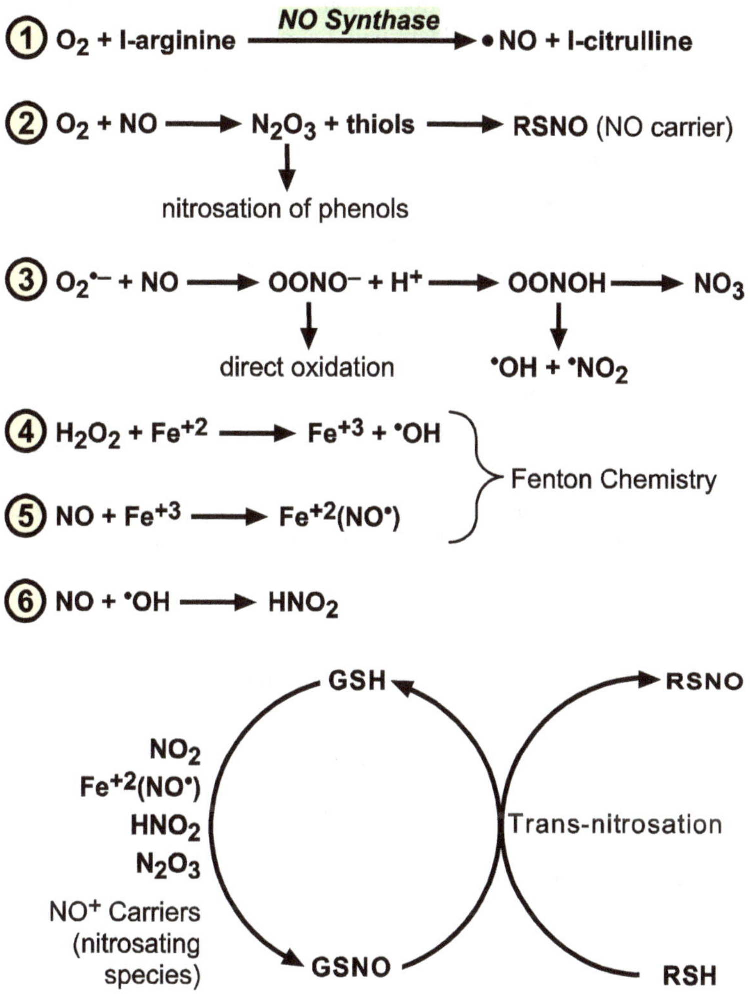

(1) O_2 + l-arginine $\xrightarrow{\textit{NO Synthase}}$ •NO + l-citrulline

(2) O_2 + NO $\longrightarrow$ N_2O_3 + thiols $\longrightarrow$ RSNO (NO carrier)

$\downarrow$

nitrosation of phenols

(3) $O_2^{•-}$ + NO $\longrightarrow$ $OONO^-$ + H^+ $\longrightarrow$ OONOH $\longrightarrow$ NO_3

$\downarrow$ direct oxidation $\downarrow$ •OH + •NO_2

(4) H_2O_2 + Fe^{+2} $\longrightarrow$ Fe^{+3} + •OH ⎫

(5) NO + Fe^{+3} $\longrightarrow$ Fe^{+2}(NO•) ⎬ Fenton Chemistry

(6) NO + •OH $\longrightarrow$ HNO_2

NO_2
Fe^{+2}(NO•)
HNO_2
N_2O_3
NO^+ Carriers
(nitrosating
species)

GSH   RSNO

Trans-nitrosation

GSNO   RSH

Fig. 1. Reactions leading to the generation of Nitric Oxide and Reactive Nitrogen Species. (Modified from Novo and Parola, 2008)

3. Use of antioxidant supplements to prevent or ameliorate diseases

Many studies have used various antioxidants at a variety of doses producing rather equivocal results. Since oxidative stress is a continuous and ongoing problem simply taking one or more antioxidants at less than optimal doses for a given situation one would not expect such a study to be effective. A scientific approach would be to use optimal doses and in a dosage form that was developed in a sustained-release, continuous -release, or time-release mode. The antioxidants should be continuously released during the disease and would be chosen based on the oxidants causing the disease. Not all oxidants are destroyed by a given antioxidant. The antioxidant potential should be matched with oxidant potential so that destruction of the oxidant is assured. Another matter of importance is to match the kinetics and half- life of a given antioxidant with the production or kinetics of the oxidant. An example of this is vitamin C - which is a water–soluble antioxidant with a short half life. It is necessary to use it in a sustained -release form and take it orally several times a day at evenly spaced times. This occurs to ensure even coverage against continuously produced oxidants e.g. peroxynitrite (OONO-) or dinitrogen trioxide (N_2O_3).

The basic problem using antioxidants as protection against oxidants is that the body produces a continuous variety of oxidants over time with differing oxidative potential (ability to oxidize). In order to counteract the effects of this multiplicity of oxidants, one would need a multiplicity of antioxidants that were available at the same time as the oxidants and at greater doses in a given area of the body. Further, since the antioxidants would not necessarily distribute evenly throughout the body, if the damage from oxidants occurred in a single location e.g., the major joint in the large toe in gout, if the antioxidants did not penetrate well into the joint where massive inflammation could occur due to uric acid crystals, negation of the inflammation probably would not occur. However by taking multiple high dose antioxidants (in time release or sustained release form except for the fat soluble antioxidants-multiple forms of vitamin E, vitamin A or its precursor beta-carotene), damage in the eye by diabetic retinopathy or macular degeneration could slow or possibly be averted.

3.1 Indications for antioxidant therapy

Some important ideas relevant to commencing antioxidant therapy should be contemplated (Rice-Evans and Diplock, 1993) and include the following:

1. Is oxidative damage associated with the disease pathology?
2. Does oxidative stress cause different diseases?
3. Does oxidative damage occur in most diseases?
4. Can the antioxidant reach the site of disease-damage in sufficient concentration?
5.   Will the antioxidant (s) utilized to prevent or stop the oxidative process in vivo?
6. Can the doses of antioxidant utilized be tolerated?
7. If the therapy is chronic, will the antioxidant supplements be safe over this time of therapy?

3.2 Examples of phenolic plant antioxidants acting as oxidant destroyers or targets of oxidation and nitration

Many of the highly colored foodstuffs (particularly vegetables and fruits) contain phenols that are highly polyphenolic in nature (Van Dyke et al., 2000). There are a series of polyphenolic compounds found in white, green and black teas. These are compounds like catechin, epicatechin, gallocatechin, epigallo catechin or their esters epicatechin gallate, or epigallocatechin gallate or even catechin gallate. Grapes possess a variety of compounds that are either monophenols or polyphenols. These compounds include phenolic acids, stilbenes, flavanols, dihydroflavanols, anthocyanins, flavanol monomers (proanthocyanidins). Flavonoids from grape skins create part of the color and taste of the wine. The non-flavonoids include the stilbenoids e.g. resveratrol, phenolic acids such as benzoic, caffeic and cinnamic acids. Resveratrol in fairly high doses has been touted as an efficient antioxidant against diseases even when people eat a diet high in fat and carbohydrate. Many Frenchmen live a long life, which is attributed to drinking red wine containing resveratrol and it is thought that this is responsible for "the French Paradox".

3.3 Major diseases or indications where antioxidant intervention is warranted

3.3.1 Atherosclerosis

Low density lipoprotein particles can be inhibited by antioxidants e.g., probucol, ascorbic acid, vitamins E, and beta carotenes as well as natural flavanoids alone were demonstrated effective against atherosclerosis which is key to inhibiting ischemic heart disease (IHD) (Ames et al., 1993). Flavanoid–rich diets appear effective as a preventative against IHD. The flavanoids which include anthocyanins and tannins are excellent antioxidants as are the non-flavanoids like resveratrol and phenolic acids e.g. benzoic, caffeic and cinnamic acid.

3.3.2 Neurological diseases

Oxidative stress occurs in the brain because of physical damage, viral, bacterial, fungal or parasitic diseases (Floyd, 1999). Different types of physical or anoxic trauma stimulate generation of oxygen and nitrogen radicals that can recombine and produce powerful oxidants or reaction with metals and create other reactive oxygen and nitrogen species. Nitric oxide and carbon monoxide are both known to create major toxicities to the brain in the various areas of the brain associated with senile dementia and Parkinson's disease as well as the other problems e.g. amyotrophic lateral sclerosis ALS and Huntington's disease. A variety of antioxidant drugs e.g. selegiline and riluzole as well as lazaroids, trilazad mesylate and tocopherol have shown promising antioxidant effects against these diseases.

3.3.3 Ischemic reperfusion injury

Reperfusion is the re-establishment of blood flow after stoppage (Dhalla et al, 2000). Once blood flow has been stopped ischemia develops, and restoration promotes production of reactive oxygen and nitrogen species causing toxicity. This phenomenon is particularly important in hypoxic states e.g. in organ transplantation i.e. heart or kidney etc. when the heart is reperfused, post ischemic stunning or contractile dysfunction, reperfusion arrhythmias and damage to the lining of vessels via endothelial injury occurs. Excessive thrombolysis associated with death occurs within 24 hours post transplantation. Antioxidants can inhibit this process associated with vascular disease and their use is mandatory. Of note, reperfusion injury can be prevented by a procedure termed ischemic preconditioning (Marin-Garcia, 2011). The practice is based on the observation that if the blood flow to an organ such as the heart is halted for a period less than five minutes and then restored, the cells downstream gain permanent protection from a second ischemic insult. For this protection to occur three cardioprotective proteins must be induced in response to the oxidative stress. 1. *iNOS*, which increases nitric oxide. NO in turn reacts with the superoxide radicals generated in response to the oxidative stress to generate the harmful nitrogen radical peroxynitrite. NO may also suppress mitochondrial respiration, while also inducing: 2. *HO-1* (heme oxygenase) which produces carbon monoxide (CO). In turn, CO induces: 3. *ecSOD* (extracellular superoxide dismutase) which inactivates the superoxide radical and peroxynitrite as well as inducing more iNOS. These types of studies further highlight the importance of endogenous antioxidants in the amelioration of oxidative stress.

3.3.4 Diabetes mellitus

Diabetes is actually caused by excessive production of excessive nitric oxide reactive oxygen and nitrogen species in or close to the pancreatic beta cells (which produce, store and release insulin) (Van Dyke et al., 2010). Once the beta cells die, the body is deficient in insulin. Therefore, insulin must be restored by injection or inhalation. Alpha cells (which produce glucagon) take the place of the beta cells and excessive glucagon causes an increase in blood glucose from sources like glycogen breakdown to exacerbate the problem. Oxidative stress from lack of the control of blood glucose occurs which depletes the antioxidant load. This creates the need for antioxidant supplementation to counteract the oxidant stress. Once the diabetic state is established, there is a deficiency of nitric oxide in the vascular system of diabetics. Since sufficient nitric oxide supply is necessary to maintain proper blood pressure - when the endothelial cells (which line the blood vessels) are damaged they produce insufficient nitric oxide to maintain proper vasodilation. Therefore blood pressure increases and hypertension occurs. Antioxidants like ascorbic acid and the various forms of vitamin E (tocopherols and tocotrienols) inhibit the auto-oxidation of glucose and glycation of blood proteins e.g. Hemoglobin A1C. Therefore antioxidant therapy can ease the toxic state which occurs in diabetes Type I or II.

3.3.5 Inflammatory diseases

Acute or chronic inflammation plays a role in most if not all disease states. This is because it is a response to damage caused by different entities be it infective (bacteria, viruses, fungus or parasites) or non-infectious – e.g., different forms of arthritis (Chade et al., 2004). Since the major inflammatory cells like neutrophils and/or macrophages generate a variety of reactive oxygen and nitrogen species as well as release various proteases tissue damage and destruction occur. Inflamed joints and even tissues e.g. inflammatory bowel diseases occur. These problems also occur in a variety of liver, kidney and pancreatic diseases etc. Preliminary clinical trials using antioxidants appear to be helpful in ameliorating the damage and pain caused by these diseases.

3.3.6 Hypertension

High blood pressure is mainly caused by having excessive vascular constrictors relative to vasodilators (Vasdev et al., 2007).The major vasoconstrictor is known to be a peptide called angiotensin II and the major vasodilator is nitric oxide. The problem is nitric oxide (.NO) has a very short half life of a few seconds or less. Therefore, it must be made produced continuously in the vasculature via endothelial cells lining the blood vessels. The body produces. NO by oxidizing the amino acid L-arginine using oxygen and the enzyme endothelial NO synthase. There must be a continuous supply of sufficient l-arginine and if there are l-arginine-like NO synthase inhibitors in the vasculature e.g., (asymmetric dimethyl arginine) less nitric oxide is made produced creating leading to hypertension (figure 2). If excessive amounts of the constrictors are generated hypertension would also develop. Therefore, long acting l-arginine supplements and angiotensin II production or receptor inhibitors are effective. Calcium channel blockers and beta receptor blocking drugs are antioxidants and have been proven effective in hypertension (Godfraind, 2005, Aruoma et al., 1991).

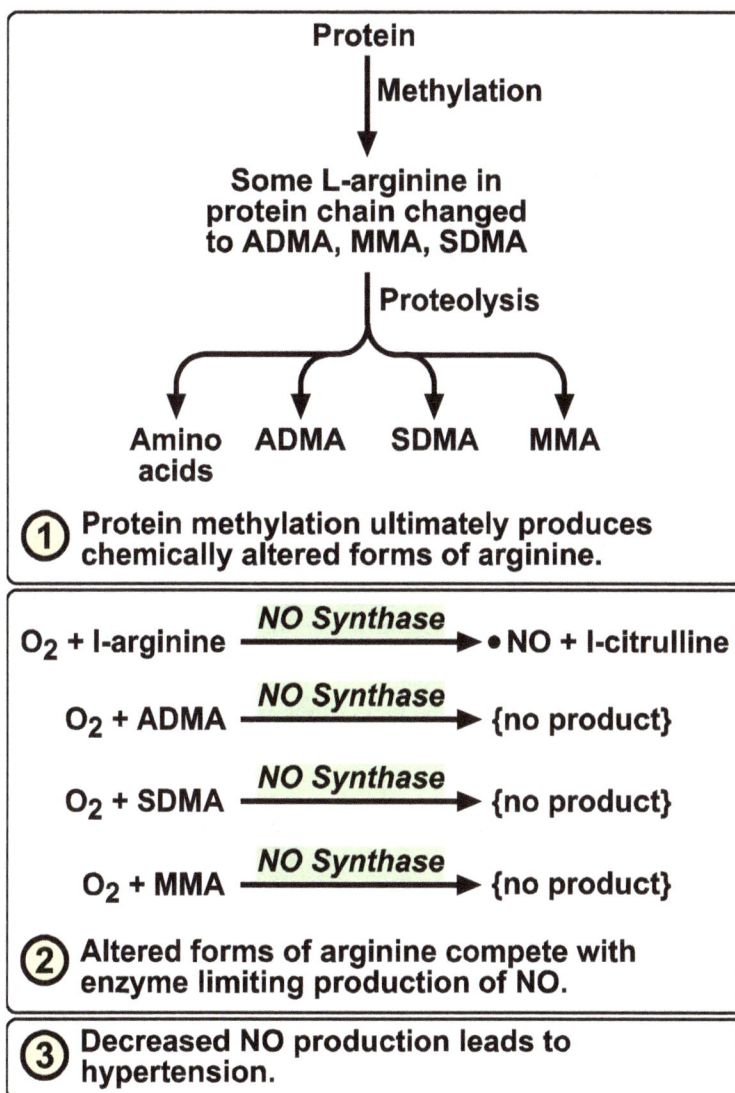

Fig. 2. Reactions leading to reduced NO production

3.3.7 Other pathological problems

Examples include sepsis and shock which cause hypoxia of vital organs, inflammation, and endothelial damage. Even with standard therapy, these patients are likely to develop adult respiratory distress syndrome (ARDS). This problem certainly is associated with activation of neutrophils and macrophages causing oxidative stress via production of oxidants and release of proteases. If these cells could be prevented from entering the areas linked to the disease, it would be effective therapy since which would control the damaging mechanisms.

Antioxidants neutralize the products generated in association with the inflammatory response and help in control the damages.

3.3.8 Cancer

Large scale epidemiological studies have indicated high dietary antioxidant diets are less likely to cause human cancer (Frei, 2004). In particular, lung cancer is an example. Particularly when people smoke, DNA damage occurs, this can often be prevented using antioxidants. However, when beta carotene was used cancer lesions increased. This is not surprising since beta carotene is not a particularly effective antioxidant unless high doses are used and these can create toxicity, even cancer.

3.3.9 HIV – Human immunodeficiency virus infection

Patients with HIV infection displaying AIDS have comprised immune systems (T lymphocyte amounts are greatly reduced) associated with both decreased reduced glutathione levels and enhanced expression of tumor necrosis factor alpha (TNF-alpha) (Fuchs et al, 1991). This stimulates reproduction of HIV while raising the antioxidant level suppresses the production of the virus. Giving a variety of antioxidants e.g. vitamin C, vitamins E, N-acetyl cysteine (reverses the rate limiting step in glutathione biosynthesis) causes a reduction in production of virus. The addition of a strong and lasting Nf-kappa β inhibitor to the mixture of antioxidants can lower the HIV viral level to undetectable. Therefore simple antioxidants can play a large role in viral reproduction. When these viruses are replicating with high efficiency, the protective antioxidant load decreases drastically. This phenomenon occurs with other viruses as well.

3.3.10 Other diseases

Vitamin E (tocopherol) appears to limit the retinopathy of prematurity consequent to the exposure of high levels of oxygen (Robertson, 2010). Reduction in the amount of intraventricular hemorrhage occurs upon treatment with vitamin E. Senile macular degeneration and cataracts may be preventable using antioxidant supplement therapy. Neonatal -distress -syndrome incidence decreases with vitamin E- (tocopherol) supplements.

3.3.11 Aging

Since ultraviolet attack and oxidative stress over time are certainly responsible in premature aging it is not surprising that antioxidants given at the correct doses and in sustained release forms several times per day would effectively slow damage over time (Ames et al., 1993) Tocopherols (vitamins E), aminoguanidine (Iqbal et al.), vitamin C, N-acetyl cysteine NAC and low dose retinoids can be helpful to prevent premature aging of the skin. Recent studies have suggested that administration of probiotics can also ameliorate inflammatory skin diseases (Hacini-Rachinel et al, 2009). Emerging studies suggest that probiotics induce regulatory T cells, which reduce inflammation by suppressing effector T cell function.

3.3.12 Drug toxicity

In cases where the major protective antioxidant glutathione is depleted by overdose of a drug -e.g. acetaminophen (Tylenol), by taking or injecting increased amounts of N-acetyl cysteine at the correct time serious toxicity to the liver can be avoided (Daly et al., 2008). This occurs because the N-acetyl cysteine (NAC) increases the amount of glutathione available and therefore prevents the toxicity of a toxic metabolite. If a drug depletes glutathione to dangerously low levels, toxicity can be avoided if N-acetyl cysteine NAC is given in the correct amount and quickly enough. Alcohol can decrease antioxidant reserves as well as acetaminophen.

An imbalance of oxidant damage / antioxidant protection has been described in congestive heart failure, and chronic renal failure. Further, eclampsia, chronic peritonitis, respiratory diseases and a variety of hematological disorders are known to involve redox similar imbalances.

4. Damage by oxidants and roles of antioxidants in animals

Animals have similar connections to oxidative stress that happens in man. However, birds have a remarkable longevity for their body size despite an increased body temperature, higher metabolic rate, and increased blood glucose concentrations compared with mammals (Holmes and Austad, 1995). Theoretically, birds should sustain a much higher degree of oxidative damage and processes leading to senescence such as glycoxidation of proteins and nucleic acids (Monnier et al., 1991). For example, a mouse that weighs approximately 20 g is equal in body size to a canary and yet the mouse will lives 3 years as with one-twentieth the oxidative burden as opposed to the canary that will lives 20 years (Holmes and Austad, 1995). This suggests that birds have developed a mechanism to cope with reactive oxygen species (ROS) assault by either reducing the amount of ROS produced or by a more efficient endogenous antioxidant defense system.

Like other species, birds rely on both exogenous and endogenous antioxidant defense systems. Exogenous antioxidants are obtained primarily from the diet and other environmental factors. For example, in another species, apes eat a diet that is generally high in fruits and vegetables. It has been stated that a chimpanzee eats about 5 grams of vitamin C a day. Generally they are quite healthy which could be attributed in part to a high antioxidant load which protects them against viruses, and other infectious diseases. In birds, endogenous antioxidant enzymes include superoxide dismutase, glutathione peroxidase and the product of the enzyme xanthine oxidase, uric acid. Uric acid is a potent antioxidant and it is arguably, the dominant antioxidant defense mechanism for birds (Seaman et al., 2008, Stinefelt et al., 2005, Machin et al., 2004, Simoyi et al., 2002, Klandorf et al., 2001). Uric acid is the end product of purine degradation. Due to the evolutionary lack of urate oxidase expression, also known as uricase, birds (comparable to reptiles, higher primates, and humans) do not convert uric acid to allantoin. It is the loss of uricase in association with increased uric acid concentrations that can be linked to a prolonged life span.

5. Future directions in antioxidant therapy

The idea that antioxidant therapy is effective against disease because it only inhibits or destroys free radicals is partially defective and incomplete (Firuzi et al., 2011. Free radical oxygen and nitrogen compounds do exist but usually are quite short-lived and are generally quite difficult to detect, especially the high energy radical compounds that cause real damage. The radicals exist for a short time and recombine with other radical compounds that have free electrons to produce oxidants that can cause major damage. A simple example is the reaction of the nitric oxide radical and oxygen radical superoxide forming peroxynitrite anion (Fig. 1).

It is the recombination of short -lived radicals- forming new strong oxidants which mainly cause the majority of the driving force for damage in oxidative and nitrosative stresses.

5.1 Limitations using antioxidant/repair therapy to treat diseases

1. It is helpful if one understands what oxidants are being generated in order to select the proper antioxidants. The doses of antioxidants must be in the therapeutic range.
2. A mixture of antioxidants should be chosen at the correct doses to eliminate the toxins formed in the disease.
3. The timing of the antioxidants has to be linked with an understanding of the half-life of each antioxidant.
4. The use of time- release or sustained -release supplements should be used so that their blood level of the different antioxidants remains relatively constant.
5. Have sufficient levels of various antioxidants at the sites of action where oxidative and nitrosative stresses occur.
6. Use increased levels of repair chemicals to prevent permanent damage (particularly for DNA) e.g. nicotinic acid.
7. It is best to use dietary antioxidant supplementation as a preventative and not as a treatment of diseases.

As Benjamin Franklin stated "An ounce of prevention is worth more than a pound of cure".

6. Conclusion

Oxidative stress represents the bodies' imbalance in the production and the utilization of reactive oxygen and nitrogen species and various reducing or antioxidant chemical systems of the body which destroy reactive intermediates and prevent/or repair the resultant damage. It is the recombination of short-lived radicals forming new strong oxidants which mainly cause the majority of the driving force for damage in oxidative and nitrosative stresses. The ability of an organism to survive these cumulative insults can be linked to both exogenous and endogenous sources of antioxidants, which can be further linked to longevity of species. Future directions in antioxidant/repair therapy will be the more effective treatment of diseases; however they are better used as a preventative strategy. Limitations to the effective management of inflammatory stress are numerous and include the nature of the oxidants generated in association with the form and half-life of the particular antioxidant administered.

7. References

Ames, B.N.; Shigenaga, M. & Hagen, T.M. (1999). Oxidants, antioxidants, and the degenerative diseases of aging, *Proc Natl Acad Sci (USA)*, 90, pp. 7915-7922

Aruoma, O.I.; Smith, C., Cecchini, R., Evans, P.J. & Halliwell, B. (1991). Free radical scavenging and inhibition of lipid peroxidation by beta-blockers and by agents that interfere with calcium metabolism. A physiologically significant process? *Biochem. Pharmacol.* 4, pp. 735-743

Chade, A.J.; Krier,J.D., Rodriguez-Porcel, M., Breen, JF., McKusick, M.A., Lerman, A & Lerman , L.O. (2004), Comparison of acute and chronic antioxidant interventions in experimental renovascular disease, *Renal Physiology*, 286, 36 F1079-F1086

Daly, F.F.; Fountain, J.D., Murray, L., Graudins, A. & Buckley, N.A (2008). Guidelines for the effective management of paracetamol poisoning in Australia and New Zeland-explanation and elaboration. A consensus statement from clinical toxicologists consulting to the Australian poisons information centres. *Med. J Aust.* 188, (5), 296-301

Dhalla, N.S.; Elmosehli, A.B., Hata,T.& Makino, N. (2000). Status of myocardial antioxidants in ischemia-reperfusion injury. *Cardiovasc Res*. 45 (3) pp. 446-456

Droge, W. (2002). Free Radicals in the physiological control of cell function. *Physiol Rev*. 82, pp. 47-95.

Firuzi, O.; Miri, R., Tavakkoli, M. & Saso, L (2011). Antioxidant Therapy : Current Status and Future Prospects. *Curr. Med. Chem*. Aug 9 [Epub ahead of print]

Floyd, R.A. (1999). Antioxidants, oxidative stress and, neurological disorders, *Proc Soc Exp Biol. Med*. 222 (3) pp. 236-245

Frei, B. (2004). Efficacy of dietary antioxidants to prevent oxidative damage and inhibit chronic disease. *Journal of Nutrition*, 134, pp. 3196S-3198S

Fuchs, J.; Janka, S., Schofer, H., Milbradt , R., Freisleben, H.J., Unkelbach,U., Oster,W, & Siems,W. (1991*). International Conference on AIDS*, Jun 16-21 &, 223 Abstract 2166

Godfraind, T. (2005). Antioxidant effects and the therapeutic mode of calcium blockers in hypertension and atherosclerosis. *Philosoph. Trans. R. Soc. Lond. B. Biol. Sci*. 360, pp. 2259-2272

Hacini-Rachinel, F.; Gheit, H., Le Luduec, J.B., Dif, F., Mancey, S. & Kaiserlian, D. (2009). Oral probiotic control skin inflammation by acting on both effector and regulatory T cells. *PLoS ONE* 4(3), e4903

Halliwell, D. & Gutteridge, J.M.C. (1989*). Free radicals in Biology and Medicine* (2nd ed.) Oxford, UK: Clarendon

Holmes, D.J. & Austad, S.N. (1995). The evolution of avian senescence patters; implications for understanding primary aging processes. *American Zoology* 3, pp. 307-317

Iqbal, M.; Probert, L.L. & Klandorf, H. (1997). Effect of dietary aminoguanidine on tissue pentosidine and reproductive performance in broiler breeder hens. *Poult. Sci*. 76, pp1574-1579.

Klandorf, H.; Rathore, D.S., Iqbal, M., Shi, X. & Van Dyke, K. (2001). Accelerated tissue aging and increased oxidative stress in broiler chickens fed allopurinol. *Comp. Biochem. and Physiol. Pt. C* 129 (2), pp. 93-104

Machin, M.; Simoyi, M., Blemings, K. & Klandorf, H. (2004). Increased dietary protein elevates plasma uric acid and is associated with decreased oxidative stress in rapidly-growing broilers. *Comparative Biochemistry and Physiology Part B*. 137, pp. 383-390

Marin-Garcia, J. (2011). Signaling in the Heart. Springer. New York, Dordrecht, Heidelberg, London

Monnier, V.M.; Sell, D.R., Ramanakoppa, H.N. & Miyata, S. (1991). Mechanisms of damage mediated by the Maillard reaction in aging. *Gerontology* 37, pp.152-165

Novo, E, & Parola, M. (2008). Redox mechanisms in hepatic chronic wound healing and fibrogenesis. *Fibrogenesis Tissue Repair* 1:5.

Rice-Evans, A. & Diplock, A.T. (1993), Current status of antioxidant therapy, *Free Radical Biology and Medicine* 15, pp 77-96

Robertson, R.P. (2010). Antioxidant Drugs for Treating Beta –Cell Oxidative Stress in Type 2 Diabetes: Glucose-centric Versus Insulin-centric Therapy, *Discovery Medicine* #044, Jan 2010, pp. 1-7

Seaman, C.; Moritz., Falkenstein, E., Van Dyke, K., Casotti, G., & Klandorf, H. (2008). Inosine ameliorates the effects of hemin-induced oxidative stress in broilers. *Comparative Biochemistry and Physiology, Part A* 151, pp. 670-675

Sies, H. (1997). Oxidative stress: oxidants and antioxidants. *Experimental Physiology*, 82, pp. 291-295

Simoyi, M.; Van Dyke, K. & Klandorf, H. (2002). Manipulation of plasma uric acid in broiler chicks and its effect on leukocyte oxidative activity. *Am. J. Physiol. Regulatory Integrative Comp Physiol.* 282, pp. 791-796

Stinefelt, B.; Leonard, S.S., Blemings, K.P., Shi, X. & Klandorf. (2005). Free Radical Scavenging, DNA protection, and inhibition of lipid peroxidation mediated by uric acid. *Ann.Clin.Lab.Sci.* 35, pp. 37-45

Van Dyke, K.; Jabbour, N., Hoeldtke, R., Van Dyke, C., & Van Dyke, M. (2010), Oxidative/nitrosative stresses trigger type I diabetes : preventable in streptozotocin rats and detectable in human disease. *Annals of the NY Acad. Sci.* 1203, pp. 138-145

Van Dyke, K.; McConnell P. & Marquardt, L. (2000). Green tea extract and its polyphenols inhibit luminol dependent chemiluminescence activated by peroxynitrite. *Luminesecence*, 15, pp. 37-43

Vasdev, S.; Gill, V, & Singal, P. (2007). Role of Advanced Glycation End Products in Hypertension and Atherosclerosis: Therapeutic Implications, *Cell Biochem. Biophys.*, 49, pp. 48-63

Wink, D.A.; Miranda, K.M. & Epsey, M.G. (2000), Effects of oxidative and nitrosative stress in cytotoxicity, *Semin. Prinatol.* 24(1), pp. 20-23

Zorov, D.B.; Bannikova, S.Y., Belousov, MY., Vyssokikh, L.D., Zorova, N.K., Isaev, N.K.,& Plotnikov, E.Y. (2005). Reactive Oxygen Species: Friends or Foes? *Biochemistry* (Moscow), 70, pp. 265-270

5

Nitric Oxide Synthase and Oxidative Stress: Regulation of Nitric Oxide Synthase

Ehab M. M. Ali[1], Soha M. Hamdy[2] and Tarek M. Mohamed[1]
[1]Biochemistry Division, Chemistry Department,
Faculty of Science, Tanta University. Tanta,
[2]Biochemistry Division, Chemistry Department,
Faculty of Science, Fayoum University, Fayoum,
Egypt

1. Introduction

Nitric oxide has been found to play an important role as a signal molecule in many parts of the organism as well as a cytotocic effector molecule of nonspecific immune response. Nitric oxide is very important functions both in helminthes and mammalian hosts. Nitric oxide may react with proteins and nucleic acids. In addition to binding to heme groups, e.g. of guanylate cyclase, hemoglobin, and cytochrome C oxidase, NO may react with nucleophilic centers like sulfur, nitrogen, oxygen and aromatic carbons. The prime target for covalent binding of NO to a functional groups in proteins under physiological condition in the presence of oxygen are SH groups. The intra-mitochondrial reaction of NO with superoxide anion yields peroxynitrite, which irreversibly modifies susceptible targets within the mitochondria, inducing oxidative and/or nitrative stresses. The signal molecule of NO is synthesized by constitutive nitric oxide synthase (cNOS). The killer molecule NO is synthesized by inducible NOS (iNOS). There is no signal or killer NO – it depends on the environments and partners involved – be very careful in that. Yes, the production is regulated in different ways. Inducible NOS is induced by numerous inflammatory stimuli, including endotoxin, cytokines and excretory/secretory products (ESP) of helminthes. ESP directly interact with the immune system and modulate host immunity. Nitric oxide is a highly reactive and unstable free radical gas that is produced by oxidation of L- arginine by oxygen and NADPH as electron donor to citrulline mediated by a family of homodimer named nitric oxide synthase. In addition to L- arginine-NO pathway, L-arginine is also metabolized to L-ornithine and urea by arginase enzyme. A side from blocking NO synthesis by depleting the cell of substrate for NOS, the arginase-mediated removal of L-arginine inhibits the expression of inducible NOS (iNOS) by repressing the translation as well as the stability of iNOS protein. Furthermore, arginase may inhibit iNOS-mediated NO production through the generation of urea.

2. Chemistry of nitric oxide

Nitrogen monoxide, called nitric oxide (NO) is an endogenous short lived free radical that freely diffuses within cells from formation to action site. Nitric oxide exhibits an enormous range of important functions in the organism. Nitric oxide interacts with other biomolecules and can combine with superoxide anion (another free radical) to form an unstable intermediate peroxynitrite which may initiate tissue injury. Peroxynitrite may also decompose to form a strong oxidant hydroxyl radical. Nitric oxide, peroxynitrite and hydroxyl radical are capable of oxidizing lipids, proteins, and nucleic acids. Nitric oxide is also a major signaling molecule in neurons and immune system, either acting on the cell in which it is produced or by penetrating cell membranes to affect adjacent cells (Zhang and Li, 2006).

Nitric oxide has been shown to be a mediator of cell injury in some pathological conditions. NO has toxic effects at high concentrations, it reacts with oxygen and superoxide. The product of the reaction with superoxide is peroxynitrite (ONOO·), also it is which decompose to form ·OH radical. Reaction of nitric oxide and H_2O_2 yields singlet oxygen ($^1O_2$). Also, the reaction pathway of NO with molecular oxygen yields nitrogen diioxide (NO_2) and dinitrogen trioxide (N_2O_3).

$$NO· + O_2 \rightarrow ONOO^-, NO_2·, N_2O_3$$

$$NO· + O_2^{-·} \rightarrow ONOO^-$$

$$NO· + H_2O_2 \rightarrow {}^1O_2$$

Nitric oxide secreted by activated cells to be a complex "cocktail" of substances. The effect of these reactive species is particularly relevant to cell injury (Rosen et al., 2002).

3. Nitric oxide synthases: Structure and function

Nitric oxide is a highly reactive and unstable free radical gas that is produced by oxidation of L-arginine by oxygen and NADPH as electron donor to citrulline mediated by a family of homodimmer named nitric oxide synthase. In addition to L-arginine-NO pathway, L-arginine is also metabolized to L-ornithine and urea by arginase enzyme (Durante et al., 2007).

The nitric oxide synthase isoforms include the neuronal type I, (nNOS), the inducible form type II (iNOS) and endothelial type III (eNOS), whereas the nNOS and eNOS are constitutively expressed enzymes (cNOS). Constitutive nitric oxide synthase produces NO for short period of time (seconds to minutes). Inducible nitric oxide synthase expression is induced by inflammatory cytokines and toxins leads to the production of much higher amounts of NO compared to the cNOS, once iNOS expressed produces NO for long period of time (hours to days). Inducible NOS typically synthesizes 100-1000 times more than constitutive nitric oxide synthase. The major differences between cNOS and iNOS activities do not reside in the concentrations of NO generated per enzyme, but rather in the duration of NO produced. In addition, iNOS protein content in fully activated cells may be higher than cNOS content. Thus, cytotoxicity usually correlates with the product of iNOS and not with the product of the two cNOS. Thus the regulated pulses versus constant unregulated NO synthesis differentiates between messenger and the killer properties of NOS (Rabelink and Luscher, 2006).

The constitutive form of NOS is anchored on the internal surface of the endothelial cell membranes continuously present, although not always active. Its activity by the endothelial cells and neurons is responsible for maintenance of physiological homeostasis such as blood pressure and blood flow, controlling leukocyte - endothelial interactions and signaling among neurons. The constitutive isoform was distinguished from the inducible form based on the dependence of the constitutive enzyme activity on calmodulin (Wendy *et al.*, 2001). Other cofactors required for all enzyme forms are flavin mononucleotide, flavin adenine dinucleotide, heme and tetrahydrobiopterin. The constitutive enzyme requires calcium ion (Ca^{2+}) for activity while the inducible enzyme dose not (Wendy *et al.*, 2001).

The inducible NOS form represents a newly synthesized enzyme, which is expressed in response to specific stimuli, such as endotoxin and cytokines leading to enhanced NO production for many hours without further stimulation. It is expressed in multiple cell types, including macrophages, vascular smooth muscle cells, vascular endothelial cells and hepatocytes. The expression of iNOS may be beneficial in host defense or in modulating the immune response; indeed, the massive NO production by iNOS from macrophages during infections inhibits the growth of many pathogens (Wendy *et al.*, 2001; Madar *et al.*, 2005).

Nitric oxide synthase protein is a dimer formed of two identical subunits. There are three distinct domains in each NOS subunit: a reductase domain, a calmodulin-binding domain and an oxygenase domain (Li and Poulos, 2005).

1. **The reductase domain**: This domain contains the flavin adenine dinucleotide (FAD) and flavin mononucleotide (FMN) moieties and acts to transfer electrons from NADPH to the oxygenase domain of the opposite subunit of the dimer, and not to the domain of the same subunit.
2. **Calmodulin-binding domain**: The binding of calmodulin is required for the activity of all the NOS isoforms. It connects NO and calcium homeostasis.
3. **The oxygenase domain**: This domain contains the binding sites for tetrahydrobiopterin, heme and arginine. The oxygenase domain catalyzes the conversion of arginine into citrulline and NO.

Several factors affect the synthesis and catalytic activity of iNOS particularly, dimerization of NOS monomers. NOS isoforms are only active as homodimers. The dimerization of NOS monomers is promoted by heme, resulting in rapid conformational changes that, by cooperative action of tetrahydrobiopetrin (BH4) and L-arginine, leads to a stable and active enzyme. Moreover, an intracellular depletion of heme, BH4 and/or L-arginine considerably contributes to decreased resistance of NOS enzymes to proteolysis (Dunbar *et al.*, 2004).

4. Molecular nitric oxide targets in cells

The broad spectrum of effects performed by NO can be exerted through two main mechanisms: the activation of guanylate cyclase (which can be soluble in the cytosol or coupled to the cell membrane) or through its interaction with the major cellular source of superoxide anion, the NO/cytochrome C oxidase, which is found in mitochondria (Gha-fourifar and Cadenas 2005; Poulos 2006).

The guanylate cyclase-dependent effects of NO mainly affect the vascular tonus thereby affecting the inflammatory reaction by increasing synthesis of guanosine 3',5'-monophosphate (cGMP), it acts as inhibitors of platelets aggregation. Other effects pertaining to mitochondrial functions involve the respiratory burst. Mitochondria can produce NO through its own Ca^{2+}-sensitive synthase (mitochondrial, mtNOS). This enzyme regulates mitochondrial oxygen consumption and transmembrane potential via a reversible reaction with cytochrome C oxidase. The intramitochondrial reaction of NO with superoxide anion yields peroxynitrite, which irreversibly modifies susceptible targets within the mitochondria, inducing oxidative and/or nitrative stress (Ghafourifar and Cadenas 2005).

In addition to their primary role in the production of energy (ATP), mitochondria generate reactive oxygen species (ROS) that can directly or indirectly affect the NO response. Since NO and ONOO- can inhibit cellular respiration at the level of cytochrome C oxidase and complexes I-III, respectively, it has been suggested that mitochondrial function can influence the balance between apoptosis and necrosis induced by NO. Nitric oxide can stimulate the biogenesis of mitochondria in a guanosine 3',5'-monophosphate (cGMP)-dependent manner (Nisoli et al. 2003; Poderoso 2009).

Nitric oxide may react with proteins and nucleic acids. In addition to binding to heme groups, e.g. of guanylate cyclase, hemoglobin, and cytochrome C oxidase, NO may react with nucleophilic center like sulfur, nitrogen, oxygen and aromatic carbons. The prime target for covalent binding of NO to a functional groups in proteins under physiological condition in the presence of oxygen is the SH group. NO has been shown to N-nitrosylate primary arylamine of nucleotides and subsequent hydrolysis yield deamianted nucleotide. NO also mediates Fe^{+2} release from target cells, destroying Fe-S clusters in enzymes, like the citric acid cycle enzyme aconitase or ferrocheletase, which catalyse Fe^{+2} into protoporphyrin. NO can inhibit several interacellular enzymes and profoundly affect the cellular gene transcription machinery (Laurent et al., 1996; Alderton et al., 2001).

The high toxicity of inducible NO comes from its high concentration and from its reactivity with oxygen and oxygen-related reactive intermediates, which yield numerous toxic species that have enzymatic and DNA-damaging properties (Alderton et al., 2001).

Depletion of glutathione, inhibition of mitochondrial superoxide dismutase (SOD), and perhaps the loss of other antioxidant defense mechanisms could permit a rise in the endogenous level of reactive oxygen species normally produced by metabolism, which is likely to enhance the toxicity of NO. By this route, reactive nitrogen and oxygen species may act in concert to inactivate the key metabolic enzymes and cause lipid peroxidation and DNA strand breaks that result in irreversible cell injury and death (Radi, 2004; (Hummel et al., 2006).

5. Protective and cytotoxic function of nitric oxide

In the cardiovascular system, nitric oxide plays a major role in the regulation of blood flow and blood pressure as well as the general homeostatic control of the vasculature. Nitric oxide also inhibits platelet aggregation and adhesion by a mechanism dependent on cyclic GMP. It may also be involved in the interaction of leucocytes with vessel walls, since it inhibits leukocytes activation (Bian et al., 2008).

NO reacts with iron in the active side of the enzyme guanylate cyclase (GC), stimulating it to produce the intracellular mediator cyclic (cGMP), that in turn enhances the release of neurotransmitters resulting in smooth muscle relaxation and vasodilation (Esplugues, 2002).

In the central nervous system, accumulating evidence indicates that nitric oxide plays a part in the formation of memory. It is also found in some peripheral nerves, where it may contribute to sensory transmission (Sunico *et al.*, 2005; Zochodne and Levy, 2005).

The effects of NO can be direct or indirect and can influence several physiological processes, ranging from DNA transcription and replication to protein synthesis and secretion. Under physiological conditions, NO mediates homeostatic anti-inflammatory reactions, such as inhibition of neutrophil adhesion, cyclooxygenase activity, cytokine production, osteoclast bone resorption, among others, in order to prevent autoimmunity (Dal -Secco *et al.*, 2006; Fukada *et al.*, 2008; Livonesi *et al.*, 2009).

Generation of NO by endothelial cells causes smooth muscle relaxation through activation of guanylate cyclase by nitrosation of its heme group. It is hypothesised that NO may have originated in host as a mechanism of first-line defense against intracellular pathogens. This theory has been confirmed by the wide occurrence of the enzyme responsible for NO production, NO-synthase, in several species, ranging from invertebrates to mammals and non-mammalian vertebrates (Ribeiro *et al.*, 1993; Fukada *et al.*, 2008).

In mammals, NO production is upregulated in response to infection by a wide range of unicellular organisms such as bacteria, yeast and parasites (Cardoni *et al.*, 1990). Evidently, evolutionary diversity has induced NO synthesis to be performed in response to different kinds of stress stimuli. In fact, several antigens derived from intracellular parasites can be recognized by innate immune receptors on macrophages, triggering NOS activity (MacMicking *et al.*, 1997; Livonesi *et al.*, 2009).

Nitric oxide plays a significant role in acute and chronic inflammation i.e. excessive production of NO contributes to vasodilation and tissue damage which characterize many inflammatory conditions. Also, NO appears to play an important role in the functions of immune system as a cytotoxic macrophage effector molecule, modulator of polymorphonuclear leucocytes chemotaxis and adhesion, mediator of tissue injury caused by adhesion of immune complexes, and a regulator of lymphocyte proliferation (Shah *et al.*, 2004).

Cytokines and NO can modulate the production of chemokines and adhesion molecules in vivo and in vitro, influencing the course of infection (Savino *et al.*, 2007; Machado *et al.*, 2008). Chemokine receptors are also involved in cellular activation during parasitic infections and this G-protein-coupled signalling pathway is implicated in NO production as well (Benevides *et al.*, 2008).

NO is perhaps the most important among the group of early mediators produced by cells of the innate immune system. Phagocytes constitute the first line of microbial defense and they function by sensing the presence of different types of infectious agents (Carneiro-Sampaio and Coutinho, 2007) through pattern recognition receptors, including Toll-like receptors (TLRs) and the most recently described NOD- (NLRs) receptors. These receptors recognize multiple microbial patterns; therefore, they are critical for triggering the production of inflammatory mediators and essential for activation of the adaptive immune response (Schnare *et al.*, 2001; Kanneganti *et al.*, 2007; Underhill, 2007).

Nitric oxide synthase is produced by antigen-presenting cells (APC) during antigen processing and presentation to T cells and it can modulate various functions of APCs. It can inhibit the expression of major histocompatibility complex class II molecules in activated macrophages and, at high concentrations, may also inhibit IL-12 synthesis, thus contributing to the desensitization of macrophages after exposure to inflammatory stimuli (Salvucci *et al.*, 1998; Pahan *et al.*, 2001; van der Veen, 2001). In chronic immune responses to intracellular pathogens, NO is reported to play a regulatory role and may promote parasite persistence. For these reasons, it is suggested that NO is cytostatic rather than cytotoxic for parasites (Jana *et al.*, 2009).

NOS enzymes or NO-activity-derived products (nitrites or nitrotyrosine) have been detected in different locations of adult worms. Neural NOS and iNOS have been found in the nervous tissue and in the parenchyma of *Schistosoma mansoni* respectively (Kohn *et al.*, 2001). The presence of NOS has been also demonstrated in *Ascaris suum; Toxocara canis; F. gigentica* (Fan *et al.*, 2004; Hamdi and Ali, 2009). Nitric oxide synthase is located in the muscular wall from adult worms in *Brugia malayi, Dirofilaria immitis* and *Acanthocheilonema vitae* filariae (Pfarr *et al.*, 2001). Expression of endothelial NOS (eNOS) has been detected in the cuticle and stichocytes from *Trichinella britovi* (Masetti *et al.*, 2004). Nitrites have been detected in the hydatid liquid of fertile *Echinococcus granulosus*. Expression of iNOS and nNOS has been detected in the parenchyma and nervous structures of the filariform larvae from *Strongyloides venezuelenesis*. Moreover, NOS expression has also been demonstrated in other phases, such as eggs, sporocysts, and cercariae of *Schistosoma* sp. (Long *et al.*, 2004) and other structures as oocytes, spermatozoids, and embryonic forms of *Brugia malayi* (Pfarr *et al.*, man 2001).

A dual role in the immunity is usually observed for NO. This well-known immune duality is usually dependent on concentration and, once dysregulated, may lead to host cell toxicity, autoimmunity or parasite persistence due to immune evasion, all of which can lead to pathology. The strength of NO toxicity is dependent on the sensitivity of the parasite, which differs among parasite strains and according to the physiological microenvironment (Gutierrez *et al.*, 2009).

Specifically, both adult worms' excretory/secretory antigens and larval somatic antigens of *T. canis* are capable of stimulating *in vitro* the production of NO at transcriptional level in rat alveolar macrophages. The stimulation of NO production by antigens of *T. canis* LII (extracted from excretory/secretory product) does not seem to play a host-defensive role. The production of NO by host cells, activated by the parasite, has negative effects not on parasite survival but on the host, and thus putatively represents a parasite evasion mechanism. Types of parasite evasion/adaptation mechanisms largely depend on the parasite's migration and definitive anatomical location. *T. canis* is characterized by its dissemination (migration) through the bloodstream until it reaches its final inside the host. This bloodstream migration would be clearly facilitated by blood-vessel dilatation described toxocariasis models. Deleterious effects were attributed on the host physiology in the release of NO by host cells, induced by the parasite itself (Espinoza *et al.*, 2002). Thus, production of NO during migration of *T. canis* LII inside their host could facilitate their migration and triggering of this production by LII may represent a parasite adaptation mechanism (Muro and Perez-Arellano, 2010).

The effect of different antigens of excretory/secretory of larval and adult worms of nematodes, on NO production from rat alveolar macrophages was observed. Excretory/secretory antigens from adult worms in *Toxocara canis* and *Strongyloides venezuelensis* stimulated the NO production from alveolar macrophages (Espinoza *et al.*,

2002). The cytoplasmatic signalling pathways involved in the NO production after stimulation with adult excretory/secretory antigen of *Toxocara canis*. It was suggested that phospholipase C macrophage pathways play an essential role in activating the production of NO triggered by this antigen. This suggests that NO production could be due to an increase of intracellular calcium and activation of the arachidonic acid pathway. Moreover, *Toxocara canis* excretory/secretory adult antigen also stimulated alveolar macrophages to produce prostaglandin E_2 (PGE$_2$). These results indicate that *Toxocara canis* can stimulate the release of vasodilatory mediators by host macrophages (Espinoza *et al.*, 2002; Hewitson *et.al*, 2009; Muro and Perez-Arellano, 2010).

6. Regulation of nitric oxide synthase

Inhibition of arginase has been shown to stimulate NO synthesis in endothelial cells. In addition, overexpression of arginase I or arginase П supresses NO generation in endothelial cells and this is associated with a significant decrease in intracellular L- arginine content (Li *et al.*, 2001). Interestingly, constitutive expression of arginase in microvascular endothelial cells counteracts NO-mediated dilation, suggesting that arginase subserves a tonic vasoconstrictor function (Chicoine *et al.*, 2004; Johnson *et al.*, 2005).

A side from blocking NO synthesis by depleting the cell of substrate for NOS, the arginase-mediated removal of L- arginine inhibits the expression of inducible NOS (iNOS) by repressing the translation as well as the stability of iNOS protein. Furthermore, arginase may inhibit iNOS-mediated NO production through the generation of urea. These findings suggest that arginase downregulates NO formation via multiple mechanisms. Interestingly, N-hydroxy-L-arginine, an intermediate formed during the catalysis of L-arginine by NOS, is a potent inhibitor of arginase, suggesting that NOS may also influence arginase activity (Johnson *et al.*, 2005).

Thus, arginase has recently been emerged as a critical regulator of NO synthesis that may contribute to the development of numerous pathologies, including vascular disease. The release of NO by endothelial NOS (eNOS) plays a crucial role in preserving vascular homeostasis. In response to changes in shear stress or receptor stimulation, NO is released from the vascular endothelium to promote blood flow by inhibiting vascular tone and platelet aggregation (Durante, 2007).

Aminoguanidine has effects on several enzyme systems. It interferes with non-enzymatic glycosylation. Aminoguanidine inhibits nitric oxide synthase (NOS), particularly its inducible form (iNOS), reducing the pathological effects due to over-activity of these enzymes and thus to over-production of NO. Aminoguanidine may affect of polyamine metabolism (Kolodziej *et al.*, 2006).

NO synthesis inhibitors have been used in vivo to evaluate their effect on parasitic infection. This strategy has mainly been used in experimental models of filariosis by *Brugia malayi*, toxocariosis, *trichinellosis*, and strongyloidiasis. The results obtained are divergent, since the use of aminoguanidin diminishes the lesions in toxocariosis, whereas it increases the parasite load in filariosis and strongyloidiasis. Moreover, mice treated with aminoguanidine, at the beginning of muscle phase of the infection, inhibit the reduction of muscle larvae number and cells of inflammatory infiltrates did not show any specific iNOS reaction]. Influence of the inhibition of the NO production by iNOS in a toxocariosis experimental model decreases the deleterious effects of the parasite upon the host, especially the lung vascular alterations (Muro and Perez-Arellano, 2010).

Oxalomalate, a tricarboxylic acid structurally related to citrate, has long been known to be a powerful competitive inhibitor of both aconitase, which is required for the first step of citric acid cycle, and NADP dependent isocitrate dehydrogenase. NO modulates also the activity of several other proteins including mitochondrial aconitase, an iron-dependent enzyme and isocitrate dehydrogenase (Irace et al., 2007).

OMA inhibits aconitase and isocitrate dehydrogenase, the Krebs cycle flux appears to be constricted at two steps with a consequent reduction of its biosynthetic ability caused by the limited availability of the successive intermediates, namely α-ketoglutarate, a precursor of L-arginine, and succinyl-CoA, a precursor of heme. In fact, exogenous treatment with α-ketoglutarate or succinyl-CoA partially or almost completely restored NO production which was inhibited by OMA. As heme is the sole cofactor absolutely required for the formation of active NOS dimers and the binding of L-arginine to iNOS facilitates its dimerization, the partial slowdown of Krebs cycle induced by OMA, leading a consequent decrease of succinyl-CoA and α-ketoglutarate formation, could explain the effect of OMA on NO production. Moreover, inhibition by OMA of the $NADP^+$-dependent isocitrate dehydrogenase reduces the availability of NADPH, a source of electrons in the NOS enzymatic reaction.This impairment, together to the depletion of L-arginine caused by α-ketoglutarate shortage, may prevent NO biosynthesis(Irace et al., 2007).

It is known that NOS are long-lived proteins and the presence of heme and L-arginine or BH4 considerably contributes to their resistance to proteolysis. Consequently, the coupled shortage of heme and L-arginine caused by OMA may determine instability of NOS dimers, and consequently, lower resistance to proteolysis. This hypothesis is supported by the observed reduction of iNOS protein content both in LPS-stimulated macrophage treated with OMA and in peritoneal macrophages recovered from LPS-stimulated rats after injection of OMA precursors (Osawa et al., 2003).

7. Conclusion

It is well known that, glutathione interacts with reactive species derived from NO oxidation and converts these species to less toxic ones Therefore, depletion of glutathione, inhibition of mitochondrial superoxide dismutase (SOD), and perhaps the loss of other antioxidant defense mechanisms could permit a rise in the endogenous level of reactive oxygen species normally produced by metabolism, which is likely to enhance the toxicity of NO development of NOS inhibitors constitutes a current strategy in the research of new compounds showing interesting properties. Inhibition of NO production would be beneficial for the host

8. References

Alderton WK, Cooper CE, Knowl RG (2001) Nitric oxide synthases: structure, function and inhibition. *Biochem J* 357: 593-615.

Benevides L, Milanezi CM, Yamauchi LM, Benjamim CF, Silva JS, Silva NM (2008) CCR2 receptor is essential to activate microbicidal mechanisms to control *Toxoplasma gondii* infection in the central nervous system. *Am J Pathol* 173: 741-751.

Bian K, Dousout M, Murad F (2008) Vascular system: role of nitric oxide in cardiovascular diseases. *J Clin Hypertens*10: 304-310.

Cardoni RL, Rottenberg ME, Segura EL (1990) Increased production of reactive oxygen species by cells from mice acutely infected with *Trypanosoma cruzi. Cell Immunol* 128: 11-21.

Carneiro-Sampaio M, Coutinho A (2007) Immunity to microbes: lessons from primary immunodeficiencies. *Infect Immun* 75: 1545-1555.

Chicoine LG, Paffet ML, Young TL Nelin LD (2004) Arginase inhibition increases nitric oxide production in bovine pulmonary arterial endothelial cells. *Am J Physiol Lung cell Mol Physiol* 287: L 60-L68.

Dal-Secco D, Moreira AP, Freitas A, Silva JS, Rossi MA, Ferreira SH, Cunha FQ (2006) Nitric oxide inhibits neutrophil migration by a mechanism dependent on ICAM-1: role of soluble guanylate cyclase. *Nitric Oxide* 15: 77-86.

Dunbar AY, Kamada Y, Jenkins GJ, Lowe ER, Billecke SS, Osawa Y (2004) Ubiquitination and degradation of neuronal nitric-oxide synthase in vitro: dimer stabilization protects the enzyme from proteolysis. *Molecular Pharmacology* 66: 964-969.

Durante W, Johnson FK, Johnson RA (2007) Arginase: A critical regulator of nitric oxide synthesis and vascular function. Clin *Exp Pharmacol Physiol* 34: 906-911.

El Kasmi KC, Qualls JE, Pesce JT, Smith AM, Thompson RW, Henao-Tamayo M, Basaraba RJ, Konig T, Schleicher U, Koo MS, Kaplan G, Fitzgerald KA, Tuomanen EI, Orme IM, Kanneganti TD, Bogdan C, Wynn TA, Murray PJ (2008) Toll-like receptor-induced arginase 1 in macrophages thwarts effective immunity against intracellular pathogens. *Nat Immunol* 9: 1399-1406.

Espinoza EY, Perez-Arellano JL, Carranza C, Collia F, Muro A (2002) In vivo inhibition of inducible nitric oxide synthase decreases lung injury induced by *Toxocara canis* in experimentally infected rats. *Parasite Immunol* 24: 511-520.

Esplugues JV (2002) NO as a signaling molecule in the nervous system. *Br J Pharmacol* 135: 1079- 95.

Fan CK, Lin YH, Hung CC, Chang SF, Su KE (2004) Enhanced inducible nitric oxide synthase expression and nitrotyrosine accumulation in experimental granulomatous hepatitis caused by *Toxocara canis* in mice. *Parasite Immunol* 26: 273-281.

Freire-de-Lima CG, Nascimento DO, Soares MB, Bozza PT, Castro-Faria-Neto HC, de Mello FG, DosReis GA, Lopes MF (2000) Uptake of apoptotic cells drives the growth of a pathogenic trypanosome in macrophages. *Nature* 403: 199-203.

Fukada SY, Silva TA, Saconato IF, Garlet GP, Avila-Campos MJ, Silva JS, Cunha FQ (2008) iNOS-derived nitric oxide modulates infection-stimulated bone loss. *J Dent Res* 87: 1155-1159.

Ghafourifar P, Cadenas E (2005) Mitochondrial nitric oxide synthase. *Trends Pharmacol Sci* 26: 190-195.

Gutierrez FRS, Mineo TWP, Pavanelli WR, Guedes PMM Silva JS (2009) The effects of nitric oxide on the immune system during Trypanosoma cruzi infection. *Mem Inst Oswaldo Cruz, Rio de Janeiro* 104: 236-245.

Hewitson JP, Grainger JR, Maizels RM (2009)Helminth immuno-regulation: The role of parasite secreted proteins in modulating host immunity. *Mol Biochem Parasitol* 167: 1-11.

Hummel SG, Fischer AJ, Martin SM, Schafer FQ Buettner GR (2006) Nitric oxide as a cellular antioxidant: a little goes along way. *Free Radic Biol Med* 40: 501- 6.

Irace C, Esposito G, Maffettone C, Rossi A, Festa M, Iuvone T, Santamaria R, Sautebin L, Carnuccio R, Colonna A. (2007) Oxalomalate affects the inducible nitric oxide synthase expression and activity. *Life Sciences* 80: 1282-1291.

Jana M, Dasgupta S, Pal U, Pahan K (2009) IL-12 p40 homodimer, the so-called biologically inactive molecule, induces nitric oxide synthase in microglia via IL-12 R beta 1. *Glia* 57: 1553-1565.

Johann AM, Barra V, Kuhn AM, Weigert A, von Knethen A, Brune B (2007) Apoptotic cells induce arginase II in macrophages, thereby attenuating NO production. *FASEB J* 21: 2704-2712.

Johnson FK, Johnson RA, Peyton K J, Durante W (2005): Arginase inhibition restores arteriolar endothelial function in Dahl rats with salt- induced hypertension. *Am J Physiol Regul Integr Comp Physiol* 288: R1057-R1062.

Kanneganti TD, Lamkanfi M, Nunez G (2007) Intracellular NOD-like receptors in host defense and disease. *Immunity* 27: 549-559.

Kohn AB, Moroz LL, Lea JM, Greenberg RM (2001) Distribution of nitric oxide synthase immunoreactivity inthe nervous system and peripheral tissues of *Schistosoma mansoni*. *Parasitol* 122: 87-92.

Kołodziej-Sobocińska M, Dziemian E, Machnicka-Rowinska B (2006) Inhibition of nitric oxide production by aminoguanidine influencesthe number of *Trichinella spiralis* parasites in infected "lowresponders" (C57BL/6) and "high responders" (BALB/c) mice. *Parasitol Res* 99: 194-196.

Laurent MM, Lepoivre M, Tenu JP (1996) Kinetic modelling of the nitric oxide gradient generated in vitro by adherent cells expressing inducible nitric oxide synthase. *Biochem J* 314: 109-113.

Li H, Poulos TL (2005) Structure, function studies on nitric oxide synthase. *J Inorg Biochem* 99: 293-305.

Li H, Meininger C J, Hawker JR (2001) Regulatory role of arginase I and ΙΙ in nitric oxide, polyamine, and proline synthesis in endothelial cells. *Am J Physiol Endocrinol Metab* 280: E75-E82.

Livonesi MC, Rossi MA, de Souto JT, Campanelli AP, de Sousa RL, Maffei CM, Ferreira BR, Martinez R, da Silva JS (2009) Inducible nitric oxide synthase-deficient mice show exacerbated inflammatory process and high production of both Th1 and Th2 cytokines during paracoccidioidomycosis. *Microbes Infect* 11: 123-132.

Long XC, Bahgat M, Chlichlia K, Ruppel A, Li YL (2004) Detection of inducible nitric oxide synthase in *Schistosoma japonicum* and *S. mansoni*. *J Helminthol* 78: 47-50.

Machado FS, Souto JT, Rossi MA, Esper L, Tanowitz HB, Aliberti J, Silva JS (2008) Nitric oxide synthase-2 modulates chemokine production by *Trypanosoma cruzi*-infected cardiac myocytes. *Microbes Infect* 10: 1558-1566.

MacMicking J, Xie QW, Nathan C (1997) Nitric oxide and macrophage function. *Annu Rev Immunol* 15: 323-350.

Madar Z, Kalet LS, Stark AH (2005) Inducible nitric oxide synthase activity and expression in liver and hepatocytes of diabetic rats. *Intern J Exp Clin Pharmacol* 73: 106- 112.

Masetti M, Locci T, Cecchettini, A., Lucchesi P, Magi M, Malvaldi G, Bruschi (2004) Nitric oxide synthase immunoreactivity in the nematode *Trichinella britovi*. Evidence for nitric oxide production by the parasite *Inter J Parasitol* 34: 715-721.

Muro A, Arellano J (2010) Nitric Oxide and Respiratory Helminthic Diseases. *J Biomedicine Biotechnoly* 10: 1-8

Nisoli E, Clementi E, Paolucci C, Cozzi V, Tonello C, Sciorati C, Bracale R, Valerio A, Francolini M, Moncada S, Carruba MO (2003) Mitochondrial biogenesis in mammals: the role of endogenous nitric oxide. *Science* 299: 896-899.

Osawa Y, Lowe ER, Everett AC, Dunbar AY, Billecke SS, (2003) Proteolytic degradation of nitric oxide synthase: effect of inhibitors and role of hsp90-based chaperones *J Pharmacol Experiment Therapeut* 304: 493-497.

Pahan K, Sheikh FG, Liu X, Hilger S, McKinney M, Petro TM (2001) Induction of nitric-oxide synthase and activation of NF-kappaB by interleukin-12 p40 in microglial cells. *J Biol Chem* 276: 7899-7905.

Pfarr KM, Qazi S, Fuhrman JA (2001) Nitric oxide synthase in filariae: demonstration of nitric oxide production by embryos in Brugia malayi and Acanthocheilonema viteae. *Experiment Parasitol* 97: 205-214.

Poderoso JJ (2009) The formation of peroxynitrite in the applied physiology of mitochondrial nitric oxide. *Arch Biochem Biophys* 484: 214-220.

Poulos TL (2006) Soluble guanylate cyclase. *Curr Opin Struct Biol* 16: 736-743.

Rabelink T, Luscher T (2006) Endothelial nitric oxide synthase: Host defense enzyme of the endothelium. *Arterioscler Thromb Vasc Biol* 26: 267-71.

Radi R (2004) Nitric oxide, oxidants and protein tyrosine nitration. *Proc Nat Acad Sci USA*; 101: 4003- 8.

Ribeiro JM, Hazzard JM, Nussenzveig RH, Champagne DE, Walker FA (1993) Reversible binding of nitric oxide by a salivary heme protein from a bloodsucking insect. *Science* 260: 539-541.

Rodriguez PC, Zea AH, DeSalvo J, Culotta KS, Zabaleta J, Quiceno DG, Ochoa JB, Ochoa AC (2003) L-arginine consumption by macrophages modulates the expression of CD3 zeta chain in T lymphocytes. *J Immunol* 171: 1232-1239.

Rosen GM, Tsai P, Pou S (2002) Mechanism of free radical generation by nitric oxide synthase . *Chem Res* 102: 1191-1199.

Salvucci O, Kolb JP, Dugas B, Dugas N, Chouaib S (1998) The induction of nitric oxide by interleukin-12 and tumor necrosis factor-alpha in human natural killer cells: relationship with the regulation of lytic activity. *Blood* 92: 2093-2102.

Savino W, Villa-Verde DM, Mendes-da-Cruz DA, Silva-Monteiro E, Perez AR, Aoki M del P, Bottasso O, Guinazu N, Silva-Barbosa SD, Gea S (2007) Cytokines and cell adhesion receptors in the regulation of immunity to *Trypanosoma cruzi*. *Cytokine Growth Factor Rev* 18: 107-124.

Schnare M, Barton GM, Holt AC, Takeda K, Akira S, Medzhitov R (2001) Toll-like receptors control activation of adaptive immune responses. *Nat Immunol* 2: 947-950.

Shah V, Lyford G, Gores G, Farrugia G. (2004) Nitric oxide in gastrointestinal health and disease. *Gastroenterol* 126: 903-913.

Sunico CR, Portillo F, Gonzalez- Forero D, Moreno - Lopez B (2005) Nitric oxide- directed synaptic remodeling in the adult mammal CNS. *J Neurosci* 25: 1448- 58.

Underhill DM (2007) Collaboration between the innate immune receptors dectin-1, TLRs, and Nods. *Immunol Rev* 219: 75-87.

van der Veen RC (2001) Nitric oxide and T helper cell immunity. *Int Immunopharmacol* 1: 1491-1500.

Wendy K., Alderton W.K., Chris E., Cooper C.E. and Knowles R.G. (2001) Nitric oxide synthases: Structure, function and inhibition. *Biochem J* 357: 593- 615.

Zhang X, Li D (2006) Peroxynitrite mediated oxidation damage and cytotoxicity in biological systems. *Life Sci* 3: 41-4.

Zochodne DW, Levy D (2005) Nitric oxide in damage, disease and repair of the peripheral nervous system. *Cell Mol Biol* 51: 255-67.

6

Assessment of the General Oxidant Status of Individuals in Non-Invasive Samples

Sandro Argüelles[1], Mercedes Cano[2], Mario F. Muñoz-Pinto[1],
Rafael Ayala, Afrah Ismaiel[1] and Antonio Ayala[1*]
*[1]Department of Biochemistry and Molecular Biology,
Faculty of Pharmacy,
[2]Department of Physiology, University of Seville,
Spain*

1. Introduction

Determination of the oxidative stress state of a person indicates the risk of suffering many disorders and diseases in humans that are a product of oxidative stress. The oxidant status of an individual is assessed by determining a group of markers in non-invasive samples. Although these biomarkers are formed by oxidation of biomolecules and are supposed to reflect changes in tissues that have been exposed to oxidants, one limitation when measuring these biomarkers in non-invasive samples is that they do not give information about the tissue localization of the oxidative stress, at the least the marker is exported into serum from the tissue. In previous work from our laboratory, we have determined that only a few generic markers of oxidation can be useful to predict the oxidant status of an individual when the markers are measured in non invasive samples. An additional aspect to consider before validating the markers is to determine how stable their levels are for the same individual throughout time. Theoretically, if these markers present a high variability, their utility to study the effects of an eventual intervention would be limited since the effects of the intervention should be clear to be observed over the basal oscillation of the marker. Results from our group show a significant intra day variation of serum biomarkers in many cases. Therefore, it is clear that more than a single measurement will be required to establish the basal status of oxidative stress of individuals and several measurements will be required for a long period of time.

2. Aging and its biomarkers

Aging can be defined as the general loss of the optimal body functions of an organism over the years. Although there are several hypotheses that attempt to explain the causes of aging, only a few are widely accepted, which does not mean they are correct. In fact, nowadays, none of these most accepted theories entirely explain the root cause of aging. Until the root cause is known, it will be difficult to design a strategy of intervention to control aging. Also,

---

* Corresponding Author

it is important to emphasize that what a theory describes as the cause of aging determines not only the design intervention strategies, but also what type of markers that can be measured to assess the rate of aging.

Among the most popular hypothesis about why we age, the theory of Harman (Harman, 1956), is 50 years-old and it is continuously being revised (Kirkwood, 2005). This theory postulates that the macromolecular damage induced by reactive oxygen species (ROS) is the main causal factor of aging and related diseases. These ERO are formed during the normal metabolism, primarily in detoxification reactions mediated by the microsomal cytochrome P-450 (Ayala and Cutler, 1997; Finkel and Holbrook, 2000; Porter and Coon, 1991) and in mitochondrial electron transport chain (Finkel and Holbrook, 2000). Also, free radicals can be induced exogenously by radiation, smoking, diet, drugs, etc.

An imbalance between oxidants and antioxidaints in favour of the oxidants, potentially leading to damage, is termed 'oxidative stress (Sies, 1997). According to free radical theory, the more oxidative stress, the higher the aging rate (biological and pathological aging) and age-related impairments. The intensity of oxidative stress depends on the amount of oxidants, available antioxidants, and the activity of the repair processes, which in turn are responsible for the elimination of oxidized molecules. As soon as oxidant products appear in the serum, urine or breath, they can be used as markers of oxidative stress. Assuming that oxidative stress is a contributing factor to the onset of age-related degenerative diseases (Halliwell et al., 1992; Halliwell, 2006; McCord, 2000), the determination of the oxidative stress state of a person not only allows comparison with the average of a population but it can also indicate the risk of suffering many disorders and diseases in humans that are a product of oxidative stress.

To assess the oxidant status of an individual under normal or pathological conditions or after any kind of intervention, a group of tests for the measurement of oxidative stress in non-invasive samples was developed in the last decade (Butler et al., 2004; Cutler, 2005; Cutler et al., 2005; Johnson, 2006). The idea being to get a complete picture of the oxidative stress state in the body by looking at the levels of markers for the ongoing oxidative damage in serum, urine, saliva and breath samples. Since most of the biological molecules can become oxidized, there are several types of biomarkers for the assessment of oxidative stress. Thus, in a biological sample, we can have oxidized DNA, proteins and lipids products and therefore several markers can be used. So, the first question we are going to analyze in this chapter is which one do we choose by considering that the blood or urinary levels of the marker must reflect tissular oxidative stress and that the marker levels must indicate only oxidation rate and not the rate of a different biological process.

3. First question about oxidative stress markers: Which one do we choose?

Traditionally, the markers of oxidative stress have been classified in three groups, depending on the macromolecule affected: markers of oxidative damage to DNA, lipids and proteins. The attack of free radicals or oxidants to DNA leads to oxidized nucleotides that can be analyzed in serum and urine. Also, the reactions of oxidant compounds with cellular membranes produce lipid peroxides and their derivatives of them and the oxidation of proteins increases the levels of carbonyl groups and the appearance of oxidized amino acids.

According to this, the panel of markers that can be determined is really large. Therefore, the first question that arises is how many of these markers should we measure in order to assess the general status of oxidation of an individual. If the answer is "let´s measure all of them", a typical scenario is that for the same person we can get contradictory results so that the marker of DNA oxidation can be high while the marker of lipid oxidation is low. Therefore, a conflict of interpretation is normal when analyzing a panel of markers. Also, the levels of the marker in serum (or urine) must reflect the oxidative damage of the tissues in general. Considering that in vivo oxidative damage is likely to occur in only a few sites or tissues at any given time, high levels of these parameters, for instance in serum, in one individual can be a consequence of: 1. - A generalized increase of oxidative damage in most of the tissues, all of them being able to export the marker into the blood proportionally to the extent of damage; 2.- The increase of oxidative damage in just a particular tissue, this being a dysfunction of the tissue, the origin of a future disease; 3. - An increase of the oxidative damage produced specifically in the circulatory system (Figure 1).

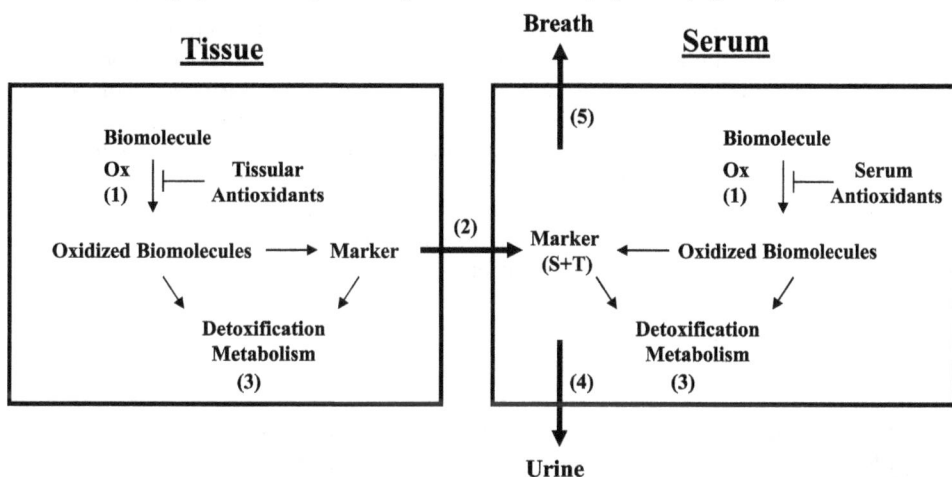

Fig. 1. **Steps involved in the levels of oxidative stress markers.** Step 1: oxidation of the biomolecule by oxidants (Ox). Step 2: Export of the tissular marker into serum. Step 3: Detoxification and metabolism of the marker. "S+T" is the sum of the amount of the marker formed in tissue and serum. Step 4 and 5: Elimination of the marker throught urine and breath.

In order to clarify whether the measurement of non invasive oxidative stress marker reflects one of the above possibilities, a study in our laboratory was undertaken a few years ago where a few generic markers of oxidation were determined simultaneously in serum and tissues of six groups of rats treated experimentally to modulate their oxidative stress status (Arguelles et al., 2004) . For each marker, the correlation between serum and tissular levels was calculated to test, first, whether changes in serum levels reflect changes in tissular levels and, second, whether these levels change concomitantly in all tissues. According to the scheme shown in Fig. 1., the levels of a particular marker in serum will depend on the rate of steps 1-5. The amount of oxidized biomolecule formed in a tissue or serum will depend on ROS levels and the levels of antioxidants (step 1). If the marker can be exported into serum from one or several tissues (pathway 2), serum determination of the biomarkers would represent the summation of the amount of modified molecules excreted from the cells and the amount of modified molecules produced in the cardiovascular system at that time ("S+T"). The amount of the marker in the serum would be increased proportionally to the degree of tissular and serum oxidative stress.

Concerning carbonyl groups (CO), a marker of protein oxidation (Levine, 2002), our results indicated that there is no significant correlations between carbonyl groups of serum and tissues, which is not surprising considering the tremendous heterogeneity of protein molecules and environments in the different tissues. It is noteworthy that there is no correlation between serum CO and CO of tissues secreting proteins that act in plasma, such as liver. Due to oxidized proteins being degraded inside the tissue, it is unlikely that they can be exported to the serum once they are degraded (Figure 2). In addition, products of protein oxidation are subject to metabolism and this metabolism can be different in the tissues studied. All these factors may play a role in determining that CO in the different tissues studied and in the serum do not correlate in general. A different situation might occur with oxidized amino acids derived from oxidized proteins, which can appear in the serum and maybe the levels correlate with the tissue levels. In any case, the results of this study showed that the measurement of CO in serum is not useful in predicting the degree of tissular protein oxidative damage and only indicates the oxidative damage to serum proteins. Because carbonyl groups in serum reflects exclusively protein oxidation in the circulatory system they might reflect the risk of cardiovascular disease. This example illustrates that only a good knowledge of the biochemical pathway of the formation of the marker allows us to draw a precise conclusion about the meaning of the different values found in serum.

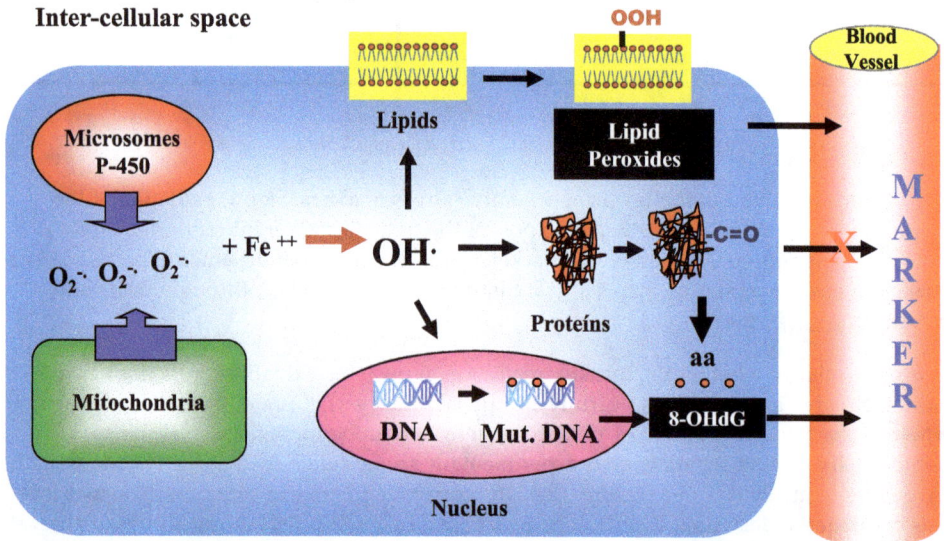

Fig. 2. **Effect of free radicals on biomolecules.** Lipid peroxides, Hydroperoxides, 8-OHdG and CO groups are formed as a consequence of free radicals attack (OH in this example). Oxidized proteins are not exported into serum because they are degraded inside the cell.

A second oxidative stress marker studied in this animal model was lipid peroxides (LP), which have been suggested as playing a role in the molecular mechanism of several pathological processes (Gutteridge and Halliwell, 2000). Contrary to the results with CO groups, LP changes concomitantly in both tissues and serum. This suggests that the measurement of LP in serum gives an indirect indication about what is happening in the tissues i.e, that it can be used as an indicator of the average amount of free radical damage to lipids in the body tissues at a given time.

Another consideration to take into account is that the biological meaning of the changes of some markers remain to be established when measured in non-invasive samples. For instance, concentration in the urine level of one of the popular markers in the field, 8-hydroxy-2'-deoxyguanosine (8-OHdG), is considered as evidence of a process of oxidative stress in DNA but also as evidence of an optimum repair level of the DNA (Halliwell, 2002). Obviously, this problem does not exist if this marker is directly determined in DNA extracted from tissues.

As can be seen, if one intends to get information about the general status of oxidation of an individual just by looking at the level of markers in the serum the problem is that tissue-specific oxidative damage does not generally cause systemic oxidative stress that can easily be measured in serum. This means that it cannot be assumed that all markers in noninvasive samples are useful for predicting the general oxidative status in all of a person´s tissues. Also, oxidative stress in circulation must be interpreted with great caution. Because of the lack of a positive correlation between oxidative stress markers in serum and tissues, whether the risk of a specific disease incidence is more associated with a given oxidative stress level of serum or tissue is yet to be determined. Also, a more comprehensive study using more biomarkers should be performed to select those whose serum levels reflect the oxidative damage in tissues.

4. Second question about oxidative stress markers: Do they change over time?

An additional aspect to consider before validating the markers is whether these markers remain stable as a function of time for each individual, not only for a long period of time but also during the day. If the source of oxidative stress is both endogenous (mitochondria and detoxification reactions) and exogenous (Finkel and Holbrook, 2000), both fractions can change as a function of many factors such us basal metabolism, medication, diet and habits, for example, smoking (Lesgards et al., 2002; Moller et al., 1996), consequently affecting levels of the markers. Theoretically, if these markers present a high variability, they would not be useful in diagnosing the basal oxidative stress of individuals. Also, their utility to study the effects of an eventual intervention would be limited since the effects of the intervention should be clear to be observed over the basal oscillation of the marker (Figure 3).

To determine the degree of variability throughout the day, and over time, in a previous work from our laboratory (Arguelles et al., 2007), three markers of oxidative stress from healthy volunteers during a period of 51 days were measured. At the same time, the variability in the levels of these markers was studied throughout the day to test whether or not they changed and if so whether the magnitude and trend of this change would be typical for every person. The aim of this second task was to determine the best moment during the day when the antioxidant or protective supplements should be administered. The results indicated that the levels of these markers can vary greatly within a person during the period studied (51 days). Figure 4 shows the results obtained in three patients. As can be seen, repeated serum lipid peroxide values on nonconsecutive days show serum lipid peroxides of patient 1 to be stable, but this is not the case for patient 3.

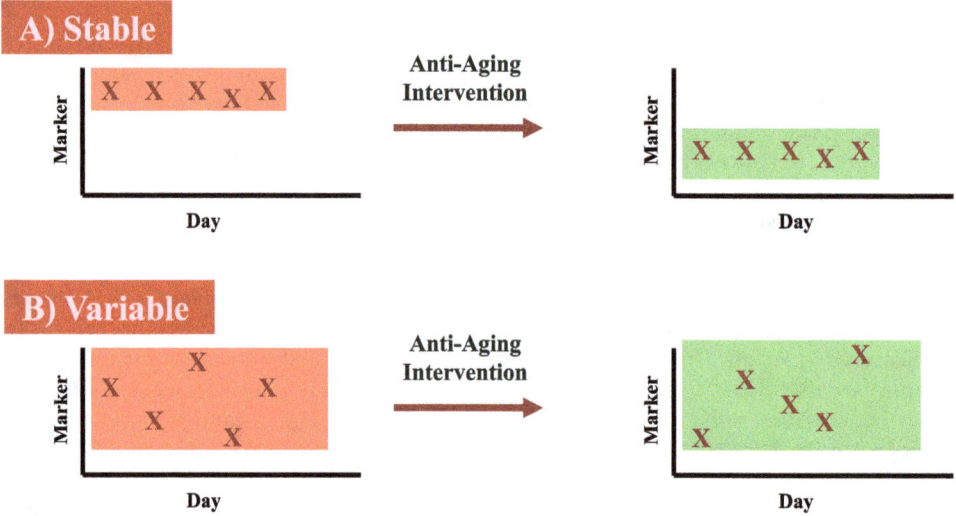

Fig. 3. Variability in the levels of a hypothetical marker. If the markers present a high variability (B), they would not be useful to study the effect of an eventual intervention.

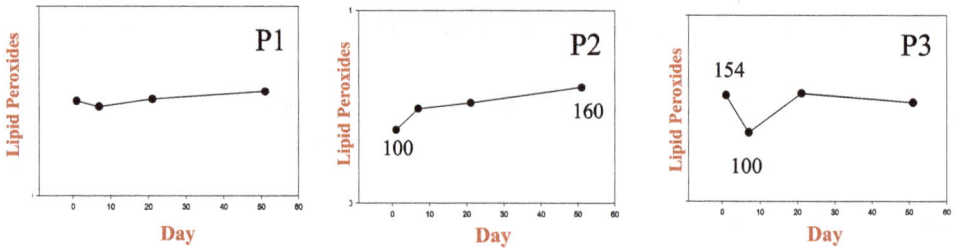

Fig. 4. Change in the morning levels of lipid peroxides in serum of three different subjects during the period studied (51 days). Results are percentage with respect to the lowest value found. (P1-P3)

As mentioned above, the two main endogenous sources of oxidative stress are mitochondrial respiration and detoxification reactions. Therefore, several factors related to basal metabolism and the constitutive content of cytochrome P-450, for example, should affect the oxidative stress of an individual. In this sense, previous works of our group have shown that people with higher amounts of hepatic p-450 present a higher oxidative stress in their proteins (Ayala and Cutler, 1997). Circadian variations in the detoxification reactions have been described and could influence (Reddy et al., 2006). Also, exogenous factors may contribute to this normal variability of the markers. Thus, both "healthy" and "unhealthy" lifestyle patterns have been described as affecting oxidative stress and antioxidant capacity (Lesgards et al., 2002; Moller et al., 1996). In our study, the influence of lifestyle factors was not considered. However, considering the variability found, it would be necessary to take a survey in order to study the relationship between habits and oxidative stress markers.

The results also show a significant intra-day variation of serum biomarkers in most of the subjects, where the general trend is that the levels of oxidative stress markers seem to increase during the day (Figure 5). If this happens every day, it is tempting to speculate that the concentration of the marker starts decreasing during the night until reaching morning values. Since the days of the assays were not consecutive , we cannot affirm that this actually happens because we do not know the night values of the day before the assay. However, we might consider the reparative aspect of sleep. In fact it is described in the paper of Lesgarsds et al. (Lesgards et al., 2002) that remaining awake all night was responsible for an important increase in urine level of TBARS. It remains to be studied whether the morning values are affected by the values reached at night the day before and/or other factors that are secreted during sleep.

Day	P1	P2	P3	P4	P5	P6	P7	P8	P9	P10
LP1	↑	↑	=	=	↑	↑	=	=		
LP2	↑		↑	=	↑	↑		↑	=	
LP3	↑	↑	↑	=	↑	↑		↑	=	
LP4	↑	↑	=	=	↑	=	↑	=	=	
CO1	↑	↑	↑	↑	↑	↑	↑	↑		
CO2	↑		↑	↑	↑	↑		↑	=	
CO3	↑	=	=	↑	=	↑		=	↑	↑
CO4	↑	↑	↑	↑	↑	↑	↑	=	=	

Fig. 5. **Summary of the Intraday variation of oxidative stress markers in different subjects (P1-P10).** Assays for lipid peroxides and carbonyl groups were measured in the same individuals three times daily on four particular days (LP1-4 and CO1-4) over a period of 51 days. The arrows show the intraday changes for each marker.

Also, it is not known whether oxidative stress increases and then the antioxidant defenses decrease as a consequence of consumption of non-enzymatic protective substances, or if the levels of antioxidants diminish first and the consequence is an elevation of oxidative stress. In any case, maybe the important item is the possible delay that can take place between the peak of oxidative stress and the participation of antioxidants, that imbalance being what cumulatively damages the cells. According to this, if the oxidative stress of an individual increases throughout the day but at the same time the subject has enough level of antioxidants, nothing would happen. But if the increase was not neutralized by the antioxidant systems, the damage to biomolecules would be possible. In fact, it has been suggested, in a new theory of longevity, that the main factor determining lifespan is not the rate of free radical production, but the cell's ability to resist short-term fluctuations in critical metabolites caused by environmental stress (Olshansky and Rattan, 2005). Besides the importance of the total antioxidant defenses, it would be important to know which compounds are the first line of scavenging antioxidant defense in neutralizing the several peaks of oxidative stress that may occurs daily and whether these compounds are the same during the 24 h-period.

As to a hypothetical intervention, ideally, it should be carried out before the increase of oxidative stress. Although we do not know the time lag between damage and increase of the markers in the serum and because for many subjects oxidative stress increases at the end of the day, it seems reasonable that the administration of the antioxidant should be distributed throughout the day instead of being administered once a day.

5. Conclusion

As a conclusion, we can say that a more comprehensive study using more biomarkers should be performed to select those whose serum levels reflect the oxidative damage in tissues. Also, It is clear that more than a single measurement will be required to establish the status of oxidative stress of individuals along with a study of lifestyle factors. In this way, a customized supplementation strategy can be recommended. Since these values present a great variability, it would be necessary to think about whether the determination of these markers could require the establishment of some indications of lifestyles previous to the assay. The observed variability of the markers does not limit their usefulness in studying the effect of intervention strategies. If the level of oxidative stress varies widely within, and between days, the markers can be useful in assessing the influence of an intervention on minimizing the height of the "oxidative stress peaks". Work is underway to investigate whether variability affects other markers. In addition, it would be interesting to try and establish a relationship between the concentration of the markers and habits because the most important factor in determining the levels of oxidative stress markers still remains to be known. Obviously, this study would require a larger population.

6. Acknowledgements

This work was supported by Spanish Ministerio de Ciencia e Innovación, BFU2010-20882.

7. References

Arguelles,S., Garcia,S., Maldonado,M., Machado,A., and Ayala,A. (2004). Do the serum oxidative stress biomarkers provide a reasonable index of the general oxidative stress status? *Biochim. Biophys. Acta*, 1674, pp. 251-259.

Arguelles,S., Gomez,A., Machado,A., and Ayala,A. (2007). A preliminary analysis of within-subject variation in human serum oxidative stress parameters as a function of time. *Rejuvenation. Res.*, 10, pp. 621-636.

Ayala,A. and Cutler,R.G. (1997). Preferential use of less toxic detoxification pathways by long-lived species. *Arch. Gerontol. Geriatr.*, 24, pp. 87-102.

Butler,R.N., Sprott,R., Warner,H., Bland,J., Feuers,R., Forster,M., Fillit,H., Harman,S.M., Hewitt,M., Hyman,M., Johnson,K., Kligman,E., McClearn,G., Nelson,J., Richardson,A., Sonntag,W., Weindruch,R., and Wolf,N. (2004). Biomarkers of aging: from primitive organisms to humans. *J. Gerontol. A Biol. Sci. Med. Sci.*, 59, pp. B560-B567.

Cutler,R.G. (2005). Oxidative stress profiling: part I. Its potential importance in the optimization of human health. *Ann. N. Y. Acad. Sci.*, 1055, pp. 93-135.

Cutler,R.G., Plummer,J., Chowdhury,K., and Heward,C. (2005). Oxidative stress profiling: part II. Theory, technology, and practice. *Ann. N. Y. Acad. Sci.*, 1055, pp. 136-158.

Finkel,T. and Holbrook,N.J. (2000). Oxidants, oxidative stress and the biology of ageing. *Nature*, 408, pp. 239-247.

Gutteridge,J.M. and Halliwell,B. (2000). Free radicals and antioxidants in the year 2000. A historical look to the future. *Ann. N. Y. Acad. Sci.*, 899, pp. 136-147.

Halliwell,B. (2002). Effect of diet on cancer development: is oxidative DNA damage a biomarker? *Free Radic. Biol. Med.*, 32, pp. 968-974.

Halliwell,B. (2006). Oxidative stress and neurodegeneration: where are we now? *J. Neurochem.*, 97, pp. 1634-1658.

Halliwell,B., Gutteridge,J.M., and Cross,C.E. (1992). Free radicals, antioxidants, and human disease: where are we now? *J. Lab Clin. Med.*, 119, pp. 598-620.

Harman,D. (1956). Aging: a theory based on free radical and radiation chemistry. *J. Gerontol.*, 11, pp. 298-300.

Johnson,T.E. (2006). Recent results: biomarkers of aging. *Exp. Gerontol.*, 41, pp. 1243-1246.

Kirkwood,T.B. (2005). Understanding the odd science of aging. *Cell*, 120, pp. 437-447.

Lesgards,J.F., Durand,P., Lassarre,M., Stocker,P., Lesgards,G., Lanteaume,A., Prost,M., and Lehucher-Michel,M.P. (2002). Assessment of lifestyle effects on the overall antioxidant capacity of healthy subjects. *Environ. Health Perspect.*, 110, pp. 479-486.

Levine,R.L. (2002). Carbonyl modified proteins in cellular regulation, aging, and disease. *Free Radic. Biol. Med.*, 32, pp. 790-796.

McCord,J.M. (2000). The evolution of free radicals and oxidative stress. *Am. J. Med.*, 108, pp. 652-659.

Moller,P., Wallin,H., and Knudsen,L.E. (1996). Oxidative stress associated with exercise, psychological stress and life-style factors. *Chem. Biol. Interact.*, 102, pp. 17-36.

Olshansky,S.J. and Rattan,S.I. (2005). At the heart of aging: is it metabolic rate or stability? *Biogerontology.*, 6, pp. 291-295.

Porter,T.D. and Coon,M.J. (1991). Cytochrome P-450. Multiplicity of isoforms, substrates, and catalytic and regulatory mechanisms. *J. Biol. Chem.*, 266, pp. 13469-13472.

Reddy, A.B., Karp, N.A., Maywood, E.S., Sage, E.A., Deery, M., O'Neill, J.S., Wong, G.K.,
    Chesham, J., Odell, M., Lilley, K.S., Kyriacou, C.P., and Hastings, M.H. (2006).
    Circadian orchestration of the hepatic proteome. *Curr. Biol.*, 16, pp. 1107-1115.
Sies,H. (1997). Oxidative stress: oxidants and antioxidants. *Exp. Physiol*, 82, pp. 291-295.

7

Heme Proteins, Heme Oxygenase-1 and Oxidative Stress

Hiroshi Morimatsu, Toru Takahashi, Hiroko Shimizu,
Junya Matsumi, Junko Kosaka and Kiyoshi Morita
Department of Anesthesiology and Resuscitology,
Okayama University Hospital,
Japan

1. Introduction

Heme is ferrous protoporphyrin-IX that is the prosthetic group of hemoproteins, such as hemoglobin, myoglobin and cytochromes that are of vital importance. In contrast, "free heme", a protein-unbound heme, that is either just synthesized but yet not incorporated into hemoproteins, or that is released from hemoprotein under oxidative conditions, is highly toxic, since it catalyzes the production of reactive oxygen species (ROS). Thus, heme proteins and free heme have an important relationship with oxidative stress.

In order to cope with this problem, the body is equipped with various defense mechanism(s) against an excessive amount of "free heme" concentrations. Heme oxygenase (HO) is one of the key players in the defense mechanism, and plays a fundamental role against the free-heme mediated oxidative process. The rate-limiting enzyme in heme catabolism, heme oxygenase-1 (HO-1), is induced by not only its substrate heme but also oxidative stress resulting from I/R injury. Heme oxygenase-1 induction leads to increased heme breakdown, resulting in the production of iron, carbon monoxide (CO), and biliverdin IXα, which is subsequently reduced to bilirubin IXα by biliverdin reductase.

Recently, large numbers of reports including ours have emerged suggesting heme proteins, HO, and its substrates such as CO, biliverdin IXα, and bilirubin IXα play important roles in pathophysiology and therapeutic implications. Here we summurize these evidences to clarify the relationship among heme proteins, HO-1, and oxidative stress.

1.1 Synthesis and degradation of heme protein 1

Heme is the prosthetic group of all heme proteins such as hemoglobin, myoglobin, cytochrome, catalase, peroxidases, nitric oxide synthase, prostaglandin synthase, and certain transcription factors. Heme is an essential molecule in all aerobic cells and plays a crucial role in physiological, pharmacological, and toxicological reactions, as well as cell differentiation and other functions. However, free heme, namely protein-unbound heme, can be toxic to cells because it results in the production of reactive oxygen species and causes cell damage (Kumar and Bandyopadhyay, 2005). To guard against this toxicity, heme levels are tightly controlled between heme biosynthesis and catabolism (Sassa, 2006).

1.1.1 Heme synthesis

The initial biosynthesis of one molecule of heme requires eight molecules of glycine and eight molecules of succinyl CoA to produce 5-aminolevulinic acid (ALA) (Sun et al., 2002) by 5-aminolevulinic acid synthase (ALAS) in mitochondria. There are two forms of ALAS, a non-tissue-specific ALAS (ALAS1) and an erythroid cell-specific ALAS (ALAS2) (Bishop et al., 1990). In the liver, heme represses the synthesis of ALAS1 mRNA at both transcriptional and translational levels (Hamilton et al., 1991) and inhibits its transfer from the cytosol into mitochondria (Ades and Harpe, 1981). In erythroid cells, heme does not inhibit ALAS2 synthesis (Sassa and Nagai, 1996) and ALAS2 activity (Ponka, 1997).

Following synthesis, mitochondrial ALA is transported to the cytosol, where ALA dehydratase (ALAD) dimerizes two molecules of ALA to produce the pyrrole ring compound porphobilinogen (PBG). The next step in the pathway involves the head-to-tail condensation of four moleclues of PBG to produce the linear tetrapyrrole intermediate hydroxymethylbilane (HMB). The enzyme for this condensation is porphobilinogen deaminase (PBG deaminase), also called hydroxymethylbilane synthase or uroporphyrinogen I synthase. Uroporphyrinogen-III synthase catalyses HMB to uroporphyrinogen III. In the absence of uroporphyrinogen-III synthase, HMB may non-enzymatically close to form uroporphyrinogen I, which cannot convert to heme.

In the next step, the acetate substituents of uroporphyrinogen III or I are all decarboxylated by the uroporphyrinogen decarboxylase in the cytosol. The resultant products are known as coproporphyrinogens, with coproporphyrinogen III being the important normal intermediate in heme synthesis.

Coproporphophyrinogen III is transported into mitochondria and is catalyzed to protoporphyrinogen IX by coproporphyrinogen oxidase. Protoporphyrinogen oxidase oxidizes protoporphyrinogen IX to protoporphyrin IX by the removal of six hydrogen atoms. Finally, ferrous iron (Fe^{2+}) is inserted into protoporphyrin IX to form heme in a reaction catalysed by ferrochelatase.

1.1.2 Heme degradation

Heme degradation starts with the reductive breakdown of the heme into carbon monoxide (CO), iron (Fe), and biliverdin in a reaction catalyzed by heme oxygenase (HO) (Tenhunen et al., 1968). Heme oxygenase exists in two isoforms, HO-1, which is inducible by heme, its substrate, and HO-2, which is constitutive and non-inducible (Shibahara et al., 1985). Heme oxygenase-1 is also known as heat shock protein 32 (Keyse and Tyrrell, 1989), as well as an acute phase reactant, and it is inducible by stressors including cytokines, heavy metals, hypoxia, and oxygen free radicals. This is the only reaction in the body that is known to produce CO. Most of the CO is excreted through the lungs, so that the CO content of expired air is a direct measure of the activity of heme oxygenase. Biliverdin is subsequently converted into bilirubin by an NAD(P)H-requiring cytosolic enzyme, biliverdin reductase

(Tenhunen et al., 1969). Bilirubin is conjugated with glucronic acid to form a more soluble bilirubin glucuronide, which is excreted in bile.

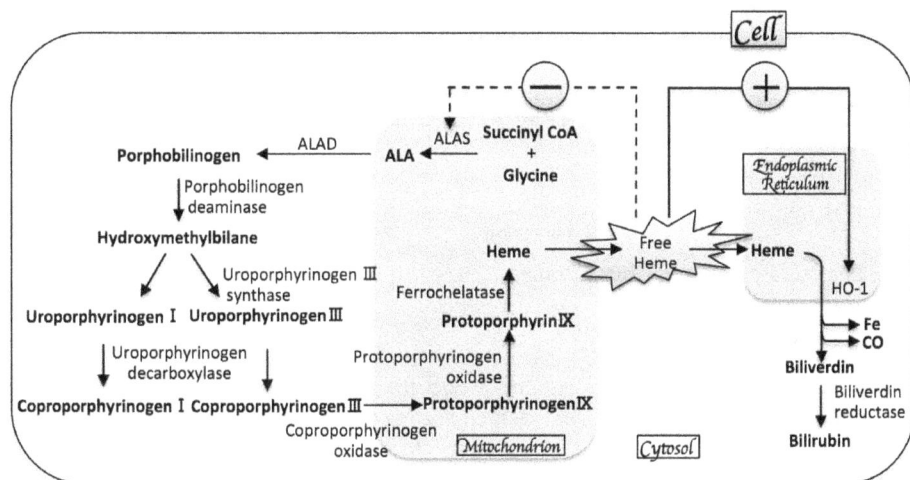

Fig. 1. Heme Metabolic Pathway

1.1.3 The regulatory effects of free heme

Free heme at low concentrations plays a beneficial regulatory role on various cellular functions. A heme concentration greater than 1 μM can be toxic to almost all cells because it catalyzes the production of reactive oxygen species (Halliwell and Gutteridge, 1990). At submicrosomal concentrations, heme is involved in regulator gene expression or repression of heme metabolism.

Heme concentrations of less than 10^{-13} M induce the synthesis of ALAS1. Repression of ALAS1 synthesis in the liver takes place at free heme concentrations of 0.1-0.3 μM, leading to decreased heme synthesis, and at 0.4-1.0 μM, HO-1 is induced in cultured chick embryo liver cells (Granick et al., 1975). In 1996, Igarashi reported two novel transcription factors, Bach1 and Bach2, as heterodimerization partners of MafK (Oyake et al., 1996). In the early 2000s, it was reported that the mammalian transcription factor Bach1, a repressor of HO-1 gene activation (Sun et al., 2002), binds with an equimolar amount of hemin (Ogawa et al., 2001). Inhibition by free heme of the DNA binding activity of Bach1 occurred at around 0.03 μM, and at 1 μM, it almost completely inhibited the DNA-binding activity of Bach1 in vitro (Ogawa et al., 2001). In heme oxygenase deficiency, hemin applied at 50 μM to the patient's plasma resulted in increased free radical generation, which was abnormal and caused varied tissue damage (Poss and Tonegawa, 1997). The products of heme degradation, CO, iron Fe, and biliverdin, contribute to cellular protection in various situations (Sassa, 2006). Bilirubin is considered a potentially important anti-oxidant and cytoprotector of physiological significance (Stocker et al., 1987) (Gopinathan et al., 1994) (Hopkins et al., 1996). Thus, heme levels are tightly controlled between heme biosynthesis and catabolism.

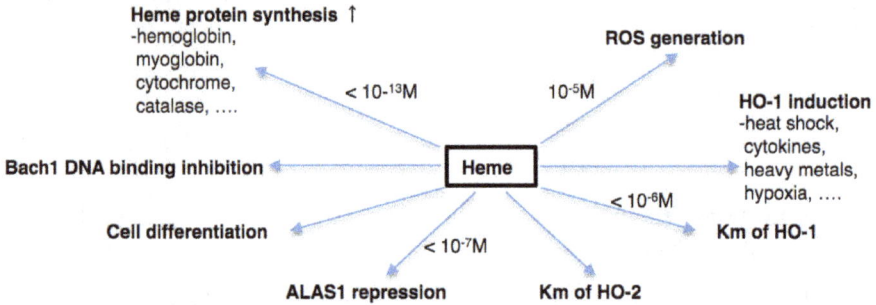

Fig. 2. The regulatory effects of free heme

2. Heme proteins as oxidants

While heme is required as the prosthetic group for heme proteins such as hemoglobin, myoglobin, and cytochrome P 450, etc., which are necessary for cellular viability, an excess amount of free heme is highly toxic to cells due to its pro-oxidant activity, driven by the divalent Fe atom contained within its protoporphyrin IX ring, which can promote the production of free radicals via Fenton chemistry (Sassa, 2006). Free heme is also highly lipophilic and readily intercalates into the lipid bilayer of adjacent cells, and it results in oxidative damage of the cytoskeleton. Furthermore, free heme that is released from methemoglobin can catalyze the oxidation of low density lipoprotein, which in turn induces lipid peroxide formation and results in endothelial cytolysis (Jeney et al., 2002).

2.1 Exacerbation of oxidative tissue injury by free heme

We have demonstrated that free heme released from heme protein plays a critical role in the development of oxidative tissue injuries by accelerating the production of reactive oxygen species (ROS) in various experimental models of oxidative tissue injuries (Takahashi et al., 2007). For instance, ROS generated by reperfusion of the kidney has been implicated in the pathogenesis of ischemic renal injury. Thus, we determined the level of microsomal heme and the gene expression of ALAS1 in the kidney following ischemia/reperfusion in rats (Shimizu et al., 2000). We found that, prior to HO-1 induction, there was a rapid and significant increase in microsomal heme concentration, which was followed by the inhibition of ALAS1 gene expression. These findings indicate that free heme concentration in the kidney increases rapidly following ischemia/reperfusion. We also found that inhibition of HO activity by tin-mesoporphyrin, a specific competitive inhibitor of HO activity, resulted in a marked increase in microsomal heme content and in the aggravation of ischemic renal injury (Shimizu et al., 2000). Thus, an enhanced and sustained increase in intracellular free heme concentration derived from cytochrome P450, a major heme protein in the kidney, may likely exacerbate the oxidative tissue injury in the kidney caused by renal ischemia/reperfusion.

2.1.2 Activation of the innate immune response by free heme

Recent studies also indicate that free heme is involved in the activation of innate immunity, which can lead to oxidative tissue injury. Exposure of endothelial cells to hemin, an oxidized form of heme that is available as a chemical, stimulates the expression of adhesion molecules such as ICAM-1, VCAM-1, and E-selectin (Wagner et al., 1997). Hemin also induces neutrophil migration *in vivo* and *in vitro*, triggers the oxidative burst, promotes cytoskeleton reorganization, and activates interleukin-8 expression in human neutrophils (Graca-Souza et al., 2002). Heme also induce TNF-α secretion by mouse peritoneal macrophages in a manner dependent on MyD88, toll-like receptor (TLR) 4, and CD14, although heme signaling through TLR4 depends on an interaction distinct from that established between TLR4 and LPS (Figueiredo et al., 2007). Moreover, free heme induces apoptotic cell death in response to pro-inflammatory agonists, as demonstrated for tumor necrosis factor (Seixas et al., 2009). Severe sepsis can develop from excessive systemic inflammatory responses to microbial infection, leading to oxidative tissue injury that ultimately results in death. Very recently, the circulating free heme released from hemoglobin during infection has been shown to contribute to the pathogenesis of severe sepsis (Larsen et al., 2010). Heme administration after low-grade polymicrobial infection induced by cecal ligation and puncture in mice promoted tissue damage and severe sepsis. Development of lethal forms of severe sepsis after high-grade infection was associated with the increase in plasma free heme concentration derived from cell-free hemoglobin and the decrease in serum concentrations of the heme sequestering protein hemopexin (HPX), whereas HPX administration after high-grade infection prevented tissue damage and lethal outcomes. Moreover, fatal septic shock in patients was associated with reduced serum HPX concentrations, suggesting that targeting free heme by HPX might be used therapeutically to prevent lethal outcomes associated with severe sepsis.

2.2 Role of HO-1 in oxidative tissue injury (liver disease and sepsis)

Oxidative stresses such as inflammation, as well as ischemia and reperfusion (I/R), injure several tissues. It has been suggested that HO-1 plays a cytoprotective role against oxidative stresses. The cytoprotective role of HO-1 influences both acute and chronic illnesses. In this chapter, we evaluate the role of HO-1 in protection against oxidative stresses at acute illnesses, mainly liver disease and sepsis.

2.2.1 Animal studies

In animal models, several reports have demonstrated the protective effect of HO-1. In the carbon tetrachloride-induced hepatotoxicity model, HO-1 expression is increased both at transcriptional and protein levels in hepatocytes. Inhibition of HO activity by tin-mesoporphyrin (Sn-MP) results in sustained liver injury, as revealed by marked increases in serum alanine transaminase (ALT), hepatic malondialdehyde formation, tumor necrosis factor-alpha (TNF-α) mRNA, inducible nitric oxide synthase (iNOS) mRNA, and DNA-binding activity of nuclear factor-kappaB (NF-κB), as well as inflammatory changes of hepatocytes (Nakahira et al., 2003). In contrast, induction of HO-1 by recombinant human interleukin-11 (rhIL-11) leads to reduced liver injury. (Kawakami et al., 2006) In the I/R liver injury model, rats pretreated with a HO-1 inducer showed greater increases in HO-1 transcriptional and protein expressions, less elevated serum ALT levels, and less increased serum TNF-α and iNOS protein and mRNA expressions than those treated with a HO-1 inhibitor. These results indicated that HO-1 overexpression protected liver against I/R injury by modulating oxidative stress and proinflammatory mediators (Yun et al., 2010). In sepsis models, lipopolysacchalide (LPS) treatment increases HO-1 at transcriptional and protein levels and decreases nonspecific delta-aminolevulinate synthase (ALAS-1), which are the rate-limiting enzymes of heme catabolism and biosynthesis, gene expression in the duodenum and the jejunum. Inhibition of HO activity by Sn-MP produces significant tissue injury (Fujii et al., 2003). LPS also induces hepatic injury as revealed by increases in serum ALT and aspartate transaminase (AST) activities, TNF-α mRNA, iNOS mRNA, and DNA-binding activity of NF-κB, and extensive hepatocyte necrosis. However, induction of HO-1 by rhIL-11 ameliorated the LPS-induced hepatic injury and decreased LPS-induced mortality (Maeshima et al., 2004) In an animal model, the cytoprotective effects of HO-1 against oxidative stress were also shown in other organs (Maeshima et al., 2005, Barreiro et al., 2002, Shimizu et al., 2000, Poole et al., 2005, Yu et al., 2009).

2.2.2 Human studies

In humans, there are some reports indicating the protective effect of HO-1. Patients with acute liver failure show increased HO-1 and decreased ALAS-1. These may indicate an increase in free heme concentration, resulting in altered heme metabolism and liver function (Fujii et al., 2004). In liver transplantation, which induces oxidative stress through I/R, a donor HO-1 genotype that modulates HO-1 induction levels is associated with outcomes, such as serum ALT and AST levels and early graft survival. This result suggests that HO-1 mediates graft survival after liver transplantation (Buis et al., 2002). In sepsis and septic shock, patients who fulfilled the criteria for severe sepsis or septic shock showed high HO-1 gene expression, and there was a positive correlation between survival and increased HO-1 concentration (Takaki et al., 2010). Patients who fulfilled the criteria for severe systemic inflammatory response syndrome and had a serum C-reactive protein level >10 mg/dL showed high HO-1 expression and serum TNF-α levels. (Mohri et al., 2006) These results indicate the relationship between inflammation and HO-1. A patient with HO-1 deficiency showed growth retardation, anemia, leukocytosis, thrombocytosis, coagulation abnormalities, elevated serum levels of haptoglobin, ferritin, and heme, a low serum bilirubin concentration, and hyperlipidemia; the patient died in childhood (Kawashima et al., 2002) This case directly shows the importance of HO-1 in homeostasis.

In summary, similar results in animal models and humans have shown the cytoprotective effect of HO-1 against oxidative stresses. The complete mechanisms related to the cytoprotective effect of HO-1 against oxidative stresses are still unknown, but several mechanisms may be involved. The major mechanism may be the removal of free heme. In oxidative stress, free heme is increased with the breakdown of hemoproteins such as hemoglobin, myoglobin, or cytochrome P450. Free heme induces the production of reactive oxygen species and low-density lipoprotein oxidation, which injures endothelial cells (Sassa, 2006). Another major mechanism may be the anti-oxidative effect of carbon monoxide and biliverdin, which are produced in heme catabolism. The detailed mechanisms related to the anti-oxidative effects of carbon monoxide and biliverdin are described in other chapters. One of the other possible mechanisms is the decrease of cytotoxic cytokines. HO-1 may affect many pathways and cytokines. For example, HO-1 inhibits macrophage activation, which triggers the inflammatory response in response to stress. In the liver, Kupffer cells, which are liver macrophages, play an important role for these reactions, such as production of TNF-α and IL-6. HO-1 inhibits the production of these cytokines by Kupffer cells and ameliorates liver damage (Babu et al., 2007, Zhong et al., 2010, Devey et al., 2009). In addition to macrophage activation, there may also be many other mechanisms, such as inactivation of the p38 mitogen-activated protein kinase pathway, which leads to a preventive effect by diminishing neutrophil infiltration. (Carchman et al., 2011, Lin et al., 2010).

In conclusion, even though the detailed mechanisms are unknown, HO-1 is one of the essential enzymes acting against oxidative stress, and its cytoprotective effect operates in many organs and probably affects patients' outcomes. More investigations into the detailed role and mechanisms of HO-1 are needed.

2.3 Bilirubin as an antioxidant

Bilirubin has been recognized as a marker of liver injury, specifically biliary obstruction. It is also well known that biliverdin is one of the metabolites catalyzed by heme oxygenase from heme proteins, and it is catalyzed by biliverdin reductase to bilirubin. An increased serum bilirubin concentration is seen as a sign of dysfunction in the hepato-billiary system or in heme protein metabolism. Free unconjugated bilirubin (UCB) can easily enter cells by passive diffusion and cause toxicity. UCB binds to discrete brain areas, such as the basal ganglia (kernicterus), and produces a wide array of neurological deficits collectively known as bilirubin encephalopathy (Shapiro, 2003). However, in 1987, it was noted that bilirubin has strong antioxidant potential *in vitro* (Stocker et al.,1987). In this study, bilirubin under 2% oxygen in liposomes had a stronger antioxidant potential than α-tocopherol known to date as the most potent protector against lipid peroxidation. This result showed that endogenous bile pigment production activated by elevated HO activity could confer antioxidative protection to cells and tissues. In another study, the potent physiologic antioxidant actions of bilirubin were reported to involve a redox cycle between bilirubin and biliverdin (Baranano et al., 2002). When bilirubin acted as an antioxidant, it was itself oxidized to biliverdin and then recycled by biliverdin reductase back to bilirubin.

2.3.1 The antioxidant and cytoprotective effects of bilirubin in animal studies

In several animal models, the antioxidant potential and cytoprotective effect of bilirubin were also reported. In an I/R heart injury model, HO-1 and bilirubin showed a protective effect with respect to postischemic myocardial performance and reduced infarct size and mitochondrial dysfunction (Clark et al., 2000). In experimental small intestinal I/R injury, bilirubin had a dose-dependent protective effect by preventing lipid peroxidation (Ceran et al., 2001). In this study, bilirubin infusion reduced the severity of postischemic intestinal injury and increased tissue malondialdehyde (MDA) levels. Malondialdehyde is a product of lipid peroxidation. Moreover, exogenous bilirubin infusion provided tissue protection in other models of hepatic (Kato et al., 2003) and renal (Adin et al., 2005) I/R injury. In an OVA-induced asthma model, the application of bilirubin inhibited airway inflammation and lung leukocyte influx (Keshavan et al., 2005). Bilirubin also inhibited vascular cell adhesion molecule 1 (VCAM-1)-mediated transendothelial lymphocyte migration *in vitro*. The authors suggested that bilirubin inhibited the cellular production of ROS in responce to VCAM-1 stimulation as an antioxidant. Furthermore, rats rendered hyperbilirubinemic by infusion of bilirubin were relatively resistant to bleomycin-induced lung injury (Wang et al., 2002). Intravenous infusion of bilirubin reduced lung fibrotic lesions and local infiltrations of inflammatory cells in histologic studies, as well as reduced levels of transforming growth factor-β (TGF-β) in the bronchoalveolar lavage fluid.

2.3.2 The relationship between serum bilirubin levels and the risk of general diseases

In several studies, mild to moderately elevated serum bilirubin levels were effective in the prevention of general diseases related to oxidative stress in humans (Ryter et al., 2007). For example, some clinical studies have indicated correlations between the serum bilirubin level and the risk of cardiovascular disease. For coronary artery disease (CAD), the relationship between serum bilirubin levels and the risk was investigated (Schwertner et al., 1994). In their study, the total bilirubin level was inversely related to the incidence of CAD independently. In the Framingham offspring study (large scale cohort study, n=5124), the relationship between serum bilirubin and myocardial infarction, coronary death, and any cardiovascular event was assessed (Djousse et al., 2001). Participants were divided into five groups by serum bilirubin level and compared. It was found that higher serum total bilirubin levels were associated with a lower risk of cardiovascular disease in men. Moreover, middle-aged patients with Gilbert syndrome (with serum bilirubin levels in the range of 20-70 µmol/l) had a lower incidence of ischemic heart disease (IHD) than healthy patients (Vitek et al., 2002). In this study, the authors referred to the total antioxidant potential of UCB. They concluded that the beneficial effect of UCB on the prevention of IHD might be important, in addition to HDL cholesterol.

The serum bilirubin level was shown to be associated with respiratory disease (Temme et al., 2001, Horsfall et al., 2011). In two studies, the relationship between serum bilirubin level and respiratory disease was examined. They reported that the serum bilirubin level was inversely correlated with the incidence of respiratory disease (lung cancer, chronic obstructive pulmonary disease) and all-cause mortality.

In conclusion, bilirubin has a strong antioxidant potential and cytoprotective effect *in vitro* and *in vivo*. The antioxidant potential of bilirubin involves a redox cycle between bilirubin and biliverdin. An elevated serum bilirubin level is associated with the incidence and the mortality of several diseases induced by oxidative stress. However, hyperbilirubinemia causes brain damage in infants and neonates. Thus, further investigations of the antioxidative and cytoprotective mechanisms of bilirubin are needed.

2.4 Carbon monoxide as an indicator of oxidative stress

Carbon monoxide (CO) is also one of the metabolites of heme proteins. It is well known that CO is a toxic gas and is used as an indicator of air pollution. Recent studies suggest that CO inhalation in very low concentration would be a therapeutic option in experimental models of sepsis, transplantation, and ischemia/reperfusion. Currently, CO concentration can be measured using two methods: CO-hemoglobin using a blood gas analyzer, and exhaled CO using a gas sampler. These new measurements will provide us important new information about patient status and underlying mechanisms of disease.

2.4.1 Increased CO concentration in exhaled air of critically ill patients

Zegdi et al. (2000) focused their attention on the exhaled CO concentration of critically ill patients, and they measured CO concentrations using an infrared CO analyzer with a sensitivity of 0.1 ppm (CO 2000, Seres, La Duranne, France). Carbon monoxide was detected in exhaled breath at a higher concentration than in inspired gas, and exhaled CO was constant at the fixed ventilator settings in hemodynamically stable patients. They suggested that the exhaled CO concentration reflects endogenous CO production and might be useful for assessing the condition of critically ill patients. Coincident with their report, Sharte and colleagues measured exhaled CO concentrations in 30 critically ill patients who underwent mechanical ventilation and compared their results to those of 6 healthy non-smoking controls without a recent history of respiratory infections who breathed spontaneously via a mouthpiece connected to a ventilator (Sharte et al., 2000). Critically ill patients showed significantly higher CO concentrations in exhaled air compared to healthy controls. Although they did not find correlations between CO concentrations in exhaled air and carboxyhemoglobin levels in arterial and central venous blood, this might be attributable to technical artifacts in the measurement of carboxyhemoglobin concentrations using an older version of the blood gas analyzer, which has a lower sensitivity. Taken together, they concluded that the increased CO concentration in exhaled air in critically ill patients suggests an induction of inducible HO-1 and might reflect the severity of illness. Since CO is one of the metabolites of heme catabolism, we also examined CO concentrations in exhaled air, carboxyhemoglobin concentrations in arterial blood, and serum levels of bilirubin, another metabolite of heme breakdown, in 29 critically ill patients with signs of systemic inflammation who were all being mechanically ventilated (Morimatsu et al., 2006). Exhaled CO concentrations were also measured in eight healthy volunteers as controls. Exhaled CO concentration was measured using the CO analyzer (CARBOLYZER mBA-2000; TAIYO Instruments, Osaka, Japan). The median exhaled CO concentration was significantly higher in critically ill patients than in controls. Of note, there was a significant correlation between CO and carboxyhemblobin, and between CO and total bilirubin levels. We also compared exhaled CO concentrations between survivors and nonsurvivors. Interestingly, survivors tended to have higher exhaled CO concentrations than nonsurvivors, but the difference was not significant because of the limited sample size, suggesting that the poorer outcome of nonsurvivors may be due to their limited capacity to produce CO or induce HO-1. Collectively, our findings suggest that there may be an increase in heme breakdown in critically ill patients, probably due to systemic oxidative stress.

2.4.2 Increased CO concentration in exhaled air in patients with systemic inflammation/sepsis

Schober et al. (2009) measured end-tidal CO concentrations and arterial CO-Hb concentrations in 20 patients undergoing cardiac surgery with cardiopulmonary bypass (CPB). They measured these indices during surgery at two time points (1 hour after induction and 1 hour after CPB). They compared pre- and post-CPB values and found that both the end-tidal CO and the arterial CO-Hb concentrations were higher post-CPB than pre-CPB. These results indicated that systemic inflammation induced by CPB resulted in oxidative stress and increased CO production. This is likely explained by specific influences of CPB on processes involved in heme degradation, such as HO-1 induction and/or hemolysis. In addition, Zegdi et al. (2002) measured the exhaled CO concentrations in 24 patients with severe sepsis or septic shock who were admitted to a medical intensive care unit and compared them to those of 5 critically ill controls. All patients were mechanically ventilated. They demonstrated for the first time that exhaled CO concentrations were greater in the septic patients than in the control group. When endogenous CO production was specifically calculated as the lung CO excretion rate at a steady state in these patients, significantly higher endogenous CO production was found in patients with severe sepsis during the first three days of treatment than in the control group, although endogenous CO production in the sepsis group decreased over time with treatment. Interestingly, survivors of sepsis had a significantly higher endogenous CO production on day 1 compared to non-survivors.

We summarized recent evidence concerning the increased exhaled CO concentrations and its significance in critically ill patients with systemic inflammation. The exhaled CO concentration could reflect endogenous HO activity and might be a useful parameter of oxidative stress. Further studies are clearly needed to elucidate whether increased endogenous CO production may predict patients' morbidity and mortality. However, techniques for monitoring CO are continuously being refined, and these techniques may eventually find their way into clinicians' offices.

3. Conclusion

In this chapeter, we showed recent evidence concerning the role of free heme in the oxidative tissue injury, and HO-1 induction as a major protective response against the free heme-mediated oxidative tissue injuries, especially focusing on acute liver injuries and septic organ damages. Preinduction of HO-1 by pharmacological modality has been shown to confer significant protection on cells, tissues and organs in these acute inflammatory disorders. We also described a novel non-invasive technology for the measurement of exhaled CO concentrations which reflect endogenous HO activity and might be a useful parameter of disease severity. In addition to the protective role of HO-1, both bile pigments and CO, the two heme metabolites by HO reaction, play critical tissue-protective roles agaisnt oxidative tissue injuries. Although the application of HO-1 and its metabolites to clinical field might be promising, further studies should clarify pending issues such as interspecies, or inter-cell type differences in ho-1 gene expression, and a cause-effect relationship between HO-1 expression and morbidity and mortality of patients.

4. References

Ades, I.Z. & Harpe, K.G. (1981). Biogenesis of mitochondrial proteins. Identification of the mature and precursor forms of the subunit of delta-aminolevulinate synthase from embryonic chick liver. *J Biol Chem*, Vol.256, No.17, (September 1981), pp. 9329-9333, ISSN 0021-9258.

Adin, C. A., Croker, B. P., & Agarwal, A. (2005). Protective effects of exogenous bilirubin on ischemia-reperfusion injury in the isolated, perfused rat kidney. *American journal of physiology. Renal physiology*, Vol.288, No.4, (April 2005), pp. F778-784, ISSN 1522-1466.

Babu, A.N., Damle, S.S., Moore, E.E., Ao, L., Song, Y., Johnson, J.L., Weyant, M., Banerjee, A., Meng, X. & Fullerton, D.A. (2007). Hemoglobin-based oxygen carrier induces hepatic heme oxygenase 1 expression in Kupffer cells. *Surgery*, Vol.142, No.2, (August 2007), pp.289-294, ISSN 0039-6060.

Baranano, D. E., Mahil, R., Christopher, D. F., & Solomon, H. S. (2002). Biliverdin reductase: a major physiologic cytoprotectant. *Proceedings of the National Academy of Sciences of the United States of America,* Vol.99, No.25, (December 2002), pp. 16093-16098, ISSN 0027-8424.

Barreiro, E., Comtois, A.S., Mohammed, S., Lands, L.C. & Hussain, S.N. (2002). Role of heme oxygenases in sepsis-induced diaphragmatic contractile dysfunction and oxidative stress. *Am J Physiol Lung Cell Mol Physiol,* Vol.283, No.2, (August 2002), pp.L476-484, ISSN 1040-0605.

Bishop, D.F., Henderson, A.S., & Astrin, K.H. (1990). Human delta-aminolevulinate synthase: assignment of the housekeeping gene to 3p21 and the erythroid-specific gene to the X chromosome. *Genomics,* Vol.7, No.2, (June 1990), pp. 207-214, ISSN 0888-7543.

Buis, C.I., van der Steege, G., Visser, D.S., Nolte, I.M., Hepkema, B.G., Nijsten, M., Slooff, M.J. & Porte, R. J. (2008). Heme oxygenase-1 genotype of the donor is associated with graft survival after liver transplantation. *Am J Transplant,* Vol.8, No.2, (February 2008), pp.377-85, ISSN 1600-6135.

Carchman, E. H., Rao, J., Loughran, P.A., Rosengart, M.R. & Zuckerbraun, B.S. Heme oxygenase-1-mediated autophagy protects against hepatocyte cell death and hepatic injury from infection/sepsis in mice. *Hepatology,* Vol.53, No.6, (June 2011), pp.2053-2062, ISSN 1600-6135.

Ceran, C., Sönmez, K., Türkyllmaz, Z., Demirogullarl, B., Dursun, A., Düzgün, E., Başaklar, A. C., & Kale, N. (2001). Effect of bilirubin in ischemia/reperfusion injury on rat small intestine. *Journal of pediatric surgery,* Vol.36, No.12, (December 2001), pp. 1764-1767, ISSN 0022-3468.

Clark, J. E., Foresti, R., Sarathchandra, P., Kaur, H., Green, C. J., & Motterlini, R. (2000). Heme oxygenase-1-derived bilirubin ameliorates postischemic myocardial dysfunction. *American journal of physiology. Heart and circulatory physiology,* Vol.278, No.2, (February 2000), pp. H643-651, ISSN 0363-6135.

Devey, L., Ferenbach, D., Mohr, E., Sangster, K., Bellamy, C.O., Hughes, J. & Wigmore, S.J. (2009). Tissue-resident macrophages protect the liver from ischemia reperfusion injury via a heme oxygenase-1-dependent mechanism. *Mol Ther,* Vol.17, No.1, (January 2009), pp.65-72, ISSN 1525-0016.

Djoussé, L., Levy, D., Cupples, L. A., Evans, J. C., D'Agostino, R. B., & Ellison, R. C. (2001). Total serum bilirubin and risk of cardiovascular disease in the Framingham offspring study. *The American journal of cardiology,* Vol.87, No.10, (May 2001), pp. 1196-1200, ISSN 0002-9149.

Figueiredo, R.T., Fernandez, P.L., Mourao-Sa, D.S., Porto, B.N., Dutra, F.F., Alves, L.S., Oliveira, M.F., Oliveira, P.L., Graca-Souza, A.V., & Bozza, M.T. (2007). Characterization of heme as activator of Toll-like receptor 4. *J Biol Chem,* Vol.282, No.28, (July 2007), pp.20221-20229, ISSN 0021-9258

Fujii, H., Takahashi, T., Nakahira, K., Uehara, K., Shimizu, H., Matsumi, M., Morita, K., Hirakawa, M., Akagi, R. & Sassa, S. (2003). Protective role of heme oxygenase-1 in the intestinal tissue injury in an experimental model of sepsis. *Critical Care Medicine,* Vol.31, No.3, (March 2003), pp.893-902, ISSN 0090-3493.

Fujii, H., Takahashi, T., Matsumi, M., Kaku, R., Shimizu, H., Yokoyama, M., Ohmori, E., Yagi, T., Sadamori, H., Tanaka, N., Akagi, R. & Morita, K. (2004). Increased heme oxygenase-1 and decreased delta-aminolevulinate synthase expression in the liver of patients with acute liver failure. *Int J Mol Med,* Vol.14, No.6, (December 2004), pp.1001-1005, ISSN 1107-3756.

Gopinathan, V., Miller, N.J., Milner, A.D. & Rice-Evans, C.A. (1994). Bilirubin and ascorbate antioxidant activity in neonatal plasma. *FEBS Lett,* Vol.349, No.2, (August 1994), pp. 197-200, ISSN 0014-5793.

Graca-Souza, A.V., Arruda, M.A., de Freitas, M.S., Barja-Fidalgo, C. & Oliveira, P.L. Neutrophil activation by heme:

implications for inflammatory processes. *Blood,* Vol.99, No.11, (June 2002), pp.4160-4165, ISSN 0006-4971.

Granick, S., Sinclair, P., Sassa, S. & Grieninger, G. (1975). Effects by heme, insulin, and serum albumin on heme and protein synthesis in chick embryo liver cells cultured in a chemically defined medium, and a spectrofluorometric assay for porphyrin composition. *J Biol Chem,* Vol.250, No.24, (December 1975), pp. 9215-9225, ISSN 0021-9258.

Halliwell, B. & Gutteridge, J.M. (1990). Role of free radicals and catalytic metal ions in human disease: an overview. *Methods Enzymol,* Vol.186, (January 1990), pp. 1-85, ISSN 0076-6879.

Hamilton, J.W., Bement, W.J., Sinclair, P.R., Sinclair, J.F., Alcedo, J.A. & Wetterhahn, K.E. (1991). Heme regulates hepatic 5-aminolevulinate synthase mRNA expression by decreasing mRNA half-life and not by altering its rate of transcription. *Arch Biochem Biophys,* Vol.289, No.2, (September 1991), pp. 387-392, ISSN 0003-9861.

Hopkins, P.N., Wu, L.L., Hunt, S.C., James, B.C., Vincent, G.M. & Williams, R.R. (1996). Higher serum bilirubin is associated with decreased risk for early familial coronary artery disease. *Arterioscler Thromb Vasc Biol,* Vol.16, No.2, (February 1996), pp. 250-255, ISSN 1079-5642.

Horsfall, L. J., Rait, G., Walters, K., Swallow, D. M., Pereira, S. P., Nazareth, I., & Petersen, I. (2011). Serum bilirubin and risk of respiratory disease and death. *JAMA : the journal of the American Medical Association.* Vol.305, No.7, (February 2011), pp. 691-697, ISSN 0098-7484.

Jeney, V., Balla, J., Yachie, A., Varga, Z., Vercellotti, G.M., Eaton, J.W. & Balla, G. Pro-oxidant and cytotoxic effects of circulating heme. *Blood,* Vol.100, No.3, (August 2002), pp.879-887, ISSN 0006-4971.

Kato, Y., Shimazu, M., Kondo, M., Uchida, K., Kumamoto, Y., Wakabayashi, G., Kitajima, M., & Suematsu, M. (2003). Bilirubin rinse: A simple protectant against the rat liver graft injury mimicking heme oxygenase-1 preconditioning. *Hepatology (Baltimore, Md.),* Vol. 38, No.2, (August 2003), pp. 364-373, ISSN 0270-9139.

Kawakami, T., Takahashi, T., Shimizu, H., Nakahira, K., Takeuchi, M., Katayama, H., Yokoyama, M., Morita, K., Akagi, R. & Sassa, S. (2006). Highly liver-specific heme oxygenase-1 induction by interleukin-11 prevents carbon tetrachloride-induced hepatotoxicity. *Int J Mol Med,* Vol.18, No.4, (October 2006), pp.537-46, ISSN 1107-3756.

Kawashima, A., Oda, Y., Yachie, A., Koizumi, S. & Nakanishi, I. (2002). Heme oxygenase-1 deficiency: the first autopsy case. *Hum Pathol,* Vol.33, No.1, (January 2002), pp.125-30, ISSN 0046-8177.

Keshavan, P., Deem, T. L., Schwemberger, S. J., Babcock, G. F., Cook-Mills, J. M., & Zucker, S. D. (2005). Unconjugated bilirubin inhibits VCAM-1-mediated transendothelial leukocyte migration. *Journal of immunology (Baltimore, Md.: 1950)*, Vol. 174, No.6, (March 2005), pp. 3709-3718, ISSN 0022-1767.

Keyse, S.M. & Tyrrell, R.M. (1989). Heme oxygenase is the major 32-kDa stress protein induced in human skin fibroblasts by UVA radiation, hydrogen peroxide, and sodium arsenite. *Proceedings of the National Academy of Sciences of the United States of America*, Vol.86, No.1, (January 1989), pp. 99-103, ISSN 0027-8424.

Kumar, S. & Bandyopadhyay, U. (2005). Free heme toxicity and its detoxification systems in human. *Toxicol Lett*, Vol.157, No.3, (July 2005), pp. 175-188, ISSN 0378-4274.

Larsen, R., Gozzelino, R., Jeney, V., Tokaji, L., Bozza, F.A., Japiassú, A.M., Bonaparte, D., Cavalcante, M.M., Chora, A., Ferreira, A., Marguti, I., Cardoso, S., Sepúlveda, N., Smith, A. & Soares, M.P. A central role for free heme in the pathogenesis of severe sepsis. *Sci. Transl. Med.*, Vol.2, No.51, (September 2010), pp.51ra71, ISSN 1946-6234.

Lin, Y.T., Chen, Y.H., Yang, Y.H., Jao, H.C., Abiko, Y., Yokoyama, K. & Hsu, C. Heme oxygenase-1 suppresses the infiltration of neutrophils in rat liver during sepsis through inactivation of p38 MAPK. *Shock*, Vol.34, No.6, (December 2010), pp.615-21, ISSN 1073-2322.

Maeshima, K., Takahashi, T., Nakahira, K., Shimizu, H., Fujii, H., Katayama, H., Yokoyama, M., Morita, K. & Akagi, R. (2004). A protective role of interleukin 11 on hepatic injury in acute endotoxemia. *Shock*, Vol.21, No.2, (February 2004), pp.134-8, ISSN 1073-2322.

Maeshima, K., Takahashi, T., Uehara, K., Shimizu, H., Omori, E., Yokoyama, M., Tani, T., Akagi, R., Morita, K. & Sassa, S. (2005). Prevention of hemorrhagic shock-induced lung injury by heme arginate treatment in rats. *Biochem Pharmacol* Vol.69, No.11, (June 2005), pp.1667-80, ISSN 0006-2952.

Mohri, T., Ogura, H., Koh, T., Fujita, K., Sumi, Y., Yoshiya, K., Matsushima, A., Hosotsubo, H., Kuwagata, Y., Tanaka, H., Shimazu, T. & Sugimoto, H. (2006). Enhanced expression of intracellular heme oxygenase-1 in deactivated monocytes from patients with severe systemic inflammatory response syndrome. *J Trauma*, Vol.61, No.3, (September 2006), pp.616-23 discussion 623, ISSN 0022-5282.

Morimatsu, H., Takahashi, T., Maeshima, K., Inoue, K., Kawakami, T., Shimizu, H., Takeuchi, M., Yokoyama, M., Katayama, H. & Morita, K. (2006) Increased heme catabolism in critically ill patients: correlation among exhaled carbon monoxide, arterial carboxyhemoglobin, and serum bilirubin IXalpha concentrations. *Am J Physiol Lung Cell Mol Physiol*, Vol.290, No.1, (January 2006), pp. L114-9, ISSN 1040-0605.

Nakahira, K., Takahashi, T., Shimizu, H., Maeshima, K., Uehara, K., Fujii, H., Nakatsuka, H., Yokoyama, M., Akagi, R. & Morita, K. (2003). Protective role of heme oxygenase-1 induction in carbon tetrachloride-induced hepatotoxicity. *Biochem Pharmacol*, Vol.66, No.6, (September 2003), pp.1091-1105, ISSN 0006-2952.

Ogawa, K., Sun, J., Taketani, S., Nakajima, O., Nishitani, C., Sassa, S., Hayashi, N., Yamamoto, M., Shibahara, S., Fujita, H. & Igarashi, K. (2001). Heme mediates derepression of Maf recognition element through direct binding to transcription repressor Bach1. *EMBO J*, Vol.20, No.11, (June 2001), pp. 2835-2843, ISSN 0261-4189.

Oyake, T., Itoh, K., Motohashi, H., Hayashi, N., Hoshino, H., Nishizawa, M., Yamamoto, M. & Igarashi, K. (1996). Bach proteins belong to a novel family of BTB-basic leucine zipper transcription factors that interact with MafK and regulate transcription through the NF-E2 site. *Mol Cell Biol,* Vol.16, No.11, (November 1996), pp. 6083-6095, ISSN 0270-7306.

Ponka, P. (1997). Tissue-specific regulation of iron metabolism and heme synthesis: distinct control mechanisms in erythroid cells. *Blood,* Vol.89, No.1, (January 1997), pp. 1-25, ISSN 0006-4971.

Poole, B., Wang, W., Chen, Y.C., Zolty, E., Falk, S., Mitra, A. & Schrier, R. (2005). Role of heme oxygenase-1 in endotoxemic acute renal failure. *Am J Physiol Renal Physiol,* Vol.289, No.6, (December 2005), pp.F1382-5, ISSN 1522-1466.

Poss, K.D. & Tonegawa, S. (1997). Reduced stress defense in heme oxygenase 1-deficient cells. *Proceedings of the National Academy of Sciences of the United States of America,* Vol.94, No.20, (September 1997), pp. 10925-10930, ISSN 0027-8424.

Ryter, S.W., Morse, D., & Choi, A. M. (2007). Carbon monoxide and bilirubin: potential therapies for pulmonary/vascular injury and disease. *American journal of respiratory cell and molecular biology,* Vol.36, No.2, (February 2007), pp. 175-182, ISSN 1044-1549.

Sassa, S. (2006). Biological implications of heme metabolism. *Journal of Clinical Biochemistry and Nutrition,* Vol.38, No.3, (May 2006), pp. 138-155, ISSN 0921-0009.

Sassa, S. & Nagai, T. (1996). The role of heme in gene expression. *International Journal of Hematology,* Vol.63, Vol.3, (April 1996), pp. 167-178, ISSN 0925-5710.

Scharte, M., Bone, H-G., Aken, HV. & Meyer, J. (2000) Increased Carbon Monoxide in Exhaled Air of Critically Ill Patients. *Biochem Biophys Res Commun,* Vol.267, No.1,(January 2000), pp.423-6, ISSN 0006-291X.

Schober, P., Kalmanowicz, M., Schwarte, LA. & Loer, SA. (2009) Cardiopulmonary bypass increases endogenous carbon monoxide production. *J Cardiothorac Vasc Anesth,* Vol.23, No.6, (April 2009), pp.802-6, ISSN 1053-0770.

Schwertner, H. A., Jackson, W. G., & Tolan, G. (1994). Association of low serum concentration of bilirubin with increased risk of coronary artery disease. *Clinical chemistry,* Vol.40, No.1, (January 1994), pp. 18-23, ISSN 0009-9147.

Seixas, E., Gozzelino, R., Chora, A., Ferreira, A., Silva, G., Larsen, R., Rebelo, S., Penido, C., Smith, N.R., Coutinho, A. &

Soares, M.P. Heme oxygenase-1 affords protection against noncerebral forms of severe malaria. *Proceedings of the National Academy of Sciences of the United States of America,* Vol. 106, No.37, (September 2009), pp.15837-15842, ISSN 0027-8424.

Shapiro, S. M. (2003). Bilirubin toxicity in the developing nervous system. *Pediatric neurology,* Vol.29, No.5, (November 2003), pp. 410-421, ISSN 0887-8994.

Shibahara, S., Muller, R., Taguchi, H. & Yoshida, T. (1985). Cloning and expression of cDNA for rat heme oxygenase. *Proceedings of the National Academy of Sciences of the United States of America,* Vol.82, No.23, (December 1985), pp. 7865-7869, ISSN 0027-8424.

Shimizu, H., Takahashi, T., Suzuki, T., Yamasaki, A., Fujiwara, T., Odaka, Y., Hirakawa, M., Fujita, H. & Akagi, R.

Protective effect of heme oxygenase induction in ischemic acute renal failure. *Critical Care Medicine,* Vol.28, No.3, (March 2000), pp.809-817, ISSN 0090-3493.

Stocker, R., Yamamoto, Y., McDonagh, A. F., Glazer, A. N., & Ames, B. N. (1987). Bilirubin is an antioxidant of possible physiological importance. *Science,* Vol.235, No.4792, (February 1987), pp. 1043-1046, ISSN 0036-8075.

Sun, J., Hoshino, H., Takaku, K., Nakajima, O., Muto, A., Suzuki, H., Tashiro, S., Takahashi, S., Shibahara, S., Alam, J., Taketo, M.M., Yamamoto, M. & Igarashi, K. (2002).

Hemoprotein Bach1 regulates enhancer availability of heme oxygenase-1 gene. *EMBO J*, Vol.21, No.19, (October 2002), pp. 5216-5224, ISSN 0261-4189.

Takahashi, T., Shimizu, H., Morimatsu, H., Inoue, K., Akagi, R., Morita, K. & Sassa, S. Heme oxygenase-1: a fundamental guardian against oxidative tissue injuries in acute inflammation. *Mini Rev Med Chem*, Vol.7, No.7, (July 2007), pp.745-753, ISSN 1389-5575.

Takaki, S., Takeyama, N., Kajita, Y., Yabuki, T., Noguchi, H., Miki, Y., Inoue, Y. & Nakagawa, T. (2010) Beneficial effects of the heme oxygenase-1/carbon monoxide system in patients with severe sepsis/septic shock. *Intensive Care Medicine*, Vol.36, No.1, (January 2010), pp.42-48, ISSN 0342-4642.

Temme, E. H., Zhang, J., Schouten, E. G., & Kesteloot, H. (2001). Serum bilirubin and 10-year mortality risk in a Belgian population. *Cancer causes & control : CCC*, Vol.12, No.10, (December 2001), pp.887-894, ISSN 0957-5243.

Tenhunen, R., Marver, H.S. & Schmid, R. (1968). The enzymatic conversion of heme to bilirubin by microsomal heme oxygenase. *Proceedings of the National Academy of Sciences of the United States of America*, Vol.61, No.2, (October 1968), pp. 748-755, ISSN 0027-8424.

Tenhunen, R., Marver, H.S. & Schmid, R. (1969). Microsomal heme oxygenase. Characterization of the enzyme. *J Biol Chem*, Vol.244, No.23, (December 1969), pp. 6388-6394, ISSN 0021-9258.

Vítek, L., Jirsa, M., Brodanová, M., Kalab, M., Marecek, Z., Danzig, V., Novotný, L., & Kotal, P. (2002). Gilbert syndrome and ischemic heart disease: a protective effect of elevated bilirubin levels. *Atherosclerosis*, Vol.160, No.2, (February 2002), pp. 449-456, ISSN 0021-9150.

Wagener, Feldman, E., de Witte, T. & Abraham, N.G.: Heme induces the expression of adhesion molecules ICAM-1, VCAM-1, and E selectin in vascular endothelial cells. *Proc Soc Exp Biol Med*, Vol.216, No.3, (December 1997), pp.456-463, ISSN 0037-9727.

Wang, H.-D., Yamaya, M., Okinaga, S., Jia, Y.-X., Kamanaka, M., Takahashi, H., Guo, L.-Y., Ohrui, T., & Sasaki, H. (2002). Bilirubin ameliorates bleomycin-induced pulmonary fibrosis in rats. *American journal of respiratory and critical care medicine*, Vol.165, No.3, (February 2002), pp. 406-411, ISSN 1073-449X.

Yu, J.B., Zhou, F., Yao, S.L., Tang, Z.H., Wang, M. & Chen, H.R. (2009). Effect of heme oxygenase-1 on the kidney during septic shock in rats. *Transl Res*, Vol.153, No.6, (June 2009), pp.283-287, ISSN 1878-1810.

Yun, N., Eum, H.A. & Lee S.M. (2010) Protective role of heme oxygenase-1 against liver damage caused by hepatic ischemia and reperfusion in rats. *Antioxid Redox Signal*, Vol.13, No.10, (November 2010), pp.1503-1512, ISSN 1523-0864.

Zegdi, R., Caid, R., Van De Louw, A., Perrin, D., Burdin, M., Boiteau, R. & Tenaillon, A. (2006) Exhaled carbon monoxide in mechanically ventilated critically ill patients: influence of inspired oxygen fraction. *Intensive Care Medicine*, Vol.26, No.9, (September 2000), pp. 1228-31, ISSN 0342-4642.

Zegdi, R., Perrin D, Burdin, M., Boiteau, R. & Tenaillon, A. (2002). Increased endogenous carbon monoxide production in severe sepsis. *Intensive Care Medicine*, Vol.28, No.6, (June 2002), pp.793-6, ISSN 0342-4642.

Zeng, Z., Huang, H.F., Chen, M.Q., Song, F. & Zhang, Y.J. (2010). Heme oxygenase-1
    protects donor livers from ischemia/reperfusion injury: The role of Kupffer cells.
    World J Gastroenterol, Vol.16, No.10, (March 2010), pp.1285-1292, ISSN 1007-9327.

8

Paraoxonase: A New Biochemical Marker of Oxidant-Antioxidant Status in Atherosclerosis

Tünay Kontaş Aşkar[1] and Olga Büyükleblebici[2]

[1]Department of Biochemistry, University of Aksaray,
[2]Vocational School, Faculty of Science, University of Çankırı Karatekin,
Turkey

1. Introduction

The paraoxonase (PON, *aryldialkyl phosphatase*, E.C. 3.1.8.1), is an Ca-dependant enzyme that is synthesised in liver. It is related to HDL and has 43-45 kDa molecular weight with glycoprotein structure. PON is an *hydrolase* that has both arylesterase (E.C. 3.1.1.2) and paraoxonase (E.C. 3.1.8.1) activity (Başkol and Köse, 2004). The name of paraoxonase is comes from its useage of organic phosphorous paraoxanes as substrate. Enzyme catalyses the hydrolysis of organic phosphorous insecticides such as diisopropyl fluorophosphate (DEP) except parathion and catalyses the hydrolysis of nerve gases such as sarin, taurin which are in the same chemical group. Howewer, natural substrate of PON is not definite (Gülcü and Gürsu, 2003).

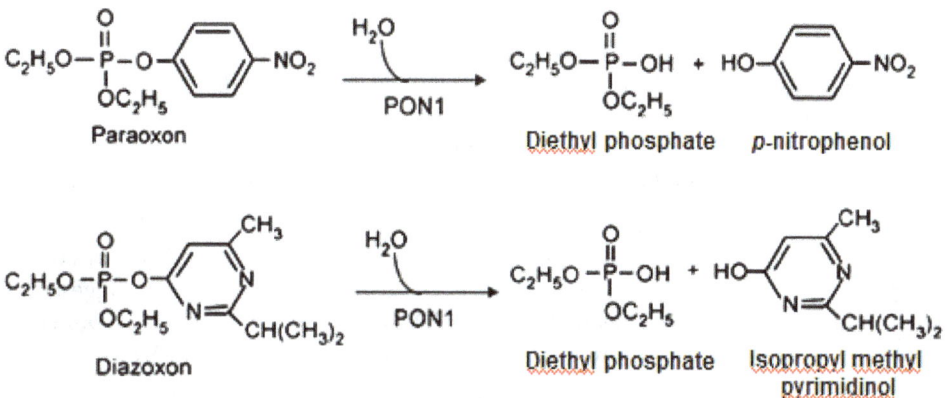

The paraoxonase was coincided initially in high density lipoproteins after electrophoresis of human serum in 1961 (Mackness and Mackness, 2004). Then, it was shown that especially purified cattle serum paraoxonase is related to lipids and has the same molecular mass with the lipid complex. With ultracentrifugation of serum, Mackness et al. (1985) reported that paraoxonase is carried in HDL structure in human blood. It is determined that the enzyme activity is related to HDL particules that contain Apo A1 in sheep (Bayrak et al., 2005). Recently, it is shown that human serum paraoxonase is related to the types of HDL which contains Apo AI and Apo J (Azarsız and Sözmen, 2000 ; Bayrak et al., 2005).

2. Izoenzymes of paraoxonase

There are three members of the PON gene family: PON1, PON2, and PON3. They all possess antioxidant properties, share 65% similarity at the amino acid level, and the genes are located in tandem on chromosome 7 in humans and on chromosome 6 in mice. Althought the amino acid chains of three PON proteins are similar 65 % in proportion, their expressions in tissues and dispersions are different from each other. PON1 and PON3 are mostly expressed in the liver and are carried in plasma bound to HDL. PON2 was found only in cells (human endothelial cells and human aortic smooth muscle cells) and may not be associated with lipoproteins (Hong-Liang et al., 2003; Aviram and Rosenblat, 2004).

Human serum paraoxonase (PON1) is the most studied family member. The 45 kDa, 354-amino acid glycoprotein encoded by the PON1 gene maps to human chromosome 7q21-22. PON1 is synthesized in liver and is found in various tissues and plasma; especially liver, kidneys and intestines. The enzyme takes place in structure of HDL in plasma (Ali et al., 2003). Calcium is required for both activity and stability of enzyme and plays a role in catalytical mechanism. The PON1 arylesterase and PON activities are calcium-dependent and can be totally and irreversibly inhibited by EDTA, while the protection of LDL against oxidation may not require calcium (Bayrak et al., 2005). Serum PON1 activity is inversely associated with oxidative stress not only in serum, but also in arterial macrophages, the hallmark of early atherogenesis, and this phenomenon is associated with enhanced atherosclerotic lesion formation. PON1 has gained currency about its antioxidant properties (Aviram and Rosenblot, 2004).

The izoenzyme PON2 is especially found in endothelial layer of liver, kidneys, heart, brain and testicular tissues (Bayrak et al., 2005). Immunohistochemical methods show that PON2 is also found in smooth muscle cells of aorta as well (Hong-Liang et al., 2003). Eventhough PON2 is defined following PON1 and less studied, it attracted more attention due to its antioxidant activity in endothelium and vascular endothelial cells (Bayrak et al., 2005).

As PON1, PON3 is basically in liver and found in plasma with HDL (Bayrak et al., 2005). PON2 and PON3 can not hydrolize paraoxonase because there is not any lysine residue in 105. position. However, the two izoenzymes (PON2 and PON3) were both suggested to possess antioxidant properties, as rabbit serum PON3 protects LDL from oxidation, and it rapidly hydrolyzes lactones. (Aviram and Rosenblat, 2004; Ekmekçi et al., 2004).

3. Structure of paraoxonase

PON1 enzyme includes 354 amino acids and three carbonhydrate chains that form 15.8 % of total mass. The gene which codes PON1 enzyme, is in q 21-22 zone of seventh chromosome.

Its isoelectric point pH is 5.1. Amino acid content does not show any specific property (Gülcü and Gürsu, 2003).

PON1 has three cysteine amino acids in its structure. While there are disulphide bonds between the two of cysteine amino acids of the enzyme (Cys 42-352), the other cysteine amino acid at 284. position is free (Figure 1). This free cysteine is present next to the active site of the enzyme and has functions on substrate recognization and binding to enzyme (Azarsız and Sözmen, 2000; Bayrak et al., 2005). Although free cysteine has significant effect on LDL protection from oxidation, it has no effect on hydrolysis of organophosphates (Ekmekçi et al., 2004). On the other hand, the disulphide bond which is found in the structure of PON1 is responsible from the cyclic form of the polypeptide chain (Memişoğulları and Orhan, 2010).

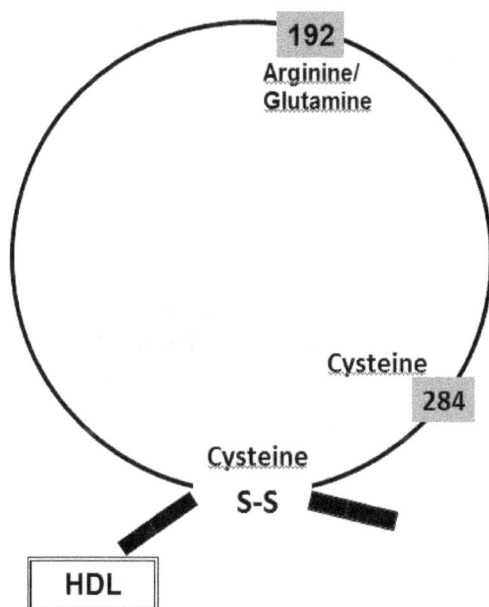

Fig. 1. The structure of paraoxonase enzyme

PON1, which is located in HDL structure, is synthesised in liver. N-terminal hydrophobic signal peptide of PON1, is required for interaction with HDL. By this signal peptid the enzyme is attached onto phospolipid and lipoproteins. Also the binding subunits of HDL with PON1, contain Apolipoprotein (Apo) A1 and Apo J (clusterin) proteins. And it is considered that Apo 1 and Apo J may affect binding (Başkol and Köse, 2004).

In three dimension structure of the enzyme, there are two calcium ions having 7.4 Å intervals in the centre of β-layers. One of these is the structural calcium and its removal from the structure causes irreversible denaturation. The other calcium takes role in catalytic action of the enzyme. This calcium ion interacts with one water molecule and the oxygene of phosphate ion (Khersonsky and Tawfik, 2005).

4. Factors affecting paraoxonase activity

Serum PON1 activity is affected by diet, pregnancy, smoking, hormones, acute phase proteins and age. PON1 activity in newborns and premature infants is half of adult PON1 activity, only after a year it reaches its adult level. PON1 activity decreases with age (Biasioli et al., 2003; Seres et al., 2004), but enzyme activity does not vary by sexes (Azarsız and Sözmen, 2000; Ekmekçi et al., 2004).

It is shown that diet which is rich in saturated fats, reduces satiety serum PON/arylesterase activity, and diet that is rich in unsaturated fats, incerases satiety serum PON/arylesterase activity (Laplaud et al., 1998). Beside this, it was shown that in adult men PON1 activity increases by moderate alcohol consumption (Azarsız and Sözmen, 2000).

It was also shown that red wine and polyphenol intake increase the PON activity in rats with Apo E deficiency, while smoking inhibits PON1 activity (Azarsız and Sözmen, 2000). The disorders affecting metabolism of Apo AI, have an effect on PON1 activity as well (Azarsız and Sözmen, 2000).

5. The functions of paraoxonase

The paraooxonase enzyme has very important functions in organism:

5.1 Protein against organophosphate (Hydrolic activity)

Paraoxonase is a calcium-dependent esterase that was initially identified by its hydrolysis of aromatic carboxylic esters and organophosphorus insecticides and nerve gases. Its name reflects its ability to hydrolyze paraoxon, a metabolite of the insecticide parathion (Cabana et al., 2003). The enzyme hydrolyses of O-P ester bond in paraoxon (Ekmekçi et al., 2004). Oxons, which do not go through detoxification in mammal liver, are hydrolysed by serum PON1 enzyme before the organophosphates display their activity (Azarsız and Sözmen, 2000).

Enzyme also catalyses the hydrolysis of various carbamates and aromatic carboxylic acid esters such as phenylacetate, 4-nitrophenylacetate, 2-naphtyl acetate (Gülcü and Gürsu, 2003). It catalyses the hydrolysis of nerve gases such as sarin, soman which are in the same chemical group.

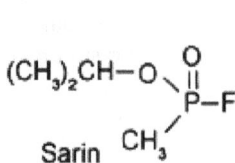

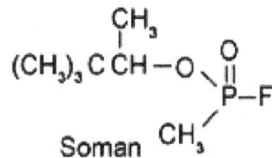

The insects are target organisms for organophosphates. Because PON1 cannot be synthesised in insects, birds, fish and reptiles. Therefore, there is a tendency for organophosphate poisoning in birds, fishes and reptiles, compared to mammals (Azarsız and Sözmen, 2000).

Earlier studies on the toxicology of organophosphate pesticides revealed that, having low serum PON1 activity increases the senstivity to the acute effects of organophosphate compounds (Costa and Furlong, 2003). Following determination of the sequence of the gene encoding PON1, identification of polymorphisms that lead to differences among people in the activity and level of expression of the enzyme, led to the idea that people with low PON1 activity may be more sensitive to organophosphate poisoning (Li and Costa, 2000)

In studies carried out in recent years, it was shown that injection of purified PON1 to animals that are exposed to organophosphate pesticides, and thus increasing serum PON1 levels artificially, it was possible to reduce the toxic effects of certain organophosphates like chlorpyrifos and diazinon; but this application was observed to be ineffective against parathion exposure (Main, 1956).

In studies carried out in recent years, it was observed that injection of purified PON1 protects against chlorpyrifos and diazinon but was not effective against parathion exposure. In order to be used as a prophylactic agent against organophosphate exposure, PON1's catalytic efficiency has to be increased and the enzyme has to be obtained in adequate amounts. In recent years, using protein engineering methods, changes were made to some amino acids of PON1, these PON1 variants were expressed in bacterial systems at sufficient levels and increase in enzyme activity against some organophosphates were obtained (Harel and Brumshtein, 2007).

These studies relevantly shows that PON1 has a role on the metabolism of some organophosphates and an artificial addition of PON1 to serum levels of these animals provides a proection against organophosphates toxicities.

5.2 Prevention of LDL oxidation (Anti-atherogenic activity)

The second important function of PON1 is its anti-atherogenic activity. Serum PON1 exists with HDL in plasma and works to prevent the oxidation of plasma lipoproteins (Aviram, 1999). PON1 is effective on lipid peroxides and also on hydrogen peroxide therefore, it is considered to have activity like peroxidase (Memişoğulları and Orhan, 2010).

Serum PON1 activity is especially important for the protection of LDL phospholipids against oxidation. In the protective effect against atherosclerosis, LDL has different anti-atherogenic features such as protection from free radical induced oxidation of LDL cholesterol in arthery wall and protection against harmful effects of oxi-LDL (Aviram, 1999). In addition to LDL, enzyme protects HDL too which is a carrier of lipid peroxide (Azarsız and Sözmen, 2000).

5.3 Protection against bacterial endotoxins(Lipopolysaccharide inactivation)

In recent studies, it was defined that HDL-associated PON1 protects the body against bacterial endotoxin via hydrolysis of bacterial lipopolysaccharides (Bayrak et al., 2005). PON1 hydrolyses bacterial lipopolysaccharides with its phosphatase effect. Also the enzyme prevents the inflammatory response by avoiding the interaction between specific macrophage binding protein and bacterial lipopolysaccharide (Aviram, 1999, Azarsız and Sözmen, 2000).

6. Role of paraoxonase in oxidative stress

Reactive oxygen species are oxygen-containing molecules that produced during normal metabolism. When the production of damaging reactive oxygene species (ROS) exceeds the capacity of the body's antioxidant defenses to detoxify them, a condition known as oxidative stress occurs. Reactive oxygene species can cause tissue damage, particularly in the endothelial tissue. Lipids and lipoproteins also affected by ROS (Altındağ et al., 2007). Therefore oxidative stress is associated with lipid peroxidation in lipoproteins and in arterial cells, including macrophages. Under oxidative stress, not only lipoproteins but lipids in cell membrane sustain to lipid peroxidation. Dense and small LDL particles are more sensitive to oxidation compared with big particles (Chait et al., 2005).These "oxidized macrophages" possess increased capability to convert native LDL into oxidized LDL. Oxidative stress has been implicated in the pathogenesis of atherosclerosis. The early atherosclerotic lesion is characterized by macrophage foam cells, filled with cholesterol, oxysterols and oxidized lipids (Fuhrman et al., 2002).

HDL associated paraoxonase (PON1) is an antioxidant enzyme that protects LDL and HDL from lipid peroxidation. And it is considered to be the main antiatherosclerotic (preclusive to atherosclerosis) component of HDL (Gülcü and Gürsu, 2003). PON1 protects lipoproteins and arterial cells against oxidation, probably by hydrolyzing lipid peroxides such as specific oxidized cholesteryl esters and phospholipids. And it was suggested to contribute to the antioxidant protection conferred by HDL on LDL oxidation (Mackness et al., 1991).

The PON1 free sulfhydryl group cysteine 284 was shown to be required for its protection against lipid peroxidation in lipoproteins (Ekmekçi et al., 2004). It was shown to be reduced in patients after myocardial infarction, in patients with familial hypercholesterolemia, and in patients with diabetes mellitus in comparison to healthy subjects (Mackness et al., 1991).

HDL-associated PON was able to hydrolyze long-chain oxidized phospholipids isolated from oxidized LDL or serve as a target for peroxides (Figure 2). PON1 through interactions between the enzyme-free sulfhydryl group and oxidized lipids. H_2O_2 is a major ROS produced by arterial wall cells during atherogenesis, and it is converted under oxidative stress into more potent hydroxyl radical leading to LDL oxidation (Aviram et al., 1998).

PON1 was found to use efficiently not only lipoprotein-associated peroxides (including cholesteryl linoleate hydroperoxides), but also hydrogen peroxide ($H2O2$). PON1 inhibits the accumulation of peroxynitrite-generated oxidized phospholipids by its ability to hydrolyze phosphatidylcholine (PC) core aldehydes and PC isoprostanes to yield lysophosphatidylcholine (Rosenberg et. al., 2003). Because of reducing hydroxide and cholesteryl linoleat hydoperoxide in LDL, it is considered that PON1 has an activity like peroxidase (Aviram et al., 1998). Thus HDL-PON may play an important role in the prevention of atherosclerosis (Aviram et al., 1998).

7. The paraoxonase activity in various diseases

In various diseases, such as obesity, menopause and malnutrition, PON1 activity is being suppressed (Yurttagül, 2007). Low PON/HDL and PON/Apo AI rates were seen at patients under high atherosclerose risk and it is concluded that the antioxidant capacity of HDL has been decreased in these cases (Azarsız and Sözmen, 2000).

In patients with coronary atherosclerose, the blocker effect of PON1 in LDL oxidation occurs with repression of inflammatory response in cells which are on the arterial wall with destroying active lipids in LDL (Yurttagül, 2007). In the studies which show PON1 protects HDL from oxidation, it is revealed that PON1 decreases 95 % rate of lipid peroxide and aldehyde accumulation. This protective effect of PON1 on HDL oxidation may be related to metal ion chelation, or to a peroxidase-like activity (Aviram et al., 1998;Yurttagül, 2007).

It has been shown that in Tropical Theileriosis disease which causes huge economic loss in stock farming PON1 activity also decreases (Turunç and Aşkar, 2012). On the other hand, PON1 activity also decreases in Dirofilariosis (Heart Worm Disease), which is a zoonotic nematode disease. The mature *Dirofilaria immitis* parasites also make oxidative damage in heart, for preventing this damage PON1 works as antioxidant (Kına, 2009).

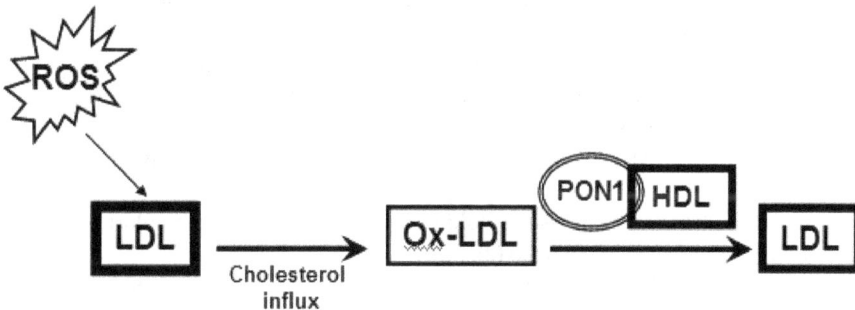

Fig. 2. The role of HDL-associated PON1 in LDL oxidation.

In Diabetes mellitus (DM), oxidative stress and depending on that, lipid peroxidation increases. Oxidised lipids are easily phagocyted by macrophages and they take part in atheropatogenesis by generating foam cells. It is determined that PON1 enzyme levels are low in DM and in patients with familial hypercholesterolemia (Aviram, 1999).

8. Conclusion

Oxidative stress, has been the centre of toxicological studies for possible mechanism of toxicity in last decades (Cochranc, 1991). PON1 is a multienzyme complex with hydrolase, arylesterase, diaoxonase, phosphatase, peroxidase and lactonase functions. PON1 has also antioxidant property, because it prevents the increase of ROS quantity by hydrolysing lipid peroxidation products. It also shows protective effect in cell membranes by neutralizing the atherogenic effects of lipid peroxides (Aviram et al., 2000).

Therefore PON1 is an important enzyme in oxidant-antioxidant status of atherosclerosis with its antioxidant effect, but more studies are required for this enzyme to prove it's role in the other diseases. With future studies, the effects of paraoxonase enzyme could be well-understood.

9. References

Ali, A.B., Zhang, Q., Lim, Y.K., Fang, D., Retnam, L., 2003. Expression of major HDL-associated antioxidant PON-1 is gender and regulated during inflammation, Free Rad Bio & Med, 34, 824-829.

Altındağ, O., Karakoc, M., Soran, N., Çelik, H., Çelik, N., Selek, Ş., 2007. Paraoxonase and arylesterase activities in patients with rheumatoid artritis, Turkish Journal of Rheumatology, 22 (4), 132-136.

Aviram, M., 1999. Does paraoxonase play a role in susceptibility to cardiovascular disease, Mol Med Tod, 5, 381-386.

Aviram, M., Hardak, E., Vaya, J., Mahmood, S., Milo, S., Hoffman, A., Billicke, S., Draganov, D., Rosenblat, M., 2000. Human serum paraoxonases (PON1) Q and R selectively decrease lipid peroxides in human coronary and carotid atherosclerotic lesions, Circulation, 101, 2510-2517.

Aviram, M., Rosenblat, M., Bisgair, C.L., 1998. Paraoxonase inhibits high density lipoprotein (HDL) oxidation and preserves its functions: a possible peroxidative role for paraoxonase, J Clin Invest, 101, 1581-1590.

Aviram, M., Rosenblat, M., 2004. Paraoxonases 1, 2, and 3, oxidative stress, and macrophage foam cell formation during atherosclerosis development. Free Radical Biology & Medicine, 37, 1304–1316

Azarsız, E., Sözmen, E.Y., 2000. Paraoksonaz ve klinik önemi, Türk Biyokimya Dergisi, 25 (3), 109-119.

Başkol, G., Köse, K., 2004. Paraoksanase: Biyokimyasal özellikleri, fonksiyonları ve klinik önemi, Erciyes Medical Journal, 26 (2), 75-80.

Bayrak, T., Bayrak, A., Demirpençe, E., Kılınç, K., 2005. Yeni bir kardiyovasküler belirteç adayı: Paraoksonaz, Hacettepe Medical Journal, 36,147-151.

Biasioli, S., Schiavon, R., Petrosino, L., De Fanti, E., Cavalcanti, G., 2003. Paraoxonase activity and paraoxonase 1 gene polymorphism in patients with uremia, ASAIO J, 49, 295-299.

Cabana, V.G., Reardon, C.A., Fenf, N., Neath, S., Lukens, J., Getz, G.S., 2003. Serum paraoxonase: effect of the apolipoprotein composition of HDL and the acute phase response, J Lipid Res, 44, 780-792.

Chait, A., Han, C.Y., Oram, J.F., Heinecke, J.W., 2005. Lipoprotein-associated inflammatory proteins: markers or mediators of cardiovascular disease? J Lipid Res, 46, 389-403.

Cochranc, C.G., 1991. Cellular injury by oxidants, Am. J. Med., 92, 235-305.

Costa L.G., Cole T.B., Furlong C.E., 2003. Polymorphisms of paraoxonase (PON1) and their significance in clinical toxicology of organophosphates. J Toxicol Clin Toxicol, 41(1): 37-45.

Ekmekçi Balcı, Ö., Donma, O., Ekmekçi, H., 2004. Paraoxonase, Cerrahpaşa J Med, 35, 78-82.

Fuhrman, B., Volkova, N., Aviram, M., 2002. Oxidative stress increases the expression of the CD36 scavenger receptor and the cellular uptake of oxidized low-density lipoprotein in macrophages from atherosclerotic mice: protective role of antioxidants and of paraoxonase, Atherosclerosis, 161, 307-316.

Gülcü, F., Gürsu, F.M., 2003. Paraoksanaz ve Aril Esteraz Aktivite Ölçümlerinin Standardizasyonu, Türk Biyokimya Dergisi, 28 (2), 45-49.

Harel M., Brumshtein B., Meged R., 2007. The 3-D structure of serum paraoxonase 1 sheds light on its activity, stability, solubility, and crystallizability. Arh Hig Rada Toksikol, 58: 347-53.

Hong-Liang, L., De-Pei, L., Chihj-Chuan, L., 2003. Paraoxonase gene polymorphisms, oxidative stress and diseases, J Mol Med, 81, 766-779.

Khersonsky, O., Tawfik, D.S., 2005. Structure-reactivity studies of serum paraoxonase PON-1 suggest that its native activity is lactonase, Biochemistry, 44, 6371-6382.

Kına, O., 2009. *Dirofilaria immitis*'li köpeklerde paraoksonaz aktivitesi ve lipid profilinin belirlenmesi, Master Thesis, University of Mustafa Kemal, Hatay, Turkey.

Li W.F., Costa L.G., Richter R.J., 2000. Catalytic efficiency determines the in vivo efficacy of PON1 for detoxifying organophosphates. Pharmacogenetics; 10: 767-79.

Mackness, M.I., Hallam, S.D., Peard, T., Warner S., Walker, C.H. 1985. The separation of sheep and human serum "A"-esterase activity into the lipoprotein fraction by ultracentrifugation. Comp. Biochem. Physiol., 82B, 675-677.

Mackness, M.I., D. Harty, D. Bhatnagar, P.H. Winocour, S. Arrol, M. Ishola, and P.N. Durrington. 1991. Serum paraoxonase activity in familial hypercholesterolaemia and insulin-dependent diabetes mellitus. *Atherosclerosis*. 86:193–199.

Mackness, M., Mackness, B., 2004. Paraoxonase 1 and atherosclerosis: is the gene or the protein more important? Free Radic Biol Med, 37, 1317–1323.

Memişoğulları, R., Orhan, N., 2010. Paraoksonaz ve Kanser, Konuralp Medical Journal, 2(2), 22-26.

Main AR. 1956. The role of A-esterase in the acute toxicity of paraoxon, TEPP and parathion. Can J Biochem Physiol,; 34: 197-216.

Seres, I., Paragh, G., Deschene, E., Fulop, T., Khalil, A., 2004. Study of factors influencing the decreased HDL associated PON1 activity with aging, Exp. Gerontol, 39, 59-66.

Turunç, V., Aşkar, T.K., 2012. The Determination of Oxidative Stress by Paraoxonase Activity, Heat Shock Protein and Lipid Profile Levels in Cattle with Theileriosisi, The Journal of the Faculty of Veterinary Medicine, University of Kafkas. (*In press*)

Yurttagül, K., 2007. Koroner arter hastalarında HDL alt grupları ile paraoksonaz enzim aktivitesi dağılımları, Doctorate thesis, University of Istanbul, Turkey. mları, Doctorate thesis, University of Istanbul, Turkey.

9

Hydrogen: From a Biologically Inert Gas to a Unique Antioxidant

Shulin Liu, Xuejun Sun and Hengyi Tao

Second Military Medical University,
China

1. Introduction

Hydrogen gas (H_2), a colorless, tasteless, odorless, non-irritating and highly flammable diatomic gas, has been used in medical applications to prevent decomposition sickness in deep divers. For a long time, H_2 was thought to be a "biologically inert gas" which could not react with biomolecules under normal pressure. In 2007, Ohsawa et al. first reported that inhalation of H_2 markedly suppressed brain injury induced by ischemia-reperfusion, which made the antioxidant properties of H_2 drew wide attention (Ohsawa et al., 2007). Soon afterwards H_2 was found to be effective for many other diseases, including hepatic and cardiac hypoxia-ischemia injury, inflammation injury caused by small intestine transplantation, neonatal hypoxia–ischemia injury, and lung allograft, (Fukuda et al., 2007; Buchholz et al., 2008; Cai et al., 2008; Hayashida et al., 2008; Kawamura et al., 2011;). Besides, other ways to administrate H_2, such as drinking H_2-saturated water, intraperitoneal and intravenous injection of H_2-saturated saline (first developed by our group), have also been proved to be effective to many disorders related with oxidative stress, such as cerebral hypoxia-ischemia injury, human type II diabetes, nephrotoxicity induced by cisplatin, Parkinson's disease and atherosclerosis in apolipoprotein (Cai et al., 2009; Chen et al., 2009; Mao et al., 2009; Sun et al., 2009; Zheng et al., 2009; Oharazawa et al., 2010). Up to now, H_2 has been proved to be effective to various disease models. Considering the unique antioxidant properties of H_2, we believe it is important to review the medical researches of this novel antioxidant in this chapter. The aim of this chapter is to summarize research findings and mechanisms concerning the therapeutic potential of H_2.

2. Physical, chemical and biological characteristics of H_2

The most lightweight gas diatomic hydrogen, the first element in the periodic table and constitutes at least 90 % of the observable universe, is rare on earth. In 1671, Robert Boyle firstly produced hydrogen when he dissolved iron in diluted hydrochloric acid. However, it was not actually discovered as a distinct gas until 1766 when Henry Cavendish originally segregated and called it "inflammable air". A few years later, Lavoisier, the father of modern chemistry, gave it a name "hydrogen". Hydrogen is a colorless, odorless, non-metallic, tasteless and highly combustible diatomic gas with the molecular formula H_2 (Huang et al., 2010). Hydrogen gas directly and violently reacts with oxidizing elements such as chlorine and fluorine and is highly flammable, a property evident in the 1937

Hindenburg zeppelin fire and its use as propellant fuel for the space shuttle. However, safe hydrogen concentrations in air and in pure oxygen gas are 4.6% and 4.1% by volume, respectively (Huang et al., 2010). On the other hand, H_2 is rather less active and behaves as an inert gas in the absence of catalysts or at body temperature (Ohta, 2011). As hydrogen is a highly potent energy source, the industrial use of hydrogen is expanding, such as the use of hydrogen fuel cells for zero-emission vehicles (Nakao et al., 2009). H_2 can be dissolved in water to 0.8 mM under atmospheric pressure, and aluminum containers are able to retain hydrogen gas for a long time (Ohta, 2011). Hydrogen gas is highly diffusible and reacts with hydroxyl radical to produce water.

3. The history of medical researches and application of H_2

For a long time, it has been believed that hydrogen did not react with biomoleculars. Hydrogen gas is routinely administered to divers as hydreliox, which contains 49% hydrogen (Abraini, 2011), to prevent decompression sickness. In 1975 and 2001, Dole and Gharib respectively reported that hydrogen under a high pressure might be a therapeutic gas for cancer and parasite-induced liver inflammation by eliminating toxic ROS (Dole et al., 1975; Gharib et al., 2001). In 2007, Ohsawa set out to see if hydrogen gas could be used as a therapeutic mitochondrial antioxidant to neutralize oxidative stress after ischemia reperfusion injury (Ohsawa et al., 2007). In these studies, Ohsawa found that hydrogen could increase cell survival significantly and hydrogen gas could reach sub-cellular compartments such as the nucleus and mitochondria. This is particularly important as mitochondria is the primary site of generation of reactive oxygen species after reperfusion and is notoriously difficult to target. To test the efficacy of hydrogen gas therapy during oxidative stress, Ohsawa et al. used a rat model of stroke, with middle cerebral artery ligation and reperfusion. Inhalation of hydrogen gas limited the stroke volume if given before the reperfusion phase of injury. Hydrogen gas treatment also reduced brain tissue lipid peroxidation and DNA oxidation, findings that were also noted in cultured cells challenged with reactive oxygen species. The decrease in reperfusion damage improved long term neurological function, such as thermoregulation and weight maintenance, at one week, implying that hydrogen gas can protect cells in vivo. Soon after that, the medical value of hydrogen rapidly drew wide attention in the world. H_2 was found to produce some positive therapeutic outcomes in other diseases of various systems, such as central nervous system, cardiovascular system, lung, renal system, liver, pancreas, and auditory system (Huang et al., 2010), as shown in Fig. 1.

4. Potential mechanisms of H_2 as an antioxidant based on informed researches

To explore the mechanism of the protective effect of hydrogen, Ohsawa and his coworkers treated PC12 cells with antimycin A (a mitochondrial respiratory complex III inhibitor) to produce hydroxyl radicals (•OH) by the Fenton reaction (Ohsawa et al., 2007). In the presence of H_2 in the medium, the intracellular levels of •OH and ONOO- were significantly reduced. Biochemical experiments using fluorescent probes and electron paramagnetic resonance spectroscopy spin traps indicated that hydrogen gas might selectively scavenge the hydroxyl radical. The authors propose this as a unique cyto-protective pathway that

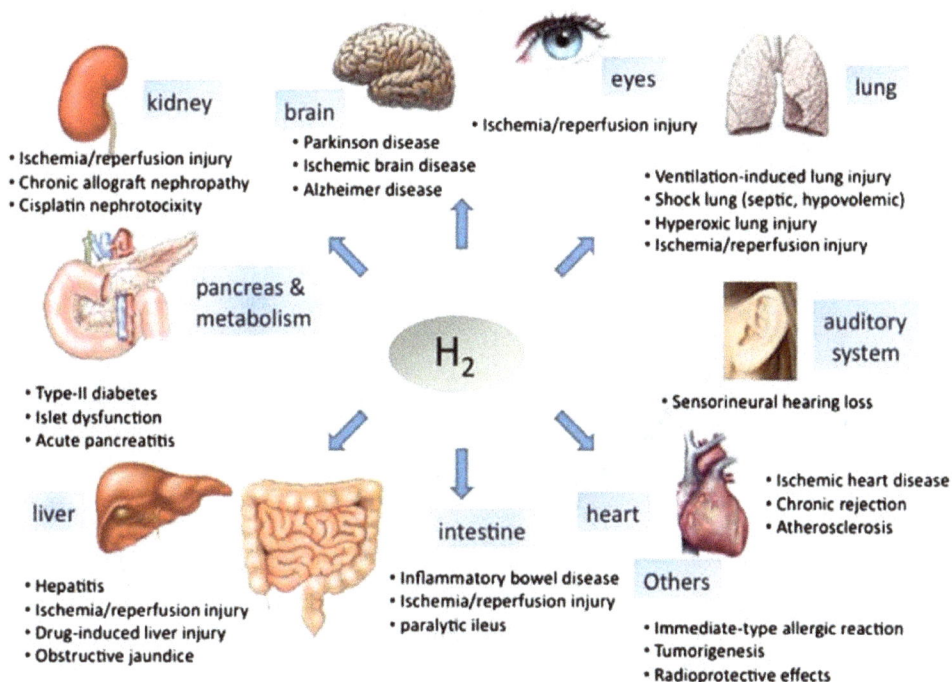

Fig. 1. Therapeutic opportunities of H_2 in a variety of disease models (Huang et al., 2010).

specifically quenches the •OH and ONOO- while preserving other reactive oxygen and nitrogen species important in signaling. Many antioxidants or enzymes that scavenge reactive oxygen species limit cyto-toxicity after ischemia and reperfusion. In the presence of catalytically active metals, however, detoxification of superoxide to hydrogen peroxide by superoxide dismutase generates the more potent hydroxyl radical. •OH and ONOO- react indiscriminately with and damage molecular targets such as nucleic acids, lipids and proteins. Based on these evidence, Ohsawa et al. drew the conclusion that hydrogen gas protected cells from oxidative damage through selective scavenging of •OH and ONOO-, as shown in Fig. 2.

However, accumulating evidence suggest that hydrogen may be not a simple ROS scavenger. There were reports that treatment with H_2 was unable to scavenge •OH (Atsushi et al., 2010). As suggested by Wood and Gladwin (2007), the concentrations of hydrogen may be insufficient to protect the numerous cellular targets of •OH. In addition, H_2 has been demonstrated to be effective for many diseases, but •OH and other ROS are not the only cause of injury for all these diseases. This indicates that •OH scavenging may not be the only underlying mechanism for the protective effects of H_2 for those diseases where oxidative stress plays a less important role in the pathogenesis. A number of more recent investigations have reported that H_2 might influence signal transduction (Itoh et al., 2009; Wang et al., 2011), suggesting that hydrogen may act as a novel signaling molecule rather than a ROS scavenger. Itoh and his associates found that H_2 could modulate the immediate-type allergic reaction, which is not causally associated with oxidative stress (Itoh et al., 2009). In another study, H_2 was found to inhibit lipopolysaccharide/interferon γ-induced

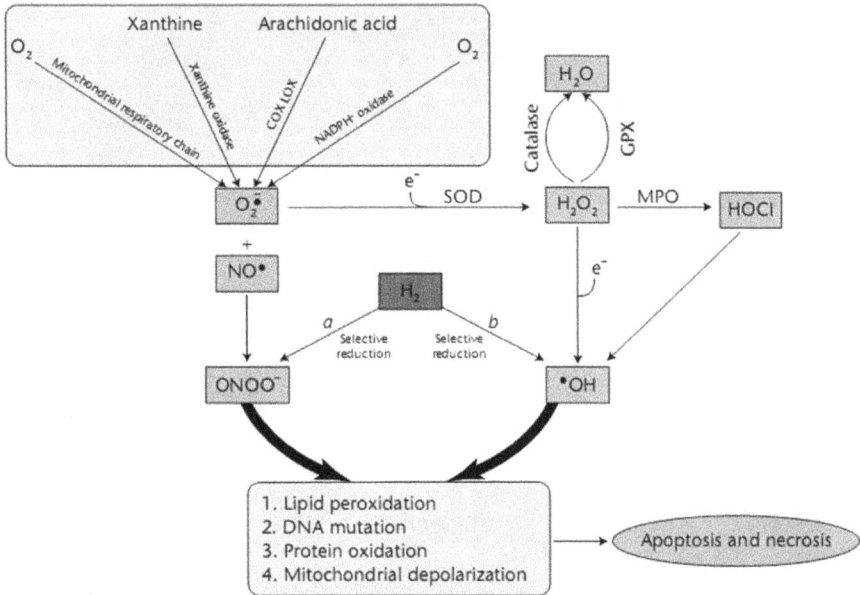

Fig. 2. The reactive oxygen species (ROS) production pathway and the selective reduction of
•OH and ONOO- by H_2 (Hong et al., 2010).

nitric oxide production through modulation of signal transduction in macrophages (Itoh et
al., 2011). All these evidences imply that the effect of hydrogen may reside not in its radical-
scavenging activity but in its ability to modulate a specific signaling pathway. Additionally,
there is evidence of endogenous hydrogen in human and animals from anaerobic
metabolism of bacteria in the large intestine (Sobko et al., 2005). In the normal terminal
breath, levels of H_2 are in the range 5-10 ppm (Maffel et al., 1977) ; in the liver, it reaches 42
µM, while in the large intestine and spleen, it is even higher (Olson & Maier, 2002) . The
existence of endogenous hydrogen also suggests that hydrogen may be capable of exerting
biological effects in physiological and pathological conditions in a manner similar to NO,
CO and H_2S. However, much additional research is required prior to verification, such as
characterizing the molecular mechanism of hydrogen's effects.

5. Methods for H_2 administration

5.1 Inhalation

For a long time, hydrogen inhalation has been applied in deep diving to prevent
decompression sickness and nitrogen narcosis, in which an exotic, breathing gas mixture of
49% hydrogen, 50% helium and 1% oxygen is used (Abraini et al., 1994), so the safety of
hydrogen for humans is fully proved. When delivered by a ventilator circuit, facemask or
nasal cannula, hydrogen can be easily delivered via inhalation (Huang et al., 2010). In the
first researches on the medical application of hydrogen, it was mainly administrated by
inhalation. There were several advantages about hydrogen inhalation. First of all, inhaled
hydrogen gas diffuses and acts more rapidly (Hong et al., 2010; Ohta, 2011). Second,

inhalation of gas does not affect blood pressure (Ohsawa et al., 2007). Third, inhalation is a way to persistently administrate hydrogen. However, safety is a big concern in hydrogen inhalation. Although hydrogen poses no risk of explosion in air and in pure oxygen when its concentration is under 4% (Ohta, 2011), there is still risk in the storage, transportation, preparation and usage of hydrogen-oxygen mixture. Furthermore, a suit of complex apparatus is required for safety, and a strict operating rule must be developed to guarantee the safety of hydrogen inhalation.

5.2 Oral intake of hydrogen-rich water

In 2008, Nagata et al. placed water saturated with hydrogen (hydrogen water) into the stomach of a rat and found that hydrogen was detected at several micromole levels in blood and drinking hydrogen water prevented cognitive impairment by reducing oxidative stress (Nagata et al., 2009). Up to now, hydrogen-rich water has proven to be effective for various disease models, including atherosclerosis (Ohsawa et al., 2008), nephrotoxicity induced by an anti-cancer drug cisplatin (Nakashima-Kamimura et al., 2009), Parkinson's disease (Fu et al., 2009), colon inflammation (Kajiya et al., 2009) and cognitive impairment (Gu et al., 2010). It is detected that H_2 can be dissolved in water up to 0.8 mM under atmospheric pressure at room temperature (Ohta, 2011). Oral intake of hydrogen-rich water seems more practical in daily life and suitable for continuous consumption for preventive or therapeutic use, and it is found to have a comparable effect to hydrogen inhalation (Nakashima-Kamimura et al., 2009). Hydrogen water can be made by several methods, including dissolving hydrogen into water under high pressure and by reaction of magnesium with water, a method developed in Japan (Ohta, 2011).

5.3 Injectable hydrogen-rich saline

In 2008, our group first injected H_2-saturated saline to rats and found that it could prevent neonatal hypoxia-ischemia injury (Cai et al., 2009). Soon after that, injected H_2-saturated saline was found to be effective against various models, including intestinal ischemia/reperfusion injury (Zheng et al., 2009), acute pancreatitis (Chen et al., 2010), experimental liver injury (Sun et al., 2010) and carbon monoxide toxicity (Sun et al., 2011). It is anticipatable that injected hydrogen saline has potential in actual clinical treatment. The unique advantage of injectable hydrogen-rich saline is that it may provide more accurate concentrations of hydrogen and there is little evaporation or other kind of hydrogen lost (Cai et al., 2009).

5.4 Production of hydrogen by intestinal bacteria

It is noteworthy that endogenous hydrogen exists in humans and animals from anaerobic metabolism of bacteria in the large intestine (Maffel et al., 1977). Studies have shown that the hydrogen level in normal terminal breath is about 5-10 ppm, but in patients with lactose intolerance and bacterial disorders, it may achieve more than 90 ppm (Olson et al., 2002). The level of hydrogen has been measured in different organs of normal mice and it is found that in the large intestine, spleen, liver and gastric mucosa the level of hydrogen is very high. Drugs inhibiting the absorption of glucose in diabetics such as acarbose have a side effect of flatulence, but they often show a heart protective effect (Hanefeld et al., 2004). Recent studies showed that the main component of these gases is hydrogen, whose increase may contribute to the heart protective effect. Oral administration of turmeric (Shimouchi et al., 2009), milk (Shimouchi et al., 2009), α-glucosidase inhibitors (Suzuki et al., 2009), coral calcium hydride (Ueda et al., 2010), and mannitol (Liu et al., 2010) could also promote the production of endogenous hydrogen, which may be one of the mechanisms for treatment of some diseases. Moreover, it is reported that oral administration of a special designed bacteria producing hydrogen gas can prevent Con A-induced hepatitis, whereas the protective effect disappeared after antibiotic treatment, indicating that improving hydrogen production of intestinal bacteria can be a treatment of diseases (Kajiya et al., 2009). It seems that oral administration of these substances is far more cheap, convenient, comfortable and simpler than other ways to administer hydrogen.

5.5 H_2-Loaded eye drops and hydrogen bath

In 2010, Oharazawa et al. dissolved H_2 in saline as H_2 loaded eye drops and found that it could ameliorate ischemia-reperfusion injury of the retina in a rat model by administering it to the ocular surface (Oharazawa et al., 2010). In 2011, Kubota used it to prevent oxidative stress-induced angiogenesis in a mouse corneal alkali-burn model with great success (Kubota et al., 2011). In Japan, taking a H_2 water bath is another method to administrate H_2 into the body in daily life. It is said that a 10 minutes bath is able to distribute H_2 throughout the whole body (Ohta, 2011).

6. Advantages and disadvantages of H_2 treatment

6.1 Advantages

Compared to traditional antioxidants, hydrogen, the newly explored antioxidant, obtains a number of advantages. First, due to its small molecular weight, hydrogen can easily penetrate biomembranes and diffuse into the cytosol, mitochondria and nucleus, which are the main site of ROS generation and DNA damage (Ohsawa et al., 2007), but difficult to targeted by other drugs. Second, as hydrogen selectively reacts with •OH and ONOO-, other important ROS (e.g. H_2O_2 and O_2-) and NO involved in cell signaling are not decreased, so the metabolic oxidation-reduction reactions are not disturbed. As we know, O_2 and H_2O_2 have important functions in neutrophils and macrophages (Ohsawa et al., 2007). Because of the hypo-reactivity of them, hydrogen does not disturb the innate immune system and allow phagocytosis of infecting organisms. Also, as endogenous NO plays an important role in vasodilation, vasoconstriction and leukocyte/endothelial interactions, it may be beneficial to spare endogenous NO (Pinsky et al., 1994; Huang et al., 2010). Third, hydrogen is better tolerated than many other antioxidants. Forth, it is found that hydrogen doesn't influence physiological parameters in the blood (temperature, blood pressure, pH, pO_2) (Hong et al., 2010). Fifth, compared to other drugs, the cost of hydrogen therapy is much lower. All of these properties of hydrogen make administration of hydrogen water a promising and treatment for various diseases.

6.2 Disadvantages

Every drug has its disadvantages; hydrogen is not an exception. For hydrogen inhalation, the main concern may be the risk of explosion. Although hydrogen won't explore under the concentration 4.6%, there is still risk of explosion during storing, transportation and dispensing. Furthermore, a complicated appliance is required for it. For hydrogen water injection, the main disadvantage may be the compliance of patients. The proper administration time, ways and dosages of hydrogen saline injection are other concerns. Oral administration of hydrogen water may be a better choice, but some evidences showed that there might be some adverse effects as well. It was found that there were a few changes in hematology and clinical chemistry parameters (Saitoh et al., 2010). In human trials, slight decreases in aspartate aminotransferase and alanine aminotransferase and increases in gammaglutamyl transferase and total bilirubin were observed (Nakao et al., 2010). Other symptoms, such as loose stools, increase in frequency of bowel movement, heartburn and headache were also experienced by some people who drunk hydrogen water (Nakao et al., 2010; Huang et al., 2010).

7. Not solved problems of H2 as a therapeutic agent

First of all, as we mentioned above, growing number of evidence suggest that ROS scavenging properties are unlikely to be the only explanation for the effects of hydrogen (Huang et al., 2010). Other not yet defined biological mechanisms of hydrogen as a signaling molecule need exploring. Some additional issues, such as, how H_2 involves the cross-talk among anti-oxidation, anti-inflammation and anti-allergy, what is the primary molecular target of H_2, remain unknown (Ohta, 2011). Secondly, although the protective effects of H_2 have been proved in clinical trials, well designed and large-scale clinical studies are required to explore ideal dose, timing and delivery methods (Huang et al., 2010). Thirdly, we need to explore the pharmacokinetics, biology and toxicity of hydrogen to well apply it clinically (Huang et al., 2010). Fourthly, although several ways to administrate hydrogen have been explored, no one has seriously compared their advantages and disadvantages in clinical trials. An ideal administration method should be selected out for clinical application.

8. References

Abraini, J.H., Gardette-Chauffour, M.C., Martinez, E., Rostain, J.C., & Lemaire, C. (1994). Psychophysiological reactions in humans during an open sea dive to 500 m with a hydrogen-helium-oxygen mixture. *J Appl Physiol*, Vol. 76, No.3, pp.1113-1118.

Atsushi, H., Inaba, H., Suzuki, E., Kasai1, K., Suzuki, H., Shinohara, A., Shirao, M., Kubo, K., & Yoshimura, Y. (2010). In vitro physicochemical properties of neutral aqueous solution systems (water products as drinks) containing hydrogen gas, 2-carboxyethyl germanium sesquioxide, and platinum nanocolloid as additives. *J Health Sci*, Vol. 56, No. 2, pp.167-174.

Buchholz, B.M., Kaczorowski, D.J., Sugimoto, R., Yang, R., Wang, Y., Billiar, T.R., McCurry, K.R., Bauer, A.J., & Nakao, A. (2008). Hydrogen inhalation ameliorates oxidative stress in transplantation induced intestinal graft injury. *Am J Transplant*, Vol.8, No.10, pp.2015-24.

Cai, J., Kang, Z., Liu, K., Liu, W., Li, R., Zhang, J.H., Luo, X., & Sun X. (2009). Neuroprotective effects of hydrogen saline in neonatal hypoxia-ischemia rat model. *Brain Res*, Vol. 1256, pp.129-137.

Chen, H., Sun, Y.P., Li, Y., Liu, W.W., Xiang, H.G., Fan, L.Y., Sun, Q., Xu, X.Y., Cai, .JM., Ruan, C.P., Su, N., Yan, R.L., Sun, X.J., & Wang, Q. (2010). Hydrogen-rich saline ameliorates the severity of l-arginine-induced acute pancreatitis in rats. *Biochem Biophys Res Commun*, Vol. 393, No. 2, pp.308-13.

Dole, M., Wilson, F.R., & Fife, W.P. (1975). Hyperbaric hydrogen therapy a possible treatment for cancer. *Science*, Vol. 190, No. 4210, pp.152-4.

Fu, Y., Ito, M., Fujita, Y., Ito, M., Ichihara, M., Masuda, A., Suzuki, Y., Maesawa, S., Kajita, Y., Hirayama, M., Ohsawa, I., Ohta, S., & Ohno, K. (2009). Molecular hydrogen is protective against 6-hydroxydopamine-induced nigrostriatal degeneration in a rat model of Parkinson's disease. *Neurosci Lett*, Vol. 453, No. 2, pp.81-85.

Fukuda, K., Asoh, S., Ishikawa, M., Yamamoto, Y., Ohsawa, I., & Ohta, S. (2007). Inhalation of hydrogen gas suppresses hepatic injury caused by ischemia/reperfusion through reducing oxidative stress. *Biochem Biophys Res Commun*; Vol. 361, No.3, pp.670-674.

Gharib, B., Hanna, S., Abdallahi, O.M., Lepidi, H., Gardette, B., & De Reggi, M. (2001). Anti-inflammatory properties of molecular hydrogen investigation on parasite-induced liver inflammation. *C R Acad Sci III*, Vol. 324, No. 8, pp.719-24.

Gu, Y., Huang, C.S., Inoue, T., Yamashita, T., Ishida, T., Kang, K.M., & Nakao, A. (2010). Drinking hydrogen water ameliorated cognitive impairment in senescence-accelerated mice. *J Clin Biochem Nutr*, Vol. 46, No. 3, pp.269-76.

Hanefeld, M., Cagatay, M., Petrowitsch, T., Neuser, D., & Petzinna, D., Rupp, M. (2004). Acarbose reduces the risk for myocardial infarction in type 2 diabetic patients meta-analysis of seven long-term studies. *Eur Heart J*, Vol. 25, No.1, pp.10-16.

Hayashida, K., Sano, M., Ohsawa, I., Shinmura, K., Tamaki, K., Kimura, K., Endo, J., Katayama, T., Kawamura, A., Kohsaka, S., Makino, S., Ohta, S., Ogawa, S., Fukuda, K. (2008). Inhalation of hydrogen gas reduces infarct size in the rat model of myocardial ischemia-reperfusion injury. *Biochem Biophys Res Commun*, Vol. 373, No.1, pp.30-35.

Huang, C.S., Kawamura, T., Toyoda, Y., Nakao, A. (2010). Recent advances in hydrogen research as a therapeutic medical gas. *Free Radic Res*, Vol. 44, No. 9, pp.971-982.

Hong, Y., Chen, S., Zhang, J.M. (2010). Hydrogen as a selective antioxidant a review of clinical and experimental studies. *J Int Med Res*, Vol. 38, No. 6, pp.1893-1903.

Itoh, T., Funjita, Y., Ito, M., Masuda, A., Ohno, K., Ichihara, M., Kojima, T., Nozawa, Y., Ito, M. (2009). Molecular hydrogen suppresses FcεRI-mediated signal transduction and prevents degranulation of mast cells. *Biochem Biophys Res Commun*, Vol. 389, No. 4, pp.651-656.

Itoh, T., Hamada, N., Terazawa, R., Ito, M., Ohno, K., Ichihara, M., Nozawa, Y. (2011). Molecular hydrogen inhibits lipopolysaccharide /interferon γ-induced nitric oxide production through modulation of signal transduction in macrophages. *Biochem Biophys Res Commun*, Vol. 411, No. 1,pp.143-149.

Kajiya, M., Sato, K., Silva, M.J., Ouhara, K., Do, P.M., Shanmugam, K.T., Kawai, T. (2009). Hydrogen from intestinal bacteria is protective for Concanavalin A-induced hepatitis. *Biochem Biophys Res Commun*, Vol. 386, No.2, pp.316-321.

Kajiya, M., Silva, M.J., Sato, K., Ouhara, K., Kawai, T. (2009). Hydrogen mediates suppression of colon inflammation induced by dextran sodium sulfate. *Biochem Biophys Res Commun*, Vol. 386, No. 1, pp.11-15.

Kawamura, T., Huang, C.S., Peng, X, Masutani, K., Shigemura, N., Billiar, T.R., Okumura, M., Toyoda, Y., Nakao, A. (2011). The effect of donor treatment with hydrogen on lung allograft function in rats. *Surgery*, Vol. 150, No. 2, pp. 240-9.

Kubota, M., Shimmura, S., Kubota, S., Miyashita, H., Kato, N., Noda, K., Ozawa, Y., Usui, T., Ishida, S., Umezawa, K., Kurihara, T., Tsubota, K. (2011). Hydrogen and N-acetyl-L-cysteine rescue oxidative stress-induced ngiogenesis in a mouse corneal alkali-burn model. *Invest Ophthalmol Vis Sci*, Vol. 52, No.1, pp.427-433.

Liu, S., Sun, Q., Tao, H., Sun, X. (2010). Oral administration of mannitol may be an effective treatment for ischemia-reperfusion injury. *Med Hypotheses*. Vol. 75, No. 6, pp.620-622.

Maffel, H.V., Metz, G., Bampoe, V., Shiner, M., Herman, S., Brook, C.G. (1977). Lactose intolerance, detected by the hydrogen breath test, in infants and children with chronic diarrhoea. *Arch Dis Child*, Vol. 52, No. 10, pp.766-771.

Mao, Y.F., Zheng, X.F., Cai, J.M., You, X.M., Deng, X.M., Zhang, J.H., Jiang, L., Sun, X.J. (2009). Hydrogen-rich saline reduces lung injury induced by intestinal ischemia/reperfusion in rats. *Biochem Biophys Res Commun*, Vol. 381, No.4, pp.602-605.

Nagata, K., Nakashima-Kamimura, N., Mikami, T., Ohsawa, I., Ohta, S. (2009). Consumption of molecular hydrogen prevents the stress-induced impairments in hippocampus-dependent learning tasks during chronic physical restraint in mice. *Neuropsychopharmacology*. Vol. 34, No. 2, pp.501-508.

Nakao, A., Sugimoto, R., Billiar, T.R., McCurry, K.R. (2009). Therapeutic antioxidant medical gas. *J Clin Biochem Nutr*. Vol. 44, No. 1, pp.1-13.

Nakao, A., Toyoda, Y., Sharma, P., Evans, M., Guthrie, N. (2010). Effectiveness of hydrogen rich water on antioxidant status of subjects with potential metabolic syndrome-an open label pilot study. *J Clin Biochem Nutr*, Vol. 46, No.2, pp.140-149.

Nakashima-Kamimura, N., Mori, T., Ohsawa, I., Asoh, S., Ohta, S. (2009). Molecular hydrogen alleviates nephrotoxicity induced by an anti-cancer drug cisplatin without compromising anti-tumor activity in mice. *Cancer Chemother Pharmacol*, Vol. 64, No.4, pp.753-761.

Oharazawa, H., Igarashi, T., Yokota, T., Fujii, H., Suzuki, H., Machide, M., Takahashi, H., Ohta, S., Ohsawa, I. (2010). Protection of the retina by rapid diffusion of hydrogen administration of hydrogen-loaded eye drops in retinal ischemia-reperfusion injury. *Invest Ophthalmol Vis Sci*, Vol. 51, No.1, pp.487-492.

Ohsawa, I., Ishikawa, M., Takahashi, K., Watanabe, M., Nishimaki, K., Yamagata, K., Katsura, K., Katayama, Y., Asoh, S., Ohta, S. (2007). Hydrogen acts as a therapeutic antioxidant by selectively reducing cytotoxic oxygen radicals. *Nat Med*, Vol.13, No.6, pp.688-694.

Ohsawa, I., Nishimaki, K., Yamagata, K., Ishikawa, M., Ohta, S. (2008). Consumption of hydrogen water prevents atherosclerosis in apolipoprotein E knockout mice. *Biochem Biophys Res Commun*, Vol. 377, No. 4, pp.1195-1198.

Ohta, S. (2011). Recent progress toward hydrogen medicine potential of molecular hydrogen for preventive and therapeutic applications. *Curr Pharm Des*, [Epub ahead of print].

Olson, J.W., Maier, R.J. (2002). Molecular hydrogen as an energy source for Helicobacter pylori. *Science*, Vol. 298, No. 5599, pp.1788-1790.

Pinsky, D.J., Naka, Y., Chowdhury, N.C., Liao, H., Oz, M.C., Michler, R.E., Kubaszewski, E., Malinski, T., Stern, D.M. (1994). The nitric oxide/cyclic GMP pathway in organ transplantation critical role in successful lung preservation. *Proc Natl Acad Sci USA*, Vol. 91, No.25, pp.12086-12090.

Sobko, T., Norman, M., Norin, E., Gustafsson, L.E., Lundberg, J.O. (2005). Birth-related increase in intracolonic hydrogen gas and nitric oxide as indicator of host-microbial interactions. *Allergy*, Vol. 60, No. 3, pp.396-400.

Sun, H., Chen, L., Zhou, W., Hu, L., Li, L., Tu, Q., Chang, Y., Liu, Q., Sun, X., Wu, M., Wang, H. (2011). The protective role of hydrogen-rich saline in experimental liver injury in mice. *J Hepatol*, Vol. 54, No. 3, pp.471-480.

Sun, Q., Cai, J., Zhou, J., Tao, H., Zhang, J.H., Zhang, W., Sun, X.J. (2011). Hydrogen-rich saline reduces delayed neurologic sequelae in experimental carbon monoxide toxicity. *Crit Care Med*, Vol. 39, No. 4, pp.765-769.

Shimouchi, A., Nose, K., Takaoka, M., Hayashi, H., Kondo, T. (2009). Effect of dietary turmeric on breath hydrogen. Dig Dis Sci. Vol. 54, No. 8, pp.1725-9.

Shimouchi, A., Nose, K., Yamaguchi, M., Ishiguro, H., Kondo, T. (2009). Breath hydrogen produced by ingestion of commercial hydrogen water and milk. *Biomark Insights*, Vol. 4, pp.27-32.

Suzuki, Y., Sano, M., Hayashida, K., Ohsawa, I., Ohta, S., Fukuda, K. (2009). Are the effects of alpha-glucosidase inhibitors on cardiovascular events related to elevated levels of hydrogen gas in the gastrointestinal tract? *FEBS Lett*, Vol. 583, No.13, pp.2157-2159.

Saitoh, Y., Harata, Y., Mizuhashi, F., Nakajima, M., Miwa, N. (2010). Biological safety of neutral-pH hydrogen-enriched electrolyzed water upon mutagenicity, genotoxicity and subchronic oral toxicity. *Toxicol Ind Health*, Vol. 26, No.4, pp.203-216.

Ueda, Y., Nakajima, A., Oikawa, T. (2010). Hydrogen-related enhancement of in vivo antioxidant ability in the brain of rats fed coral calcium hydride. *Neurochem Res*, Vol. 35, No. 10, pp.1510-1515.

Wood, K.C., Gladwin, M.T. (2007). The hydrogen highway to reperfusion therapy. *Nat Med*, Vol. 13, No. 6, pp.673-674.

Wang, C., Li, J., Liu, Q., Yang, R., Zhang, J.H., Cao, Y.P., Sun, X.J. (2011). Hydrogen-rich saline reduces oxidative stress and inflammation by inhibit of JNK and NF-κB activation in a rat model of amyloid-beta-induced Alzheimer's disease. *Neurosci Lett*, Vol. 491, No. 2, pp.127-132.

Zheng, X., Mao, Y., Cai, J., Li, Y., Liu, W., Sun, P., Zhang, J.H., Sun, X., Yuan, H. (2009). Hydrogen-rich saline protects against intestinal ischemia/reperfusion injury in rats. *Free Radic Res*, Vol. 43, No. 5, pp.478-84.

Section 3

Cellular and Molecular Targets

10

Renal Redox Balance and Na⁺, K⁺-ATPase Regulation: Role in Physiology and Pathophysiology

Elisabete Silva and Patrício Soares-da-Silva
Department of Pharmacology and Therapeutics,
Faculty of Medicine, Porto University,
Portugal

1. Introduction

An imbalance between oxidant production and antioxidant defences has been associated with the development of several conditions such as hypertension, obesity-associated hypertension and diabetes, as well as during the ageing process (Bedard & Krause, 2007; Makino et al., 2003; Touyz & Schiffrin, 2004; Valko et al., 2007; Wilcox, 2005). This association is achieved through direct reactive oxygen species damage upon biomolecules or reactive oxygen species-induced alterations in gene and protein regulation (Finkel, 2003; Gill & Wilcox, 2006; McCubrey et al., 2006; Valko et al., 2007).

Being the kidney an organ severely affected in the above-mentioned conditions, most likely tissue redox balance has implications in renal physiology and pathophysiology. In this Na⁺,K⁺-ATPase is especially important since it is a key molecule in renal electrolyte regulation. As such, this chapter concerns the regulation of Na⁺,K⁺-ATPase by reactive products of oxygen metabolism and its physiological and pathophysiological implications. Focus is given to NADPH-oxidase derived reactive oxygen species, since they appear to be important for the redox-signal, and to superoxide dismutases, for being the anti-oxidant molecules that most efficiently scavenge and remove reactive oxygen species.

This chapter begins with a brief introduction to the basic principles of renal function, Na⁺,K⁺-ATPase structure and mechanisms of regulation, followed by a short review on renal reactive oxygen species production and anti-oxidant defence. An exploration of the new findings and ideas on the dynamic interplay between renal redox balance, the molecular effects at the cellular level and Na⁺,K⁺-ATPase function is approached more deeply in the following section. Finally, experimental animal models supporting that loss of redox balance and altered Na⁺,K⁺-ATPase function contribute to the development of renal associated pathologies is addressed. The chapter ends with a broad overview, given in the conclusion section.

2. The kidney function and Na⁺, K⁺-ATPase

In an adult organism the kidney plays an important role in the regulation of blood pressure, nutrient and electrolyte reabsorption and drug and metabolite excretion. This is achieved due to the presence of specialized proteins that are distributed into specific domains of the apical or basolateral membrane of the distinct nephron segments (Abdolzade-Bavil et al., 2004).

Na^+,K^+-ATPase is the major transporter of sodium ions in renal basolateral epithelia throughout the nephron and one of the most important renal transporters (Jaitovich & Bertorello, 2010). Na^+,K^+-ATPase is an oligomeric transmembrane protein composed of two main subunits, α and β (Figure 1). The α-subunit is the catalytic domain of Na^+,K^+-ATPase and contains the binding site for sodium ions, potassium ions, ATP, steroid hormones and phosphorylation sites for protein kinase A and protein kinase C (Aperia, 2001; Bertorello et al., 1991; Ewart & Klip, 1995; Feraille & Doucet, 2001; Schwartz et al., 1988). The β-subunit is involved in enzyme maturation, localization to the plasma membrane and stabilization of the potassium-occluded intermediate (Geering, 2008). A third subunit has been recently described to bind α and β complex in some tissues, such as heart, kidney and brain. This subunits belongs to the FXYD proteins, a group of structurally similar polypeptides expressed in a tissue-specific manner, and modulates cation binding affinity to Na^+,K^+-ATPase (Crambert & Geering, 2003; Geering, 2006; Geering et al., 2003).

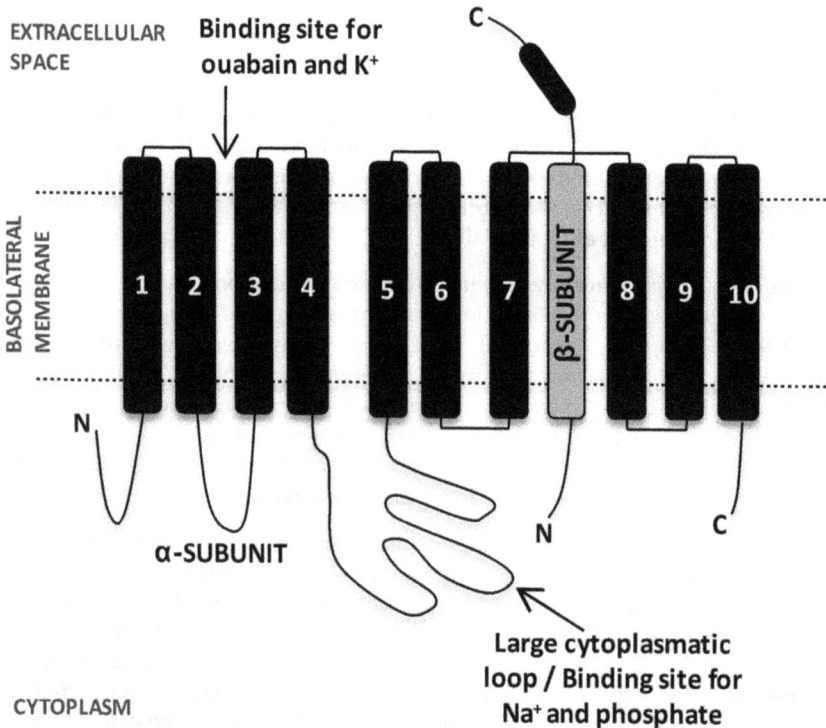

Fig. 1. Schematic representation of Na^+,K^+-ATPase. Na^+,K^+-ATPase is composed of two main subunits: a catalytic α-subunit (black) and a glycosylated β-subunit (grey).

There are 4 known isoforms of the α-subunit: $α_1$, $α_2$, $α_3$ and $α_4$, all with a unique tissue distribution. The $α_1$-isoform is expressed ubiquitously (Blanco & Mercer, 1998), and it is the major isoform expressed in the kidney (Kaplan, 2002).

The β-subunit has 3 known isoforms: β_1, β_2 and β_3. Detection of the tissue distribution of the β-subunit isoforms has been more difficult due to the lack of specific antibodies. However, antibody sensitivity has been improved by deglycosylation of the β-subunit. Current knowledge is that the β_1-isoform is expressed in most tissues, including the kidney (Vagin et al., 2007). The tissue specific distribution of α and β subunits indicates that each combination exhibits unique cellular functions.

In the kidney, Na⁺,K⁺-ATPase catalyzes ATP-dependent transport of three sodium ions in exchange for two potassium ions, maintains intracellular ion balance and membrane potential and is also responsible for maintaining sodium gradient. Thus providing the driving force for nutrients, electrolytes and water reabsorption (Aperia, 2001; Feraille et al., 2001; Kaplan, 2002; Skou, 1957). In the renal proximal tubules Na⁺,K⁺-ATPase plays an essential role in the bulk reabsorption of sodium and sodium-dependent reabsorption of nutrients and other electrolytes (Feraille et al., 2001) (Figure 2). Despite approximately 70% of sodium and potassium being reabsorbed in the proximal tubules the final adjustment is made in the distal tubules and the collecting ducts, where Na⁺,K⁺-ATPase also plays a crucial role.

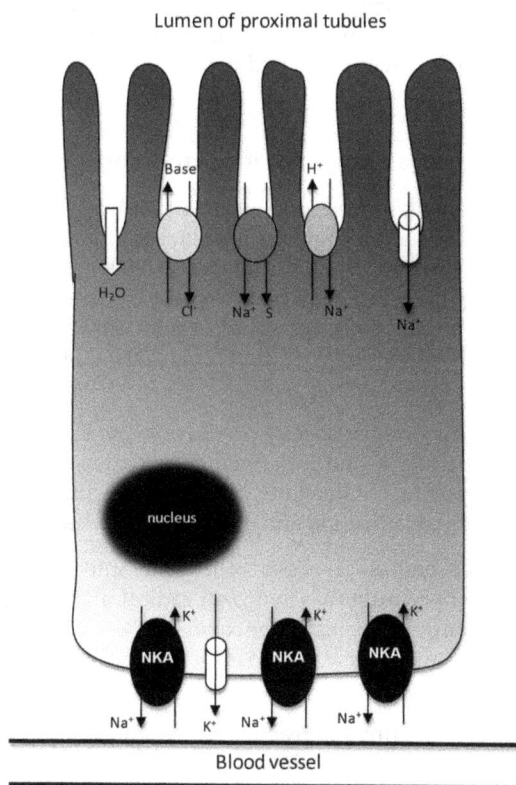

Fig. 2. Schematic representation of solute transport in the proximal tubules. A sodium gradient produced by the Na⁺,K⁺-ATPase allows sodium and organic substrates (S –sugars; amino acids; organic anions) to be transported into the cell. K⁺ is recycled across basolateral membrane through K⁺ channels.

Given the key role of Na^+,K^+-ATPase for normal renal function changes in Na^+,K^+-ATPase regulation are associated with the development of several conditions such as hypertension, obesity-associated hypertension and diabetes as well as during the ageing process (Jaitovich et al., 2010; Silva et al., 2010; Vague et al., 2004; Wang et al., 2009). As such understanding the mechanisms involved in regulation of Na^+,K^+-ATPase throughout the nephron is of major importance.

Renal Na^+,K^+-ATPase is regulated by several hormones such as dopamine, noradrenaline, aldosterone and ouabain, growth factors, peptides, several cytoskeleton proteins such as ankyrins, spectrins, adducins, actin and moesin (Aperia, 2001; Cantiello, 1997; Devarajan et al., 1994; Kraemer et al., 2003; Nelson & Veshnock, 1987; Therien & Blostein, 2000; Tripodi et al., 1996; Zhang et al., 1998) and more directly by ionic distribution across the membrane (Aperia, 2001; Feraille et al., 2001; Haber et al., 1987; Therien et al., 2000; Xie & Askari, 2002; Xie & Cai, 2003; Zhou et al., 2003). These regulatory factors may alter Na^+,K^+-ATPase function through interference with protein synthesis, insertion in membrane compartments, enzyme internalization and substrate affinity. Protein kinases, calcium, cAMP and reactive oxygen species are known secondary messengers involved in Na^+,K^+-ATPase regulation. In comparison to the other secondary messengers, little was known about reactive oxygen species-mediated Na^+,K^+-ATPase regulation and much information has been gathered in the last decade.

3. Renal reactive oxygen species generation and anti-oxidant defence

Reactive oxygen species are now looked at as normal products of cell metabolism used in various physiological functions and recognised for playing a dual role as both harmful and beneficial to the organism (Valko et al., 2007; Valko et al., 2006). Produced in low/moderate concentrations reactive oxygen species may be important mediators in cellular responses to noxia, in the defence against infectious agents, in cellular signalling pathways and in the induction of a mitogenic response (Finkel & Holbrook, 2000; Gill et al., 2006; McCubrey et al., 2006; Valko et al., 2007).

An imbalance between reactive oxygen species production and anti-oxidant defence leads to the disruption of redox homeostasis and is defined as oxidative stress. Oxidative stress is a deleterious process that can induce damage to cell structures (lipids, membranes, proteins and DNA) and lead to cellular dysfunction and eventually cell death (Harman, 1956).

Reactive oxygen species encompass a series of oxygen intermediates that include the superoxide anion, hydrogen peroxide, the hydroxyl radical and hypochlorous acid. In the organism they can be produced by xanthine oxidase, NADPH-oxidase, mitochondrial oxidative phosphorylation, lypoxygenase, cytochrome P450 mono-oxygenase and heme-oxygenase 1 (Figure 3).

Despite the existence of several sources of reactive oxygen species, NADPH-oxidase appears to be especially important for the redox-signal (Gill et al., 2006; Lassegue et al., 2001). Seven NADPH-oxidase isoforms with tissue specific distribution have been identified: NOX1, NOX2, NOX3, NOX4, NOX5, DUOX1 and DOUX2. NOX1, NOX2, NOX3 and NOX4 form heterodimers with membrane p22phox, that stabilizes the NOX proteins and docks cytosolic specific regulatory and activator subunits needed for NADPH-oxidase function (Bedard et al., 2007; Gill et al., 2006; Nauseef, 2008). NOX1 activity requires cytosolic NOXO1, in some cases p47phox, NOXA1, and Rac. NOX2 activity requires cytosolic p47phox, p67phox, and Rac. Moreover, p40phox may also contribute to activation of NOX2. NOX3 activity requires NOXO1. NOX4 is active without cytosolic subunits. NOX5, DUOX1, and DUOX2 are activated by calcium and do not appear to require cytosolic subunits (Bedard et al., 2007).

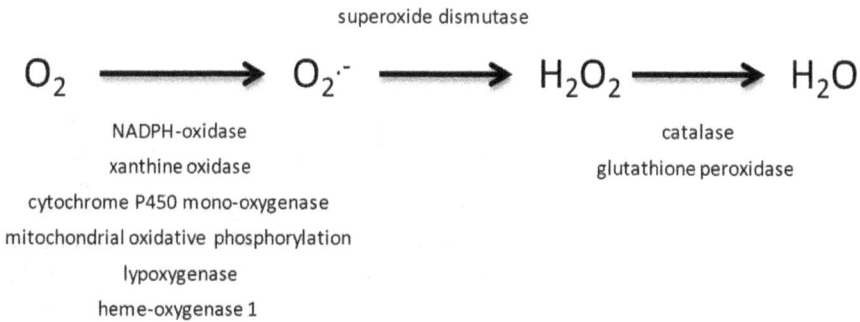

superoxide dismutase

$$O_2 \longrightarrow O_2^{\cdot-} \longrightarrow H_2O_2 \longrightarrow H_2O$$

NADPH-oxidase
xanthine oxidase
cytochrome P450 mono-oxygenase
mitochondrial oxidative phosphorylation
lypoxygenase
heme-oxygenase 1

catalase
glutathione peroxidase

Fig. 3. Schematic representation of the balance oxidant and anti-oxidant enzymes. Multiple enzymes may induce reactive oxygen species generation that in the organism is efficiently detoxified by anti-oxidant enzymes.

Most of NADPH-oxidase isoforms are responsible for the generation of superoxide anion. However, there is still some debate on whether NOX4 generates hydrogen peroxide or superoxide anion that rapidly dismutates into hydrogen peroxide (Bedard et al., 2007; Gill et al., 2006).

The kidney is known to express at least 4 NADPH-oxidase isoforms: NOX1, NOX2, NOX4, NOX5 and NADPH-oxidase regulatory subunits (Bedard et al., 2007; Gill et al., 2006). Despite receiving considerable attention, little is known about NADPH-oxidase function in normal renal physiology (Bedard et al., 2007; Lambeth, 2007). It has been suggested that NADPH-oxidase family may play a role in secretion of erythropoietin, regulation of blood pressure by reaction of superoxide with nitric oxide limiting nitric oxide relaxing effect on afferent arterioles, alteration of cell fate through MAPK pathways activation, induction of apoptosis or cell hypertrophy through ERK 1/2 activation, regulation of gene expression by activation of transcription factors such as NF-kB or c-jun, and innate immunity (Cui & Douglas, 1997; Dorsam et al., 2000; Gorin et al., 2004; Lodha et al., 2002; Lopez et al., 2003; Rhyu et al., 2005; Wilcox, 2003). Renal NADPH-oxidase activity is influenced by diverse stimuli such as angiotensin II, chemokine receptors and aldosterone (Bedard et al., 2007; Cave et al., 2006; Dworakowski et al., 2006; Gill et al., 2006; Lambeth, 2007; Lambeth et al., 2007).

Under physiological conditions renal reactive oxygen species production is largely contained by a complex and efficient array of antioxidant defence systems. These include antioxidant free radical scavengers such as ascorbate, vitamin E, C and A and antioxidant enzymes such as superoxide dismutase, catalase and glutathione peroxidase. Renal superoxide dismutase can rapidly dismutate cellular superoxide anion into hydrogen peroxide that is converted to water and molecular oxygen by catalase or glutathione peroxidase (Figure 3).

Transduction of the chemical reactive oxygen species signal into biological relevant events can occur through a stable sulfenic acid modification of cysteine residues in selected proteins, resulting in protein function alterations (Cave et al., 2006; Finkel, 2003; McCubrey et al., 2006). Once oxidized, proteins can undergo spontaneous or enzymatic reduction back to the initial conformation. This mechanism represents a form of signal transduction similar to phosphorylation.

A large number of proteins have been identified as specific targets of reversible oxidation, including structural proteins, transcription factors, membrane receptors, ion channels, protein kinases and protein phosphatases (Bedard et al., 2007; Cave et al., 2006) (Figure 4). Protein tyrosine phosphatases are probably the best studied, since they control the phosphorylation status of numerous signal-transducing proteins (Finkel, 2003; Meng et al., 2002). Reactive oxygen species-induced oxidation of protein tyrosine phosphatases decreases phosphatase activity by altering the tyrosine/phosphatase balance and thereby influencing signal transduction (Figure 4). This mechanism constitutes an indirect way of reactive oxygen species-mediated activation of the mitogen activated protein kinases signal pathway. However, a direct mechanism of activation of these pathways is also possible through reactive oxygen species-induced activation of membrane receptors, such as

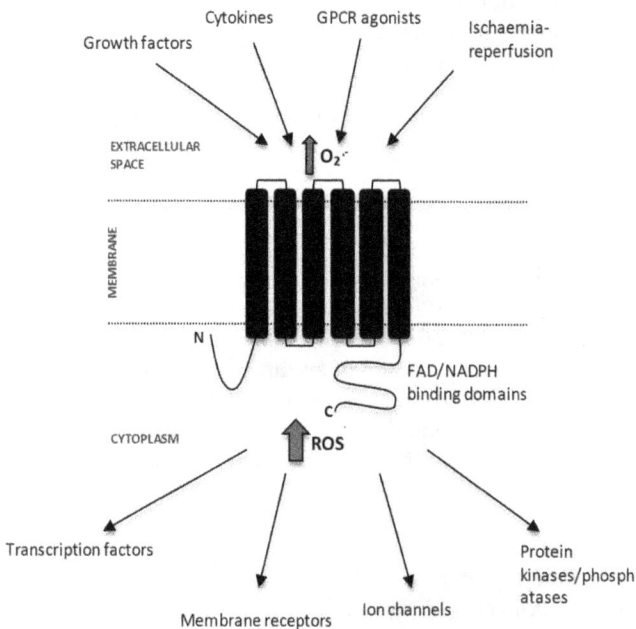

Fig. 4. Schematic representation of known activators of NADPH-oxidase isoforms and downstream effects of NADPH-oxidase derived reactive oxygen species. Diverse stimuli activate NADPH-oxidase isoforms including G-protein coupled receptor (GPCR) agonists, cytokines, growth factors and ischemia-reperfusion. NADPH-oxidase derived reactive oxygen species may influence several signalling pathways through changes in the activity of structural proteins, transcription factors, membrane receptors, ion channels and protein kinases/phosphatases.

endothelial growth factor receptor and platelet-derived growth factor receptor (McCubrey et al., 2006). As such, superoxide anion and hydrogen peroxide can be cell damaging or cell signalling molecules and play an important role in the renal physiology, as well as in the development of renal associated conditions such as diabetes and hypertension (Bedard et al., 2007; Makino et al., 2003; Touyz et al., 2004; Valko et al., 2007; Wilcox, 2005).

4. Redox balance and renal Na⁺, K⁺-ATPase regulation

Reactive oxygen species-mediated molecular mechanisms underlying Na⁺,K⁺-ATPase regulation may rely on an oxidative modification of the enzyme. In opossum kidney cells Na⁺,K⁺-ATPase activity is inhibited by acute incubation with 500 mM of hydrogen peroxide (unpublished results). This finding is in accordance with the works performed by Boldyrev *et al* (Boldyrev & Kurella, 1996) addressed at studying the kinetic parameters of the Na⁺,K⁺-ATPase after its partial inhibition by hydrogen peroxide. They used hydrogen peroxide in the mM concentration range and suggest that oxidized SH-groups of Na⁺,K⁺-ATPase interfered with the capacity of the enzyme to form active oligomers which are essential for higher Na⁺,K⁺-ATPase activity (Figure 5).

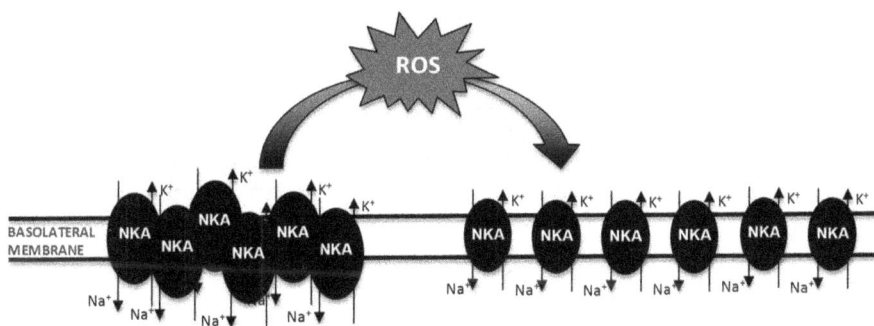

Fig. 5. Schematic representation of direct modulation of Na⁺,K⁺-ATPase by reactive oxygen species (ROS). ROS oxidize Na⁺,K⁺-ATPase interfering with the capacity of the enzyme to form active oligomers and thus decreasing Na⁺,K⁺-ATPase activity.

In cells reactive oxygen species are produced in specific domains, such as membrane microdomains, where NADPH-oxidase isoforms are present, or in the mitochondria. Depending on the cellular location where up-regulation of reactive oxygen species production takes place different modifications in cellular receptors and signalling pathways that alter phosphorylation, gene or protein expression can be achieved. Thus, also different outcomes regarding Na⁺,K⁺-ATPase activity and expression may be observed.

Reactive oxygen species-mediated alteration of cell signalling events that are participate in Na⁺,K⁺-ATPase regulation have been described both *in vivo* and *in vitro*. In Madin-Darby canine kidney cells it has been demonstrated that low potassium-induced an increase in Na⁺,K⁺-ATPase protein content and cell surface Na⁺,K⁺-ATPase expression (Zhou et al., 2003). The signal pathway that transduces the stimulatory effect of low potassium onto up-regulation of Na⁺,K⁺-ATPase is dependent on NADPH-oxidase-derived reactive oxygen species production, Sp1 up-regulation and enhanced transcription of both α and β-subunits of Na$^+$,K$^+$-ATPase (Yin et al., 2008). Also, in opossum kidney cell line increased levels of reactive oxygen species production in cells with 80 passages in culture were responsible for up-regulation of Na$^+$,K$^+$-ATPase (Silva et al., 2006; Silva & Soares-da-Silva, 2007). Opossum kidney cells with 80 passages in culture were shown to have overexpression of NOX1, superoxide dismutase 1, superoxide dismutase 2 and superoxide dismutase 3 isoforms (Silva et al., 2007) and decreased availability to catalyze hydrogen peroxide degradation (unpublished results). When opossum kidney cells with 80 passages in culture were treated with antioxidants (apocynin or TEMPOL) Na$^+$,K$^+$-ATPase activity was found to be significantly decreased (Silva et al., 2007) (Figure 6).

In the proximal tubules of aged rats age-related increase in reactive oxygen species levels were found to be responsible for decreased basal Na$^+$,K$^+$-ATPase activity (Asghar et al., 2001; Silva et al., 2010). The molecular mechanism responsible for the observed decrease in Na$^+$,K$^+$-ATPase activity was a higher basal Na$^+$,K$^+$-ATPase phosphorylation due to reactive oxygen species-mediated increase in protein kinase C activity (Asghar et al., 2003; Asghar et al., 2001; Asghar & Lokhandwala, 2004) (Figure 6).

In the renal medulla where the final regulation of sodium and potassium in the urine takes place, ageing was accompanied by a significant increase in Na$^+$,K$^+$-ATPase activity and expression of the α$_1$-subunit (Silva et al., 2010). Furthermore, not only was Na$^+$,K$^+$-ATPase activity increased in renal medulla of aged Wistar Kyoto rats but also, in this part of the kidney, hydrogen peroxide production was increased with age and in comparison with renal cortex (Silva et al., 2010). Given that Na$^+$,K$^+$-ATPase regulation differs between the proximal and distal nephron segments it is possible that, in the renal medulla, increased reactive oxygen species may activate cell specific signal pathways that up-regulate Na$^+$,K$^+$-ATPase activity. In fact, a study performed by Beltowski and co-workers (Beltowski et al., 2004) investigating whether superoxide anion was involved in the regulation of renal Na$^+$,K$^+$-ATPase support this hypothesis (Figure 6). In this study, infusion of compounds modulating superoxide anion concentration into the abdominal aorta proximally to the renal rat arteries increased medullary Na$^+$,K$^+$-ATPase activity but had no effect on cortical Na$^+$,K$^+$-ATPase activity. Both apocynin and TEMPOL decreased the medullary Na$^+$,K$^+$-ATPase activity. The inhibitory effect of apocynin and TEMPOL was abolished by inhibitors of nitric oxide synthase, soluble guanylate cyclase and protein kinase G. The suggested mechanisms of Na$^+$,K$^+$-ATPase regulation is that NADPH-oxidase-derived superoxide anion increases Na$^+$,K$^+$-ATPase activity in the renal medulla by reducing the availability of nitric oxide. Other mechanism of reactive oxygen species-mediated regulation of Na$^+$,K$^+$-ATPase has been described by the same group. They reported that leptin-induced stimulation of renal Na$^+$,K$^+$-ATPase involves hydrogen peroxide generation, Src kinase, transactivation of the EGF receptor, and stimulation of ERK (Beltowski et al., 2006; Wojcicka et al., 2008) (Figure 6). However, the mechanism of Na$^+$,K$^+$-ATPase regulation in leptin-treated rats shifts from hydrogen peroxide/ERK-dependent to superoxide anion/nitric oxide-dependent after 8 days of treatment, due to a decrease in superoxide dismutase activity and as a consequence higher cellular levels of superoxide anion (Beltowski, 2010; Beltowski et al., 2008).

Finally, reactive oxygen species have also been demonstrated to alter Na$^+$,K$^+$-ATPase activity due to interference with membrane receptor function (Figure 6). Asghar et al (2004) demonstrated that dopamine was unable to inhibit Na+,K+-ATPase activity in old rats and treatment with antioxidants restored the coupling of dopamine type 1 receptor to G proteins. Further studies allowed the identification of the mechanism responsible for decoupling of dopamine type 1 receptor from G proteins (Asghar & Lokhandwala, 2006; Fardoun et al., 2007). In renal proximal tubules increased reactive oxygen species production activates NF-KB and promotes its translocation to the nucleus, where it increases transcription of protein kinase C. This causes an increase in protein kinase C expression and activity, which leads to GRK-2 translocation to the membrane and subsequent dopamine type 1 receptor hyper-phosphorylation and uncoupling from protein G.

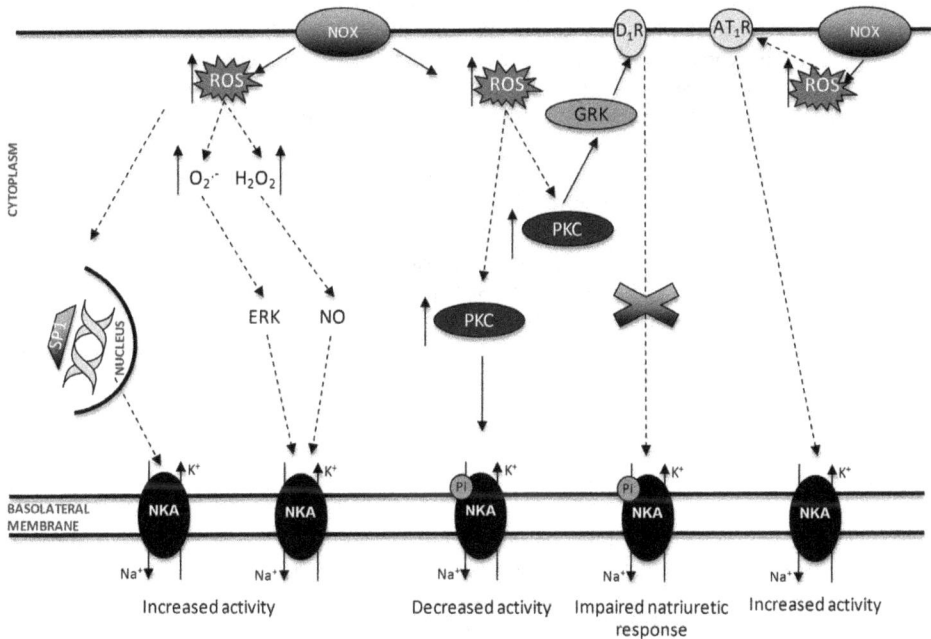

Fig. 6. Schematic representation of reactive oxygen species-mediated modulation of Na+,K+-ATPase. An increase in cellular levels of reactive oxygen species, mainly due to increase activity of NADPH-oxidase, activates specific cell signal cascades that interfere with Na+,K+-ATPase function. Filled arrows indicate a direct action. Dashed arrows indicate that the pathway is not known or was simplified.

5. Loss of redox balance: Functional consequences in renal physiology

Na+,K+-ATPase function in the renal proximal tubules has been a target of extensive research due to the fact that, as mentioned before, it plays a crucial role in the bulk of sodium and nutrient reabsorption. Moreover, as recently reviewed by Wang and co-workers (Wang et al., 2009) it is well known that altered renal proximal tubular sodium reabsorption is implicated in the development of essential hypertension.

In the renal proximal tubules it is well established that dopamine promotes natriuresis while angiotensin II increases sodium retention. Dopamine promotes natriureis via the activation of dopamine type 1 receptors. The dopamine type 1 receptors couple to G_s-proteins and activate the adenylate cyclase-cAMP-protein kinase A signaling pathway. In the kidney dopamine type 1 receptors can also couple to $G_{q/11}$ and activate the phospholipase C-diacylglycerol-protein kinase C pathway. Both a cross-talk between the protein kinase A and protein kinase C signaling pathways or an activation of the phospholipase C-protein kinase C pathway by protein kinase A have been described (Aperia, 2000; Brismar et al., 2000; Gomes & Soares-da-Silva, 2002; Hussain & Lokhandwala, 1998; Jose et al., 2000). Thus, activation of plasma membrane dopamine type 1 receptors stimulates a tissue specific signal cascade that leads to the activation of the protein kinase C δ-isoform. Protein kinase C δ-isoform phosphorylates the $α_1$-subunit of Na+,K+-ATPase producing a conformational change of amino-terminal, which through interaction with other domains of the $α_1$-subunit of Na+,K+-ATPase exposes the binding domains for phosphoinositide 3-kinase and adaptor protein-2. Binding of these proteins induces the activation of Na+,K+-ATPase endocytosis in the proximal tubules and promotes natriuresis (Cinelli et al., 2008; Efendiev et al., 2003; Pedemonte et al., 2005). Angiotensin II exerts an anti-natriuretic effect via the activation of angiotensin II type 1 receptors. Angiotensin II type 1 receptors are predominantly coupled to G-proteins and signal through phospholipases, inositol-phosphatases, calcium channels and serine/threonine and tyrosine kinases. Activation of plasma angiotensin II type 1 receptors stimulates a tissue-specific signaling cascade that leads to the activation of the protein kinase C β-isoform. The protein kinase C β-isoform phosphorylates the $α_1$-subunit of Na+,K+-ATPase producing a conformational change that increases the interaction between the $α_1$-subunit of Na+,K+-ATPase and adaptor protein-1, which results in the recruitment of the enzyme to the plasma membrane (Efendiev et al., 2000; Efendiev et al., 2003).

A reactive oxygen species-associated defect in renal dopamine type 1 receptor function has been observed not only in experimental models of hypertension (Banday et al., 2008; Hussain et al., 1999; Hussain & Lokhandwala, 1997a; Hussain & Lokhandwala, 1997b) but also in diabetes (Banday et al., 2005; Marwaha & Lokhandwala, 2006) and ageing (Fardoun et al., 2006; Hussain et al., 1999; Vieira-Coelho et al., 1999). Failure of dopamine to modulate sodium reabsorption results in diminished natriuresis and blood pressure elevation. More recently it was also demonstrated that reactive oxygen species-mediated angiotensin type 1 receptor up-regulation increases sodium transporters and subsequently contributes to sodium retention and blood pressure elevation (Banday & Lokhandwala, 2008a; Banday & Lokhandwala, 2008b). Angiotensin type 1 receptor up-regulation has been observed in experimental models of hypertension (Reja et al., 2006).

Interest in the regulation of sodium transport in renal medulla is more recent and mainly due to the existence of a possible role in the initiation and development of several forms of experimental hypertension (Cowley & Roman, 1996; Cowley et al., 1992). In renal medulla reactive oxygen species appear to directly alter sodium reabsorption and indirectly alter medullar blood flow, contributing to the development of hypertension (Cowley, 2008; Taylor et al., 2006a; Taylor et al., 2006b). One of the main sources of superoxide anion in this part of the kidney is NADPH-oxidase. NADPH-oxidase–derived superoxide anion was found to contribute to the development of salt-induced hypertension in Dahl salt-sensitive rats (Taylor et al., 2006a). A functional consequence of elevations of superoxide anion within the renal medulla was found to be an immediate reduction of sodium excretion. In this setting, Na+,K+-ATPase may play a role since renal medullary superoxide anion can increase

Na⁺,K⁺-ATPase activity by reducing availability of nitric oxide, as previously described (Beltowski et al., 2004). This mechanism may contribute to an increase in sodium reabsorption and the development of hypertension.

It is now evident that by disrupting several mechanisms responsible for maintenance of sodium homeostasis reactive oxygen species can contribute to the development of renal pathologies and hypertension.

6. Conclusion

Reactive oxygen species-mediated regulation of Na⁺,K⁺-ATPase has been receiving considerable attention. There is now an increasing number of publications addressing multiple mechanisms by which reactive oxygen species may alter Na⁺,K⁺-ATPase function and contribute to the development of several conditions, with special focus being given to hypertension. However, many intermediate events in the mechanisms of reactive oxygen species-mediated regulation of Na⁺,K⁺-ATPase are still unknown and much work needs to be done. Moreover, given that normalization of the redox imbalance in the proximal and distal nephron segments may require the use of specific anti-oxidant molecules and/or pharmacological modulation of different signaling pathways attention should be paid in future therapeutic approaches.

7. References

Abdolzade-Bavil A, Hayes S, Goretzki L, Kroger M, Anders J & Hendriks R (2004). Convenient and versatile subcellular extraction procedure, that facilitates classical protein expression profiling and functional protein analysis. *Proteomics* 4(5): 1397-1405. ISSN 1615-9853

Aperia A (2001). Regulation of sodium/potassium ATPase activity: impact on salt balance and vascular contractility. *Current hypertension reports* 3(2): 165-171. ISSN 1522-6417

Aperia AC (2000). Intrarenal dopamine: a key signal in the interactive regulation of sodium metabolism. *Annual review of physiology* 62: 621-647. ISSN 0066-4278

Asghar M, Hussain T & Lokhandwala MF (2003). Overexpression of PKC-betaI and -delta contributes to higher PKC activity in the proximal tubules of old Fischer 344 rats. *American journal of physiology. Renal physiology* 285(6): F1100-1107. ISSN 1931-857X

Asghar M, Kansra V, Hussain T & Lokhandwala MF (2001). Hyperphosphorylation of Na-pump contributes to defective renal dopamine response in old rats. *Journal of the american society of nephrology* 12(2): 226-232. ISSN 1046-6673

Asghar M & Lokhandwala MF (2004). Antioxidant supplementation normalizes elevated protein kinase C activity in the proximal tubules of old rats. *Experimental biology and medicine (Maywood)* 229(3): 270-275. ISSN 1535-3702

Asghar M & Lokhandwala MF (2006). Antioxidant tempol lowers age-related increases in insulin resistance in Fischer 344 rats. *Clinical and experimental hypertension* 28(5): 533-541. ISSN 1064-1963

Banday AA, Lau YS & Lokhandwala MF (2008). Oxidative stress causes renal dopamine D1 receptor dysfunction and salt-sensitive hypertension in Sprague-Dawley rats. *Hypertension* 51(2): 367-375. ISSN 1524-4563

Banday AA & Lokhandwala MF (2008a). Loss of biphasic effect on Na/K-ATPase activity by angiotensin II involves defective angiotensin type 1 receptor-nitric oxide signaling. *Hypertension* 52(6): 1099-1105. ISSN 1524-4563

Banday AA & Lokhandwala MF (2008b). Oxidative stress-induced renal angiotensin AT1 receptor upregulation causes increased stimulation of sodium transporters and hypertension. *American journal of physiology. Renal physiology* 295(3): F698-706. ISSN 1931-857X

Banday AA, Marwaha A, Tallam LS & Lokhandwala MF (2005). Tempol reduces oxidative stress, improves insulin sensitivity, decreases renal dopamine D1 receptor hyperphosphorylation, and restores D1 receptor-G-protein coupling and function in obese Zucker rats. *Diabetes* 54(7): 2219-2226. ISSN 0012-1797

Bedard K & Krause KH (2007). The NOX family of ROS-generating NADPH oxidases: physiology and pathophysiology. *Physiological reviews* 87(1): 245-313. ISSN 0031-9333

Beltowski J (2010). Leptin and the Regulation of Renal Sodium Handling and Renal Na-Transporting ATPases: Role in the Pathogenesis of Arterial Hypertension. *Current cardiology reviews* 6(1): 31-40. ISSN 1875-6557

Beltowski J, Jamroz-Wisniewska A, Wojcicka G, Lowicka E & Wojtak A (2008). Renal antioxidant enzymes and glutathione redox status in leptin-induced hypertension. *Molecular and cellular biochemistry* 319(1-2): 163-174. ISSN 1573-4919

Beltowski J, Marciniak A, Jamroz-Wisniewska A & Borkowska E (2004). Nitric oxide -- superoxide cooperation in the regulation of renal Na(+),K(+)-ATPase. *Acta biochimica polonica Pol* 51(4): 933-942. ISSN 0001-527X

Beltowski J, Wojcicka G, Trzeciak J & Marciniak A (2006). H2O2 and Src-dependent transactivation of the EGF receptor mediates the stimulatory effect of leptin on renal ERK and Na+,K+-ATPase. *Peptides* 27(12): 3234-3244. ISSN 0196-9781

Bertorello AM, Aperia A, Walaas SI, Nairn AC & Greengard P (1991). Phosphorylation of the catalytic subunit of Na+,K+-ATPase inhibits the activity of the enzyme. *Proceedings of the national academy of sciences of the United States of America* 88(24): 11359-11362. ISSN 0027-8424

Blanco G & Mercer RW (1998). Isozymes of the Na+,K+-ATPase: heterogeneity in structure, diversity in function. *American journal of physiology* 275(5 Pt 2): F633-650. ISSN 0002-9513

Boldyrev A & Kurella E (1996). Mechanism of oxidative damage of dog kidney Na/K-ATPase. *Biochemical and biophysical research communications* 222(2): 483-487. ISSN 0006-291X

Brismar H, Holtback U & Aperia A (2000). Mechanisms by which intrarenal dopamine and ANP interact to regulate sodium metabolism. *Clinical and experimental hypertension* 22(3): 303-307. ISSN 1064-1963

Cantiello HF (1997). Changes in actin filament organization regulate Na+,K+-ATPase activity. Role of actin phosphorylation. *Annals of the New York academy of sciences* 834: 559-561. ISSN 0077-8923

Cave AC, Brewer AC, Narayanapanicker A, Ray R, Grieve DJ, Walker S, et al. (2006). NADPH oxidases in cardiovascular health and disease. *Antioxidants & redox signaling* 8(5-6): 691-728. ISSN 1523-0864

Cinelli AR, Efendiev R & Pedemonte CH (2008). Trafficking of Na+,K+-ATPase and dopamine receptor molecules induced by changes in intracellular sodium concentration of renal epithelial cells. *American journal of physiology. Renal physiology* 295(4): F1117-1125. ISSN 0363-6127

Cowley AW, Jr. (2008). Renal medullary oxidative stress, pressure-natriuresis, and hypertension. *Hypertension* 52(5): 777-786. ISSN 1524-4563

Cowley AW, Jr. & Roman RJ (1996). The role of the kidney in hypertension. *The journal of the american medical association* 275(20): 1581-1589. ISSN 0098-7484

Cowley AW, Roman RJ, Fenoy FJ & Mattson DL (1992). Effect of renal medullary circulation on arterial pressure. *Journal of hypertension* 10(7): S187-193. ISSN 0952-1178

Crambert G & Geering K (2003). FXYD proteins: new tissue-specific regulators of the ubiquitous Na⁺,K⁺-ATPase. *Science signaling - The signal transduction knowledge environment* 2003(166): RE1. ISSN 1525-8882

Cui XL & Douglas JG (1997). Arachidonic acid activates c-jun N-terminal kinase through NADPH oxidase in rabbit proximal tubular epithelial cells. *Proceedings of the National Academy of Sciences of the United States of America* 94(8): 3771-3776. ISSN 0027-8424

Devarajan P, Scaramuzzino DA & Morrow JS (1994). Ankyrin binds to two distinct cytoplasmic domains of Na⁺,K⁺-ATPase a-subunit. *Proceedings of the National Academy of Sciences of the United States of America* 91(8): 2965-2969. ISSN 0027-8424

Dorsam G, Taher MM, Valerie KC, Kuemmerle NB, Chan JC & Franson RC (2000). Diphenyleneiodium chloride blocks inflammatory cytokine-induced up-regulation of group IIA phospholipase A(2) in rat mesangial cells. *Journal of pharmacology and experimental therapeutics* 292(1): 271-279. ISSN 0022-3565

Dworakowski R, Anilkumar N, Zhang M & Shah AM (2006). Redox signalling involving NADPH oxidase-derived reactive oxygen species. *Biochemical Society transactions* 34(Pt 5): 960-964. ISSN 0300-5127

Efendiev R, Bertorello AM, Pressley TA, Rousselot M, Feraille E & Pedemonte CH (2000). Simultaneous phosphorylation of Ser11 and Ser18 in the alpha-subunit promotes the recruitment of Na⁺,K⁺-ATPase molecules to the plasma membrane. *Biochemistry* 39(32): 9884-9892. ISSN 0006-2960

Efendiev R, Budu CE, Cinelli AR, Bertorello AM & Pedemonte CH (2003). Intracellular Na⁺ regulates dopamine and angiotensin II receptors availability at the plasma membrane and their cellular responses in renal epithelia. *The journal of biological chemistry* 278(31): 28719-28726. ISSN 0021-9258

Ewart HS & Klip A (1995). Hormonal regulation of the Na⁺,K⁺-ATPase: mechanisms underlying rapid and sustained changes in pump activity. *American journal of physiology* 269(2 Pt 1): C295-311. ISSN 0002-9513

Fardoun RZ, Asghar M & Lokhandwala M (2007). Role of nuclear factor kappa B (NF-kappaB) in oxidative stress-induced defective dopamine D1 receptor signaling in the renal proximal tubules of Sprague-Dawley rats. *Free radical biology & medicine* 42(6): 756-764. ISSN 0891-5849

Fardoun RZ, Asghar M & Lokhandwala M (2006). Role of oxidative stress in defective renal dopamine D1 receptor-G protein coupling and function in old Fischer 344 rats. *American journal of physiology. Renal physiology* 291(5): F945-951. ISSN 1931-857X

Feraille E & Doucet A (2001). Sodium-potassium-adenosinetriphosphatase-dependent sodium transport in the kidney: hormonal control. *Physiological reviews* 81(1): 345-418. ISSN 0031-9333

Finkel T (2003). Oxidant signals and oxidative stress. *Current opinion in cell biology* 15(2): 247-254. ISSN 0955-0674

Finkel T & Holbrook NJ (2000). Oxidants, oxidative stress and the biology of ageing. *Nature* 408(6809): 239-247. ISSN 0028-0836

Geering K (2008). Functional roles of Na⁺,K⁺-ATPase subunits. *American journal of physiology. Renal physiology* 17(5): 526-532. ISSN 1062-4821

Geering K (2006). FXYD proteins: new regulators of Na⁺,K⁺-ATPase. *American journal of physiology. Renal physiology* 290(2): F241-250. ISSN 1931-857X

Geering K, Beguin P, Garty H, Karlish S, Fuzesi M, Horisberger JD, et al. (2003). FXYD proteins: new tissue- and isoform-specific regulators of Na⁺,K⁺-ATPase. *Annals of the New York academy of sciences* 986: 388-394. ISSN 0077-8923

Gill PS & Wilcox CS (2006). NADPH oxidases in the kidney. *Antioxidants & redox signaling* 8(9-10): 1597-1607. ISSN 1523-0864

Gomes P & Soares-da-Silva P (2002). Role of cAMP-PKA-PLC signaling cascade on dopamine-induced PKC-mediated inhibition of renal Na⁺,K⁺-ATPase activity. *American journal of physiology. Renal physiology* 282(6): F1084-1096. ISSN 0363-6127

Gorin Y, Ricono JM, Wagner B, Kim NH, Bhandari B, Choudhury GG, et al. (2004). Angiotensin II-induced ERK1/ERK2 activation and protein synthesis are redox-dependent in glomerular mesangial cells. *Biochemical journal* 381(Pt 1): 231-239. ISSN 1470-8728

Haber RS, Pressley TA, Loeb JN & Ismail-Beigi F (1987). Ionic dependence of active Na-K transport: "clamping" of cellular Na⁺ with monensin. *American journal of physiology* 253(1 Pt 2): F26-33. ISSN 0002-9513

Harman D (1956). Aging: a theory based on free radical and radiation chemistry. *Journal of gerontology* 11(3): 298-300. ISSN 0022-1422

Hussain T, Kansra V & Lokhandwala MF (1999). Renal dopamine receptor signaling mechanisms in spontaneously hypertensive and Fischer 344 old rats. *Clinical and experimental hypertension* 21(1-2): 25-36. ISSN 1064-1963

Hussain T & Lokhandwala MF (1997a). Dopamine-1 receptor G-protein coupling and the involvement of phospholipase A2 in dopamine-1 receptor mediated cellular signaling mechanisms in the proximal tubules of SHR. *Clinical and experimental hypertension* 19(1-2): 131-140. ISSN 1064-1963

Hussain T & Lokhandwala MF (1997b). Renal dopamine DA1 receptor coupling with G(S) and G(q/11) proteins in spontaneously hypertensive rats. *The American journal of physiology* 272(3 Pt 2): F339-346. ISSN 0002-9513

Hussain T & Lokhandwala MF (1998). Renal dopamine receptor function in hypertension. *Hypertension* 32(2): 187-197. ISSN 0194-911X

Jaitovich A & Bertorello AM (2010). Salt, Na⁺,K⁺-ATPase and hypertension. *Life sciences* 86(3-4): 73-78. ISSN 1879-0631

Jose PA, Eisner GM & Felder RA (2000). Renal dopamine and sodium homeostasis. *Current hypertension reports* 2(2): 174-183. ISSN 1522-6417

Kaplan JH (2002). Biochemistry of Na⁺,K⁺-ATPase. *Annual review of biochemistry* 71: 511-535. ISSN 0066-4154

Kraemer DM, Strizek B, Meyer HE, Marcus K & Drenckhahn D (2003). Kidney Na⁺,K⁺-ATPase is associated with moesin. *European journal of cell biology* 82(2): 87-92. ISSN 0171-9335

Lambeth JD (2007). Nox enzymes, ROS, and chronic disease: an example of antagonistic pleiotropy. *Free radical biology & medicine* 43(3): 332-347. ISSN 0891-5849

Lambeth JD, Kawahara T & Diebold B (2007). Regulation of Nox and Duox enzymatic activity and expression. *Free radical biology & medicine* 43(3): 319-331. ISSN 0891-5849

Lassegue B, Sorescu D, Szocs K, Yin Q, Akers M, Zhang Y, et al. (2001). Novel gp91(phox) homologues in vascular smooth muscle cells : nox1 mediates angiotensin II-induced superoxide formation and redox-sensitive signaling pathways. *Circulation research* 88(9): 888-894. ISSN 1524-4571

Lodha S, Dani D, Mehta R, Bhaskaran M, Reddy K, Ding G, et al. (2002). Angiotensin II-induced mesangial cell apoptosis: role of oxidative stress. *Molecular medicine* 8(12): 830-840. ISSN 1076-1551

Lopez B, Salom MG, Arregui B, Valero F & Fenoy FJ (2003). Role of superoxide in modulating the renal effects of angiotensin II. *Hypertension* 42(6): 1150-1156. ISSN 1524-4563

Makino A, Skelton MM, Zou AP & Cowley AW, Jr. (2003). Increased renal medullary H_2O_2 leads to hypertension. *Hypertension* 42(1): 25-30. ISSN 1524-4563

Marwaha A & Lokhandwala MF (2006). Tempol reduces oxidative stress and restores renal dopamine D1-like receptor- G protein coupling and function in hyperglycemic rats. *American journal of physiology. Renal physiology* 291(1): F58-66. ISSN 1931-857X

McCubrey JA, Lahair MM & Franklin RA (2006). Reactive oxygen species-induced activation of the MAP kinase signaling pathways. *Antioxidants & redox signaling* 8(9-10): 1775-1789. ISSN 1523-0864

Meng TC, Fukada T & Tonks NK (2002). Reversible oxidation and inactivation of protein tyrosine phosphatases in vivo. *Molecular cell* 9(2): 387-399. ISSN 1097-2765

Nauseef WM (2008). Biological roles for the NOX family NADPH oxidases. *The Journal of biological chemistry* 283(25): 16961-16965. ISSN 0021-9258

Nelson WJ & Veshnock PJ (1987). Ankyrin binding to Na⁺,K⁺-ATPase and implications for the organization of membrane domains in polarized cells. *Nature* 328(6130): 533-536. ISSN 0028-0836

Pedemonte CH, Efendiev R & Bertorello AM (2005). Inhibition of Na⁺,K⁺-ATPase by dopamine in proximal tubule epithelial cells. *Seminars in nephrology* 25(5): 322-327. ISSN 0270-9295

Reja V, Goodchild AK, Phillips JK & Pilowsky PM (2006). Upregulation of angiotensin AT1 receptor and intracellular kinase gene expression in hypertensive rats. *Clinical and experimental pharmacology & physiology* 33(8): 690-695. ISSN 0305-1870

Rhyu DY, Yang Y, Ha H, Lee GT, Song JS, Uh ST, et al. (2005). Role of reactive oxygen species in TGF-beta1-induced mitogen-activated protein kinase activation and epithelial-mesenchymal transition in renal tubular epithelial cells. *Journal of the american society of nephrology* 16(3): 667-675. ISSN 1046-6673

Schwartz A, Grupp G, Wallick E, Grupp IL & Ball WJ, Jr. (1988). Role of the Na⁺,K⁺-ATPase in the cardiotonic action of cardiac glycosides. *Progress in clinical and biological research* 268B: 321-338. ISSN 0361-7742

Silva E, Gomes P & Soares-da-Silva P (2006). Overexpression of Na⁺,K⁺-ATPase parallels the increase in sodium transport and potassium recycling in an in vitro model of proximal tubule cellular ageing. *The Journal of membrane biology* 212(3): 163-175. ISSN 0022-2631

Silva E, Pinto V, Simao S, Serrao MP, Afonso J, Amaral J, et al. (2010). Renal aging in WKY rats: changes in Na+,K+ -ATPase function and oxidative stress. *Experimental gerontology* 45(12): 977-983. ISSN 1873-6815

Silva E & Soares-da-Silva P (2007). Reactive oxygen species and the regulation of renal Na+-K+-ATPase in opossum kidney cells. *American journal of physiology. Regulatory, integrative and comparative physiology* 293(4): R1764-1770. ISSN 0363-6119

Skou JC (1957). The influence of some cations on an adenosine triphosphatase from peripheral nerves. *Biochimica et biophysica acta* 23(2): 394-401. ISSN 0006-3002

Taylor NE, Glocka P, Liang M & Cowley AW, Jr. (2006a). NADPH oxidase in the renal medulla causes oxidative stress and contributes to salt-sensitive hypertension in Dahl S rats. *Hypertension* 47(4): 692-698. ISSN 1524-4563

Taylor NE, Maier KG, Roman RJ & Cowley AW, Jr. (2006b). NO synthase uncoupling in the kidney of Dahl S rats: role of dihydrobiopterin. *Hypertension* 48(6): 1066-1071. ISSN 1524-4563

Therien AG & Blostein R (2000). Mechanisms of sodium pump regulation. *American journal of physiology. Cell physiology* 279(3): C541-566. ISSN 0363-6143

Touyz RM & Schiffrin EL (2004). Reactive oxygen species in vascular biology: implications in hypertension. *Histochemistry and cell biology* 122(4): 339-352. ISSN 0948-6143

Tripodi G, Valtorta F, Torielli L, Chieregatti E, Salardi S, Trusolino L, et al. (1996). Hypertension-associated point mutations in the adducin a and b subunits affect actin cytoskeleton and ion transport. *The Journal of clinical investigation* 97(12): 2815-2822. ISSN 0021-9738

Vagin O, Sachs G & Tokhtaeva E (2007). The roles of the Na$^+$,K$^+$-ATPase b$_1$ subunit in pump sorting and epithelial integrity. *Journal of bioenergetics and biomembranes* 39(5-6): 367-372. ISSN 0145-479X

Vague P, Coste TC, Jannot MF, Raccah D & Tsimaratos M (2004). C-peptide, Na+,K(+)-ATPase, and diabetes. *Experimental diabesity research* 5(1): 37-50. ISSN 1543-8600

Valko M, Leibfritz D, Moncol J, Cronin MT, Mazur M & Telser J (2007). Free radicals and antioxidants in normal physiological functions and human disease. *The international journal of biochemistry & cell biology* 39(1): 44-84. ISSN 1357-2725

Valko M, Rhodes CJ, Moncol J, Izakovic M & Mazur M (2006). Free radicals, metals and antioxidants in oxidative stress-induced cancer. *Chemico-biological interactions* 160(1): 1-40. ISSN 0009-2797

Vieira-Coelho MA, Hussain T, Kansra V, Serrao MP, Guimaraes JT, Pestana M, et al. (1999). Aging, high salt intake, and renal dopaminergic activity in Fischer 344 rats. *Hypertension* 34(4 Pt 1): 666-672. ISSN 0194-911X

Wang X, Armando I, Upadhyay K, Pascua A & Jose PA (2009). The regulation of proximal tubular salt transport in hypertension: an update. *Current opinion in nephrology and hypertension* 18(5): 412-420. ISSN 1535-3842

Wilcox CS (2005). Oxidative stress and nitric oxide deficiency in the kidney: a critical link to hypertension? *American journal of physiology* 289(4): R913-935. ISSN 0363-6119

Wilcox CS (2003). Redox regulation of the afferent arteriole and tubuloglomerular feedback. *Acta physiologica scandinavica* 179(3): 217-223. ISSN 0001-6772

Wojcicka G, Jamroz-Wisniewska A, Widomska S, Ksiazek M & Beltowski J (2008). Role of extracellular signal-regulated kinases (ERK) in leptin-induced hypertension. *Life sciences* 82(7-8): 402-412. ISSN 0024-3205

Xie Z & Askari A (2002). Na$^+$,K$^+$-ATPase as a signal transducer. *European journal of biochemistry / FEBS* 269(10): 2434-2439. ISSN 0014-2956

Xie Z & Cai T (2003). Na$^+$,K$^+$-ATPase-mediated signal transduction: from protein interaction to cellular function. *Molecular interventions* 3(3): 157-168. ISSN 1534-0384

Yin W, Yin FZ, Shen WX, Cai BC & Hua ZC (2008). Requirement of hydrogen peroxide and Sp1 in the stimulation of Na,K-ATPase by low potassium in MDCK epithelial cells. *The international journal of biochemistry & cell biology* 40(5): 942-953. ISSN 1357-2725

Zhang Z, Devarajan P, Dorfman AL & Morrow JS (1998). Structure of the ankyrin-binding domain of a-Na$^+$,K$^+$-ATPase. *The journal of biological chemistry* 273(30): 18681-18684. ISSN 0021-9258

Zhou X, Yin W, Doi SQ, Robinson SW, Takeyasu K & Fan X (2003). Stimulation of Na$^+$,K$^+$-ATPase by low potassium requires reactive oxygen species. *American journal of physiology. Cell physiology* 285(2): C319-326. ISSN 0363-6143

11

Protection of Mouse Embryonic Stem Cells from Oxidative Stress by Methionine Sulfoxide Reductases

Larry F. Lemanski[1,2,*], Chi Zhang[*,3], Andrei Kochegarov[1],
Ashley Moses[1], William Lian[1], Jessica Meyer[1], Pingping Jia[4],
Yuanyuan Jia[5], Yuejin Li[6,7], Keith A. Webster[5],
Xupei Huang[6], Michael Hanna[1], Mohan P. Achary[8],
Sharon L. Lemanski[2] and Herbert Weissbach[9]
[1]Department of Biological and Environmental Sciences,
Texas A&M University-Commerce, Commerce, Texas
[2]Department of Anatomy and Cell Biology
and The Cardiovascular Research Center,
School of Medicine, Temple University,
Philadelphia, Pennsylvania,
USA

1. Introduction

Methionine sulfoxide reductases (Msr) belong to a family of enzymes that includes one MsrA and three MsrBs (MsrB1, MsrB2 and MsrB3) [Zhang et al., 2010, 2011]. We have identified all four enzymes expressed in mouse embryonic stem cell cultures. In addition, we have found a truncated form of MsrA transcript that could have easier access to oxidize methionine residues on proteins than the longer form of the MsrA protein, thus possibly providing an evolutionary selective advantage [Jia et al., 2011].

In this chapter, we will review and summarize the findings from our recent studies. We have stated previously [Zhang et al., 2010, 2011] that the vital cellular functions of the Msr

[3]Department of Radiation Oncology, New York Presbyterian Hospital,
Columbia University Medical Center, New York, New York
[4]Miami Project to Cure Paralysis, University of Miami Miller School of Medicine, Miami, Florida
[5]Department of Molecular and Cellular Pharmacology,
University of Miami, Miller School of Medicine, Miami, Florida
[6]Department of Biomedical Science, Florida Atlantic University; Boca Raton, Florida
[7]Department of Pediatrics, Division of Cardiology, Johns Hopkins School of Medicine, Baltimore, Maryland
[8]Department of Radiation Oncology, School of Medicine, Temple University, Philadelphia, Pennsylvania
[9]Center for Molecular Biology and Biotechnology, Charles E. Schmidt College of Science, Florida Atlantic
University, Jupiter, Florida
USA
[*]Dr. Lemanski and Dr. Zhang share equally as first authors on this review chapter.

family of enzymes are to protect cells from oxidative damage by enzymatically reducing the oxidized sulfide groups of methionine residues in proteins from the sulfoxide form (-SO) back to sulfide thus restoring normal protein functions as well as reducing intracellular reactive oxygen species (ROS). Studies have been performed on the Msr family genes to examine the regulation of gene expression. Using real-time PCR, we have shown that expression levels of the four Msr family genes are under differential regulation by anoxia/reoxygenation treatment, acidic culture conditions and interactions between MsrA and MsrB. Results from these *in vitro* experiments suggest that although these genes function as a whole in oxidative stress protection, each of the Msr genes could be responsive to environmental stimuli differently at the tissue level [Zhang et al., 2011]. We have further shown that one member of the Msr gene family, methionine sulfoxide reductase A (MsrA) can reduce methionine sulfoxide residues in proteins formed by oxidation of methionine by reactive oxygen species (ROS). Msr is an important protein repair system that can also function to scavenge ROS. Our studies have confirmed the expression of MsrA in mouse embryonic stem cells (ESCs) in culture conditions [Zhang et al., 2010]. A cytosol-located and mitochondria-enriched expression pattern has been observed in these cells. To confirm the protective function of MsrA in ESCs against oxidative stress, a siRNA approach was used to knockdown MsrA expression in ESCs. MsrA siRNA treated cells showed less resistance than control ESCs to hydrogen peroxide treatment. Overexpression of MsrA gene products in ES cells showed improved survivability to hydrogen peroxide treatment. This indicates that MsrA plays an important role in cellular defenses against oxidative stress in ESCs. Thus, Msr genes may provide a new target in stem cells to increase their survivability during therapeutic applications [Zhang et al., 2010].

We confirmed that oxidative stress was induced by exposing cells for increasing time periods to anoxia and reoxygenation in a hypoxia chamber [Zhang et al., 2010, 2011]. The expression of MsrA mRNA and protein expression decreased after 4 hours of oxygen depletion. Localization of MsrA proteins in the cytosol and mitochondria of mouse ESCs was evaluated by confocal microscopy and showed a differential distribution with more concentrated levels in the mitochondria. In further studies, knockdown of MsrA expression in mouse embryonic stem cells using siRNAs reduced resistance of the cells to H_2O_2 mediated oxidative stress. In these studies we also found that overexpression of an MsrA-eGFP fusion protein in mouse ESCs provided additional protection against H_2O_2 induced oxidative stress [Zhang et al., 2011].

Based on our studies, we believe that by overexpressing MsrA in stem cells, we can provide significant protection against harsh environments such as ischemia/reperfusion, for example during stroke or heart attack. Currently new approaches to treat disease, such as neurodegenerative disease or ischemia/reperfusion-induced brain or heart damage, are being developed using adult or embryonic stem cell transplantation. One severe limitation in using these stem cells is the low survival of the cells after surgery partly due to high levels of oxidative stress. Results of our recent studies [Zhang et al., 2010, 2011; Jia et al., 2011] suggest a very promising approach to solving this problem by enhancing the resistance of stem cells to oxidative damage after transplantation into a hostile environment caused by ischemia-reperfusion. It will be important to further investigate the differentiation capability of stem cells overexpressing MsrA. Due to the antioxidant actions of these proteins and their reparative properties in recovering target protein function and indirect control of cellular reactive oxygen species (ROS) levels, we believe that overexpressing Msr genes in stem cells could reduce ROS damage to the cells without losing the cells' sensitivity to ROS as a differentiation signal. Thus, the Msr gene family could potentially serve a key role in engineering stem cells to obtain higher resistance to oxidative damage, while retaining the cells' potential for differentiation into adult cell types [Zhang et al., 2010].

2. Background and overview of oxidative stress

ROS are chemically-reactive molecules containing molecular oxygen. They are free radicals, containing an unpaired shell electron: singlet oxygen, superoxide radical, superoxide anion, hydrogen peroxide and hypochlorite ion. ROS are formed by an incomplete one-electron reduction of oxygen. The other non-oxygen chemical radicals are also highly chemically reactive, however, ROS are most abundant in biochemical reactions. They are capable of damaging all bio-molecules: lipids, proteins and nucleic acids.

There are several cellular organelles that generate ROS. Mitochondria are the energy powerhouses of cells producing the major biological energy carrier ATP. In mitochondria, certain enzymes of the cycle transfer electrons from the substrates to carriers such as NAD(P) and FAD which bring electrons to the electron transport (or respiratory) chain. The electron transport chain is generally described as four protein complexes that include NADH oxidase, succinic dehydrogenase, cytochome c reductase and cyotchrome c oxidase; ATP synthase is sometimes referred to as a fifth complex. A small percentage of electrons leak from complex I or III to oxygen resulting in the formation of superoxide anion radicals instead of water. Superoxide anion radicals may spontaneously, or in the presence of metals, turn into hydrogen peroxide and further into a hydroxyl radical: the most chemically active reactive oxygen species.

There are defensive enzymes that are capable of catalyzing chemical reactions of these reactive species to less harmful molecules. Protein oxidation potentially compromises many protein functions, including inhibition of enzymatic and binding activities, increased susceptibility to aggregation and proteolysis, increased or decreased uptake by cells, and altered immunogenicity which then interrupts normal cell functions and induces significant biological damage. Cells have multiple antioxidant systems to scavenge free radicals, including superoxide dismutases (SODs), peroxidases, and catalases. SODs catalyze the dismutation of superoxide anion radicals into molecular oxygen and the less toxic hydrogen peroxide. Peroxidases are a group of enzymes that catalyze the conversion of multiple peroxides (including hydrogen peroxide) into hydroxides. Catalase is an enzyme responsible for conversion of hydrogen peroxide to water and oxygen. The other enzymes, including methionine sulfoxide reductases, reduce oxidized protein molecules.

Methionine sulfoxide reductase is an important enzyme which catalyzes the reduction of free and protein-bound methionine sulfoxide to methionine (Fig. 1). Depending on the nature of the oxidizing species, methionine may undergo a two-electron oxidation to methionine sulfoxide or a one-electron oxidation to the methionine radical cation. This reduction is a repair mechanism for oxidatively damaged proteins. There are two enzymes responsible for reduction of oxidized methionine in proteins: Methionine sulfoxide reductase (Msr) A and B. Most organisms, from bacteria to humans possess MsrA and MsrB genes. The MsrA and MsrB genes exhibit no sequence similarity. MsrA and MsrB genes in several bacterial species are clustered as operons or are fused with each other via connecting domains [Lowther et al., 2002]. In eukaryotes, the MsrA and MsrB genes are typically single.

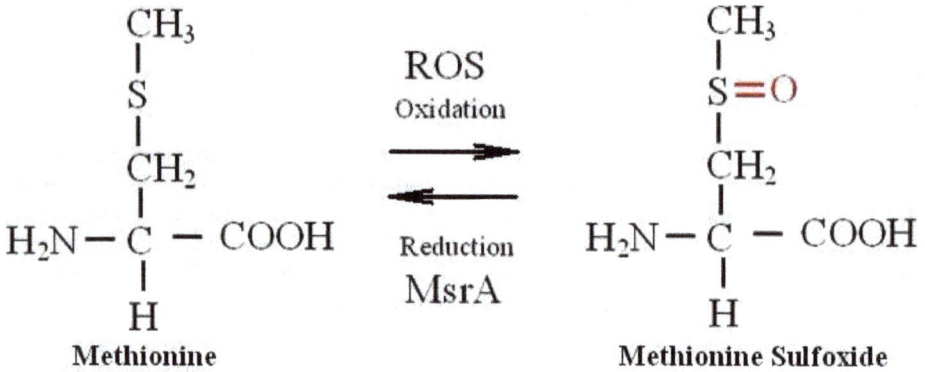

$$
\underset{\text{Methionine}}{
\begin{array}{c}
CH_3 \\
| \\
S \\
| \\
CH_2 \\
| \\
H_2N - C - COOH \\
| \\
H
\end{array}
}
\quad
\begin{array}{c}
\text{ROS} \\
\text{Oxidation} \\
\longrightarrow \\
\longleftarrow \\
\text{Reduction} \\
\text{MsrA}
\end{array}
\quad
\underset{\text{Methionine Sulfoxide}}{
\begin{array}{c}
CH_3 \\
| \\
S = O \\
| \\
CH_2 \\
| \\
H_2N - C - COOH \\
| \\
H
\end{array}
}
$$

Fig. 1. Methionine Sulfoxide Reductase reduces oxidized methionine.

In mammals, three MsrB genes (MsrB1, B2 and B3) and one MsrA gene have been identified. MsrB1 is distributed in the cytosol and nucleus of a cell while MsrB2 is localized in mitochondria [Kim and Gladyshev, 2004]. In humans there are four alternatively spliced isoforms of MSRB3 [Ahmed et al, 2011]. MsrB3 is targeted to the endoplasmic reticulum (ER) and mitochondria [Kim and Gladyshev, 2004].

Although only a single MsrA gene is found in mammals, the corresponding protein is localized in multiple cellular compartments [Kim and Gladyshev, 2005]. Studies on human MsrA gene structures have identified two distinct putative promoters that generate three transcripts. The main MsrA transcript (MsrA1, GenBank: mouse, NM026322; human, NM012331) has been translated into the longest protein. The long MsrA transcript consists of six exons separated by introns (Fig. 2). In the first exon, MsrA has a mitochondrion localization signal peptide (Fig. 3) that targets it to mitochondria [Kim & Gladyshev, 2005]. There is an alternative splicing form of MsrA (S) with short exon 1 missing a mitochondrion localization signal peptide (Fig. 2 and 3) (GenBank: mouse, AK018338; human, AY690665). MsrA (S) is targeted to the cytosol and nucleus. Expression of the MsrA (S) was mostly found in the brain, especially in an early developmental stage.

Alternative forms without exon 3 were identified (Figure 2, GenBank: mouse, BC014738; human, CK819754) which do not have enzymatic activity. MsrA shortform (-3) is present in the cytosol and nucleus. An additional alternative form has also been detected, in which exon 6, the last exon, is replaced with a segment of an unknown gene [GenBank: CD 365491]. This form shared the first exon with the mitochondrial MsrA form, but it is probably catalytically inactive due to an absence of the last exon which contains two cysteine residues critical for catalysis [Kim & Gladyshev, 2005].

Deletion of the fifth exon (113bp) was found in the smaller MsrA form that generates a frame shift in the sixth exon directly attached to the fourth exon thereby forming a premature stop codon (Figure 2). A c-terminal truncated form of MsrA also has been cloned from mouse ESCs due to the skipping of exon 5 and subsequent frame shift in exon 6, also generating a premature stop codon (Figure 2). The total length of the truncated form protein is 148 amino acids, compared to the full length long form protein that has 233 amino acids; both contain a mitochondrial signal peptide at the N-terminus. The truncated form still retains the GCFWG functional motif (catalytic active site) but contains neither of the two cysteines at the c-terminus. The truncated protein shows a different subcellular localization and altered response to anoxia/reoxygenation. These different methionine sulfoxide reductases exhibit different substrate specificities. They may target different methionine residues depending on the protein sequence. Due to the functional importance of the two c-terminal cysteines in the redox reaction, it is likely that the enzyme activity of the variants will be dramatically decreased.

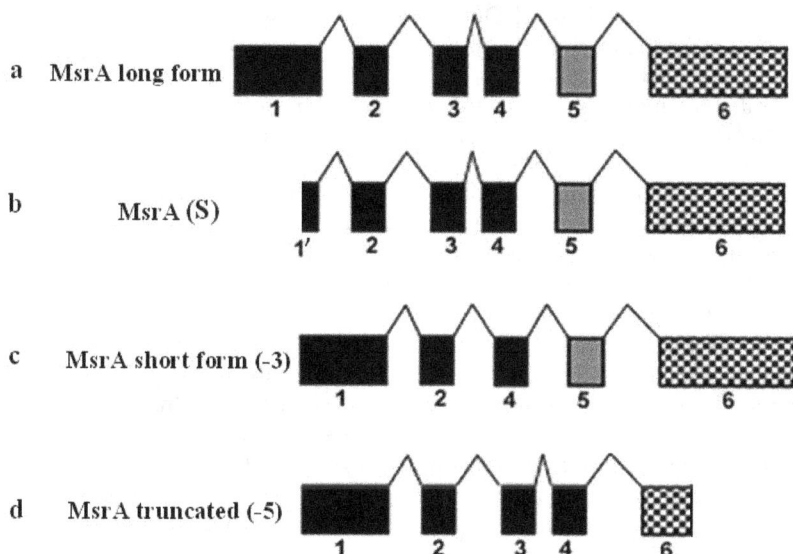

Fig. 2. Alternative splicing isoforms of MsrA gene. a. The long form of MsrA has six exons. b. The one short form MsrA (S) has a short exon 1. c. The short form lacks exon 3. d. The truncated form lacks exon 5, and exon 6 is truncated. (Reprinted with permission of the authors from Jia et al., 2011 J. Biomed. Sci. 18:46).

Fig. 3. Variants of mouse and human MsrAs. The mitochondrial signal peptide is indicated in italic letters in the box. The arrows show the last amino acid residue of exon 1.

3. Methionine sulfoxide reductase

3.1 Protection of embryonic stem cells

As described earlier, methionine sulfoxide reductase A (MsrA), a member of the Msr gene family, can reduce methionine sulfoxide residues in proteins formed by oxidation of methionine by reactive oxygen species (ROS). Msr can also function to scavenge ROS [Levine et al., 1996]. We have confirmed the expression of MsrA in cultured mouse ESCs with a mitochondria-enriched expression pattern [Zhang et al., 2010].

It is well known that oxidative stress plays a key role in cellular injury of patients suffering acute ischemia/reperfusion of multiple tissues including acute myocardial infarction (AMI, heart attack) or chronic insufficient blood perfusion (ischemia) caused by atherosclerosis or vascular stenosis [Zhang et al., 2010]. During oxidative stress, excess reactive oxygen species (ROS) induce damage to proteins, lipids, and nucleic acids that can lead to cell death [Honig & Rosenberg, 2000; Boldyrev et al., 2004; Onyango et al., 2005]. Stem cell transplantation and subsequent differentiation into mature tissues may be able to repair cells and tissues lost to oxidative stress. Cells have multiple anti-oxidation mechanisms to protect against the oxidative insults induced by ischemia/reperfusion [Boldyrev et al., 2004]. Enzymes including superoxide dismutase, catalase, and glutathione peroxidase scavenge the superoxide anion and H_2O_2 to prevent ROS-induced damage. Cells that are modified to over-express such genes are expected to be more resistant to oxidative stress [Blass, 2001]. However, if the levels of ROS become too low, this could have a deleterious effect because ROS are required as signaling molecules to promote differentiation of ESCs, for example into cells of the cardiovascular system [Boldyrev et al., 2004; Sauer et al., 2000; Thiruchelvam et al., 2005; Wo et al., 2008]. Thus, reducing ROS inside cells by using direct anti-oxidants such as vitamin E could interfere with normal cell function. Clearly, this also could interfere with stem cell therapy since the ROS level must be balanced so that cells can undergo normal differentiation without resulting in damage or cell death. Thus, if indeed this hypothesis is correct, then an indirect mechanism to reduce ROS involving the Msr pathways would be advantageous for stem cell therapy by retaining normal ROS basal levels.

ESCs, with their remarkable property of totipotency, have the potential to produce new adult cells to replace those in damaged tissues. Tissue damage and cell death can be caused by reperfusion and ischemia from an accumulation of free radicals in the cells which could result in oxidation and functional impairment directly or through signal transduction pathways such as c-Jun N-terminal kinase (JNK) and p38 mitogen-activated protein kinase (MAPK) [Ueda et al., 2002]. In patients suffering from myocardial infarct or stroke, the infarcted organ is stressed from acidosis due to lack of an adequate blood supply resulting in hypoxia. Kubasiak et al. [2002] have proposed that hypoxia and acidosis are the two conditions that can dramatically induce cell death in cardiomyocytes when present simultaneously. Improvement in the survival of transplanted stem cells in harsh hypoxic and acidic environments is a critical issue for the success of therapeutic tissue repair. Stem cells under conditions of oxidative stress and in acidic environments are an area that requires further study.

The Msr system [Weissbach et al., 2002] is an important self-defense mechanism that protects cells against free radical damage. Proteins, lipids, nucleic acids and other cellular components can be oxidized by free radicals leading to impaired cellular function and/or cell death. Methionine (met), either as a free amino acid molecule or as a peptide, can be oxidized to methionine sulfoxide (Met-(O)). This kind of structural change can cause impaired function of a variety of proteins [Abrams et al., 1981; Taggart et al., 2000; Jones et al., 2008; Shao et al., 2008]. There is a family of enzymes, encoded by methionine sulfoxide reductases (Msr) genes that can reduce Met-(O) to Met. MsrA and MsrB genes have been identified that are specific for reducing the epimers, Met(O)-S and Met(O)-R [Weissbach et al., 2002]. Both MsrA and MsrBs have been implicated in the protection of cells against oxidative damage that in turn have been attributed to MsrA and MsrB genes, suggesting that they may be involved in various age related diseases [Gabbita et al., 1999; Pal et al., 2007; Brennan & Kantorow, 2008; Kim & Gladyshev, 2005].

With respect to protective mechanisms against oxidative damage, it will be important to determine whether the local environment conditions within damaged tissues modulate Msr expression in adjacent stem cells. Cross talk between family members is not well understood and it is not clear whether altered expression levels of one enzyme in the Msr family affects other(s), whether there might exist a co-regulation between different Msr genes; or what controls Msr gene expression. In recent studies, we have described how individual Msr genes respond to varying regulatory processes [Zhang et al., 2010, 2011; Jia et al., 2011]. An acidic culture environment and depletion of oxygen affect Msr gene expression, at the mRNA level; the most significant response was observed in MsrB3, indicating a non-housekeeping activity for this particular gene [Zhang et al., 2011]. Interestingly, MsrB3 showed downregulation of transcription concomitant with MsrA mRNA knockdown by MsrA-specific siRNA [Zhang et al., 2011].

Free radical damage to cellular components, including proteins, is commonly observed under disease conditions and in aging processes. One of the cellular defense mechanisms to reduce oxidation of proteins, thus reducing oxidative stress and restoring their functions, relies on the methionine sulfoxide reductase (msr) family of genes in mammals. Although both MsrA and MsrB genes conduct the redox reactions using a similar chemical reaction, MsrBs convert the R epimer of Met-(O) (Met-R-(O)) back to methionine while MsrA reduces the S epimer of Met-(O)(Met-S-(O)) [Kryukov et al., 2002; Kim & Gladyshev, 2005]. MsrBs localize in different cellular compartments; the MsrB1 protein is cytosolic and nuclear while the MsrB2 is localized in the mitochondria. Human MsrB3, by alternative first exon splicing, produces two forms which targets the endoplasmic reticulum (ER) and the mitochondria of the cell [Kim & Gladyshev, 2004, 2005, 2006].

Even though there is only a single MsrA gene in mammals, its corresponding protein is found in multiple cellular compartments [Vougier et al., 2003]. It turns out that there are two distinct putative promoters of the MsrA gene that generate three transcripts. MsrA1 transcript forms the longest protein that targets the mitochondria. MsrA2 and MsrB3 are formed from a second promoter and localize in the cytosol and nuclei [Lee et al., 2006; Pascual et al., 2009]. Two novel splice forms: MsrA2a and MsrA2b were found recently in rat smooth muscle cells [Haenold et al., 2007]. Alternative splicing occurred in the second exon with the MsrA2, a functional isoform. Thus, alternative promoters and alternative splicing both contribute to the variety of MsrA isoforms that are responsible for methionine sulfoxide reduction in the different cellular compartments.

Currently, most studies on MsrA isoforms emphasize the 5' terminus where different isoforms for mitochondrial signal peptide are present and dictates where protein products are localized within the mitochondria [Kim & Gladyshev, 2006; Lee et al., 2006; Haenold et al., 2007]. There is evidence that transcripts of MsrA from alternative splicings at the 3' end of the MsrA gene are present in the mammalian EST database, however, no detailed studies on these transcripts have been reported [Kim & Gladyshev, 2006].

In our most recent studies in cultured mouse embryonic stem cells, we discovered an MsrA transcript from alternative splicing at the 3'end, that skips exon 5, and produces a shortened isoform with a truncated protein product containing the conserved catalytic active site [Jia et al., 2011]. We performed studies on this truncated isoform's expression pattern under normal culture conditions and in response to oxygen depletion/reoxygenation conditions in mouse embryonic stem cells. The MsrA/B system does not function to eliminate ROS directly, but rather to reverse its damaging effects in the cells. Moreover, methionine is oxidized by ROS in proteins, causing the formation of the R (Met-R-O) and S (Met-S-O) epimers of methionine sulfoxide (met-O) and potential diminution of protein function. MsrA and MsrB are able to reduce protein bound Met-S-O and Met-R-O, respectively, and restore their function. MsrA has been shown to reduce Met-S-O in several oxidized proteins. Scavenging ROS by the Msr system may result from methionine residues in proteins acting as catalytic anti-oxidants. Strong support for a scavenger role of the Msr system is derived from our recent studies on the overexpression of the Msr system, or knocking out the Msr system in cultured cells [Zhang et al., 2010, 2011]. Overexpression of MsrA using adenovirus delivery significantly lowers the hypoxia-induced increase in ROS ordinarily induced by hypoxia and promotes cell survival of PC12 cells in culture [Yermolaieva et al., 2004]. The siRNA knockdown of MsrA and MsrBs in human lens cells also reveals that the MsrA/B system has a protective influence on hydrogen peroxide-induced cellular injury [Kantorow et al., 2004; Yermolaieva et al., 2004].

MsrB is the only Msr family gene whose protein can reduce the R form of methionine sulfoxide unlike the MsrA that is able to reduce the S form of methionine sulfoxide [Kim & Gladyshev, 2004, 2005]. In our studies using siRNA to knockdown MsrA expression and MsrA-GFP fusion protein overexpression, we demonstrated that MsrA plays an important role in protecting stem cells against oxidative damage which has important implications in stem cell therapy.

3.2 MsrA expression patterns in mouse ESCs under differing levels of oxidative stress

We have confirmed MsrA gene expression in mouse embryonic stem cells and have performed studies on MsrA gene expression during hypoxia/reoxygenation using real-time RT-PCR for mRNA transcription, as well as Western blotting with anti-MsrA antibody for MsrA protein expression. This work was published originally in detail in our recent paper [Zhang et al., 2010] and will be summarized below.

Using real-time RT-PCR, mRNA transcription level regulation of the MsrA gene under different levels of oxidative stress was analyzed. Oxidative stress levels were induced by exposing cells to increased time periods of anoxia (in 90% N_2, 5% H_2 and 5% CO_2) and reoxygenation combinations in a hypoxia chamber. It was found that the MsrA mRNA levels decreased with increased exposure to hypoxic and reoxygenation treatments. After 4 hours of oxygen depletion, decreases of MsrA mRNA could be observed; however, the original expression level returned by 8 hours of reoxygenation. One explanation for this observation is that it was a transient, instant response of stem cells to oxygen depletion, since mRNA transcription reductions were apparent for MsrA, MsrBs and MMP9 (Matrix Metalloproteinase-9) genes.

Mouse embryonic stem cell protein samples were collected after the above treatments. Proteins from each sample were loaded in equal amounts onto 4-12% gradient NuPAGE polyacrylamide gels (Invitrogen, Carlsbad, CA). Anti-MsrA antibodies (Abcam, Cambridge, MA) were used in Western blot analysis that showed a gradual decrease of the MsrA protein, with extending treatments of anoxia/reoxygenation, consistent with the mRNA level results.

3.3 Laser confocal microscopic analysis of MsrA protein localization in mouse ESCs

A MsrA-eGFP fusion plasmid was constructed in the pEGFP-N1 vector (Clontech, CA) and transfected into the plasmids into ES cells. 48 to 72 hours after transfection, the ES cells were observed in order to study the subcellular localizations of MsrA in the expression of MsrA-eGFP fusion proteins [Zhang et al., 2010]. Cells then were incubated with MitoTracker (Molecular Probes, WI) for mitochondrial staining before fixation and later mounted in DAPI to stain the nuclei. Confocal microscopy demonstrated a strong colocalization of the green fluorescence signals (MsrA-eGFP) with red MitoTracker signals (mitochondria). Also observed was a general cytosol distribution of MsrA-eGFP outside of the nuclei and what appeared to be between mitochondria. Cytosolic localization with concentrated mitochondrial localization of MsrA-eGFP was not seen in stem cells transfected with the pEGFP-N1 vector control. GFP (eGFP) expression in stem cells showed signals in both nuclei and cytosolic domains suggesting nonspecificity (Figure 4).

3.4 Knock down of MsrA expression using siRNAs reduced resistance of mouse embryonic stem cells to H₂O₂-mediated oxidative stress

To determine whether MsrA can protect stem cells from oxidative stress, we designed and commercially synthesized three siRNAs for MsrA knockdown experiments (Invitrogen, CA). A negative control siRNA and a FITC-labeled siRNA also were synthesized. Using FITC-labeled siRNA, we optimized transfection conditions so that nearly 100% of the stem cells were transfected with siRNA [Zhang et al., 2010].

3.4.1 Real-time RT-PCR confirms knock down of the MsrA mRNA

siRNA transfection in stem cells was performed and total RNA collected at one, two and three days after transfection. The siRNA designed for the MsrA gene (168) and a negative control siRNA, with randomized sequence (882c), were transfected into the stem cells using identical methods. Real-time RT-PCR experiments were conducted with MsrA specific primers designed against the c-teminal end coding sequence, in order to avoid the alternative spliced 5′ end transcripts [Kim & Gladyshev, 2006]. Comparing MsrA mRNA levels after MsrA specific siRNA transfections with the negative control siRNA, we found that one siRNA, from the three designed in our laboratory reduced the mRNA level of the MsrA in the stem cells to only 20% of the control cells (Figure 5A). The reduced levels of MsrA mRNA were maintained for at least 72 hours post-transfection [Zhang et al., 2010].

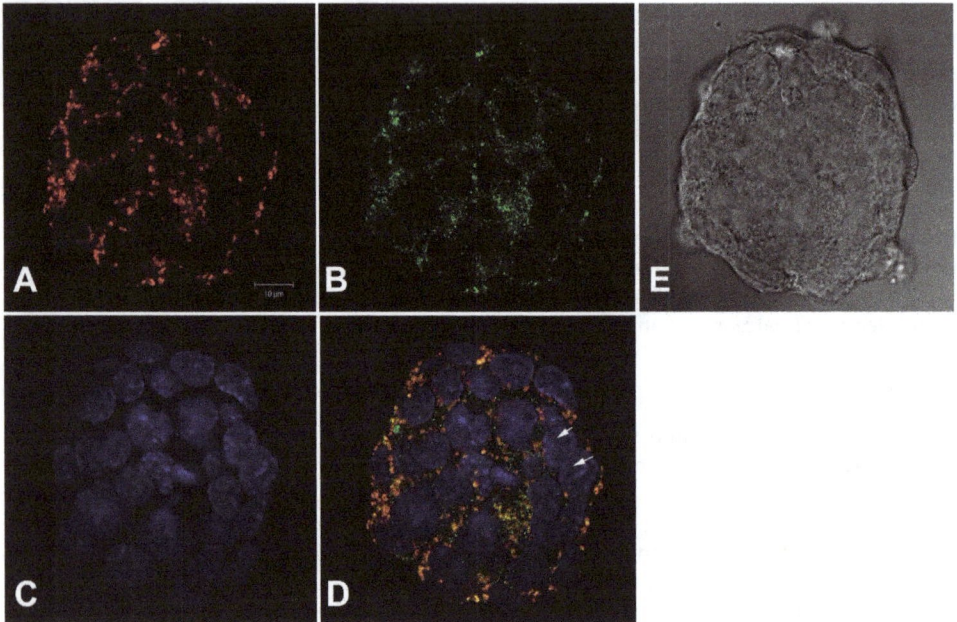

Fig. 4. Confocal microscopy of the subcellular localizations of MsrA-eGFP fusion protein in mouse embryonic stem cells after pEGFP-N1-MsrA fusion expression plasmid transfection for 3 days. A: Stem cells scanned for Mitotracker staining for mitochondria. B: Green fluorescence shows the MsrA-eGFP fusion protein. C: DAPI staining shows nuclei. D: Merged image of A–C showing that MsrA-eGFP localizes in both mitochondria and cytosol, occasionally in nuclei (arrows). E: Phase contrast image of a colony of stem cells. Scalebar is 10 μm. (Reprinted with permission of the authors from Zhang et al., 2010 J. Cell. Biochem. 111(1):94-103).

3.4.2 Western blots confirm knock down of the MsrA proteins

Protein samples from stem cells were collected post-siRNA transfection at 24 hours, 48 hours and 72 hours. Western blotting experiments were conducted using anti-MsrA antibody which binds to a 24kD MsrA protein band (Figure 5B). Western membranes were stripped and rehybridized with β-actin antibody to normalize the MsrA concentrations in each sample (Figure 5D). From the Western blotting data as well as the densitometry of blots (Figure 5E) scanned by a gel imaging system (Alpha Innotech Corp., San Leandro, CA), we unequivocally demonstrate a significant reduction of MsrA protein 24-hours after transfection of MsrA-specific siRNA in the stem cells as compared to the negative control siRNA. It is notable that the MsrA protein levels quickly return to normal levels on the second day after transfection (Figure 5B and 5E), in spite of the mRNA levels still remaining low (see Figure 5A). We also detected a faint band corresponding to MsrA protein dimers (~50kD) from Western blotting after the protein samples were run in reducing polyacrylamide gels (Figure 5B). To verify the existence of the dimer form of the MsrA protein in mouse embryonic stem cells, we repeated the same Western blotting protocol but

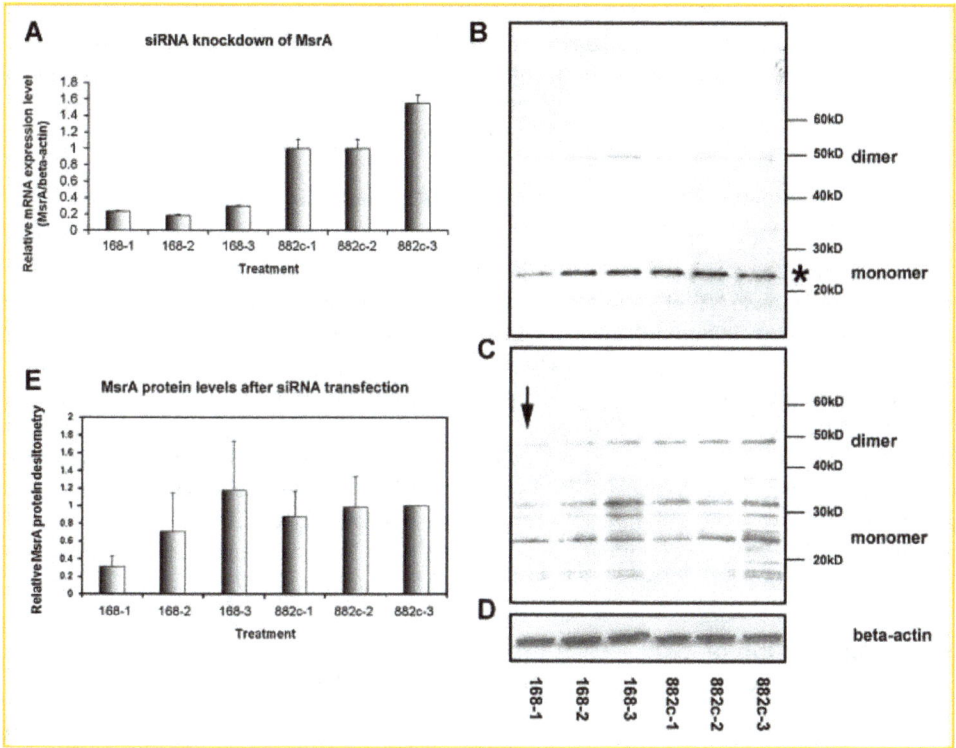

Fig. 5. A: Confirmation of the knockdown of MsrA mRNA by MsrA-specific siRNA transfection using real-time RT-PCR. MsrA-specific siRNA (168) successfully reduces MsrA mRNA levels to only ~20% compared to the negative control siRNA (882c) transfected cells 1 day post-transfectionally. mRNA levels remain low even 3 days after transfection. 168-1 stands for MsrA-specific siRNA (168) transfection for 1 day. B: Confirmation of the knockdown of MsrA protein by MsrA-specific siRNA transfection using Western blot analysis. Western blotting was done with protein samples from stem cells transfected with MsrA-specific siRNA (168) or negative control siRNA (882c) for various days (days 1, 2, and 3) in a reducing PAGE gel. C: Western blot after running the same protein samples as in panel B in a non-reducing PAGE gel. Arrow points indicate significantly reduced MsrA dimers after 1 day transfection of MsrA-specific siRNA (168). D: The same blot as used in B was reprobed for β-actin protein showing equal loading of protein samples in each lane. E: Densitometry studies after scanning the monomer bands shown in panel B (shown with an asterisk) a confirmed significantly decreased expression level of MsrA after 1 day transfection of MsrA-specific siRNA compared to the negative control (168-1 vs. 882c-1). Data were pooled from four separate experiments. 168-1 and 882c-1 stand for transfection of MsrA-specific siRNA (168) or a negative control siRNA (882c) for 1 day. (Reprinted with permission of the authors from Zhang et al., 2010 J. Cell. Biochem. 111(1):94-103).

used a non-reducing PAGE. It clearly shows that the dimer form is present in significant amounts in the stem cells (Figure 5C). These data suggest that it is the dimer form, instead of monomer form, that is reduced after siRNA knockdown of the MsrA mRNA (arrow, Figure 5C) [Zhang et al., 2010]. Densitometry in Figure 5E illustrates the mean values using data pooled from four separate experiments with reducing PAGE (i.e., from four blots as shown in Figure 5B). Only densitometry of the monomer forms (indicated by * in Figure 5B) after reducing PAGE was performed and shown in Figure 5E.

3.4.3 Stem cells show a lowered resistance to H_2O_2 treatments after MsrA siRNA transfection

After examining the effects of siRNA in the MsrA knockdown experiments, we transfected the siRNA into mouse ES cells and treated the cells with hydrogen peroxide to evaluate their resistance to hydrogen peroxide after maximal MsrA protein downregulation after 24 hours. Hydrogen peroxide was diluted into the culture medium at differing concentrations and incubated with the cells overnight prior to MTT assays to compare cell numbers in each survival group. Results clearly showed a significantly reduced resistance of stem cells to hydrogen peroxide treatment with less cell survival after MsrA knock down (Figure 6). Importantly, differences, between MsrA-specific siRNA and the negative control siRNA transfected cells were most significant at the higher hydrogen peroxide concentrations used (i.e., between 100 μM and 300 μM). No differences were observed at lower concentrations when cells begin to die (i.e., between 25 μM to 50 μM) [Zhang et al., 2010].

4. Overexpression of MsrA-eGFP provides additional protection against hydrogen peroxide–induced oxidative stress to mouse embryonic stem cells

We performed experiments in which MsrA-eGFP fusion constructs were transfected into cultured stem cells for two days [Zhang et al., 2010]. With the confirmation of expression of fusion proteins by the appearance of green fluorescence under a fluorescence microscope, hydrogen peroxide at a concentration of 75 μM was added to the medium for 2 hours before the cells were incubated with propidium iodide to stain for non-living cells. At 75 μM concentration, about 50% of the cells survived from previous serial dilution assays (Figure 6). In further studies (illustrations not included in this review — see Zhang et al., 2011), cells were trypsinized and flow cytometry was performed to compare the differences of the cell death ratios in cells with or without MsrA-eGFP fusion protein expression [Zhang et al., 2010]. Results from flow cytometry confirm the expression of the MsrA-eGFP fusion protein, when comparing the cell numbers of EGFP channels (excitation wavelength at 488 nm) between transfected cells and nontransfected cells [Zhang et al., 2010]. Based on cell flow cytometry, the cells were subdivided into four groups: 1. dead cells with no eGFP expression; 2. dead cells with eGFP or MsrA-GFP expression; 3. live cells with no eGFP expression; 4. dead cells with eGFP or MsrA-GFP expression. These experiments showed that of those cells expressing MsrA-GFP fusion protein only 8.67% were dead cells [Zhang et al., 2010]. However, in cells expressing eGFP only, 12.14% of the cells show propidium iodide positive staining (dead). Therefore MsrA-eGFP protected the cells from hydrogen peroxide-mediated oxidative damage,and thus, decreased cell death by 28.6% compared to the eGFP control there are no significant differences in cell death ratios among nontransfected cells within DMEM medium (no DNA transfection), eGFP transfected cells and MsrA-eGFP transfected cells without hydrogen peroxide treatment [Zhang et al., 2010].

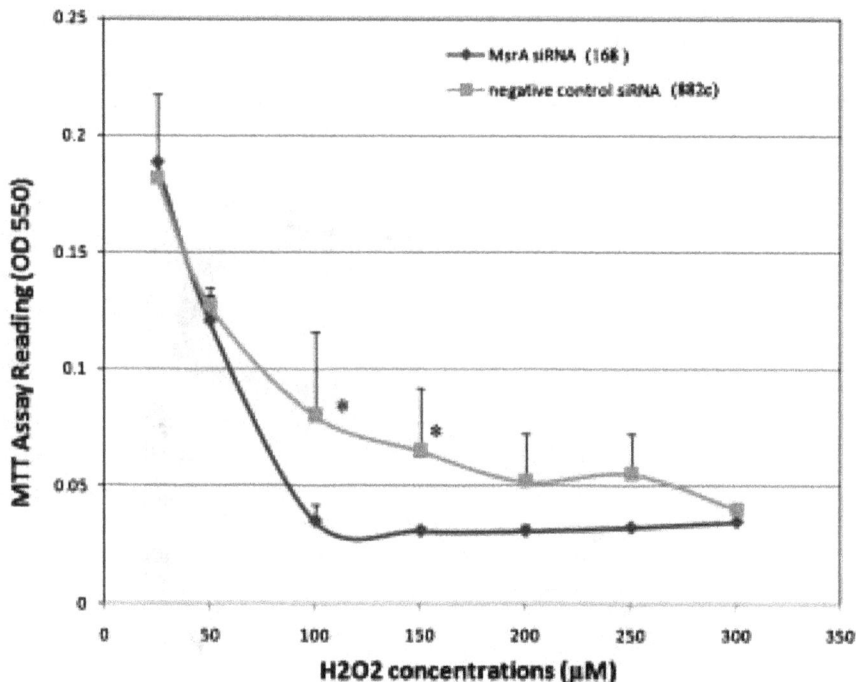

Fig. 6. MTT assays on stem cells treated by MsrA siRNA (168) versus a negative control siRNA (882c). After MsrA-specific siRNA (168) transfection for 24 h, the cells are significantly more prone to oxidative damage mediated by high concentrations of H_2O_2 treatment compared to a negative control siRNA (882c) transfected cells. The asterisks indicate $P < 0.05$. (Reprinted with permission of the authors from Zhang et al., 2010 J. Cell. Biochem. 111(1):94-103).

5. Confirmation of the maintained potency to differentiate into multiple adult cell types from mouse embryonic stem cells in feeder-layer-free culture conditions

To avoid interference of feeder layer embryonic fibroblast cells in our earlier stem cell studies, we employed a feeder-layer-free culture system for our mouse embryonic stem cell culture [Narayanan et al., 1993]. Cells were maintained at a low passage number (less than 20) and cultured on gelatin-coated culture dishes with Leukemia Inhibitory Factors (LIF) added to the medium under conditions previously published [Zhang et al., 2011]. Cells using these conditions maintain their typical stem cell morphology, growing into round colonies with smooth edges (Figure 7A).

To confirm that cells used for subsequent studies would carry genuine stem cell characteristics, we routinely stained the cells for embryonic stem cell specific markers and tested the capability of the cells to differentiate into various adult cell types at the same cell passages we used for the studies of Msr gene expression regulation [Zhang et al., 2011]. The cells under these culture conditions remain undifferentiated and positively stain for Stage-specific Embryonic Antigen-1 (SSEA-1) antibody (Abcam, MA) (Figure 7B). From immunostaining, we have determined that there are 90-95% SSEA-1 stainable cells in the population. After induction of embryoid bodies from the stem cell cultures and subsequent cell differentiation for 2-4 weeks [Muller et al., 2000], we have been able to stain differentiated cells by various cell linage markers and confirmed multi-potency of the cells using our feeder-layer-free culture conditions [Zhang et al., 2011]. Capability of these cells to

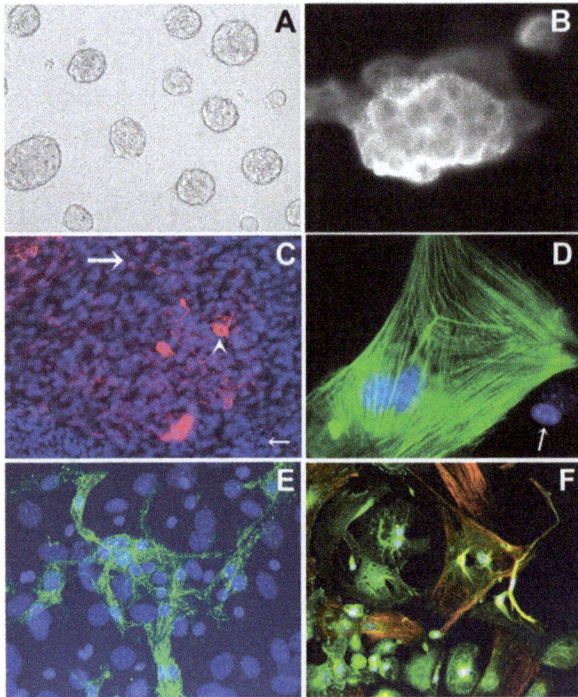

Fig. 7. Confirmation of stem cell identity and multipotent differentiation capability of the cultured mouse embryonic stem cells (MESCs) in feeder-layer-free culture system. A: Phase-contrast image of cultured MESCs without a feeder layer maintaining their typical round colonies with smooth edges indicating lack of differentiation. B: Positive staining with stem cell specific marker SSEA-1. C: Positive staining for the neuronal cell marker: neurofilament (NF). Antibodies to NF stain both cell bodies (arrowhead) and axons (large arrow). Nuclei are stained blue with DAPI (small arrow). D: Positive staining (green stain) for the cardiac cell marker: cardiac Troponin T (cTnT). Organized myofibrils are seen. E: Positive staining for the skeletal muscle cell marker, fast skeletal Troponin T (fsTnT) (green stain). DAPI, a blue nuclear fluorescent dye shows nuclear staining. F: Positive staining for both a-actinin (green) and desmin (red), markers for mesoderm-derived cell types. Magnifications for Figure 1: (A) 100x; (B) 250x; (C) 100x; (D) 600x; (E) 250x; (F) 250x. (Reprinted with permission of the authors from Zhang et al., 2011 J. Cell Biochem. 112(1):98-106).

differentiate into neurons and cardiac muscle cells were of particular interest to us since strokes and heart attacks are most likely the two most prominent health problems resulting from ischemia. It is evident that both areas might benefit from advancing stem cell therapy. Differentiated cells stained with neurofilament antibody were used to identify neuronal cells (Figure 7C), while cardiac troponin T is specific for cardiomyocytes (Figure 7D). In addition, fast skeletal muscle troponin T monoclonal antibodies stain only skeletal muscle troponin T (Figure 7E), and desmin and α-actinin strongly suggest, but are not an absolute conformation of general muscle cell types (Figure 7F).

5.1 Responses of Msr gene expression to anoxia/reoxygenation treatments

Using real-time RT-PCR, we have compared the regulation of methionine sulfoxide reductase gene expression at the transcription level under oxidative stress [Zhang et al., 2011]. Different levels of oxygen depletion and oxidative stress were induced by treating the embryonic mouse stem cells with increasing time periods of anoxia (in 90% N2, 5% H2 and 5% CO2) and reoxygenation (in air) in combined treatments. Results indicate that MsrA mRNA levels were gradually decreased with prolonged anoxic and reoxygenation treatments. RT-PCR results of MsrB1 and MsrB2 do not show an oxygen-dependent or oxidative stress regulated expression pattern, however, an oxygen-dependent expression pattern for MsrB3 is evident [Zhang et al., 2011]. Sensitivity of the MsrB3 mRNA expression level to oxygen depletion has been unequivocally demonstrated with significantly decreased mRNA levels at four, eight, and twelve hours of anoxic conditions to nearly half of the normal levels of normal oxygen levels. We found that the mRNA levels return to normal values after prolonged reoxygenation (> 4 hours) [Zhang et al., 2011]. The same oxygen-dependent expression pattern has been observed in matrix metalloprotein 2 and matrix metalloprotein 9 (MMP9). All real-time RT-PCR results have been normalized to β-actin levels for comparison.

5.2 Knockdown of MsrA expression at the mRNA level has no direct impact on MsrB1 or B2 gene expression but significantly decreases MsrB3 expression at the mRNA level

In view of earlier reports showing the potential for gene expression level interactions between MsrA and MsrB [Moskovitz, 2007], we repeated real-time RT-PCR experiments on all three reported MsrB genes (MsrB1, MsrB2 and MsrB3) using cells with MsrA mRNA downregulated by the siRNA transfection approach [Zhang et al., 2011]. Results confirm that all three MsrBs are expressed in mouse embryonic stem cells. Neither MsrB1 nor MsrB2 expression at the mRNA level were altered in the cells with MsrA mRNA downregulation (Figure 8).

Nonetheless, MsrB3 expression was significantly decreased in MsrA-specific siRNA transfected cells compared to the cells transfected with negative control siRNA. On the basis of our data, we propose that there are no direct interactions between MsrA and MsrB1 or MsrB2 gene expression at the mRNA level in mouse embryonic stem cells. However, MsrB3 mRNA expression is shown to be influenced by the levels of MsrA mRNA concentrations in the cells. A reduction of about 50% MsrB3 mRNA expression was noted as early as day 2 after MsrA gene knockdown. It is noteworthy that MsrA-specific siRNA downregulates MsrA mRNA for several days while its protein levels in the cell start to return towards normal after only 24 hours post-transfection (Figure 8E and 8F) [Zhang et al., 2010, 2011].

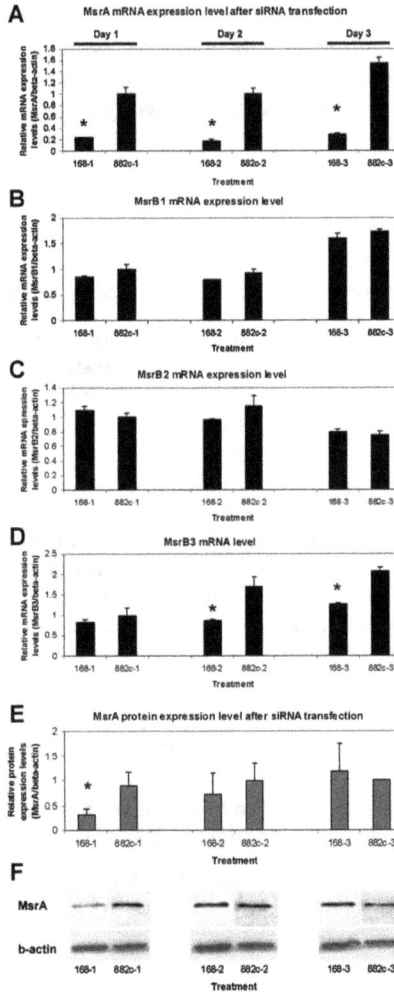

Fig. 8. Real-time RT-PCR studies on MsrB gene expression at the mRNA level after downregulation of MsrA expression by siRNA (168) transfection. A: MsrA mRNA levels; B: MsrB1 mRNA levels; C: MsrB2 mRNA levels. D: MsrB3 RNA levels. MsrB3 is the only one showing significantly increased mRNA expression after MsrA knockdown. E: MsrA protein levels from densitometry of Western blots shown in F. F: Western blotting analysis of MsrA protein and β-actin after treating MESCs with MsrA-specific siRNA(168) or negative control siRNA(882) for days 1, 2, and 3. Triplets were used in each experiment. Two independent experiments were performed for real-time RT-PCRs and three Western blotting experiments were done. Results were averaged with standard errors of mean (SEM) presented with error bars. 168-1 represents samples collected on day 1 with siRNA (168) treatment. Results from MsrA-specific siRNA (168) treated samples labeled with asterisks (*) show statistically significant differences from negative control siRNA (882) treated groups (P≤0.05). (Reprinted with permission of the authors from Zhang et al., 2011 J. Cell Biochem. 112(1):98-106).

5.3 The pH of the culture medium influences Msr gene expression in cultured mouse embryonic stem cells

In our published work on MsrA, it was observed that prolonged culture of mouse embryonic stem cells (MES) in the same medium, without refreshing, promotes MsrA mRNA expression [Zhang et al., 2010, 2011]. To further explore this phenomenon, we cultured the cells in media pre-adjusted to pH 7.5, 6.8 or 6.4 by acetic acid or hydrochloric acid, with either acid showing similar results. Media was changed every 24 hours for three days and cells were collected just before the next culture media change. Total RNA and proteins were extracted for real-time RT-PCR and Western blotting studies. When comparing the different pH conditions on the same days, we found that MsrA, B1 and B3 genes have significantly increased their mRNA expression in cells cultured in the more acidic media on day 3. Significant increases of MsrB3 mRNA were also found in pH 6.8 media on day 2. The most significant change was noticed in MsrB3 which had a 4-6 fold increase of its mRNA in either pH 6.4 or 6.8 culture media compared to pH 7.5 on day 3. However, even with the significant increase in MsrA mRNA expression in pH 6.4 medium on day 3, no increase of protein level in stem cells was observed using MsrA antibody, the only antibody available for Western blotting on mouse Msr family genes (Figure 9E and 9F).

6. Evidence of the existence of a truncated form of MsrA in mouse embryonic stem cells

From our previous studies on MsrA [Zhang et al., 2010], we have consistently found that there is an observable protein band with a molecular weight (MW) of about 16 kD from Western blotting experiments using anti-MsrA antibody (Figure 10A, thick arrow), and we examined this in much more detail, recently publishing our study in which we will summarize the results and discussion in the following section of this review chapter [Jia et al., 2011]. The hybridization signal is low, and becomes evident on the X-ray film only after the longer form MsrA bands are overexposed (Figure 10A, thin arrow). There are also protein bands with molecular weights of ~46 and ~48 kD indicating homodimers of MsrA long form proteins (Figure 10A, arrow head), even with protein samples thoroughly treated by reducing agents before SDS-PAGE, as has been reported by other laboratories [Lee et al., 2006]. On our films, there is also a band with a molecular weight of 39 kD, possibly heterodimers formed by a long form protein of MsrA (MW: 23 kD) and a smaller form of MsrA protein (MW 16 kD) (Figure 10A, *). In addition, a minor band with a molecular weight of ~32 kD, possibly formed by homodimers of two molecules of the smaller form of MsrA proteins (Figure 10A, **). The 32kD and 39kD bands cannot readily be explained by the dimerization from cytosolic isoforms of MsrA which are 19-20kD in size [Kim & Gladyshev, 2006; Lee et al., 2006]. To confirm the existence of this smaller form of protein, we have carried out RT-PCR using total RNA extracted from mouse embryonic stem cells to amplify the full cDNA [Jia et al., 2011]. The forward and reverse primers were designed based on the 5'-UTR and 3'-UTR of the known full length MsrA cDNA respectively (Genebank#: NM 026322.3). Products from RT-PCR were loaded onto an agarose gel for electrophoresis; and two bands are readily visible on the gel, with the smaller one (Figure 10B, thick arrow) showing approximately 1/20th of the intensity and being about 100bp smaller than the larger band (Figure 10B, thin arrow).

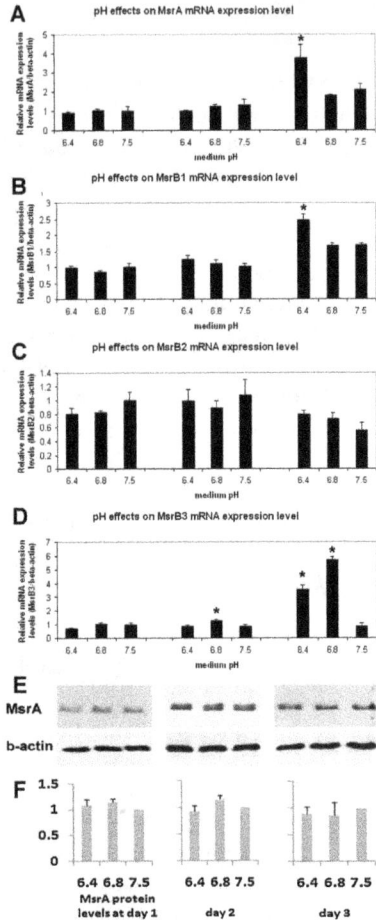

Fig. 9. Real-time RT-PCR studies on Msr gene expression in culture media with different pH conditions (pH 6.4, 6.8, and 7.5) at 1, 2, and 3 days in culture. All Msr genes, except for MsrB2, show increased expression at the mRNA level after 3 days in culture with acidic media with MsrB3 showing the most dramatic responses. A: MsrA mRNA levels; B: MsrB1 mRNA levels; C: MsrB2 mRNA levels; D: MsrB3 mRNA levels. E: Western blotting assays on MsrA protein and β-actin after culturing MESCs in media with different pH for days1, 2, and3. F: Densitometry of MsrA bands normalized by β-actin after Western blotting shows equal levels of MsrA protein expression in cells cultured in media with different pH on the same days. In all comparisons, data from cells cultured at pH 7.5 are arbitrarily set as unit 1. Results with statistically significant differences from control groups (pH 7.5) were labeled with asterisks (*) (P≤0.05). Triplets were used in each experiment and at least two independent experiments were performed. Results were averaged with standard errors of mean (SEM) presented as error bars. Results are labeled with asterisks (*) if there are statistically significant differences (P≤0.05) while comparing samples treated with acidic culture media and cells cultured at pH 7.5. (Reprinted with permission of the authors from Zhang et al., 2011 J. Cell Biochem. 112(1):98-106).

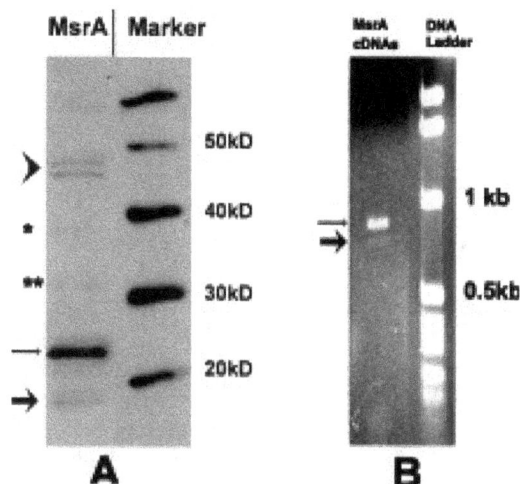

Fig. 10. Western blotting experiment using MsrA antibody reveals a smaller MsrA protein in cultured mouse embryonic stem cells. A. Thick arrow: the smaller form (~16 kD); thin arrow: the large form (~23kD); arrowhead: possible dimers from the long form MsrA proteins; *: possible heterodimers from a large form and a shorter form protein; **: possible homodimers from two shorter form proteins. B. Agarose gel electrophoresis of the amplified cDNA of MsrA showing a smaller band. PCR bands: long form: 857bp (thin arrow); short form: 744bp (thick arrow). (Reprinted with permission of the authors from Jia et al., 2011 J. Biomed. Sci. 18:46).

6.1 Cloning of the truncated cDNA form of MsrA

RT-PCR products of the smaller band described above were recovered from agarose gels and ligated to the pGEM-T-easy vector (Invitrogen, CA). After determining the DNA sequence, the smaller form sequence was aligned and compared to the full length GenBank cDNA. A deletion of the fifth exon (113bp) was found in the smaller form which ends up with a frame shift in the sixth exon directly attached to the fourth generating a new premature stop codon (Figure 11 A and B). The total length of the truncated protein is 148 amino acids, compared to the full length protein of 233 amino acids in length, both containing a mitochondrial signal peptide at the N-terminus (Figure 2). The truncated form still retains the GCFWG functional motif (catalytic active site) but contains neither of the two cysteines at the c-terminus (Figure 11B). Due to the functional importance of the two c-terminal cysteines in the redox reaction, it is reasonable to believe that the enzyme activity for methionine sulfoxide reduction will decrease dramatically, which requires confirmation by studies on purified proteins translated from this truncated template. It is interesting that we have also identified a virtually identical mouse EST sequence from a kidney cDNA library in GenBank (GenBank ID: BG970953.1) with the same intron splicing pattern as the truncated form of MsrA cloned from embryonic stem cells, indicating that this isoform might not be stem cell specific. The comparison between the EST sequence and truncated MsrA is illustrated in Figure 12. Except for missing ten nucleotides at the end of the third exon (boxes in Figure 12 and Figure 11A), the EST sequence showed 100% identity to the truncated cDNA from the 113th nucleotide (at the 5' UTR) to the very end of truncated form of the Msr cDNA.

A

```
1   : CCTGGCTGCGGAGGTGGAGAAAC GGCAGCAACTCTGACCTCCCGCGGTGGC

52  : GCCCAGCCAGCATCAATTTTGGCCCGCGCGGCTTCACGTACTCTGGGAACT

103 : TTGTCCCTCAGAGACAGCCAGGCCTGGAGGTCTGCCCCGCGAGACGGACAC

154 : GCCGCCCATGCTCTCCGCCTCTAGAAGGGCTCTCCAGCTCCTCTCTAGCGC
                    M   L   S   A   S   R   R   A   L   Q   L   L   S   S   A

205 : CAACCCGGTACGGAGGATGGGCGACTCAGCTTCGAAAGTCATCTCCGCAGA
       N   P   V   R   R   M   G   D   S   A   S   K   V   I   S   A   E

256 : GGAAGCCTTGCCGGGACGCACCGAGCCGATCCCTGTAACACCCAAACACCA
       E   A   L   P   G   R   T   E   P   I   P   V   T   A   K   H   H

307 : TGTCAGTGGCAACAGAACGGTTGAACCTTTCCCAGAGGGAACACAGATGGC
       V   S   G   N   R   T   V   E   P   F   P   E   G   T   Q   M   A

358 : TGTATTTGGAATGGGCTGCTTCTGGGGAGCTGAGCGCAAGTTCTGGGTCTT
       V   F   G   M   G   C   F   W   G   A   E   R   K   F   W   V   L

409 : GAAAGGAGTGTACTCAACCCAAGTGGGCTTTGCAGGAGGCCACACACGCAA
       K   G   V   Y   S   T   Q   V   G   F   A   G   G   H   T   R   N

460 : TCCCACCTACAAAGAGGTCTGCTCAGAAAAAAACTGGCCACGCCGAAGTCGT
       P   T   Y   K   E   V   C   S   E   K   T   G   H   A   E   V   V

511 : CCGGGTTGTGTACCGGCCAGAGCACATCAGCTTTGAGGAACTGCTCAAGGT
       R   V   V   Y   R   P   E   H   I   S   F   E   E   L   L   K   V

562 : CTTCTGGGAGAATCACGACCCGACCCAAG GTTCTTTCAAAGCATAACTTTG
       F   W   E   N   H   D   P   T   Q   G   S   F   K   A   *

613 : GCCCCATCACCACCGACATCCGAGAGGGACAAGTGTTCTACTATGCCGAAG

664 : ACTACCACCAGCAATACCTGAGCAAGAACCCAGACGGCTACTGTGGCCTCG

715 : GGGGTACTGGAGTTTCCTGCCCGATGGCCA
```

B

```
                    Mitochondria signal peptide
long  1   MLSASRRALQLLS SANPVRRMGD SASKVISAEEALPGRTEP IPVTAKHHVSGNRTVEPFP   60
short 1   MLSASRRALQLLS SANPVRRMGD SASKVISAEEALPGRTEP IPVTAKHHVSGNRTVEPFP   60

                    GCFWG motif
long  61  EGTQMAVFGM GCFWG AERKFWVLKGVYS TQVGFAGGHT RNPTYKEVCS EKTGHAEVVRVV  120
short 61  EGTQMAVFGM GCFWG AERKFWVLKGVYS TQVGFAGGHT RNPTYKEVCS EKTGHAEVVRVV  120

long  121 YRPEHISFEELLKVFWENHDPTQGMRQCNDFGTQYRSAVYPTSAVQMEAALRSKEEYQKV  180
short 121 YRPEHISFEELLKVFWENHDPTQGSFKA*  148
long  181 LSKHNFGPITTDIREGQVFYYAEDYHQQYLSKNPDGYCGLGGTGVSCPMAIKK*  233
                                     *                        *
```

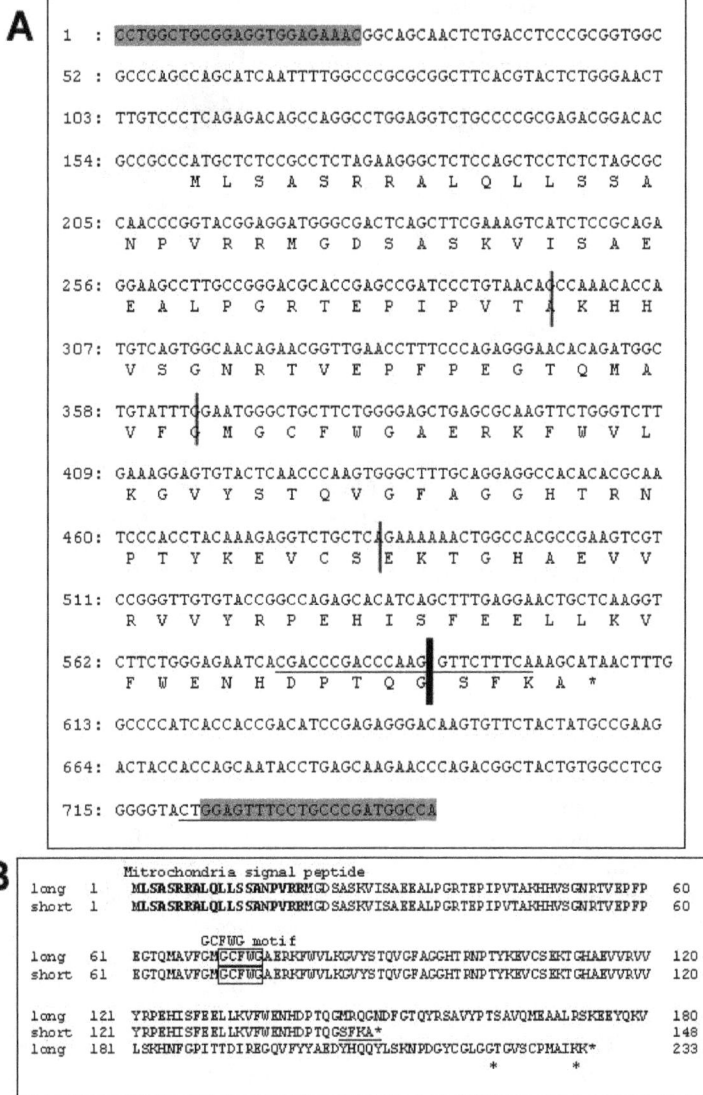

Fig. 11. The cDNA sequence of the truncated form of MsrA with the predicted open reading frame. A. Thin lines indicate exon-exon junctions. The thick line points to the junction of exon 4 and 6 skipping exon 5 in the truncated form but is included in the long form of MsrA. Shaded sequences are the two primers used to amplify the whole cDNA. Sequences underlined show the primer pairs for real time PCR to specifically amplify the truncated form with the forward primer designed on the junction of exon 4 and 6. B. The truncated form contains 148 amino acids with a mitochondrial signal peptide at the N-terminus and a GCFWG motif but no c-terminal cysteines. (Reprinted with permission of the authors from Jia et al., 2011 J. Biomed. Sci. 18:46).

```
EST      1    GAGACAGCCAGGCCTGGAGGTCTGCCCCGCGAGACGGACACGCCGCCCATGCTCTCCGCC    60
              ||||||||||||||||||||||||||||||||||||||||||||||||||||||||||||
Trunc  113    GAGACAGCCAGGCCTGGAGGTCTGCCCCGCGAGACGGACACGCCGCCCATGCTCTCCGCC   172

EST     61    TCTAGAAGGGCTCTCCAGCTCCTCTCTAGCGCCAACCCGGTACGGAGGATGGGCGACTCA   120
              ||||||||||||||||||||||||||||||||||||||||||||||||||||||||||||
Trunc  173    TCTAGAAGGGCTCTCCAGCTCCTCTCTAGCGCCAACCCGGTACGGAGGATGGGCGACTCA   232

EST    121    GCTTCGAAAGTCATCTCCGCAGAGGAAGCCTTGCCGGGACGCACCGAGCCGATCCCTGTA   180
              ||||||||||||||||||||||||||||||||||||||||||||||||||||||||||||
Trunc  233    GCTTCGAAAGTCATCTCCGCAGAGGAAGCCTTGCCGGGACGCACCGAGCCGATCCCTGTA   292

EST    181    ACAGCCAAACACCATGTCAGTGGCAACAGAACGGTTGAACCTTTCCCAGAGGGAACACAG   240
              ||||||||||||||||||||||||||||||||||||||||||||||||||||||||||||
Trunc  293    ACAGCCAAACACCATGTCAGTGGCAACAGAACGGTTGAACCTTTCCCAGAGGGAACACAG   352

EST    241    ATGGCTGTATTTGGAATGGGCTGCTTCTGGGGAGCTGAGCGCAAGTTCTGGGTCTTGAAA   300
              ||||||||||||||||||||||||||||||||||||||||||||||||||||||||||||
Trunc  353    ATGGCTGTATTTGGAATGGGCTGCTTCTGGGGAGCTGAGCGCAAGTTCTGGGTCTTGAAA   412

EST    301    GGAGTGTACTCAACCCAAGTGGGCTTTGCAGGAGGCCACACACGCAATCCCACCTACAAA   359
              ||||||||||||||||||||||||||||||||||||||||||||||||||||||||||||
Trunc  413    GGAGTGTACTCAACCCAAGTGGGCTTTGCAGGAGGCCACACACGCAATCCCACCTACAAA   472

EST    360    G┌─────────┐AGAAAAAACTGGCCACGCCGAAGTCGTCCGGGTTGTGTACCGGCCAGAG   410
             |└         │|||||||||||||||||||||||||||||||||||||||||||||||
Trunc  473    G│AGGTCTGCTC│AGAAAAAACTGGCCACGCCGAAGTCGTCCGGGTTGTGTACCGGCCAGAG   532
              └─────────┘

EST    411    CACATCAGCTTTGAGGAACTGCTCAAGGTCTTCTGGGAGAATCACGACCCGACCCAAGGT   470
              ||||||||||||||||||||||||||||||||||||||||||||||||||||||||||||
Trunc  533    CACATCAGCTTTGAGGAACTGCTCAAGGTCTTCTGGGAGAATCACGACCCGACCCAAGGT   592

EST    471    TCTTTCAAAGCATAACTTTGGCCCCATCACCACCGACATCCGAGAGGGACAAGTGTTCTA   530
              ||||||||||||||||||||||||||||||||||||||||||||||||||||||||||||
Trunc  593    TCTTTCAAAGCATAACTTTGGCCCCATCACCACCGACATCCGAGAGGGACAAGTGTTCTA   652

EST    531    CTATGCCGAAGACTACCACCAGCAATACCTGAGCAAGAACCCAGACGGCTACTGTGGCCT   590
              ||||||||||||||||||||||||||||||||||||||||||||||||||||||||||||
Trunc  653    CTATGCCGAAGACTACCACCAGCAATACCTGAGCAAGAACCCAGACGGCTACTGTGGCCT   712

EST    591    CGGGGGTACTGGAGTTTCCTGCCCGATGGCCA    622
              ||||||||||||||||||||||||||||||||
Trunc  713    CGGGGGTACTGGAGTTTCCTGCCCGATGGCCA    744
```

Fig. 12. Comparison between truncated form of MsrA cDNA (Trunc) and a mouse EST sequence from a kidney cDNA library in Genbank (EST). The EST sequence lacks ten nucleotides comparing with the truncated cDNA which is shown in the box. The same area is also shown in a box in Figure 2A, which is located at the very end of the third exon. (Reprinted with permission of the authors from Jia et al., 2011 J. Biomed. Sci. 18:46).

6.2 Confocal microscopy reveals different subcellular localizations for the truncated MsrA protein compared to the full length using eGFP fusion constructs

The long form of the MsrA full length protein conjugated with the eGFP tag has been studied previously in our laboratory and was found to be localized predominantly in mitochondria with some staining in the cytosol [Zhang et al., 2010, 2011]. Further studies confirmed this finding with the MsrA long form-eGFP fluorescence signals (green, figure 13B); the mitochondria stained by MitoTracker (red, figure 13A) mostly overlap each other and show an orange color (figure 13G).

After our discovery of the truncated form of the MsrA, we did a detailed study of this form and have published these results [Jia et al., 2011], from which we describe and summarize these recent findings in the following paragraphs. The MsrA truncated form-eGFP fusion protein shows a more nonspecific localization, mostly in the cytosol, although green fluorescent signals in mitochondria are detectable (Figure 13 D, E, F and H). To confirm this observation, we have generated a combination of three single slices of confocal scans top view (A), upper-side view (B) and right-side view (C) on a stem cell colony with the truncated-MsrA-eGFP transfection (Figure 14A, B and C) [Jia et al., 2011]. This provides a clear three dimensional image showing the localization of the truncated protein. On the focal point of the scans (cross-point of the horizontal green line and the vertical pink line), the green fluorescent signal is excluded from nuclei stained by DAPI. Most of the green signals are not overlapping with the red mitochondria in all three view angles (arrowhead, Figure 14C) although there are some detectable co-localizations evident (arrow, Figure 14B) [Jia et al., 2011].

Studies reported by Lee et al. [2006] and Pascual et al. [2009] show the existence of two alternative promoters for the MsrA gene that encode different isoforms of MsrA proteins in mitochondria or cytosol/nuclei due to the presence or absence of a N-terminal mitochondrial signal peptide. Studies from Kim and Gladyshev [2006], using GFP fusion techniques and deletion mutagenesis, reveal functional domains in the MsrA peptide sequence, including sequences close to the c-terminus, that may also direct the specific locations of the protein in subcellular compartments. Localization of the mitochondrial form of MsrA in the cytosol and nuclei were also noted by Kim and Gladyshev [2005] in MsrA overexpression experiments. Although syntheses of different isoforms with or without the N-terminal signal peptide might be the optimal way for cells to direct protein sorting, it should not be overlooked that the same isoform might be able to locate to multiple cellular compartments. While we do not rule out the possibility that the altered localization pattern of the truncated protein, compared to the long form, is due to GFP fusion interference, it is very unlikely considering the fact this same method has been used successfully to reveal subcellular localization of MsrA in all of our cell lines [Zhang et al., 2010, 2011; Jia et al., 2011]. Previous studies from our laboratory on the truncated MsrA-eGFP fusion protein suggest a necessary domain at the c-terminal sequence for permanently docking of the protein on mitochondria. In addition to the mitochondrial signal peptide, there may exist another essential domain at the c-terminal end of the full length protein, without which, the truncated proteins are able to be sorted to the mitochondria but will eventually leak out back into the cytosol. In the deletion mutagenesis studies of Kim and Gladyshev [2005], the deletion is limited only to the very end or very middle of the N-terminus, not totally overlapping the portion omitted in this truncated form which could harbor more functional domain units [Kim & Gladyshev, 2005].

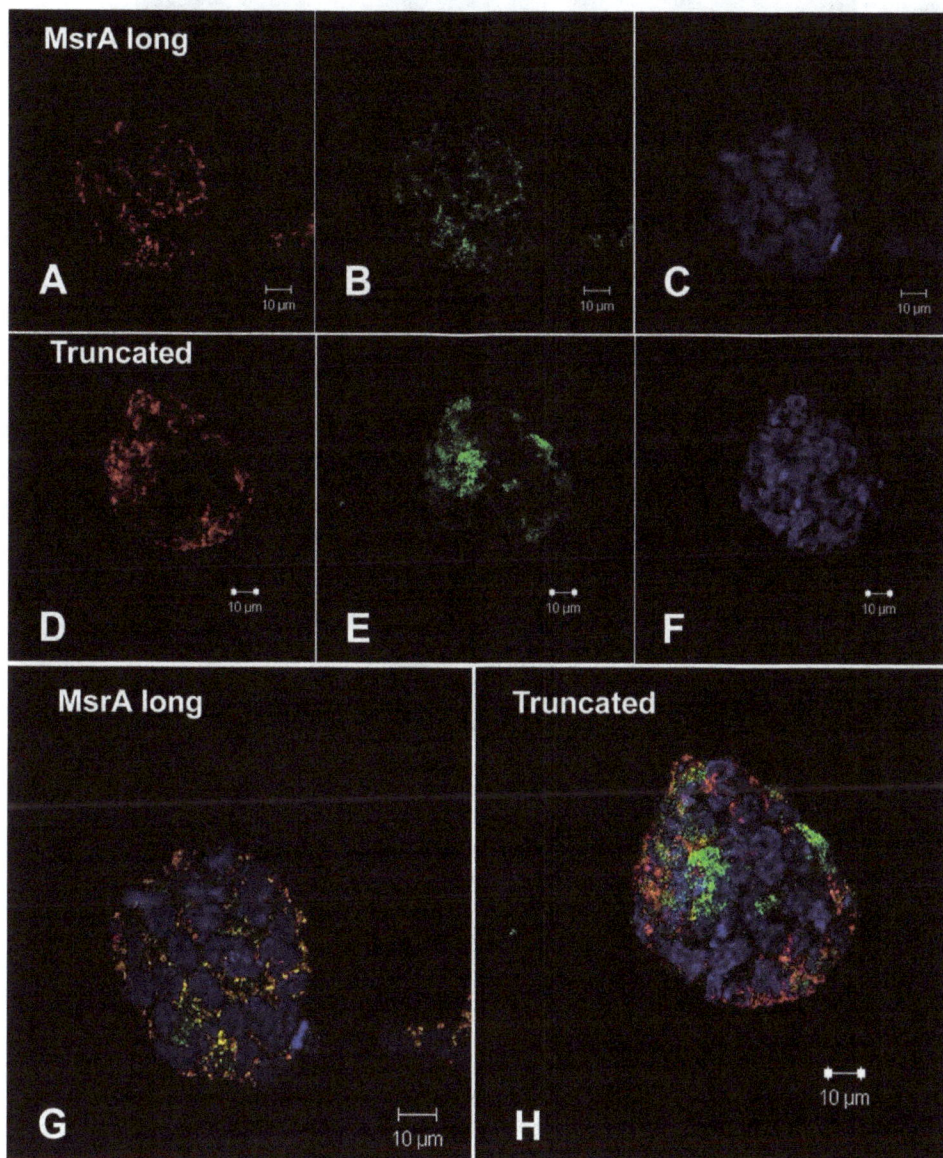

Fig. 13. Confocal microscopy on the long form MsrA-eGFP (A,B, C,G) and truncated MsrA-eGFP (D,E,F,H) transfected mouse embryonic stem cells. A,D: mitochondria stained by propidium iodide; B,E: green fluorescence showing GFP tags; C,F: nuclei stained with DAPI; G: overlapped image from A, B and C, but with higher magnification; H: image overlapped from D, E and F; Bars: 10 μm. (Reprinted with permission of the authors from Jia et al., 2011 J. Biomed. Sci. 18:46).

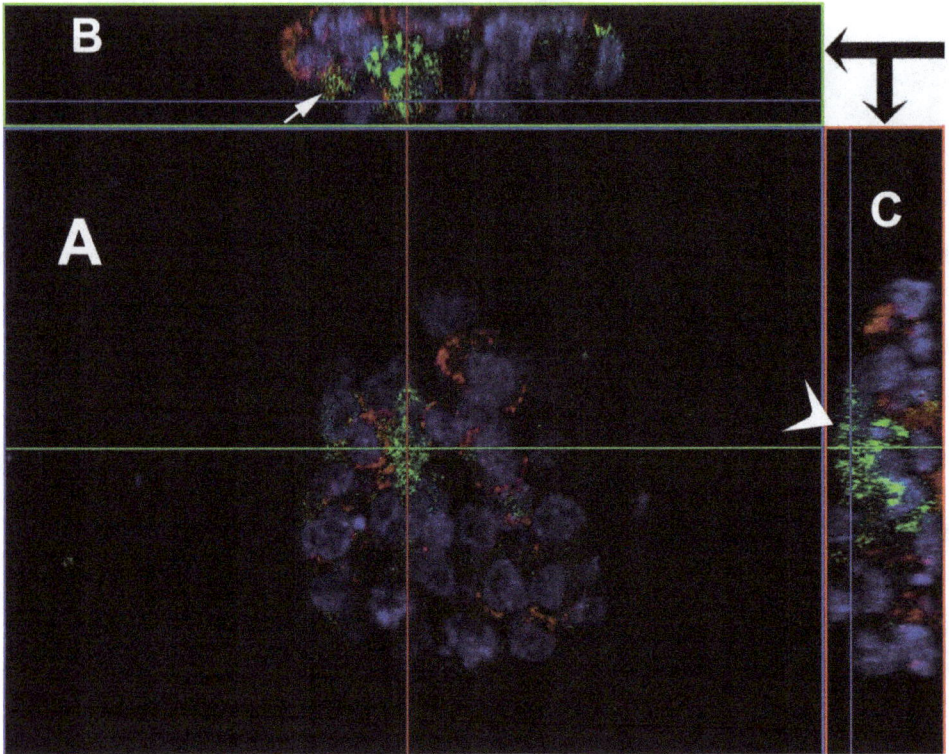

Fig. 14. Confocal microscopy of a single colony of mouse embryonic stem cells. A: top view of a single slice of scanning showing most eGFP signals are not overlapping with mitochondria or nuclei. B: single slice scanning from upper-side view; J-2: single slice scanning from rightside view. The arrow in B indicates some overlapping signal of truncated Msr-eGFP and mitochondria. The arrowhead in C points to the area where the truncated Msr-eGFP signals are not overlapping with mitochondria, but mainly in cytosol. The cross-point of the horizontal green line and vertical red line points to the area we are observing. The blue lines in B and C show the current slice position for this confocal scanning (Reprinted with permission of the authors from Jia et al., 2011 J. Biomed. Sci. 18:46).

6.3 Real time RT-PCR shows a different response of mRNA expression levels for the trunctated form compared to the full length MsrA

Real time RT-PCR studies done in our laboratory on the long form MsrA expression responses to oxygen deprivation and reoxygenation show that the expression levels decrease along with longer anoxia/reoxygenation treatment combinations [Zhang et al., 2010, 2011]. Similar studies on the truncated MsrA transcripts show different responses compared to the long form except during 4 hours of anoxia treatment in which both truncated and long forms show decreases at the mRNA level, likely an initial response of the stem cells to oxygen deprivation as observed in all Msr genes as well as in other genes including metalloproteinases (MMP2 and MMP9) [Zhang et al., 2010, 2011; Jia et al., 2011]. mRNA expression of the truncated form decreases dramatically when a short period of reoxygenation (4 hours) was given after a long period (12 hours) of anoxia (12+4 hrs). The level of reactive oxygen species (ROS) in the cells are expected to rise substantially. Expression level partially recovers after 12 hours of reoxygenation following the 12 hours of anoxia treatment (12+12 hrs) to the same level as 12 hours of reoxygenation following 8 hours of anoxia (8+12 hrs), at which point the level of reactive oxygen species might have decreased compared to the point of 12+4 [Jia et al., 2011]. We suggest that the truncated form of the MsrA transcript is responsive to the cellular level of ROS. By comparing expression levels at 12+4 and 12+12 time points for long and truncated forms of MsrA mRNA, it is evident that the truncated form is more sensitive to oxidative stress changes than the long form.

6.4 The c-terminal truncated form of MsrA was cloned from mouse embryonic stem cells by skipping exon 5 and having a frame shift mutation in exon 6, generating a premature stop codon

The truncated MsrA protein displays a different subcellular localization pattern expression in response to anoxia/reoxygenation treatment of the stem cells. Further studies on the enzymatic activity of this shortened peptide is required to classify it as a functional isoform. One possibility for the evolutionary advantage of retaining such a truncated form might be that producing a smaller size protein while containing an active GCWFG site, may allow this "mini-protein" to readily access the oxidized methionine residues on proteins, while the larger structure hinders access of the long form protein. Since the truncated protein does not contain c-terminal cysteines, perhaps the final relieving of the oxidation step needs the long form MsrA. Interestingly, heterodimers between long form and truncated proteins were observed in our Western blotting experiments [Jia et al., 2011].

7. Implications and conclusions on protection of mouse embryonic stem cells from oxidative stress by methionine sulfoxide reductase A (MsrA)

Previous studies on methionine sulfoxide reductase (Msr) have been performed in various mammalian cell lines [Kantorow et al., 2004; Yermolaieva et al., 2004; Petropoulos et al., 2001] as well as animal models [Ruan et al., 2002; Moskovitz et al., 2003]. Regulation of the Msr system as an antioxidant mechanism for stem cells, however, was totally unknown before our studies [Zhang et al., 2010, 2011; Jia et al., 2011]. It is critical to understand more about stem cell anti-oxidative stress capabilities in order to develop protocols for therapeutic applications of stem cells stressed by high oxidative environments. Our published *in vitro* studies using mouse embryonic stem cell cultures demonstrate the vital importance of the MsrA protein protecting these stem cells from hydrogen peroxide mediated oxidative damage [Zhang et al., 2010]. This same protection may exist *in vivo* and potentially may improve the cell viability of stem cells in patients having ischemia/reperfusion oxidative damage, such as in an infarcted myocardium. Msr genes also could play a major role in shielding adult stem cells or embryonic stem cells stored in adult tissues, from oxidative damage. Such cells apparently are maintained in a quiescent state in adult tissues for long periods that probably extend throughout the entire lifetime of the individual [Ding, 2003, Ryu et al., 2006].

Three MsrBs have been cloned (MsrB1, B2 and B3) from mammals, which are localized in subcellular compartments [Kim and Gladyshev, 2004, 2005, 2006] and show variable expression patterns in different tissues. There is only a single known mammalian MsrA gene and its proteins are localized in both the cytosol and in mitochondria [Vougier et al., 2003]. Our published studies using the eGFP fusion protein to track MsrA expression show that the majority of MsrA proteins are present in the cytosol and the mitochondria components of cells [Zhang et al., 2010]; a similar localization pattern has been shown for rat liver cells [Vougier et al., 2003]. In our experiments, limited fluorescent signals were observed in the cell nuclei, indicating low concentrations of MsrA-eGFP [Zhang et al., 2010]. It is possible that these low eGFP fluorescent levels were due to steric molecular interference being conjugated to MsrA [Hansel et al., 2002]. Studies on human MsrA showed the exclusive mitochondrial protein localization [Hansel et al., 2002]. The mouse MsrA protein, using the eGFP construct, revealed localization of MsrA in mitochondria, cytosol and nuclei [Kim & Gladyshev, 2005]. Kim and Gladyshev [2005] proposed that the N-terminal mitochondrial signal peptide directs the MsrA protein to the mitochondria. It appears there is localization of MsrA variations according to species and type of cell.

In mammalian cells, MsrA is a gene lacking numerous exogenous influences unlike reports for insects suggesting that ecdysone can induce MsrA expression in Drosophila melanogaster during development [Cherbas et al., 1986; Roesijadi et al., 2007]. We have demonstrated that combinations of anoxia and reoxygenation to treat mouse embryonic stem cells have shown a unique result in which MsrA mRNA and protein expression decreases with increased periods of anoxia or reoxygenation treatment [Zhang et al., 2010]. Increasing the periods of anoxia or reoxygenation results in elevated levels of reactive oxygen species expression and MsrA mRNA and protein expression in stem cells. These results are consistent with reports [Petropoulos et al., 2001] that rat MsrA gene expression decreases with age and during replicative senescence of human WI-38 fibroblasts. Both scenarios correlate aging with higher levels of oxidative stress. It is relevant to note that all these studies show the MsrA expression levels to decrease, not increase, with prolonged oxidative stress exposure/aging. Increasing MsrA expression would hypothetically provide more protection for cells when exposed to higher levels of oxidative stress [Zhang et al., 2010]. It remains unknown whether this mechanism exists in nature as a protective mechanism to prevent stem cells in the body from turning into cancerous tissue by allowing cells to die by ROS damage once the hypoxic environment is established in a growing tumor [Zhang et al., 2010]. It is also possible that by decreasing MsrA via oxidative stress, embryonic stem cells are allowing higher and longer periods of ROS spikes to signal for differentiation initiation in which case overexpression of MsrA in stem cells will be detrimental to their plasticity. It would be very interesting to further investigate this phenomenon, for future guidance in manipulating MsrA expression in stem cells in response to oxidative damage to increased cell survivability and differentiation [Zhang et al., 2010]. We have found that after siRNA knockdown of MsrA, its mRNA remains at low levels for 72 hours or more. The MsrA protein levels, however, return to normal within 48 hours. Our laboratory was the first to report translational regulation of the MsrA protein in stem cells [Zhang et al., 2010]. The mechanism of regulation remains unknown. It seems evident, however, that the stringent control in mouse ESC to keep the MsrA protein level high, even with low levels of mRNA present, suggests that this protein is essential to cellular function and survivability [Zhang et al., 2010].

We confirmed the successful knockdown of MsrA mRNA levels in the stem cells after 24 hours following MsrA siRNA transfection. There was almost an 80% reduction of mRNA levels and a 70% reduction of protein in the cells at this time point. In addition we added H_2O_2 to the cell cultures after 24 hours to determine a possible loss of the protective effect after MsrA knockdown at high H_2O_2 concentrations. We found that the protected loss after MsrA knockdown was not significant if H_2O_2 concentrations are low (< 50μM), suggesting that the protective effects from other methionine sulfoxide reductases, such as MsrBs, are intact in the stem cells to fight against oxidative damage [Zhang et al., 2010]. In order to test our hypothesis and show that loss of protective effects is from MsrA alone rather than concurrent loss of MsrBs' expression after MsrA siRNA transfection, we performed experiments using real-time RT-PCR on the three MsrB genes (MsrB1, MsrB2 and MsrB3) using samples with MsrA siRNA transfection. These experiments proved that all three MsrBs are present in stem cells after 24 hours of siRNA post-transfection, without any changes in MsrB mRNA expression levels. Our earlier work confirms that there is no direct interaction between MsrA and MsrBs gene expression in mouse embryonic stem cells.

These results reiterate the existence of additional protective effects of MsrA overexpression in mouse embryonic stem cells. We conclude, based on our studies, that overexpression of MsrA in stem cells provides major protection to stem cells in harsh environments such as ischemia/reperfusion during strokes or heart attacks. New approaches to treat diseases, including neurodegenerative diseases and ischemia/reperfusion-induced brain or heart damage using adult or embryonic stem cell transplantation are being developed. A major roadblock of using stem cells is their low survivability after surgery due to the harsh oxidative environments into which these cells are placed. Without preparing and protecting stem cells to resist high oxidative stress prior to transplanting them into damaged tissue areas dictates likely failure and death of the implanted cells. In our recent studies, we believe we have found a very promising approach to solving this problem by enhancing the resistance of stem cells to oxidative damage once they are transplanted into an oxidatively stressing environment [Zhang et al., 2010]. We believe it will be very important to further investigate the differentiation capability of stem cells overexpressing MsrA. Since the Msr genes can directly function to reduce oxidized proteins back to their functional forms by indirectly controlling the ROS levels in these cells, it seems reasonable that overexpressing Msr genes in stem cells might function to reduce ROS damage to the cells without losing the cells' sensitivity to ROS as a differentiation signal [Zhang et al., 2010]. Thus the Msr gene family could potentially serve a key role in engineering stem cells to obtain higher resistance to oxidative damage, while retaining the cells' potential for differentiation into the adult cell types described in a given clinical application.

8. Potential regulatory interactions of the Msr genes

In our earlier studies, we demonstrated that knockdown of the MsrA gene in mouse embryonic stem cells causes a loss of protection against H_2O_2-induced oxidative damage [Zhang et al., 2010]. Studies from other laboratories have shown a potential intergenic relationship of gene expression between the MsrA and MsrB genes from the MsrA knockout mouse model; MsrA knockout mice showed parallel losses at the levels of the MsrB1 mRNA and the MsrB1 protein [Moskovitz et al., 2003]. Subsequent studies on MsrA knockout mice placed on selenium deficient diets showed a reduction of MsrB activity in a tissue specific manner [Moskovitz, 2007]. In our studies [Zhang et al., 2011], a short term (three days) loss of MsrB1 or MsrB2 expression was not detected and the long term effect of MsrA knockdown of MsrB1 and MsrB2 expressions is not known. MsrB3, however, shows decreased expression at days 2 and 3. Although the mRNA of MsrA is still down 70% compared to negative control siRNA transfected cells at day 3, the MsrA protein expression has returned to normal levels. It is possible that the signal to decrease MsrB3 expression is at the level of MsrA mRNA, and not its protein. The mechanism by which MsrA regulates MsrB expression remains to be elucidated and will form a most interesting topic for further studies, considering the fact that MsrB3 and MsrA genes are located on different chromosomes in both human and mouse. Earlier studies from our laboratory, together with previous findings from other laboratories, strongly suggest that MsrA plays a significant role in MsrB gene expression [Zhang et al., 2010, 2011].

There have been a few studies on regulatory factors that influence Msr gene expression. In prokaryotes, studies on *H. pylori* demonstrate that certain stress conditions such as treatment with peroxide, peroxynitrite, or iron starvation, results in a 3- to 3.5-fold transcriptional up-regulation of the Msr gene [Alamuri & Maier, 2006]. In *H. pylori*, Msr codes for a 42-kDa protein with fused MsrA- and MsrB-like domain [Weissbach et al., 2002]. The only available evidence that Msr gene expression is influenced by pH stems from studies on *Streptococcus gordonii* which when entering the blood stream (pH 7.3) from the oral cavity (pH 6.2) promotes MsrA expression and possibly protects and increases survivability of the bacteria [Vriesema et al., 2000]. In Drosophila, ecdysone was found to be effective in promoting MsrA but not MsrB expression [Roesijadi et al., 2007]. In mouse, a selenium deficient diet results in decreased enzymatic activities for both MsrA and MsrB [Uthus & Moskovitz, 2007]. MsrA is also found downregulated in human hepatitis B positive hepatocellular carcinoma (HCC) with metastasis compared to HCCs without metastasis [Lei et al., 2007]. Recent studies on insulin/IGF receptor (IIR)/FOXO pathways indicate that downregulation of signaling in this pathway has been shown to extend lifespan in worms and flies [Minniti et al., 2009]. FOXO-mediated transcription is required for the long lifespan, thus there is great interest in identifying FOXO target genes. Also, it was reported recently that methionine sulfoxide reductase A expression is regulated by the DAF-16/ FOXO pathway in Caenorhabditis elegans [Minniti et al., 2009]. Moreover, another study has shown that Spx, a global transcriptional regulator of the disulfide specific oxidative stress response in B. subtilis plays a central role in the paraquat-specific induction of MsrA and MsrB expression [You et al., 2008].

Previously, we have demonstrated, for the first time that Msr genes are responsive to environmental culture condition changes that mimic pathological situations [Zhang et al., 2010, 2011; Jia et al., 2011]. Since MsrA is the only enzyme currently known for Met-S-(O) reduction and since MsrB1 has the highest enzymatic activity for Met-R-(O) reduction [Kim & Gladyshev, 2004], and both show only slight responses to oxygen level and media pH changes, MsrA and MsrB1 (and possibly also MsrB2) likely function as housekeeping enzymes for normalizing oxidative status in the cells [Zhang et al., 2011]. However, MsrA and MsrB1 do possess the capability of being regulated in a severely harsh environment. MsrB3, on the other hand, shows the lowest expression level in real time RT-PCR studies compared to MsrB1 and B2, but the most dramatic response to both oxygen level and media pH among all the Msr genes [Zhang et al., 2011]. Thus, MsrB3 could be a major player responding to increased cellular oxidative stress. It seems also reasonable to hypothesize that an acidic pH environment is an important signal for MsrB3 expression induction based on current results, while oxygen depletion, although also a tissue damaging signal, shuts down MsrB3 transcription [Zhang et al., 2011].

Using real time RT-PCR, we have shown that MsrB3 responds most dramatically, among all Msr genes, to oxygen deprivation, culture media pH changes and MsrA knockdown [Zhang et al., 2010, 2011]. MsrB3 expression decreases significantly under anoxic conditions but increases dramatically after two days of culture in acidic medium, suggesting that MsrB3 is a major player in response to changes of tissue oxidative stress in embryonic stem cells. Knockdown of MsrA by siRNA in these cells also has shown a parallel decrease of MsrB3 transcription (30-50%) compared with the negative controls, but not for MsrB1 and B2, indicating different intergenic interactions between MsrA and members of the MsrB group. In the present studies, we suggest new evidence to examine the expressional regulation of Msr genes. The flexibility of MsrB3 expression levels provides a potential target for future research to improve oxidative stress resistance in therapeutic stem cells or even to reduce this resistance in harmful tissues such as in cancer cells [Zhang et al., 2011].

9. Truncated form of methionine sulfoxide reductase A

Methionine sulfoxide reductase A (MsrA), an enzyme in the Msr gene family, is important in the cellular anti-oxidative stress defense mechanism. It acts by reducing the oxidized methionine sulfoxide in proteins back to sulfide and by reducing the cellular level of reactive oxygen species. MsrA, the only enzyme in the Msr gene family that can reduce the S-form epimers of methionine sulfoxide, has been located in different cellular compartments including mitochondria, cytosol and nuclei of various cell lines [Zhang et al., 2010, 2011]. One possibility for the existence of a truncated form of the MsrA transcripts could be that with a smaller protein size, yet retaining a GCWFG action site, this protein might have easier access to oxidize methionine residues on proteins than the longer form of the MsrA protein, thus having an evolutionary selection advantage. This research opens the door for further study on the role and function of the truncated MsrA embryonic mouse stem cells [Jia et al., 2011].

10. Conclusions

Studies from our laboratories and others support the concept that the Msr family of enzymes protects cells from oxidative sulfide groups of methionine residues in proteins. As a consequence, normal protein functions are restored and there is a reduction of intracellular reactive oxygen species. Further experiments suggest that the four Msr family genes are under differential regulation by anoxia/reoxygenation, acidic culture conditions and MsrA and MsrB interactions. Results of experiments on mouse embryonic stem cells reveal that the gene family functions in oxidative stress protection with each of the Msr genes responding differently to environmental stimuli at the tissue level. Based upon our experiments using mouse embryonic stem cell cultures (ESC), it is clear that one member of the Msr gene family, methionine sulfoxide reductase A (MsrA) reduces methionine sulfoxide residues in proteins formed by oxidation of methionine by reactive oxygen species (ROS). In these mouse ESC cultures, knockdown of MsrA expression resulted in a significantly lowered resistance of the mouse ESCs to hydrogen peroxide treatment, while overexpression of the MsrA gene resulted in increased survivability of the ESCs. Thus,

MsrA appears to play a significant role in the resistance of ESCs to oxidative stress. As future studies go forward on the use of embryonic stem cells in therapeutic applications, the Msr family of genes may prove to be important in producing stem cells that have the ability to differentiate into the desired adult end tissues, while at the same time possessing a high resistance to the severe oxidative environment that stem cells are subjected to, especially at the early phases of transplantation. Further studies on the methionine sulfoxide reductases (Msr) at the enzyme as well as molecular genetic levels may be a key to future successful stem cell therapeutic applications.

11. References

Abrams WR, G. Weinbaum, L. Weissbach, H. Weissbach, and N. Brot. Enzymatic reduction of oxidized alpha-1-proteinase inhibitor restores biological activity. Proc Natl Acad Sci U S A. 1981;78(12):7483-6.

Ahmed Z.M., R. Yousaf, B.C. Lee, S.N. Khan, S. Lee, K. Lee, T. Husnain, A.U. Rehman, S. Bonneux, M. Ansar, W. Ahmad, S.M. Leal, V.N. Gladyshev, I.A. Belyantseva, G. Van Camp, S. Riazuddin, T.B. Friedman, and S. Riazuddin. Functional null mutations of MSRB3 encoding methionine sulfoxide reductase are associated with human deafness DFNB74. Am J Hum Genet. 2011, 88(1):19-29.

Alamuri, P, and R.J. Maier. Methionine sulfoxide reductase in helicobacter pylori: Interaction with methionine-rich proteins and stress-induced expression. J Bacteriol. 2006 Aug;188(16):5839-50.

Blass, J.P. Brain metabolism and brain disease: is metabolic deficiency the proximate cause of Alzheimer dementia? J. Neurosci. Res. 2001:66:851-856.

Boldyrev, A., E. Bulygina, and A. Makhro. Glutamate receptors modulate oxidative stress in neuronal cells. A mini-review. Neurotox Res. 2004:6:581-587.

Brennan LA, and M. Kantorow. Mitochondrial function and redox control in the aging eye: Role of MsrA and other repair systems in cataract and macular degenerations. Exp. Eye Res., 2009 Feb;88(2):195-203. Epub 2008 Jun 7.

Cherbas, C.L., R.A. Schultz, M.M.D. Koehler, C. Savakis, P. Cherbas. Structure of the EIP28/29 gene, an ecdysone-inducible gene from Drosophila. J. Mol. Biol. 1986: 189: 617-631.

Choi, D., H.J. Lee, S. Jee, S. Jin, S.K. Koo, S.S. Paik, S.C. Jung, S.Y. Hwang, K.S. Lee, and B. Oh. In vitro differentiation of mouse embryonic stem cells: enrichment of endodermal cells in the embryoid body. Stem Cells. 2005:23:817-827.

Ding, S., T.Y. Wu, A. Brinker, E.C. Peters, W. Hur, N.S. Gray, and P.G. Schultz. Synthetic small molecules that control stem cell fate. Proc. Natl. Acad. Sci. U.S.A. 2003:100:7632-7637.

Gabbita, SP, M.Y. Aksenov, M.A. Lovell, and W.R. Markesbery. Decrease in peptide methionine sulfoxide reductase in alzheimer's disease brain. J. Neurochem. 1999 Oct ;73(4):1660-6.

Haenold, R, R. Wassef, A. Hansel, S.H. Heinemann, and T. Hoshi. Identification of a new functional splice variant of the enzyme methionine sulphoxide reductase A (MSRA) expressed in rat vascular smooth muscle cells. Free Radic. Res. 2007 Nov;41(11):1233-45.

Hamada, H., M. Kobune, K. Nakamura, Y. Kawano, K. Kato, O. Honmou, K. Houkin, T. Matsunaga, and Y. Niitsu. Mesenchymal stem cells (MSC) as therapeutic cytoreagents for gene therapy. Cancer. Sci. 2005:96:149-156.

Hansel, A., L. Kuschel, S. Hehl, C. Lemke, H.J. Agricola, T. Hoshi, and S.H. Heinemann. Mitochondrial targeting of the human peptide methionine sulfoxide reductase (MSRA), an enzyme involved in the repair of oxidized proteins. FASEB J. 2002:16:911-913.

Hansel, A., S. Jung, T. Hoshi, and S.H. Heinemann. A second human methionine sulfoxide reductase (hMSRB2) reducing methionine-R-sulfoxide displays a tissue expression pattern distinct from hMSRB1. Redox Rep. 2003:8:384-388.

Honig, L.S., and R.N. Rosenberg. Apoptosis and neurologic disease. Am. J. Med. 2000:108:317-330.

Huang, J.Q., J.M. Trasler, S. Igdoura, J. Michaud, N. Hanal, and R.A. Gravel. Apoptotic cell death in mouse models of GM2 gangliosidosis and observations on human Tay-Sachs and Sandhoff diseases. Hum. Mol. Genet. 1997:6:1879-1885.

Jia , P, C. Zhang, Y. Jia, K.A. Webster, X. Huang, A.A. Kochegarov, S.L. Lemanski, and L.F. Lemanski Identification of a truncated form of methionine sulfoxide reductase A expressed in mouse embryonic stem cells J. Biomed. Sci. 2011:18:46.

Jones, E.M., T.C. Squier, C.A. Sacksteder. An altered mode of calcium coordination in methionine-oxidized calmodulin. Biophys J. 2008 Dec;95(11):5268-80.

Kantorow, M., J.R. Hawse, T.L. Cowell, S. Benhamed, G.O. Pizarro, V.N. Reddy, and J.F. Hejtmancik. Methionine sulfoxide reductase A is important for lens cell viability and resistance to oxidative stress. Proc. Natl. Acad. Sci. U.S.A. 2004:101:9654-9659.

Kim, H.Y., and V.N. Gladyshev. Methionine sulfoxide reduction in mammals: characterization of methionine-R-sulfoxide reductases. Mol. Biol. Cell. 2004:15:1055-1064.

Kim, H.Y., and V.N. Gladyshev. Role of structural and functional elements of mouse methionine-S-sulfoxide reductase in its subcellular distribution. Biochemistry. 2005:44:8059-8067.

Kim, H.Y., and V.N. Gladyshev. Alternative first exon splicing regulates subcellular distribution of methionine sulfoxide reductases. BMC Mol. Biol. 2006:7:11

Kryukov G.V., R.A. Kumar, A. Koc, Z. Sun, and V.N. Gladyshev. Selenoprotein R is a zinc-containing stereo-specific methionine sulfoxide reductase. Proc. Natl. Acad. Sci. USA. 2002, 99:4245-4250.

Kubasiak L.A., O.M. Hernandez, N.H. Bishopric, and K.A. Webster. Hypoxia and acidosis activate cardiac myocyte death through the bcl-2 family protein BNIP3. Proc. Natl. Acad. Sci. USA. 2002;99(20):12825-30.

Lee J.W., N.V. Gordiyenko, M. Marchetti, N. Tserentsoodol, D. Sagher, S. Alam, H. Weissbach, M. Kantorow, and I.R. Rodriguez. Gene structure, localization and role in oxidative stress of methionine sulfoxide reductase A (MSRA) in the monkey retina, Exp. Eye Res. 2006; 82:816-827.

Lei , K.F., Y.F. Wang, X.Q. Zhu, P.C. Lu, B.S. Sun, H.L. Jia, N. Ren, Q.H. Ye, H.C. Sun, L. Wang, Z.Y. Tang, and L.X. Qin. Identification of MSRA gene on chromosome 8p as a candidate metastasis suppressor for human hepatitis B virus-positive hepatocellular carcinoma. BMC Cancer. 2007 Sep 4 ;7:172.

Levine, R.L., L. Mosoni, B.S. Bertlett, E.R. Stadtman. Methionine residues as endogenous antioxidants in proteins. Proc. Natl. Acad. Sci. USA. 1996: 96: 15036-15040.

Lowther, W.T, H. Weissbach, F. Etienne, N. Brot, B.W. Mathews. The mirrored methionine sulphide reductases of Neisseria gonorrhoeae pilB. Nature Structural Biology. 2002: 9(5): 348-352.

Minniti A.N., R. Cataldo , C. Trigo , L. Vasquez , P. Mujica , F. Leighton , N.C. Inestrosa , R. Aldunate . Methionine sulfoxide reductase A expression is regulated by the DAF-16/FOXO pathway in Caenorhabditis elegans. Aging Cell. 2009: 8(6):690-705.

Moskovitz, J., and E.R. Stadtman. Selenium-deficient diet enhances protein oxidation and affects methionine sulfoxide reductase (MsrB) protein level in certain mouse tissues. Proc. Natl. Acad. Sci. USA. 2003:100:7486-7490.

Moskovitz, J. Roles of methionine suldfoxide reductases in antioxidant defense, protein regulation and survival. Curr. Pharm. Des. 2005:11:1451-1457.

Moskovitz , J. Prolonged selenium-deficient diet in MsrA knockout mice causes enhanced oxidative modification to proteins and affects the levels of antioxidant enzymes in a tissue-specific manner. Free Radic. Res. 2007 Feb;41(2):162-71.

Narayanan, R, K.A. Higgins, J.R. Perez, T.A. Coleman, and C.A. Rosen. Evidence for differential functions of the p50 and p65 subunits of NF-kappa B with a cell adhesion model. Mol . Cell. Biol. 1993 Jun;13(6):3802-10.

Onyango, I.G., J.P. Bennett Jr, and J.B. Tuttle. Endogenous oxidative stress in sporadic Alzheimer's disease neuronal cybrids reduces viability by increasing apoptosis through pro-death signaling pathways and is mimicked by oxidant exposure of control cybrids. Neurobiol. Dis. 2005:19:312-322.

Pal, R, D.B. Oien, F.Y. Ersen, and J. Moskovitz. Elevated levels of brain-pathologies associated with neurodegenerative diseases in the methionine sulfoxide reductase A knockout mouse. Exp. Brain Res. 2007 Jul; 180(4):765-74.

Pascual, I, I.M. Larrayoz, and I.R. Rodriguez. Retinoic acid regulates the human methionine sulfoxide reductase A (MSRA) gene via two distinct promoters, Genomics 2009; 93:62-71

Petropoulos, I., J. Mary, M. Perichon, and B. Friguet. Rat peptide methionine sulphoxide reductase: cloning of the cDNA, and down-regulation of gene expression and enzyme activity during aging. Biochem.J. 2001:355:819-825.

Roesijadi, G., S. Rezvankhah, D.M. Binninger, and H. Weissbach. Ecdysone induction of MsrA protects against oxidative stress in Drosophila. Biochem. Biophys. Res. Commun. 2007:354:511-516.

Ruan, H., X.D. Tang, M.L. Chen, M.L. Joiner, G. Sun, N. Brot, H. Weissbach, S.H. Heinemann, L. Iverson, C.F. Wu, and T. Hoshi. High-quality life extension by the enzyme peptide methionine sulfoxide reductase. Proc. Natl. Acad. Sci.USA. 2002:99:2748-2753.

Ryu, B.Y., K.E. Orwig, J.M. Oatley, M.R. Avarbock, and R.L. Brinster. Effects of aging and niche microenvironment on spermatogonial stem cell self-renewal. Stem Cells. 2006:24:1505-1511.

Sauer, H., G. Rahimi, J. Hescheler, and M. Wartenberg. Role of reactive oxygen species and phosphatidylinositol 3-kinase in cardiomyocyte differentiation of embryonic stem cells. FEBS Lett. 2000:476:218-223.

Shao, B, G. Cavigiolio, N. Brot, M.N. Oda, and J.W. Heinecke. Methionine oxidation impairs reverse cholesterol transport by apolipoprotein A-I. Proc. Natl. Acad. Sci. USA. 2008 Aug 26;105(34):12224-9.

Taggart, C, D. Cervantes-Laurean, G. Kim, N.G. McElvaney, N. Wehr, J. Moss, et al. Oxidation of either methionine 351 or methionine 358 in alpha 1-antitrypsin causes loss of anti-neutrophil elastase activity. J Biol Chem. 2000 Sep 1;275(35):27258-65.

Thiruchelvam, M., O. Prokopenko, D.A. Cory-Slechta, B. Buckley, and O. Mirochnitchenko. Overexpression of superoxide dismutase or glutathione peroxidase protects against the paraquat + maneb-induced Parkinson disease phenotype. J. Biol. Chem. 2005:280:22530-22539.

Ueda, S, H. Masutani, H. Nakamura, T. Tanaka, M. Ueno, and J. Yodoi. Redox control of cell death. Antioxid. Redox Signal. 2002;4(3):405-14.

Uthus, E.O., and J. Moskovitz . Specific activity of methionine sulfoxide reductase in CD-1 mice is significantly affected by dietary selenium but not zinc. Biol. Trace Elem. Res. 2007 Mar ;115(3):265-76.

Vougier, S., J. Mary, and B. Friguet. Subcellular localization of methionine sulphoxide reductase A (MsrA): evidence for mitochondrial and cytosolic isoforms in rat liver cells. Biochem. J. 2003:373:531-537.

Vriesema, A.J., J. Dankert, S.A. Zaat. A shift from oral to blood pH is a stimulus for adaptive gene expression of streptococcus gordonii CH1 and induces protection against oxidative stress and enhanced bacterial growth by expression of msrA. Infect. Immun. 2000 Mar;68(3):1061-8.

Weissbach, H, F. Etienne, T. Hoshi, S.H. Heinemann, W.T. Lowther, B. Matthews, et al. Peptide methionine sulfoxide reductase: Structure, mechanism of action, and biological function. Arch. Biochem. Biophys. 2002;397(2):172-8.

Wo, Y.B., D.Y. Zhu, Y. Hu, Z.Q. Wang, J. Liu, and Y.J. Lou. Reactive oxygen species involved in prenylflavonoids, icariin and icaritin, initiating cardiac differentiation of mouse embryonic stem cells. J. Cell. Biochem. 2008:103:1536-1550.

Wodarz, D. Effect of stem cell turnover rates on protection against cancer and aging. J. Theor. Biol. 2007:245:449-458.

Yermolaieva, O., R. Xu, C. Schinstock, N. Brot, H. Weissbach, S.H. Heinemann, and T. Hoshi. Methionine sulfoxide reductase A protects neuronal cells against brief hypoxia/reoxygenation. Proc. Natl. Acad. Sci.USA. 2004:101:1159-1164.

You, C., A. Sekowska , O. Francetic , I. Martin-Verstraete , Y. Wang , A. Danchin . Spx mediates oxidative stress regulation of the methionine sulfoxide reductases operon in Bacillus subtilis. BMC Microbiol. 2008: 8:128.

Zhang, C., P. Jia, Y. Jia, H. Weissbach, K.A. Webster, X. Huang, S.L. Lemanski, M. Achary, and L.F. Lemanski. Methionine Sulfoxide Reductase A (MsrA) Protects Cultured Mouse Embryonic Stem Cells from H2O2-mediated Oxidative Stress, J. Cell. Biochem. 2010: 111(1):94-103.

Zhang, C, P. Jia, Y. Jia, Y. Li, K.A. Webster, X. Huang, M. Achary, S.L. Lemanski and L.F.
    Lemanski. Anoxia, acidosis, and intergenic interactions selectively regulate
    methionine sulfoxide reductase transcriptions in mouse embryonic stem cells J. Cell
    Biochem. 2011:112(1):98-106.

12

Effects of Oxidative Stress and Antenatal Corticosteroids on the Pulmonary Expression of Vascular Endothelial Growth Factor (VEGF) and Alveolarization

Ana Remesal, Laura San Feliciano and Dolores Ludeña
Department of Paediatrics (A.R., L.SF.),
Department of Cellular Biology and Pathology (D.L.),
School of Medicine, University of Salamanca,
Salamanca University Hospital,
Spain

1. Introduction

The lung is the human organ most susceptible to oxidative damage. The transition from fetal, a low-oxygen environment, to FiO_2: 0.21, induces a relative degree of oxidative stress for all newborns, but especially in the case of preterm neonates whose lung development has been interrupted. A supplementary supply of oxygen (hyperoxia) is often used to treat preterm newborns and these infants may not yet be prepared to protect their lungs from oxidative injury.

One of the most important advances in neonatal care has been the introduction of antenatal corticosteroid therapy for preventing respiratory distress syndrome and improving the survival of preterm infants (Liggins et al., 1972; 1995). Foetal and postnatal lung development is regulated by glucocorticoids (Speirs et al., 2004).

During lung development, vascular endothelial growth factor (VEGF) is an important growth factor in vasculogenesis and angiogenesis (Voelkel et al., 2006), and it plays a central role in epithelial-endothelial interactions, which are critical for normal lung development (Zhao et al., 2005). VEGF is also involved in alveolarization (Thebaud et al., 2005), and it has been reported that the inhibition of VEGF impairs this process (Zhao et al., 2005; Van Tuyl et al. 2005).

Situations such as hypoxia, hyperoxia, and the administration of antenatal corticosteroids may regulate VEGF expression and may interfere with the alveolarization process (Remesal et al. 2009, 2010; San Feliciano et al., 2011).

Among the mechanisms implicated in lung damage due to hyperoxia, alterations in the expression of pulmonary VEGF have been described (Roberts et al., 1983; Maniscalco et al., 2005; Remesal et al., 2009). In our previous experimental studies in rats, we have also observed that dexamethasone and hyperoxia have an additive effect on the inhibition of VEGF, with a decrease in alveolarization (Remesal et al., 2010).

In this chapter we shall discuss the importance of vasculogenesis and angiogenesis in lung development, the role of VEGF in lung development, the effect of oxidative stress on the pulmonary expression of VEGF and alveolarization, the effects of antenatal glucocorticoids added to oxidative stress on the pulmonary expression of VEGF and alveolarization and, finally, the role of VEGF in the pathogenesis of bronchopulmonary dysplasia (BPD).

2. Angiogenesis and vasculogenesis in lung development

Vascular development occurs during all stages of lung development. The formation of the pulmonary vasculature includes three processes: angiogenesis, which leads to central vessels by the budding of new vessels from previous ones; vasculogenesis, which results in peripheral vessels by in situ differentiation of mesenchymal cells into hemangioblasts, and the fusion between the central and peripheral systems, creating the pulmonary circulation (Papaioannou et al., 2006; Akeson et al., 2000).

Vascular development in the lung has been shown to be a determinant of the maturation of lung structure, and angiogenesis and vasculogenesis are necessary for the successful development of the organ (Voelkel et al., 2006).

Interactions between the airways and blood vessels are critical for normal lung development, suggesting that a coordinated and timely release of vascular growth factors would promote alveolar development (Thebaud et al., 2005).

Jakkula et al. (Jakkula et al., 2000) administered antiangiogenic agents, SU-5416, thalidomide and fumagillin, to neonatal rats and observed a decrease in alveolarization and lung growth, with a histological pattern similar to BPD.

Van Tuyl et al.(Van Tuyl et al., 2005) found that the inhibition of vascularization in vitro resulted in a significant decrease in the morphogenesis of the airways, suggesting that pulmonary vascular development would be a factor in lung morphogenesis.

Schwarz et al.(Schwarz et al., 2000) reported that the inhibition of neovascularization with endothelial monocyte-activating polypeptide II (EMAP II) resulted in an arrest in the morphogenesis of the airways. It has also been reported that pulmonary vascular development is dependent on reciprocal interactions with the lung epithelium. Gebb and Shannon (Gebb et al., 2000) showed that mesenchymal cells cultured in the absence of epithelial cells degenerate, and they also observed significantly fewer cells positive for VEGFR-2 (VEGF receptor 2). In contrast, lung mesenchyme recombined with lung epithelial cells contained abundant cells positive for VEGFR-2, and their spatial distribution was similar to that observed in vivo in the lungs of foetal and neonatal rats. In an in vivo study of the lung, they found, as from foetal day 11, precursor cells with a positive expression of VEGFR-2 in the mesenchyme in the developing epithelium.

Studies carried out by Burri (Burri, 2006) have shown that in the human lung microvascular maturation begins very early and partly takes place during alveolarization, ending at an age of 2-3 years.

Genetic analyses have shown that cell-cell interactions and cell-extracellular matrix, growth factors and transcription factors are involved in vascular development (Roth-Kleiner et al., 2003).

One of the most important factors related to the processes described above is clearly VEGF (Thebaud et al., 2005).

3. The role of VEGF in lung development

Vascular endothelial growth factor (VEGF) is a potent mitogen for endothelial cells (Ferrara et al., 1997; Lassus et al., 1999) as well as being chemotactic for these cells, and it is one of the most potent mediators of vascular regulation (Bhatt et al., 2000). Regarding water and proteins, VEGF exerts a potent effect on vascular permeability (Dvorak et al., 1995). VEGF is also necessary for the survival of endothelial cells, especially in hyperoxic environments (Watkins et al., 1999; Maniscalco et al., 2002).

In 1996, the studies of Ferrara et al (Ferrara et al., 1996) and Carmeliet el al. (Carmeliet et al., 1996) revealed the critical role of VEGF in embryonic vasculogenesis and angiogenesis. Those authors observed the death of mouse embryos, with the inactivation of a single VEGF allele.

The VEGF signalling pathway is important in vasculogenesis, particularly during foetal lung development (Gebb et al., 2000; Voelkel et al., 2006) . The expression of VEGF and VEGFR-2 has been demonstrated in airway tube and vascular mesenchymal cells during foetal development and also in vitro in cultures containing epithelial and mesenchymal elements of foetal lung (Gebb et al., 2000).

In baboons, Maniscalco et al. (Maniscalco et al., 2002) found an association between increased PECAM-1 (platelet endothelial cell adhesion molecule) and increased VEGF levels, suggesting a role of VEGF in pulmonary vasculogenesis.

Lassus et al (Lassus et al., 2001) found a higher concentration of VEGF in tracheal aspirates of preterm infants than in those from term infants. Levy et al (Levy et al., 2005) reported that in foetal lungs VEGF expression was higher in the canalicular and saccular stages.

In a murine model, Bhatt et al. (Bhatt et al., 2000) observed that VEGF mRNA levels increased three-fold during the first two weeks of postnatal age: i.e., the alveolarization phase and expansion of the microvasculature; VEGFR-2 mRNA increased in parallel. In those studies, performed by in situ hybridization, the authors demonstrated that VEGF mRNA was mainly located in the epithelial cells of the distal air spaces.

The mRNA expression of VEGFR-1 and VEGFR-2 also increases during normal lung development in mice (Costa et al., 2001; Gebb et al., 2000), and these receptors are located on the endothelial cells of pulmonary vessels in proximity to the epithelium in development. This spatial relationship suggests that VEGF would play an important role in the development of the alveolar capillary bed.

Del Moral et al. (Del Moral et al., 2006) described that exogenous VEGF$_{164}$ induces bronchial morphogenesis in cultured embryonic mouse lung. The effect of VEGF on the epithelium would be indirect, through an interaction with the mesenchyme, because these authors only found VEGFR-2 to be expressed in mesenchymal cells.

The expression of VEGF mRNA and VEGF protein is localized to distal epithelial cells in human foetal lung during the second half of pregnancy and its levels increase with time (Brown et al., 2001).

More recently, in humans Groenman et al. (Groenman et al., 2007) observed the expression of VEGF in the epithelium during the first trimester, as well as the expression of VEGFR-2 in mesenchymal cells adjacent to the epithelium and mesenchyme in vascular structures.

In a mouse model with lung renal capsule grafts, Zhao el al (Zhao et al., 2005) identified the role of VEGF in lung development. By inhibiting VEGF they observed the inhibition of vascular development and a significant alteration in epithelial development.

The absence of $VEGF_{164}$ and $VEGF_{188}$ isoforms results in a decrease in peripheral vascular development and a delayed formation of air spaces in mouse lungs (Van Tuyl et al., 2005).

It has been shown that the inhibition of VEGF results in the inhibition of angiogenesis and alveolarization in lung development in rats. The inhibition of VEGF for even short periods of time may also reduce the number of blood islands and endothelial cells expressing VEGFR-2 (Van Tuyl et al., 2005).

Thebaud B et al. (Thebaud et al., 2005) performed in vitro and in vivo studies in rat lung, reporting that the inhibition of the VEGF signal stopped alveolarization, offering a model similar to BPD and emphysema.

Use of SU-5416, which blocks the VEGF receptor, before or after birth results in a reduction in pulmonary vascularization and alveolarization (Jakkula et al., 2000).

Reports have been made of the existence of a regulatory loop in which epithelium-derived VEGF induces vascular development and endothelial relayed signals directly or indirectly, via the mesenchymal compartment, and stimulates epithelial differentiation and branching (Van Tuyl et al., 2005).

According to the studies carried out by Yamamoto et al. (Yamamoto et al., 2007) primary septum formation depends on interactions between the respiratory epithelium and the underlying vessels, suggesting a dependence of the development of pulmonary capillaries on the VEGF-A derived from the epithelium.

In human foetal lung cell cultures in the second half of gestation, Brown et al (Brown et al., 2001) found the VEGFR-2 receptor and neurofilin I in epithelial cells, suggesting a possible autocrine role of VEGF in the proliferation and differentiation of human alveolar epithelial cells. They also observed that exogenously administered VEGF increased tissue differentiation parameters and that this led to a proliferation of epithelial cells in the distal airways with the morphology of type II pneumocytes and, additionally, it increased the production of several components of surfactant.

VEGF stimulates the production of pulmonary surfactant by type II pneumocytes (Voelkel et al., 2006; Papaioannou et al., 2006; Ferrara et al., 2003), leading to lung maturation and preventing the development of respiratory distress syndrome in newborns.

Compernolle et al (Compernolle et al., 2002) performed a study in premature mice in which they observed an increase in the production of surfactant proteins B and C after treatment with VEGF, showing that type II cells express VEGFR-2. Although the precise mechanisms by which VEGF stimulates the synthesis of surfactant have not been clearly defined, it appears that VEGF stimulates the synthesis of platelet-activating factor (PAF), a potent inducer of glycogenolysis in the foetal lung, and activates kinase C protein, a central regulator of surfactant secretion and glycogen metabolism.

It has been reported that the inhibition of VEGF results in endothelial cell death, suggesting that VEGF could also be a survival factor for endothelial cells (Maniscalco et al., 2005).

4. Effects of oxidative stress on pulmonary VEGF expression and alveolarization

Some studies have shown the involvement of reactive oxygen species (ROS) in the mitogenic cascade initiated by tyrosine kinase receptors of many growth factors. Colavitti et al (Colavitti et al., 2002) identified reactive oxygen species as mediators involved in the signal transduction of VEGFR-2; these findings were paradoxical, however, because it is well known that angiogenesis occurs at low oxygen concentrations. Nevertheless, situations of complete anoxia are very rare under physiological conditions. Other studies have shown that low oxygen concentrations give rise to oxygen free radicals through a mechanism involving an abnormal flow of electrons in the respiratory mitochondrial chain (Chandel et al., 1998).

Preterm birth in ambient air, before the full development of antioxidant defences, elicits a relative degree of hyperoxia, which blocks normal lung vascular development and alveologenesis (Massaro et al., 2004; Thebaud, 2007).

The perinatal period is critical for proper adaptation to postnatal lung life. Lung damage during the perinatal period could interfere with lung growth, leading to abnormalities in lung structure and function that persist in children and adults. Premature infants with immature lungs often require high inspired oxygen levels, which are toxic to alveolar and endothelial cells (Crapo, 1986, 2003a, 2003b). Studies carried out by our group and other researchers have shown that hyperoxia alters alveolar growth by interfering with essential growth factors such as VEGF (Maniscalco et al., 1997; Remesal et al., 2009, 2010).

In our studies (Remesal et al., 2009, 2010) we evaluated pulmonary VEGF expression by RT-PCR and immunohistochemistry with densitometry analysis in Wistar rats at 0, 4 and 14 days of life. In order to analyse alveolarization, quantitative morphometric assessment was carried out by superimposing a sample over a square grid pattern (model CPLW 1018, Zeiss Optical, Hannover Md) for conventional haematoxylin-eosin lung preparations (Blanco et al., 1994), and the mathematical model of Weibel (Weibel et al., 1966) was applied, measuring Lm (mean air space size), Nv (number of alveoli) and ISA (mean internal surface area). All procedures were approved by the Animal Health Care Committee of the University of Salamanca. The experiments were performed following the regulations of the Directive of the Council of the European Community (DOCE L 222; 24/08/1999). The rats were born at 21-22 days of gestation by natural delivery and were mixed and randomly distributed in a litter size adjusted to 8 pups in order to control for the effects of this on nutrition and growth (Crnic, L. S. et al., 1978). Rat pups were exposed to hypoxia for two hours (0.10 FiO2) at 4-8 hours of life in a sealed chamber with continuous O_2 monitoring (SERVOMEX 1440 gas analyser) and then recovered for a further 2 hours under hyperoxia (>0.95 FiO$_2$) (hyperoxia after hypoxia group) or normoxia (room air, 0.21 FiO$_2$) (normoxia after hypoxia group) or not recovered (hypoxia group). The rat pups studied at 4 and 14 days were later maintained in room air. The animals of the control group were under normoxic conditions throughout the study.

In our studies (Remesal et al., 2009, 2010), we observed a decrease in pulmonary VEGF expression in rats that were exposed to hyperoxia after hypoxia and this finding correlated with less alveolar septation. We observed that pulmonary VEGF expression remained decreased at 14 days after recovery with oxygen at birth (Fig.1).

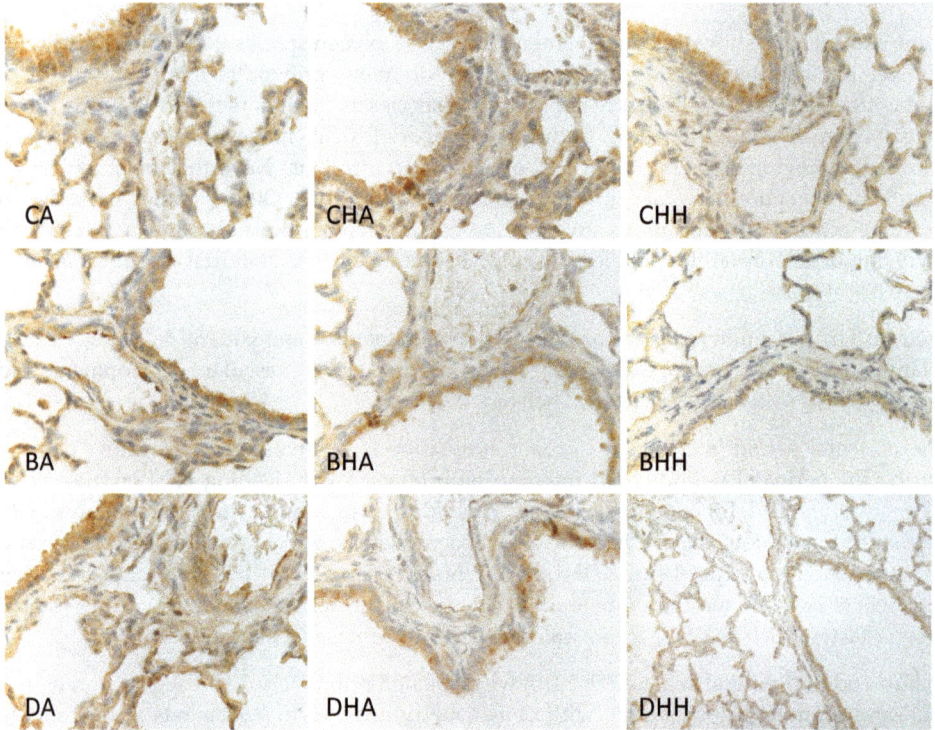

C: Control. B: Betamethasone. D: Dexamethasone. A: Air. HA: Hypoxia + Air. HH: Hypoxia + Hyperoxia

Fig. 1. Immunohistochemistry. Lung sections for each group.

The effect of hyperoxia on the postnatal lung varied, depending on the length of exposure, the degree of hyperoxia, and the age and animal species used in the model. In mice, McGrath-Morrow et al. (Grath-Morrow et al., 2005) found differences between the lung damage caused by hyperoxia and damage due to blocking VEGFR-2. In lungs exposed to hyperoxia the damage was more persistent; the animals had a more reduced mitotic index and showed increased apoptosis.

Roberts et al.(Roberts et al., 1983) reported that newborn rats exposed to an FiO$_2$ of 1 exhibited a decrease in the number of capillaries due to the toxic effect of oxygen on endothelial cells. Kunig et al. (Kunig et al., 2005) found that lungs subjected to hyperoxia had a decreased number of alveoli and showed reduced vascular growth.

Although hyperoxia could inhibit lung growth through several mechanisms, clinical and experimental studies have suggested that damage to VEGF signalling plays an important role in the pathogenesis of BPD (Thebaud et al., 2005,; Kunig et al., 2005; Maniscalco et al., 2002;Lin et al., 2005; Thebaud, 2007). In many studies it has been shown that hyperoxia decreases VEGF levels.

Thebaud B et al.(Thebaud et al., 2005) described irreversible hypoalveolarization induced by oxygen in rat lung development, and this was associated with decreased VEGF expression and reduced vascular growth.

Wageenaar et al (Wagenaar et al., 2004) investigated gene expression by analysing "microarray" DNA in the lungs of preterm rats exposed to prolonged hyperoxia, observing a decrease in VEGF levels and those of its receptor VEGFR-2.

Endothelial cells are particularly sensitive to oxidative stress and decreases in VEGF levels due to hyperoxia further increase the susceptibility of the epithelium to hyperoxic injury and a greater loss of these cells after acute hyperoxia (Kunig et al., 2005; Lin et al., 2005). During exposure to hyperoxia, a destruction of the lung microvascular system was observed, accompanied by a cessation of endothelial regeneration, which is essential for the repair of lung injury induced by oxygen (Watkins et al., 1999).

Klekamp et al. (Klekamp et al., 1999) reported a reduction in VEGF in the lungs of rats exposed to hyperoxia ($FiO_2 > 0.95$ between postnatal days 6 and 14) associated with alveolar cell apoptosis and a reduced expression of VEGFR-2 and VEGFR- 1.

Maniscalco et al. (Maniscalco et al., 2002) found that preterm baboons treated with oxygen had a 70% decrease in PECAM-1 protein and a 27% decrease in capillary density in the presence of dysmorphic capillaries in comparison with a control group. They observed a decrease in VEGF mRNA, supporting the hypothesis that the development of BPD could be the result of a disruption of the genetic program of angiogenic factors and receptor expression in endothelial cells.

Watkins et al. (Watkins et al., 1999) found that in damage due to hyperoxia in newborn and adult rabbits the proportion of $VEGF_{189}$ decreased, normal values being re-established during recovery under normoxic conditions.

In a model of hyperoxia exposure in mice, Zimova et al (Zimova-Herknerova et al., 2008) observed a decrease in VEGF mRNA levels that was lower when the animals were given retinoic acid.

The decrease in VEGF expression in hyperoxia could be related to the suppression of the expression of hypoxia-induced factor 2-α (HIF2-α or HLF) (Maniscalco et al., 2002). Moreover, hypoxia-induced factor (HIF) is inhibited by increasing levels of oxygen (Thebaud, 2007).

In a study carried out on newborn rats, Hosford et al (Hosford et al., 2003) showed that the expression of VEGF, VEGFR-1 and VEGFR-2 was decreased at 12 and 14 days after exposure to a hyperoxic environment during the critical period of alveolar development (4 to 14 days). They observed that under conditions of normoxia there was a strong correlation between VEGF and HLF. This correlation disappeared after hyperoxia, suggesting that low levels of HLF, after exposure to high concentrations of O_2, would not stimulate the expression of VEGF.

Roper et al. (Roper et al., 2004) found that alveolar type II epithelial cells exposed to hyperoxia showed DNA damage (broken strands, modifications of bases, changes in sister chromatids and the oxidation of guanine to 8-oxoG), even though they were morphologically intact. Hyperoxic injury was produced by the interaction of ROS with macromolecules such as DNA, lipids and proteins.

Oxidative damage to tissues could be exacerbated by damage to VEGF signalling. It has been observed that VEGF induces manganese superoxide dismutase (MnSOD) and nitric oxide (NO), both of which are scavengers of reactive oxygen species (Maniscalco et al., 2005). In VEGF-overexpressing transgenic mice, Siner et al (Siner et al., 2007) have shown that VEGF induces cytoprotection via the induction of heme oxygenase-1 (an antioxidant) and also that VEGF has the capacity to decrease levels of apoptosis markers.

In addition VEGF conferred cytoprotection through an A1-dependent mechanism, a critical regulator in lung injury and cell death induced by hyperoxia. He et al. (He et al., 2005) showed that transgenic VEGF$_{165}$-overexpressing mice had increased A1 protein levels and A1 mRNA and that they survived longer under conditions of hyperoxia with a FiO$_2$ of 1.

In a study of preterm baboons subjected to hyperoxia for 6 to 10 days, Maniscalco et al. (Maniscalco et al., 2005) found an increase in the levels of p53, a transcription factor that represses VEGF transcription, in distal epithelial cells. They also found oxidant DNA damage with increased 8-oxoG levels, which would be the mechanism responsible for the increase in p53 level.

In a study performed with mice to detect alveolar type II cells, Yee et al. (Yee et al., 2006) observed a loss of these cells after exposure to hyperoxia; this would be related to the decrease in VEGF levels in hyperoxia.

In studies by Maniscalco et al. (Maniscalco et al., 1995, 2002) it has been reported that VEGF is expressed in type II cells with low surfactant protein C levels. Thus, the change in the distal epithelial cell phenotype was seen in situations of hyperoxia, with more cells with high amounts of surfactant protein C, resulting in a decrease in the production of VEGF.

It has also been shown that oxidative stress inactivates the survival signal of VEGF in endothelial cells through the action of peroxynitrite (Acarregui et al., 1999) .

Furthermore, in vitro and in vivo studies have described the induction of VEGF by reactive oxygen species. In ferrets, Becker et al (Becker et al., 2000) reported an increase in VEGF in a model of lung ischemia, regardless of exposure to oxygen, and in a mouse model Corne et al. (Corne et al., 2000) observed an increase in VEGF levels in bronchoalveolar lavage fluid after 72 hours of exposure to an FiO$_2$ concentration of 1.

In our studies (Remesal et al., 2009, 2010) the animals were previously exposed to acute hypoxia. It has been described that early exposure to hypoxia predisposes the animals to an increased response to any further damage (Haworth et al., 2003). Saugstad (Saugstad, 1988) reported that reoxygenation following hypoxia generated an increase in oxygen free radicals levels that could not be neutralized by antioxidant defences and that resulted in injury to cellular structures. Ekekezie et al. (Ekekezie et al., 2003) and Lin et al. (Lin et al., 2005) reported that hyperoxia negatively regulates the expression of VEGF, despite recovery in air. These findings are similar to our own.

5. Antenatal glucocorticoids

One of the most important advances in neonatal care has been the introduction of antenatal glucocorticoid therapy for preventing respiratory distress syndrome and improving the survival of preterm infants (1995). Foetal and postnatal lung development is regulated by glucocorticoids (Speirs, H. J. et al., 2004). There are many reports showing that corticoids can alter the normal stages of lung development during the period of alveolarization. Corticoids are known to trigger the structural maturation of mesenchymal cells and the functional maturation of the surfactant system (Jobe, A. H., 2001).

Dexamethasone and Betamethasone are the only corticosteroids recommended for antenatal therapy (National Institute of Health NIH, 1995), and as yet there are no indications favouring the use of one preparation over the other (Crowley, 2007), although there are some reports giving preference to bethametasone (Lee et al., 2006; Miracle et al., 2008; Remesal, et al., 2010; San Feliciano et al., 2011). The dosage and number of treatment cycles in gestating women at risk of preterm delivery are still under debate because of the long-term effects of these drugs on growth (Sweet et al., 2010).

5.1 Effects of betamethasone vs dexamethasone on the expression of VEGF and alveolarization

Massaro et al. (Massaro et al., 2004) reported that in all species septation, whether antenatal or postnatal, occurs during a period in which the plasma concentration of glucocorticoids is low and that it ends when the concentration of glucocorticoids increases. Also, in studies performed by Liu et al. (Liu et al., 2004) the authors showed that dexamethasone inhibits IGF-I, this having an impact on alveolarization.

Studies by Massaro (Massaro et al., 2004) and Clerch (Clerch et al., 2004) have shown that glucocorticoid administration to rats or mice during the period of septation, when the plasma concentration of steroids is usually low, damages spontaneous septation and the development of the pulmonary vasculature. There was no further spontaneous septation and vasculogenesis when steroid treatment was withdrawn (tested in rats up to 95 days). This observation suggests that there would be a critical period for the development of these events. In those studies, using microarray analysis of pulmonary gene expression the authors identified a negative regulation of VEGFR-2 by dexamethasone, resulting in the inhibition of septation.

Furthermore, in a study in newborn rats treated with steroids during the first 4 days of life Tschanz et al. (Tschanz et al., 2003) have shown that the lung has the ability to recover from the damage caused by steroids when these are suspended.

In our studies (Remesal et al., 2010; San Feliciano et al., 2011), in order to evaluate possible differences between the effect of both antenatal glucocorticoids, dexamethasone, betamethasone or saline solution were administered intravenously to pregnant Wistar rats on the 20th and 21st days of gestation. The newborn rats were exposed to the different experimental situations described previously.

We observed a decreased pulmonary VEGF expression that was correlated with a decrease in alveolarization in the animals that received antenatal dexamethasone. This effect on VEGF and alveolarization was not found in the animals that received antenatal betamethasone (Fig.1). We observed a negative effect on lung maturation and VEGF expression caused by antenatal dexamethasone lasting from the saccular phase until the end of the alveolarization period (Remesal et al., 2010; San Feliciano et al., 2011).

The only structural difference in the molecule of these two corticosteroids is the orientation of the methyl group at position 16, and this small difference seems to have important consequences. Other authors have reported the different biological effects of both corticosteroids. Rayburn et al. (Rayburn et al., 1997) observed better memory in rats treated with antenatal betamethasone and poorer memory in the group treated with antenatal dexamethasone.

In an in vitro study of thymocytes, Buttgereit et al (Buttgereit et al., 1999) observed that different glucocorticoids, including dexamethasone and betamethasone, differed in the strength of their genomic and non-genomic effects; they found that dexamethasone was five times more potent in inhibiting cellular respiration.

Other authors have reported differences in genomic and non-genomic effects between the two molecules and the expression of NMDA receptors (McGowan et al., 2000; Setiawan et al., 2007) and in the expression of the sodium channels of respiratory epithelial cells. Although we have found no reports comparing the effects of betamethasone or dexamethasone on lung VEGF, our results indicate that the molecules do have different actions (Remesal et al., 2010; San Feliciano et al., 2011).

In clinical studies, different findings have been reported with the use of either molecule (dexamethasone or betamethasone). Baud et al (Baud et al., 1999) published a multicenter study with a cohort of 883 children with gestational ages of 24 to 31 weeks over a period of 4 years. The mothers of 361 infants had received betamethasone; the mothers of 165 infants had received dexamethasone, and the mothers of 357 children had not received glucocorticoids. The authors compared the frequency of cystic periventricular leukomalacia among the three groups using a multivariate analysis adjusted for confounding factors such as sex, chorioamnionitis, infection, multiple gestation, and other relevant factors. 8.4% of children in the group that had not received antenatal steroids developed cystic periventricular leukomalacia as compared to 4.4% of the children in the antenatal betamethasone-treated group and 10.9% of those in the antenatal dexamethasone-treated group.

In the study by Lee et al. (Lee et al., 2006), which included 3600 children with birth weights of less than 1500 grams, the authors found a decrease in mortality in the children who had received antenatal betamethasone as compared with the group that had not received antenatal steroids; they failed to find this decrease in mortality in the antenatal dexamethasone-treated group. Also, dexamethasone was associated with an increase in neonatal mortality as compared with betamethasone. They found a lower incidence of severe retinopathy in children who had received antenatal betamethasone and this would be related to the inhibition of TNF-α. This factor was also involved in angiogenesis. In that study, the authors also observed a lower frequency of intraventricular hemorrhage in children who had received antenatal betamethasone. Similar results have also been published in the Cochrane meta-analysis reported by Crowley (Crowley, 2007).

Feldman et al. (Feldman et al., 2007) reported a study of 334 preterm infants with birth weights of less than 1500 grams at birth: 186 children had received antenatal betamethasone and 148 had received antenatal dexamethasone. The most important finding observed in that study was that the children who had received antenatal betamethasone had a lower incidence of respiratory distress syndrome and BPD.

In cultured embryonic rat lung, Oshika E et al (Oshika et al., 1998) showed that treatment with dexamethasone resulted in growth retardation, abnormal branching, dilated proximal tubules, and a suppression of the proliferation of the epithelial cells of the distal tubules.

Schellenberg and Liggins (Schellenberg et al., 1987) administered dexamethasone to pregnant rats and studied its effects on lung development in the offspring during foetal and postnatal life. The results revealed an inhibition of lung growth and body growth, and a reduction in DNA contents in the animals treated with dexamethasone. Dexamethasone appeared to affect the populations of cells that produce elastin and cells that produce collagen during foetal lung development.

In in vitro studies of human lung fibroblast cultures, it was observed that dexamethasone inhibited fibroblast proliferation and chemotactic activity in a dose-dependent manner (Brenner et al., 2001).

There are different studies, such as ours, describing a relationship between steroids and VEGF. In studies carried out by Vento et al.(Vento et al., 2002) on preterm infants, the authors found that lung VEGF levels were decreased in the dexamethasone-treated group as compared with the untreated group.

Many in vitro studies (Nauck et al., 1997; Nauck et al., 1998; Harada et al., 1994; Tanabe et al., 2006) have described a strong negative regulation of the induction of VEGF expression by dexamethasone in different cell types, including alveolar epithelial cells.

Hewitt et al (Hewitt et al., 2006) showed that VEGF expression was decreased in the placentas of pregnant rats treated with dexamethasone.

In an in vitro study, Gille et al (Gille et al., 2001) found that in human keratinocyte cells treated with glucocorticoids VEGF mRNA was rapidly degraded within 2 hours. The decrease in mRNA stability could be one mechanism of glucocorticoid-induced inhibition of VEGF.

In an in vitro study of human foetal lungs, Acarregui et al. (Acarregui et al., 1999) found increased levels of VEGF mRNA when the lung tissue was maintained in dexamethasone for 4 days as from the onset of incubation in 0.20 FiO$_2$. However, when the epithelium was first incubated in FiO$_2$ 0.20 and then treated with dexamethasone, mRNA levels were not increased. According to those authors, dexamethasone induced differentiation in type II cells and this was responsible for the variations in the levels of VEGF.

In studies of preterm infants, Lassus et al. (Lassus et al., 1999, 2001) found no differences in the concentration of VEGF in tracheal aspirates from children who had received antenatal steroids and those who had not. They also failed to find any differences in the concentration of VEGF in preterm infants treated early on with postnatal dexamethasone.

Moreover, D'Angio et al. (D'Angio et al., 1999) have described higher levels of VEGF in tracheal aspirates in preterm infants treated with dexamethasone.

In a study carried out in mice, Compernolle et al. (Compernolle et al., 2002) observed that antenatal dexamethasone stimulated the expression of foetal VEGF when administered at low doses (0.8 mg / kg), but suppressed the production of VEGF when administered at high doses (2.4 mg / kg). Thus, excessive amounts of glucocorticoids could neutralize the beneficial effects of VEGF in the lung.

Conversely, a study in mice conducted by Bhatt et al. (Bhatt et al., 2000), who administered dexamethasone (0.1-5 mg/ Kg/ day) from postnatal day 6 to 9, analyzed VEGF mRNA and VEGFR-2 mRNA levels in lungs, and found that VEGF and VEGFR-2 increased with increasing doses of dexamethasone. They also analyzed the effects of dexamethasone on the amount of mRNA HLF, which has been associated with an increased transcription of VEGF in normoxia and hypoxia. In the group treated with 5mg/Kg/day of dexamethasone the amount of HLF-α in the lung tripled and this could be suggested as a possible mechanism accounting for the effects of dexamethasone on lung VEGF mRNA levels. There were no differences between dexamethasone and control groups in the study of the VEGF protein. Dexamethasone treatment did not alter the pattern of VEGF-expressing cells; mainly distal alveolar epithelial cells.

The different experimental models used, or the fact that the effect of dexamethasone on the differentiation of type II cells might lead to a relative increase in VEGF mRNA (Acarregui et al., 1999) could explain the discrepancies found in the effect of dexamethasone on VEGF expression.

In our studies (Remesal et al., 2010; San Feliciano et al., 2011) we observed that dexamethasone inhibited septation in rats (with postnatal alveolarization), as reported by Massaro and Massaro (Massaro et al., 1986). We believe that the effect of dexamethasone on the decrease in VEGF may be related to the impairment in alveolarization.

Also in our studies (Remesal et al., 2010; San Feliciano et al., 2011), we found no decrease in VEGF in the group treated with antenatal betamethasone (Fig.1). We have found few studies in the literature addressing the effect of betamethasone and VEGF.

In a study by Aida et al. (Aida et al., 2004) it was reported that VEGF stimulates the expression of glucocorticoid receptors, although these might possibly have a more reduced function.

In an experimental study on sheep, Suzuki et al. (Suzuki et al., 2006) found no differences in the levels of VEGF mRNA between the group treated with antenatal betamethasone and the group that received no treatment.

Roubliova et al. (Roubliova et al., 2008) conducted a study that included 112 rabbit foetuses. The mothers received 0.05 or 0.1 mg/ Kg/ day of betamethasone at 25 and 26 days of gestation, but the mothers of the control group only received saline. The authors studied lung VEGF expression and found that VEGF was increased in endothelial cells, epithelial cells, and smooth muscle cells, these results being similar to those found in our study. They also observed that this effect was dose-dependent.

5.2 Effects of antenatal glucocorticoids added to oxidative stress damage on the pulmonary expression of VEGF and alveolarization

In our studies (Remesal et al., 2010) we have observed that dexamethasone and hyperoxia have an additive effect on the inhibition of VEGF, with a decrease in alveolarization (Fig.1).

It has been reported that both factors (dexamethasone and hyperoxia) arrest alveolarization (Frank , 1992; Veness-Meehan et al., 2000)

There are other studies that have reported that hyperoxia and dexamethasone jointly enhance decreases in the expression of VEGF (Ozaki et al., 2002; Edelman et al., 1999).

In the study performed by Ozaki et al. (Ozaki et al., 2002) the authors assessed the response of VEGF in the retinas of newborn rabbits after treatment with dexamethasone, exposure to hyperoxia with FiO_2 0.8 to 1 for 4 days, and exposure to hyperoxia with subsequent recovery in air for 5 days. They found that VEGF mRNA was decreased after hyperoxia and recovery in air in the animals that had received dexamethasone as compared to animals that had received dexamethasone and were in ambient air.

Edelman et al. (Edelman et al., 1999) used a model of corneal neovascularization induced in rat cornea by cauterization. They observed that after cauterization an increase occurred in VEGF mRNA and protein produced by leukocytes and macrophages adjacent to the lesion. This was a model in which hypoxic zones also occurred around the lesion due to the cauterization, and hypoxia could induce the expression of VEGF. They found that treatment with dexamethasone or systemic hyperoxia inhibited the increase in VEGF and that combined treatment with both dexamethasone and hyperoxia had an additive effect.

HGF (hepatocyte growth factor) is a growth factor with mitogenic activity against epithelial cells. Dexamethasone suppresses the gene expression of HGF and inhibits the growth factors responsible for the induction of HGF mRNA expression (Lassus et al., 2002). It has also been shown that glucocorticoids inhibit the inflammatory cytokine production. It has been shown that interleukin 1 and 6 induce the expression of HGF in vitro (Tamura et al., 1993; Zarnegar, 1995). Through its effect on the HGF, dexamethasone exerts an adverse influence on lung development and the repair of acute lung injury in the lungs of preterm infants (Lassus et al., 2002).

There is a balance between retinoic acid and glucocorticoids during normal lung development. Alterations in this balance produce abnormal alveolarization (Jobe, 2003). It has been observed that the survival of newborn rats exposed to oxygen improves with concurrent treatment with retinoic acid and dexamethasone, suggesting a possible complementary effect (Veness-Meehan et al., 2000). Retinoic acid could compensate the acceleration of septal maturation due to the administration of steroids, the benefit of treatment with these drugs persisting during epithelial maturation (Bourbon et al., 2005). It has also been observed that vitamin A levels in the blood of ventilated premature children whose lung function improved with glucocorticoid treatment were higher (Shenai et al., 2000).

According to our own results (Remesal et al., 2010) it would appear that betamethasone inhibits the negative action of hyperoxia on VEGF.

Chandrasekar et al. (Chandrasekar et al., 2008) found that betamethasone reduced intensity of oxidative stress and improved the response of the pulmonary arteries to vasodilators in lambs with pulmonary hypertension. They observed that betamethasone increased the levels of eNOS and MnSOD.

6. VEGF and BPD

The first definition of BPD was given by Northway in 1967 (Northway et al., 1967) as chronic lung disease in childhood as a result of therapy with mechanical ventilation and oxygen for respiratory distress syndrome after premature birth. Traditionally, BPD has been defined by the persistence of respiratory signs and symptoms, the need for supplemental oxygen to treat hypoxemia, and radiographic abnormalities at 36 weeks corrected age.

In light of the new therapeutic possibilities available today, BPD has been seen to affect very immature preterm infants. In the pathogenesis of BDP, the interruption of lung development plays a more important role than volutrauma or inflammation (Jobe, 1999). Alveolar hypoplasia and dysmorphic changes of the pulmonary microvasculature have been consistent findings in animal models of BPD and in the autopsies of children who died of BPD (Maniscalco et al., 2002; Coalson et al., 1988, 1999; Bhatt et al., 2001; De Paepe et al., 2006).

Among the mechanisms that arrest alveolar development in BPD, the following have been implicated: "oxidative stress", mechanical ventilation, proinflammatory factors, glucocorticoids, bombesin-like peptides, and poor nutrition (Jobe, 1999; Jobe et al., 2001).

Premature infants have a non-developed microvasculature and suffer from lung damage due to hyperoxia. (Crapo,1986, 2003a, 2003b ; Spyridopoulos et al., 1997). Given the critical role of endothelial cells during lung development, strategies that alter the signal of VEGF during the foetal and perinatal periods would lead to BPD (Voelkel et al., 2006; Lin et al., 2005).

The importance of VEGF in foetal lung growth has led to the vascular theory of BPD, which has considerable importance in lung injury due to prematurity, mechanical ventilation, and hyperoxia treatment during perinatal life (Abman, 2001; Voelkel et al., 2006).

The predominant lung histology at autopsy in children with bronchopulmonary dysplasia is characterized by an arrest in lung development, including alveolar development and vascular growth (Kunig et al., 2005; Lin et al., 2005; Jobe, 1999; Jobe et al., 2001; Abman, 2001; Kunig et al., 2006).

Bhatt et al. (Bhatt et al., 2001) reported a decrease in the expression of VEGF, VEGFR-1 and TIE-2 in the lungs of children who had died of BPD and had dysmorphic capillaries, suggesting an arrest of the development of the pulmonary vasculature.

Ambalavanan and Novak (Ambalavanan et al., 2003) found lower levels of VEGF in the tracheal aspirates of preterm infants upon mechanical ventilation in the first 24 hours of life, such infants developing BPD in comparison with those who did not develop BPD. The results of experimental studies carried out on rats suggest that the inhibition of VEGF receptor activity reduces alveolarization and vascular growth, leading to histological changes in the lung that resemble those found in BPD (Lin et al., 2005).

The decrease of VEGF in BPD has implications for the survival of endothelial cells. In addition, VEGF inhibits tumour necrosis factor (TNF-α) and hence endothelial cell apoptosis (Bhatt et al., 2001). It has also been reported that the inhibition of VEGF receptors in immature lungs reduces the expression of nitric oxide synthase (NOS) and the bioactivity of NO and later contributes to the development of structural damage and functional BPD (Papaioannou et al., 2006; Lin et al., 2005; Tang et al., 2004).

The baud et al. (Thebaud et al., 2005) performed intratracheal treatment with gene therapy with VEGF to rats exposed to hyperoxia during lung development and observed a longer survival and preservation and restoration of normal alveolarization, even when treatment was performed with already established BPD. VEGF gene transfer also increased the expression of NOS, suggesting that some of the beneficial effects of VEGF could be mediated by NO.

Kunig et al. (Kunig et al., 2005, 2006) treated newborn rats with human recombinant VEGF during and after exposure to hyperoxia, observing an increase in vascular growth and improved alveolarization. VEGF treatment after the damage caused by hyperoxia in neonatal rats improved alveolarization, vascular growth and lung growth, and prevented the development of BPD.

In the above studies (Thebaud et al., 2005; Kunig et al., 2005), VEGF led to highly permeable immature capillaries and pulmonary oedema. However, in a combination of gene transfer of VEGF and angiopoietin-1, the alveolarization and angiogenesis were preserved and improved, with more mature and fewer permeable capillaries.

These studies would increase the possibilities for the treatment of children with severe BPD, although caution should be exercised on attempting to extrapolate findings from animal models for the treatment of human disease.

In the case of children with BPD, De Paepe et al. (De Paepe et al., 2006) have suggested that changes in lung architecture would not simply be due to a decrease in angiogenesis, describing an increase in pulmonary capillary density in children who had been subjected to mechanical ventilation for long periods.

Akeson et al. (Akeson et al., 2003) studied transgenic $VEGF_{164}$-overexpressing mice. They noted that when the increased expression of $VEGF_{164}$ was in the proximal airways the abnormalities did not occur at the junction of the vascular network of the lung. When the expression of this molecule was higher in the distal airways, there was an alteration of the junction of the vascular network, demonstrating that normal pulmonary vascular development requires a precise spatial expression of VEGF-A for proper morphogenesis, and indicating that caution should be exercised in the use of VEGF as a therapeutic agent in neonates.

Changes in the lungs of patients with BPD are similar to those seen in our dexamethasone and hyperoxia groups, with fewer and larger alveoli (Remesal et al., 2010; San Feliciano et al., 2011). Betamethasone does not decrease VEGF expression or alveolarization and seems to inhibit the negative action of hyperoxia on VEGF (Remesal et al., 2010).

7. Conclusions

VEGF is one of the most important growth factors related to lung development and it is negatively regulated by oxidative stress.

With the limitation that we did not use an experimental model to reproduce BPD, the results of our studies suggest that repeated administration of antenatal dexamethasone, mainly associated with supplemental oxygen therapy, might worsen the morphological changes seen in BPD, with a negative effect on lung maturation and VEGF expression until the end of the alveolarization period.

Our studies also support the notion that betamethasone could be the drug of choice for treating pregnant women at risk of preterm delivery.

8. Acknowledgments

We wish to dedicate this chapter to the memory of Prof. Carmen Pedraz

9. References

Abman, S. H. (15-11-2001). Bronchopulmonary dysplasia: "a vascular hypothesis". Am.J.Respir.Crit Care Med. 164: 1755-1756.

Acarregui, M. J., Penisten, S. T., Goss, K. L., Ramirez, K., and Snyder, J. M. (1999). Vascular endothelial growth factor gene expression in human fetal lung in vitro. Am.J.Respir.Cell Mol.Biol. 20: 14-23.

Aida, K., Shi, Q., Wang, J., VandeBerg, J. L., McDonald, T., Nathanielsz, P., and Wang, X. L. (2004). The effects of betamethasone (BM) on endothelial nitric oxide synthase (eNOS) expression in adult baboon femoral arterial endothelial cells. J.Steroid Biochem.Mol.Biol. 91: 219-224.

Akeson, A. L., Greenberg, J. M., Cameron, J. E., Thompson, F. Y., Brooks, S. K., Wiginton, D., and Whitsett, J. A. (15-12-2003). Temporal and spatial regulation of VEGF-A controls vascular patterning in the embryonic lung. Dev.Biol. 264: 443-455.

Akeson, A. L., Wetzel, B., Thompson, F. Y., Brooks, S. K., Paradis, H., Gendron, R. L., and Greenberg, J. M. (2000). Embryonic vasculogenesis by endothelial precursor cells derived from lung mesenchyme. Dev.Dyn. 217: 11-23.

Ambalavanan, N. and Novak, Z. E. (2003). Peptide growth factors in tracheal aspirates of mechanically ventilated preterm neonates. Pediatr.Res. 53: 240-244.

Baud, O., Foix-L'Helias, L., Kaminski, M., Audibert, F., Jarreau, P. H., Papiernik, E., Huon, C., Lepercq, J., Dehan, M., and Lacaze-Masmonteil, T. (14-10-1999). Antenatal glucocorticoid treatment and cystic periventricular leukomalacia in very premature infants. N.Engl.J.Med. 341: 1190-1196.

Becker, P. M., Alcasabas, A., Yu, A. Y., Semenza, G. L., and Bunton, T. E. (2000). Oxygen-independent upregulation of vascular endothelial growth factor and vascular barrier dysfunction during ventilated pulmonary ischemia in isolated ferret lungs. Am.J.Respir.Cell Mol.Biol. 22: 272-279.

Bhatt, A. J., Amin, S. B., Chess, P. R., Watkins, R. H., and Maniscalco, W. M. (2000). Expression of vascular endothelial growth factor and Flk-1 in developing and glucocorticoid-treated mouse lung. Pediatr.Res. 47: 606-613.

Bhatt, A. J., Pryhuber, G. S., Huyck, H., Watkins, R. H., Metlay, L. A., and Maniscalco, W. M. (15-11-2001). Disrupted pulmonary vasculature and decreased vascular endothelial growth factor, Flt-1, and TIE-2 in human infants dying with bronchopulmonary dysplasia. Am.J.Respir.Crit Care Med. 164: 1971-1980.

Blanco, L. N. and Frank, L. (1994). Development of gas-exchange surface area in rat lung. The effect of alveolar shape. Am.J.Respir.Crit Care Med. 149: 759-766.

Bourbon, J., Boucherat, O., Chailley-Heu, B., and Delacourt, C. (2005). Control mechanisms of lung alveolar development and their disorders in bronchopulmonary dysplasia. Pediatr.Res. 57: 38R-46R.

Brenner, R. E., Felger, D., Winter, C., Christiansen, A., Hofmann, D., and Bartmann, P.
    (2001). Effects of dexamethasone on proliferation, chemotaxis, collagen I, and
    fibronectin-metabolism of human fetal lung fibroblasts. Pediatr.Pulmonol. 32: 1-7.
Brown, K. R., England, K. M., Goss, K. L., Snyder, J. M., and Acarregui, M. J. (2001). VEGF
    induces airway epithelial cell proliferation in human fetal lung in vitro.
    Am.J.Physiol Lung Cell Mol.Physiol. 281: L1001-L1010.
Burri, P. H. (2006). Structural aspects of postnatal lung development - alveolar formation
    and growth. Biol.Neonate. 89: 313-322.
Buttgereit, F., Brand, M. D., and Burmester, G. R. (15-7-1999). Equivalent doses and relative
    drug potencies for non-genomic glucocorticoid effects: a novel glucocorticoid
    hierarchy. Biochem.Pharmacol. 58: 363-368.
Carmeliet, P., Ferreira, V., Breier, G., Pollefeyt, S., Kieckens, L., Gertsenstein, M., Fahrig, M.,
    Vandenhoeck, A., Harpal, K., Eberhardt, C., Declercq, C., Pawling, J., Moons, L.,
    Collen, D., Risau, W., and Nagy, A. (4-4-1996). Abnormal blood vessel development
    and lethality in embryos lacking a single VEGF allele. Nature. 380: 435-439.
Chandel, N. S., Maltepe, E., Goldwasser, E., Mathieu, C. E., Simon, M. C., and Schumacker,
    P. T. (29-9-1998). Mitochondrial reactive oxygen species trigger hypoxia-induced
    transcription. Proc.Natl.Acad.Sci.U.S.A. 95: 11715-11720.
Chandrasekar, I., Eis, A., and Konduri, G. G. (2008). Betamethasone attenuates oxidant stress
    in endothelial cells from fetal lambs with persistent pulmonary hypertension.
    Pediatr.Res. 63: 67-72.
Clerch, L. B., Baras, A. S., Massaro, G. D., Hoffman, E. P., and Massaro, D. (2004). DNA
    microarray analysis of neonatal mouse lung connects regulation of KDR with
    dexamethasone-induced inhibition of alveolar formation. Am.J.Physiol Lung Cell
    Mol.Physiol. 286: L411-L419.
Coalson, J. J., Kuehl, T. J., Prihoda, T. J., and deLemos, R. A. (1988). Diffuse alveolar damage
    in the evolution of bronchopulmonary dysplasia in the baboon. Pediatr.Res. 24:
    357-366.
Coalson, J. J., Winter, V. T., Siler-Khodr, T., and Yoder, B. A. (1999). Neonatal chronic lung
    disease in extremely immature baboons. Am.J.Respir.Crit Care Med. 160: 1333-
    1346.
Colavitti, R., Pani, G., Bedogni, B., Anzevino, R., Borrello, S., Waltenberger, J., and Galeotti,
    T. (1-2-2002). Reactive oxygen species as downstream mediators of angiogenic
    signaling by vascular endothelial growth factor receptor-2/KDR. J.Biol.Chem. 277:
    3101-3108.
Compernolle, V., Brusselmans, K., Acker, T., Hoet, P., Tjwa, M., Beck, H., Plaisance, S., Dor,
    Y., Keshet, E., Lupu, F., Nemery, B., Dewerchin, M., Van, V. P., Plate, K., Moons, L.,
    Collen, D., and Carmeliet, P. (2002). Loss of HIF-2alpha and inhibition of VEGF
    impair fetal lung maturation, whereas treatment with VEGF prevents fatal
    respiratory distress in premature mice. Nat.Med. 8: 702-710.
Corne, J., Chupp, G., Lee, C. G., Homer, R. J., Zhu, Z., Chen, Q., Ma, B., Du, Y., Roux, F.,
    McArdle, J., Waxman, A. B., and Elias, J. A. (2000). IL-13 stimulates vascular
    endothelial cell growth factor and protects against hyperoxic acute lung injury.
    J.Clin.Invest. 106: 783-791.
Costa, R. H., Kalinichenko, V. V., and Lim, L. (2001). Transcription factors in mouse lung
    development and function. Am.J.Physiol Lung Cell Mol.Physiol. 280: L823-L838.

Crapo, J. D. (1986). Morphologic changes in pulmonary oxygen toxicity. Annu.Rev.Physiol. 48:721-31.: 721-731.

Crapo, J. D. (2003a). Oxidative stress as an initiator of cytokine release and cell damage. Eur.Respir.J.Suppl. 44:4s-6s.: 4s-6s.

Crapo, J. D. (1-11-2003b). Redox active agents in inflammatory lung injury. Am.J.Respir.Crit Care Med. 168: 1027-1028.

Crnic, L. S. and Chase, H. P. (1978). Models of infantile undernutrition in rats: effects on milk. J Nutr. 108: 1755-1760.

Crowley, P. (18-7-2007). WITHDRAWN: Prophylactic corticosteroids for preterm birth. Cochrane.Database.Syst.Rev. CD000065.

D'Angio, C. T., Maniscalco, W. M., Ryan, R. M., Avissar, N. E., Basavegowda, K., and Sinkin, R. A. (1999). Vascular endothelial growth factor in pulmonary lavage fluid from premature infants: effects of age and postnatal dexamethasone. Biol.Neonate. 76: 266-273.

De Paepe, M. E., Mao, Q., Powell, J., Rubin, S. E., DeKoninck, P., Appel, N., Dixon, M., and Gundogan, F. (15-1-2006). Growth of pulmonary microvasculature in ventilated preterm infants. Am.J.Respir.Crit Care Med. 173: 204-211.

Del Moral, P. M., Sala, F. G., Tefft, D., Shi, W., Keshet, E., Bellusci, S., and Warburton, D. (1-2-2006). VEGF-A signaling through Flk-1 is a critical facilitator of early embryonic lung epithelial to endothelial crosstalk and branching morphogenesis. Dev.Biol. 290: 177-188.

Dvorak, H. F., Brown, L. F., Detmar, M., and Dvorak, A. M. (1995). Vascular permeability factor/vascular endothelial growth factor, microvascular hyperpermeability, and angiogenesis. Am.J.Pathol. 146: 1029-1039.

Edelman, J. L., Castro, M. R., and Wen, Y. (1999). Correlation of VEGF expression by leukocytes with the growth and regression of blood vessels in the rat cornea. Invest Ophthalmol.Vis.Sci. 40: 1112-1123.

Ekekezie, I. I., Thibeault, D. W., Rezaiekhaligh, M. H., Norberg, M., Mabry, S., Zhang, X., and Truog, W. E. (2003). Endostatin and vascular endothelial cell growth factor (VEGF) in piglet lungs: effect of inhaled nitric oxide and hyperoxia. Pediatr.Res. 53: 440-446.

Feldman, D. M., Carbone, J., Belden, L., Borgida, A. F., and Herson, V. (2007). Betamethasone vs dexamethasone for the prevention of morbidity in very-low-birthweight neonates. Am.J.Obstet.Gynecol. 197: 284.

Ferrara, N., Carver-Moore, K., Chen, H., Dowd, M., Lu, L., O'Shea, K. S., Powell-Braxton, L., Hillan, K. J., and Moore, M. W. (4-4-1996). Heterozygous embryonic lethality induced by targeted inactivation of the VEGF gene. Nature. 380: 439-442.

Ferrara, N., Gerber, H. P., and LeCouter, J. (2003). The biology of VEGF and its receptors. Nat.Med. 9: 669-676.

Ferrara, N. and vis-Smyth, T. (1997). The biology of vascular endothelial growth factor. Endocr.Rev. 18: 4-25.

Frank, L. (1992). Prenatal dexamethasone treatment improves survival of newborn rats during prolonged high O2 exposure. Pediatr.Res. 32: 215-221.

Gebb, S. A. and Shannon, J. M. (2000). Tissue interactions mediate early events in pulmonary vasculogenesis. Dev.Dyn. 217: 159-169.

Gille, J., Reisinger, K., Westphal-Varghese, B., and Kaufmann, R. (2001). Decreased mRNA
stability as a mechanism of glucocorticoid-mediated inhibition of vascular
endothelial growth factor gene expression by cultured keratinocytes. J.Invest
Dermatol. 117: 1581-1587.

Grath-Morrow, S. A., Cho, C., Cho, C., Zhen, L., Hicklin, D. J., and Tuder, R. M. (2005).
Vascular endothelial growth factor receptor 2 blockade disrupts postnatal lung
development. Am.J.Respir.Cell Mol.Biol. 32: 420-427.

Groenman, F., Rutter, M., Caniggia, I., Tibboel, D., and Post, M. (2007). Hypoxia-inducible
factors in the first trimester human lung. J.Histochem.Cytochem. 55: 355-363.

Harada, S., Nagy, J. A., Sullivan, K. A., Thomas, K. A., Endo, N., Rodan, G. A., and Rodan, S.
B. (1994). Induction of vascular endothelial growth factor expression by
prostaglandin E2 and E1 in osteoblasts. J.Clin.Invest. 93: 2490-2496.

Haworth, S. G. and Hislop, A. A. (2003). Lung development-the effects of chronic hypoxia.
Semin.Neonatol. 8: 1-8.

He, C. H., Waxman, A. B., Lee, C. G., Link, H., Rabach, M. E., Ma, B., Chen, Q., Zhu, Z.,
Zhong, M., Nakayama, K., Nakayama, K. I., Homer, R., and Elias, J. A. (2005). Bcl-2-
related protein A1 is an endogenous and cytokine-stimulated mediator of
cytoprotection in hyperoxic acute lung injury. J.Clin.Invest. 115: 1039-1048.

Hewitt, D. P., Mark, P. J., and Waddell, B. J. (2006). Glucocorticoids prevent the normal
increase in placental vascular endothelial growth factor expression and placental
vascularity during late pregnancy in the rat. Endocrinology. 147: 5568-5574.

Hosford, G. E. and Olson, D. M. (2003). Effects of hyperoxia on VEGF, its receptors, and
HIF-2alpha in the newborn rat lung. Am.J.Physiol Lung Cell Mol.Physiol. 285:
L161-L168.

Jakkula, M., Le Cras, T. D., Gebb, S., Hirth, K. P., Tuder, R. M., Voelkel, N. F., and Abman, S.
H. (2000). Inhibition of angiogenesis decreases alveolarization in the developing rat
lung. Am.J.Physiol Lung Cell Mol.Physiol. 279: L600-L607.

Jobe, A. H. (2001). Glucocorticoids, inflammation and the perinatal lung. Semin.Neonatol. 6:
331-342.

Jobe, A. H. (2003). Antenatal factors and the development of bronchopulmonary dysplasia.
Semin.Neonatol. 8: 9-17.

Jobe, A. H. and Bancalari, E. (2001). Bronchopulmonary dysplasia. Am.J.Respir.Crit Care
Med. 163: 1723-1729.

Jobe, A. J. (1999). The new BPD: an arrest of lung development. Pediatr.Res. 46: 641-643.

Klekamp, J. G., Jarzecka, K., and Perkett, E. A. (1999). Exposure to hyperoxia decreases the
expression of vascular endothelial growth factor and its receptors in adult rat
lungs. Am.J.Pathol. 154: 823-831.

Kunig, A. M., Balasubramaniam, V., Markham, N. E., Morgan, D., Montgomery, G., Grover,
T. R., and Abman, S. H. (2005). Recombinant human VEGF treatment enhances
alveolarization after hyperoxic lung injury in neonatal rats. Am.J.Physiol Lung Cell
Mol.Physiol. 289: L529-L535.

Kunig, A. M., Balasubramaniam, V., Markham, N. E., Seedorf, G., Gien, J., and Abman, S. H.
(2006). Recombinant human VEGF treatment transiently increases lung edema but
enhances lung structure after neonatal hyperoxia. Am.J.Physiol Lung Cell
Mol.Physiol. 291: L1068-L1078.

Lassus, P., Nupponen, I., Kari, A., Pohjavuori, M., and Andersson, S. (2002). Early postnatal dexamethasone decreases hepatocyte growth factor in tracheal aspirate fluid from premature infants. Pediatrics. 110: 768-771.

Lassus, P., Ristimaki, A., Ylikorkala, O., Viinikka, L., and Andersson, S. (1999). Vascular endothelial growth factor in human preterm lung. Am.J.Respir.Crit Care Med. 159: 1429-1433.

Lassus, P., Turanlahti, M., Heikkila, P., Andersson, L. C., Nupponen, I., Sarnesto, A., and Andersson, S. (15-11-2001). Pulmonary vascular endothelial growth factor and Flt-1 in fetuses, in acute and chronic lung disease, and in persistent pulmonary hypertension of the newborn. Am.J.Respir.Crit Care Med. 164: 1981-1987.

Lee, B. H., Stoll, B. J., McDonald, S. A., and Higgins, R. D. (2006). Adverse neonatal outcomes associated with antenatal dexamethasone versus antenatal betamethasone. Pediatrics. 117: 1503-1510.

Levy, M., Maurey, C., Chailley-Heu, B., Martinovic, J., Jaubert, F., and Israel-Biet, D. (2005). Developmental changes in endothelial vasoactive and angiogenic growth factors in the human perinatal lung. Pediatr.Res. 57: 248-253.

Liggins, G. C. and Howie, R. N. (1972). A controlled trial of antepartum glucocorticoid treatment for prevention of the respiratory distress syndrome in premature infants. Pediatrics. 50: 515-525.

Lin, Y. J., Markham, N. E., Balasubramaniam, V., Tang, J. R., Maxey, A., Kinsella, J. P., and Abman, S. H. (2005). Inhaled nitric oxide enhances distal lung growth after exposure to hyperoxia in neonatal rats. Pediatr.Res. 58: 22-29.

Liu, H., Chang, L., Rong, Z., Zhu, H., Zhang, Q., Chen, H., and Li, W. (2004). Association of insulin-like growth factors with lung development in neonatal rats. J.Huazhong.Univ Sci.Technolog.Med.Sci. 24: 162-165.

Maniscalco, W. M., Watkins, R. H., D'Angio, C. T., and Ryan, R. M. (1997). Hyperoxic injury decreases alveolar epithelial cell expression of vascular endothelial growth factor (VEGF) in neonatal rabbit lung. Am.J.Respir.Cell Mol.Biol. 16: 557-567.

Maniscalco, W. M., Watkins, R. H., Finkelstein, J. N., and Campbell, M. H. (1995). Vascular endothelial growth factor mRNA increases in alveolar epithelial cells during recovery from oxygen injury. Am.J.Respir.Cell Mol.Biol. 13: 377-386.

Maniscalco, W. M., Watkins, R. H., Pryhuber, G. S., Bhatt, A., Shea, C., and Huyck, H. (2002). Angiogenic factors and alveolar vasculature: development and alterations by injury in very premature baboons. Am.J.Physiol Lung Cell Mol.Physiol. 282: L811-L823.

Maniscalco, W. M., Watkins, R. H., Roper, J. M., Staversky, R., and O'Reilly, M. A. (2005). Hyperoxic ventilated premature baboons have increased p53, oxidant DNA damage and decreased VEGF expression. Pediatr.Res. 58: 549-556.

Massaro, D. and Massaro, G. D. (1986). Dexamethasone accelerates postnatal alveolar wall thinning and alters wall composition. Am.J.Physiol. 251: R218-R224.

Massaro, D. and Massaro, G. D. (2004). Critical period for alveologenesis and early determinants of adult pulmonary disease. Am.J.Physiol Lung Cell Mol.Physiol. 287: L715-L717.

McGowan, J. E., Sysyn, G., Petersson, K. H., Sadowska, G. B., Mishra, O. P., ivoria-Papadopoulos, M., and Stonestreet, B. S. (1-10-2000). Effect of dexamethasone treatment on maturational changes in the NMDA receptor in sheep brain. J.Neurosci. 20: 7424-7429.

Miracle, X., Di Renzo, G. C., Stark, A., Fanaroff, A., Carbonell-Estrany, X., and Saling, E. (2008). Guideline for the use of antenatal corticosteroids for fetal maturation. J Perinat.Med. 36: 191-196.

Nauck, M., Karakiulakis, G., Perruchoud, A. P., Papakonstantinou, E., and Roth, M. (12-1-1998). Corticosteroids inhibit the expression of the vascular endothelial growth factor gene in human vascular smooth muscle cells. Eur.J.Pharmacol. 341: 309-315.

Nauck, M., Roth, M., Tamm, M., Eickelberg, O., Wieland, H., Stulz, P., and Perruchoud, A. P. (1997). Induction of vascular endothelial growth factor by platelet-activating factor and platelet-derived growth factor is downregulated by corticosteroids. Am.J.Respir.Cell Mol.Biol. 16: 398-406.

NIH. Effect of corticosteroids for fetal maturation on perinatal outcomes. NIH Consensus Development Panel on the Effect of Corticosteroids for Fetal Maturation on Perinatal Outcomes. JAMA. 273: 413-418.

Northway, W. H., Jr., Rosan, R. C., and Porter, D. Y. (16-2-1967). Pulmonary disease following respirator therapy of hyaline-membrane disease. Bronchopulmonary dysplasia. N.Engl.J.Med. 276: 357-368.

Oshika, E., Liu, S., Ung, L. P., Singh, G., Shinozuka, H., Michalopoulos, G. K., and Katyal, S. L. (1998). Glucocorticoid-induced effects on pattern formation and epithelial cell differentiation in early embryonic rat lungs. Pediatr.Res. 43: 305-314.

Ozaki, N., Beharry, K., Nishihara, K. C., Akmal, Y., Ang, J. G., and Modanlou, H. D. (2002). Differential regulation of prostacyclin and thromboxane by dexamethasone and celecoxib during oxidative stress in newborn rabbits. Prostaglandins Other Lipid Mediat. 70: 61-78.

Papaioannou, A. I., Kostikas, K., Kollia, P., and Gourgoulianis, K. I. (17-10-2006). Clinical implications for vascular endothelial growth factor in the lung: friend or foe? Respir.Res. 7:128.: 128.

Rayburn, W. F., Christensen, H. D., and Gonzalez, C. L. (1997). A placebo-controlled comparison between betamethasone and dexamethasone for fetal maturation: differences in neurobehavioral development of mice offspring. Am.J.Obstet.Gynecol. 176: 842-850.

Remesal, A., Pedraz, C., San Feliciano, L., and Ludeña, D. (2009). Pulmonary expression of vascular endothelial growth factor (VEGF) and alveolar septation in a newborn rat model exposed to acute hypoxia and recovered under conditions of air or hyperoxia. Histol.Histopathol. 24: 325-330.

Remesal, A., San Feliciano, L., Isidoro-Garcia, M., and Ludeña, D. (4-5-2010). Effects of Antenatal Betamethasone and Dexamethasone on the Lung Expression of Vascular Endothelial Growth Factor and Alveolarization in Newborn Rats Exposed to Acute Hypoxia and Recovered in Normoxia or Hyperoxia. Neonatology. 98: 313-320.

Roberts, R. J., Weesner, K. M., and Bucher, J. R. (1983). Oxygen-induced alterations in lung vascular development in the newborn rat. Pediatr.Res. 17: 368-375.

Roper, J. M., Mazzatti, D. J., Watkins, R. H., Maniscalco, W. M., Keng, P. C., and O'Reilly, M. A. (2004). In vivo exposure to hyperoxia induces DNA damage in a population of alveolar type II epithelial cells. Am.J.Physiol Lung Cell Mol.Physiol. 286: L1045-L1054.

Roth-Kleiner, M. and Post, M. (2003). Genetic control of lung development. Biol.Neonate. 84: 83-88.

Roubliova, X. I., Van der Biest, A. M., Vaast, P., Lu, H., Jani, J. C., Lewi, P. J., Verbeken, E. K., Tibboel, D., and Deprest, J. A. (2008). Effect of maternal administration of betamethasone on peripheral arterial development in fetal rabbit lungs. Neonatology. 93: 64-72.

San Feliciano L., Remesal, A., Isidoro-Garcia, M., and Ludeña, D. (9-2-2011). Dexamethasone and Betamethasone for Prenatal Lung Maturation: Differences in Vascular Endothelial Growth Factor Expression and Alveolarization in Rats. Neonatology. 100: 105-110.

Saugstad, O. D. (1988). Hypoxanthine as an indicator of hypoxia: its role in health and disease through free radical production. Pediatr.Res. 23: 143-150.

Schellenberg, J. C., Liggins, G. C., and Stewart, A. W. (1987). Growth, elastin concentration, and collagen concentration of perinatal rat lung: effects of dexamethasone. Pediatr.Res. 21: 603-607.

Schwarz, M. A., Zhang, F., Gebb, S., Starnes, V., and Warburton, D. (2000). Endothelial monocyte activating polypeptide II inhibits lung neovascularization and airway epithelial morphogenesis. Mech.Dev. 95: 123-132.

Setiawan, E., Jackson, M. F., MacDonald, J. F., and Matthews, S. G. (15-6-2007). Effects of repeated prenatal glucocorticoid exposure on long-term potentiation in the juvenile guinea-pig hippocampus. J.Physiol. 581: 1033-1042.

Shenai, J. P., Mellen, B. G., and Chytil, F. (2000). Vitamin A status and postnatal dexamethasone treatment in bronchopulmonary dysplasia. Pediatrics. 106: 547-553.

Siner, J. M., Jiang, G., Cohen, Z. I., Shan, P., Zhang, X., Lee, C. G., Elias, J. A., and Lee, P. J. (2007). VEGF-induced heme oxygenase-1 confers cytoprotection from lethal hyperoxia in vivo. FASEB J. 21: 1422-1432.

Speirs, H. J., Seckl, J. R., and Brown, R. W. (2004). Ontogeny of glucocorticoid receptor and 11beta-hydroxysteroid dehydrogenase type-1 gene expression identifies potential critical periods of glucocorticoid susceptibility during development. J.Endocrinol. 181: 105-116.

Spyridopoulos, I., Brogi, E., Kearney, M., Sullivan, A. B., Cetrulo, C., Isner, J. M., and Losordo, D. W. (1997). Vascular endothelial growth factor inhibits endothelial cell apoptosis induced by tumor necrosis factor-alpha: balance between growth and death signals. J.Mol.Cell Cardiol. 29: 1321-1330.

Suzuki, K., Hooper, S. B., Wallace, M. J., Probyn, M. E., and Harding, R. (2006). Effects of antenatal corticosteroid treatment on pulmonary ventilation and circulation in neonatal lambs with hypoplastic lungs. Pediatr.Pulmonol. 41: 844-854.

Sweet, D. G., Carnielli, V., Greisen, G., Hallman, M., Ozek, E., Plavka, R., Saugstad, O. D., Simeoni, U., Speer, C. P., and Halliday, H. L. (2010). European consensus guidelines on the management of neonatal respiratory distress syndrome in preterm infants - 2010 update. Neonatology. 97: 402-417.

Tamura, M., Arakaki, N., Tsubouchi, H., Takada, H., and Daikuhara, Y. (15-4-1993). Enhancement of human hepatocyte growth factor production by interleukin-1 alpha and -1 beta and tumor necrosis factor-alpha by fibroblasts in culture. J.Biol.Chem. 268: 8140-8145.

Tanabe, K., Tokuda, H., Takai, S., Matsushima-Nishiwaki, R., Hanai, Y., Hirade, K., Katagiri, Y., Dohi, S., and Kozawa, O. (1-9-2006). Modulation by the steroid/thyroid hormone superfamily of TGF-beta-stimulated VEGF release from vascular smooth muscle cells. J.Cell Biochem. 99: 187-195.

Tang, J. R., Markham, N. E., Lin, Y. J., McMurtry, I. F., Maxey, A., Kinsella, J. P., and Abman, S. H. (2004). Inhaled nitric oxide attenuates pulmonary hypertension and improves lung growth in infant rats after neonatal treatment with a VEGF receptor inhibitor. Am.J.Physiol Lung Cell Mol.Physiol. 287: L344-L351.

Thebaud, B. (2007). Angiogenesis in lung development, injury and repair: implications for chronic lung disease of prematurity. Neonatology. 91: 291-297.

Thebaud, B., Ladha, F., Michelakis, E. D., Sawicka, M., Thurston, G., Eaton, F., Hashimoto, K., Harry, G., Haromy, A., Korbutt, G., and Archer, S. L. (18-10-2005). Vascular endothelial growth factor gene therapy increases survival, promotes lung angiogenesis, and prevents alveolar damage in hyperoxia-induced lung injury: evidence that angiogenesis participates in alveolarization. Circulation. 112: 2477-2486.

Tschanz, S. A., Makanya, A. N., Haenni, B., and Burri, P. H. (2003). Effects of neonatal high-dose short-term glucocorticoid treatment on the lung: a morphologic and morphometric study in the rat. Pediatr.Res. 53: 72-80.

Van Tuyl, T. M., Liu, J., Wang, J., Kuliszewski, M., Tibboel, D., and Post, M. (2005). Role of oxygen and vascular development in epithelial branching morphogenesis of the developing mouse lung. Am.J.Physiol Lung Cell Mol.Physiol. 288: L167-L178.

Veness-Meehan, K. A., Bottone, F. G., Jr., and Stiles, A. D. (2000). Effects of retinoic acid on airspace development and lung collagen in hyperoxia-exposed newborn rats. Pediatr.Res. 48: 434-444.

Vento, G., Matassa, P. G., Ameglio, F., Capoluongo, E., Tortorolo, L., and Romagnoli, C. (2002). Effects of early dexamethasone therapy on pulmonary fibrogenic mediators and respiratory mechanics in preterm infants. Eur.Cytokine Netw. 13: 207-214.

Voelkel, N. F., Vandivier, R. W., and Tuder, R. M. (2006). Vascular endothelial growth factor in the lung. Am.J.Physiol Lung Cell Mol.Physiol. 290: L209-L221.

Wagenaar, G. T., ter Horst, S. A., van Gastelen, M. A., Leijser, L. M., Mauad, T., van, d., V, de, H. E., Hiemstra, P. S., Poorthuis, B. J., and Walther, F. J. (15-3-2004). Gene expression profile and histopathology of experimental bronchopulmonary dysplasia induced by prolonged oxidative stress. Free Radic.Biol.Med. 36: 782-801.

Watkins, R. H., D'Angio, C. T., Ryan, R. M., Patel, A., and Maniscalco, W. M. (1999). Differential expression of VEGF mRNA splice variants in newborn and adult hyperoxic lung injury. Am.J.Physiol. 276: L858-L867.

Weibel, E. R., Kistler, G. S., and Scherle, W. F. (1966). Practical stereological methods for morphometric cytology. J.Cell Biol. 30: 23-38.

Yamamoto, H., Yun, E. J., Gerber, H. P., Ferrara, N., Whitsett, J. A., and Vu, T. H. (1-8-2007). Epithelial-vascular cross talk mediated by VEGF-A and HGF signaling directs primary septae formation during distal lung morphogenesis. Dev.Biol. 308: 44-53.

Yee, M., Vitiello, P. F., Roper, J. M., Staversky, R. J., Wright, T. W., Grath-Morrow, S. A., Maniscalco, W. M., Finkelstein, J. N., and O'Reilly, M. A. (2006). Type II epithelial cells are critical target for hyperoxia-mediated impairment of postnatal lung development. Am.J.Physiol Lung Cell Mol.Physiol. 291: L1101-L1111.

Zarnegar, R. (1995). Regulation of HGF and HGFR gene expression. EXS. 74:33-49.: 33-49.

Zhao, L., Wang, K., Ferrara, N., and Vu, T. H. (2005). Vascular endothelial growth factor co-ordinates proper development of lung epithelium and vasculature. Mech.Dev. 122: 877-886.

Zimova-Herknerova, M., Myslivecek, J., and Potmesil, P. (2008). Retinoic acid attenuates the mild hyperoxic lung injury in newborn mice. Physiol Res. 57: 33-40.

13

Structural and Activity Changes in Renal Betaine Aldehyde Dehydrogenase Caused by Oxidants

Jesús A. Rosas-Rodríguez[1], Hilda F. Flores-Mendoza[2],
Ciria G. Figueroa-Soto[2], Edgar F. Morán-Palacio[1]
and Elisa M. Valenzuela-Soto[2]*
[1]*Departamento de Ciencias Químico Biológicas y Agropecuarias,
Universidad de Sonora Unidad Regional Sur, Navojoa, Sonora,*
[2]*Centro de Investigación en Alimentación y Desarrollo A.C., Hermosillo, Sonora,*
México

1. Introduction

Oxidative stress has been implicated in a variety of diseases such as glomerulonephritis and tubulointerstitial nephritis, renal insufficiency, proteinuria, Alzheimer's and Parkinson's disease, diabetes and hypertension, as well as contributing to the pathogenesis of ischemia reperfusion injury in the kidney (Banday & Lokhandwala, 2011; Martin & Goeddeke-Merickel, 2005; Touyz, 2004; Vaziri, 2004). One of the most important functions of kidney is the regulation of liquid volume. Broad shifts in osmolality, high urea concentrations and low oxygen tension are required by the urine concentrating mechanism to produce concentrated urine (Kwon et al., 2009; Burg & Ferraris, 2008; Neuhofer & Beck, 2006). There is evidence indicating that the osmotic stress and low oxygen tension produced in medullary cells generates an increase in the concentration of reactive oxygen species (ROS), which triggers damage to kidney cells.

The source of ROS in kidney is (1) the hyperosmolality of the environment where the medulla cells are found, (2) NADH oxidase activity and (3) the mitochondrial respiratory chain (Banday & Lokhandwala, 2011; Harper et al., 2004; Mori & Crowley, 2003; Zou et al., 2001). The most abundant ROS in renal cells are the superoxide radical (O_2^-) and hydrogen peroxide (H_2O_2). Hydrogen peroxide is produced during the incomplete reduction of oxygen to H_2O, and it is biologically important because it is reduced to the highly reactive hydroxyl radical (Rhee et al., 2003). The hydroxyl radical has been implicated in endothelial dysfunction (Coyle et al. 2006). The amount of ROS produced is in balance with the quantity eliminated by the antioxidants. However, when this balance is upset the possibility of cellular damage by ROS increases (Haliwell, 2007; Lushchak 2011).

Metal ions such as iron, cadmium and copper have been found to play a role in several human diseases owing to their capacity to generate oxidative stress (Bonda et al., 2011;

---

* Corresponding Author

Butterworth, 2010; Jomova & Valko, 2011; Jomova et al., 2010). Copper generates ROS (hydroxyl •OH and superoxide anion radical $O_2^{\bullet-}$) by two mechanisms: Fenton and/or Haber-Weiss chemistry (Jomova & Valko, 2011; Prousek, 2007) and by decreasing glutathione levels (Speisky et al., 2009). Glutathione has several functions including acting as a powerful cell antioxidant and as a copper chelating agent; therefore decreased glutathione levels can enhance oxidative stress (Mattie & Freedman, 2004; Steinebach & Wolterbeek, 1994).

Copper ions display a high affinity to thiol groups in proteins and can directly bind to them causing enzyme inactivation or altered protein conformation (Giles et al., 2003; Letelier et al., 2005; Prudent & Girault, 2009; Zhang et al., 2010). For example, the enzyme glutathione S-transferase (GST) from rat liver, crab and shrimp is inhibited by copper (Elumalai et al., 2002; Letelier et al., 2006; Salazar-Medina et al., 2010), and proteins such as LDL and HDL are also oxidized by copper (Burkitt, 2001). The thiol group exhibits a variety of oxidation states which allow for nucleophilic attack, electron transfer, hydride transfer and oxygen atom transfer (Giles et al., 2003). The binding between copper and the thiol group can oxidize the sulphur, producing sulfenic and sulfinic acids or a disulfide bridge between vicinal cysteines (Giles et al., 2003).

Identifying the proteins that are susceptible to oxidative damage could lead to a better understanding of the complexity of the alterations caused by oxidative stress, and allow us to counteract the injury caused by ROS. A decrease in the activity of the enzymes involved in osmolyte synthesis and stress response are associated with renal injury, and as such could generate other diseases in the organism. Further study of the regulation and biochemical characteristics of those enzymes is necessary in order to identify and understand the effects of ROS in kidney and their relationship to the osmolyte system. One of those enzymes is betaine aldehyde dehydrogenase (BADH EC 1.2.1.8), which catalyzes the oxidation of betaine aldehyde to synthesize glycine betaine.

Glycine betaine (GB) is one of the major non-perturbing osmolytes that are actively accumulated by plant, bacteria and mammalian cells under hypertonic conditions, and this process is important in the regulation of cell volume (Chen & Murata, 2011; Burg & Ferraris, 2008, Burg et al., 2007). Glycine betaine has also been found to suppress increases in oxidative stress in old rats (Go et al. 2005), prevent isoprenaline-induced myocardial infarction due to its antioxidant properties (Ganesan et al. 2009), prevent taurolithocholate 3-sulfate-induced oxidative stress in rat hepatocytes (Graf et al. 2002), activate chaperone-mediated disaggregation—suggesting that GB has a specific interaction of activation with ClpB and/or DnK (Umenishi et al., 2005), and to promote aquaporine PI (AQPI) expression under severe hypertonic conditions in kidney (Diamant et al., 2001). Glycine betaine has been described as an osmoprotectant against deleterious effect of urea (Burg et al. 1996).

Betaine aldehyde dehydrogenase (BADH, EC 1.2.1.8) belongs to the superfamily of aldehyde dehydrogenases (ALDH9) (Julián-Sanchez et al., 2007; Vasiliou et al., 1999) and has been found in many prokaryotic and eukaryotic organisms. The enzyme has been purified from and characterized for a number of these sources: plants (Arakawa et al., 1987; Figueroa-Soto & Valenzuela-Soto, 2001; Livingstone et al., 2003; Oishi & Ebina, 2005; Pan, 1988; Valenzuela-Soto & Muñoz-Clares, 1994; Weretylnik & Hanson, 1989), microorganisms (Boch et al., 1997; Falkenberg & Strom, 1990; Mori et al., 1992; Mori et al., 1980; Nagasawa et al., 1976; Velasco-Garcia et al., 1999) and animals (Chern & Pietruszko, 1999; Guzman-Partida & Valenzuela-Soto, 1998; Hjelmquist et al., 2003; Kurys et al., 1989; Rothschil & Guzman-Barron, 1954).

Studies of steady-state kinetics have shown that BADH from swine kidney and amaranth plants follows an iso bi-bi ordered mechanism (Figueroa-Soto & Valenzuela-Soto, 2000; Valenzuela-Soto & Muñoz-Clares, 1993), that of *Pseudomonas aeruginosa* follows a random steady-state, with a much preferred route in which the nucleotide first links to the enzyme (Velasco-Garcia et al., 1999), and a ping-pong mechanism for the enzymes from the fungus *Cylindrocarpon didymum* (Mori et al., 1980), and the bacterium *Escherichia coli* (Falkenberg & Strom, 1990). In an iso bi-bi ordered mechanism, NAD+ is the first substrate to combine with the enzyme and NADH is the last product released.

The enzyme betaine aldehyde dehydrogenase from swine kidney (skBADH) requires renal physiological ionic strength to maintain its tetrameric conformation; at low ionic strength the enzyme dissociates, forming dimers which are inactive and very stable (Valenzuela-Soto et al., 2003). Additionally, the enzyme requires the physiological ionic strength provided by a monovalent cation for maximum thermostability (Valenzuela-Soto et al., 2005).

SkBADH, like the majority of aldehyde dehydrogenases, has a catalytic cysteine (Cys 288) (González-Segura et al., 2002; Muñoz-Clares et al., 2003), and also has a vicinal cysteine (Cys 289). The chemical mechanism of BADH involves the following steps: 1) There is a nucleophilic attack by the thiolate group of the catalytic cysteine on the carbon of carbonyl group of the aldehyde, forming the thiohemiacetal intermediate; 2) The transfer of the hydride to the C-4 position of the nicotinamide portion of NAD(P)+, thus reducing the nucleotide; 3) A water molecule acting as a nucleophile breaks the thioester bond, thus forming the acid product of the reaction (Muñoz-Clares & Valenzuela-Soto, 2008; Muñoz-Clares et al., 2010).

In previous studies, we found that in swine kidney BADH the thiolate from catalytic cysteine can be oxidized by hydrogen peroxide, forming a disulfide bond between catalytic and vicinal cysteine (Rosas-Rodríguez & Valenzuela-Soto, 2011). With the aim of better understanding how ROS affect the stability and activity of the enzymes involved in osmolyte synthesis, in this study we examined the impact of hydrogen peroxide and copper on the activity, structure and stability of renal BADH. Physiological conditions such as pH and ionic strength were analyzed and contrasted with the conditions that are optimal for enzyme activity.

2. Materials and methods

2.1 Chemicals

Betaine aldehyde, DTT, EDTA, glycine betaine, GSH, HEPES, 2-mercaptoethanol, NAD+ (sodium salt), KCl and $CuCl_2$ were obtained from Sigma-Aldrich SA de CV, México. All other chemicals and solvents used in this study were of analytical grade.

2.2 Enzyme purification and activity assay

BADH was purified from porcine kidney tissue and its activity was assayed at 30 °C as previously reported (Guzman-Partida & Valenzuela-Soto, 1998) with a modification in the affinity chromatographic step, where an N-6-Hexyl-AMP-sepharose matrix was used. The pure enzyme was stored at –20 °C in a 10 mM potassium phosphate buffer, pH 6.8, 1 mM EDTA, 14 mM 2-mercaptoethanol, 0.2 M KCl and 10% glycerol. The standard assay system contained 0.1 M Hepes-KOH, pH 8.0, 0.5 mM betaine aldehyde (BA) and 1.0 mM NAD$^+$ in a final volume of 0.4 mL. SkBADH activity was assayed spectrophotometrically measuring NAD$^+$ reduction by the increase in extinction at 340 nm using an Ultrospec 4000 (Pharmacia) spectrophotometer.

Prior to treatments skBADH (0.15 mg/mL) was dialyzed overnight at 4 °C against 10 mM Hepes-KOH buffer, pH 7.0, 1 mM EDTA, 10 mM 2-mercaptoethanol (buffer A, low ionic strength), and against buffer A plus 150 mM KCl (buffer B, physiological ionic strength).

2.3 Effect of hydrogen peroxide in enzyme activity

SkBADH activity in the presence of hydrogen peroxide was evaluated at distinct pH values and ionic strengths. The enzyme was assayed in the presence of 0.1 mM of hydrogen peroxide in 10 mM Hepes buffer, pH 6.8, 7.0, 7.2, 7.4 and 8.0 and enzyme activity was measured. The effect of ionic strength was evaluated by incubating the enzyme at each pH tested with 0.15 M KCl for 60 minutes prior to the enzyme activity assay.

2.4 Kinetics of inactivation by copper

The enzyme was dialyzed with 10 mM Hepes buffer, pH 7 or pH 8.0, 10% glycerol (v/v), and 1 mM 2-mercaptoethanol. The enzyme was incubated for 120 min with 10 μM CuCl$_2$, in the presence or absence of 0.15 M KCl, an aliquot was taken at different times and residual enzyme activity was measured by the standard assay. Rate constants for inactivation were calculated by fitting the data to a single exponential decay equation (Eq. 1) using OriginPro 8.0 (OriginLab, Northampton, MA, USA).

$$A_t = A_\infty + A_0 e^{-k_{obs} \cdot t} \tag{1}$$

Where A_t, A_0 and A_∞ are enzyme activity at time t, zero, and infinite, respectively, expressed as a percentage of the initial activity; k_{obs} is the observed pseudo-first order constant in monophasic inactivation.

2.5 Enzyme reactivation kinetics

The reactivation of inactivated enzyme was carried out with 5 mM dithiothreitol (DTT), 10 mM glutathione (GSH) or 10 mM 2-mercaptoethanol. An aliquot was taken at different times and enzyme activity was measured using the standard assay. Activity data were fitted to a double exponential growth equation (Eq. 2) using OriginPro 8.0.

$$A_t = A_1 (1 - e^{-k_{obs} \cdot t}) + A_2 (1 - e^{-k_{obs}2 \cdot t}) \tag{2}$$

Where A_t, A_0 and A_∞ are enzyme activity at time t, zero, and infinite, respectively, expressed as a percentage of the initial activity; k_{obs} is the observed pseudo-first order constant in monophasic inactivation.

2.6 Enzyme fluorescence assay

Intrinsic tryptophan fluorescence was used to estimate skBADH fluorescence quenching by enzyme substrates (BA and NAD$^+$), products (GB and NADH) and inhibitors (H$_2$O$_2$ and copper) at the concentrations given in each figure legend. Another set of experiments was carried out in which the enzyme was incubated for 120 min in the presence of 100 μM H_2O_2 or 10 μM $CuCl_2$. Fluorescence spectra were recorded at an excitation wavelength of 296 nm and emission wavelength of 300-400 nm (5 nm bandwidth) using a QM-2003 fluorometer (Photon Technology International) with a 75-W xenon lamp as the light source. The spectra were corrected by subtracting the spectrum obtained for the solvent under identical conditions. Three readings of emission data were accumulated. The assay was done using 2 μM of enzyme monomer in 100 mM Hepes-KOH buffer at pH 7.0 and then at pH 8.0. Fluorescence spectral centers of mass (intensity-weighted average emission wavelengths, λ_{av}) were calculated according to the following equation:

$$\lambda_{av} = \Sigma \, \lambda I \, (\lambda)/ \, \Sigma I \, (\lambda) \qquad (3)$$

where λ is the emission wavelength and $I(\lambda)$ represents the fluorescence intensity at wavelength λ.

2.7 Enzyme structural models

The structural model of the skBADH tetramer was constructed using MOE v2009.10 software (Chemical Computing Group). The homology model was built using the cDNA-deduced swine kidney BADH amino acid sequence. The crystallographic structures from cod liver (1A4S, DOI:10.2210/pdb1a4s/pdb) and *Escherichia coli* BADH (1WNB, DOI:10.2210/pdb1wnb/pdb) were used as templates to generate the final homology model.

The hydrogen peroxide and copper interaction models with the skBADH active site were built using the PyMOL Molecular Graphics System, Version 1.3, Schrödinger, LLC. The model was designed based on previously described experimental evidence of the alteration of skBADH active site (Rosas-Rodríguez & Valenzuela-Soto, 2011).

3. Results

3.1 Effect of pH and hydrogen peroxide on enzyme activity

To test the influence of pH on skBADH oxidation provoked by hydrogen peroxide, enzyme activity was assayed from pH 6.8 to 8.0. Enzyme activity was higher at pH 8.0 in the absence or presence of hydrogen peroxide (Fig. 1). SkBADH activity decreased 50% in the presence of 100 μM H_2O_2 at all pH values tested relative to the enzyme assays done with no hydrogen peroxide (Fig. 1A). Chemical modification studies carried out with BADH from *Pseudomonas aeruginosa* indicated that the catalytic cysteine residue exists as thiolate at pH values of 5.5 to 9.0 (González-Segura et al., 2002). Therefore, H_2O_2 is able to oxidize the catalytic cysteine in the skBADH at the pH values assayed in this work.

Physiological ionic strength obtained using 0.15 M KCl was tested to analyze its influence on skBADH oxidation by 100 μM H_2O_2. Enzyme activity decreased 50% when hydrogen peroxide was included in the activity assay medium (Fig. 1B). Physiologic ionic strength was not able to maintain enzyme activity in the presence of hydrogen peroxide at any of the pH values tested (Fig. 1B).

Previous studies indicated that enzyme exposure to Na^+ or K^+ ions maintains more than 50% of skBADH activity and that physiological ionic strength is necessary to maintain the tetrameric structure of skBADH (Guzman-Partida & Valenzuela-Soto, 1998; Valenzuela-

Soto et al., 2003). SkBADH is inhibited in the presence of 100 μM H_2O_2, a concentration that may be found in medulla cells during concentrated urine formation, and which requires high osmolarity and low oxygen tension (Burg & Ferraris, 2008; Neuhofer & Beck, 2006). In addition, these peroxide concentrations may be found in the medulla cells of hypertensive (Touyz & Briones, 2011; Thengchaisri & Kuo, 2003) and diabetic patients (Coughlan et al., 2009; Forbes et al., 2008), i.e. pathologies related to high ROS concentrations.

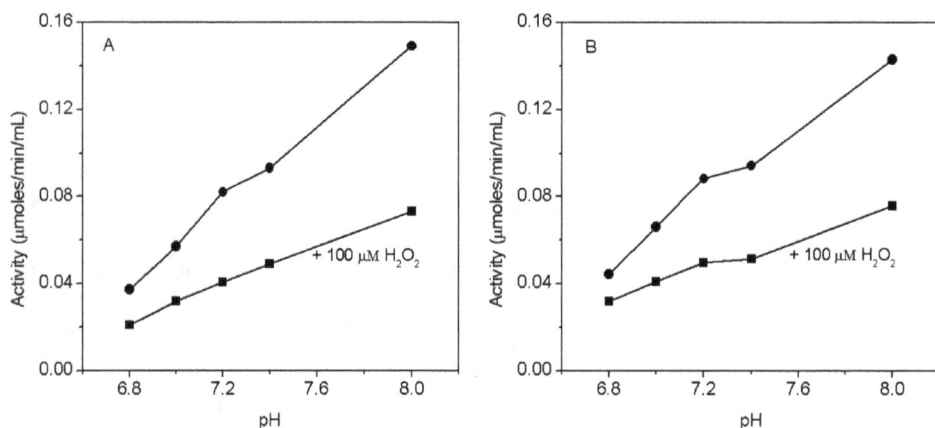

Fig. 1. Effect of pH and ionic strength on skBADH inactivation caused by hydrogen peroxide. Enzyme activity measured at: (A) low ionic strength; (B) physiological ionic strength (0.15 M KCl).

3.2 The effect of hydrogen peroxide on enzyme fluorescence

A methodical evaluation of the effect of hydrogen peroxide on enzyme tertiary structure was performed by incubating the enzyme in the presence of 0-0.5 mM H_2O_2. SkBADH maximum emission was detected at 333 nm (Fig. 2). Enzyme fluorescence was quenched by all hydrogen peroxide concentrations tested (Fig. 2), however fluorescence spectral centers of mass did not change and no blue or red shift was found.

A hydrogen peroxide concentration of 100 μM was chosen to analyze the effect of incubation time on enzyme quenching, because this concentration can be found in kidney. When the enzyme was incubated for 120 min in the presence of 100 μM H_2O_2, the fluorescence of skBADH was quenched in a time dependent manner (Fig. 3). The presence of hydrogen peroxide decreased the skBADH maximum emission by more than 50% at 120 min of incubation. Changes in enzyme emission fit a single decay equation with an R-square of 0.9893 (Fig. 3 Inset). It is noteworthy that skBADH fluorescence quenching was only detected in the presence of peroxide and the shape of the fluorescence spectra did not change. The fluorescence spectral centers of mass did not change either.

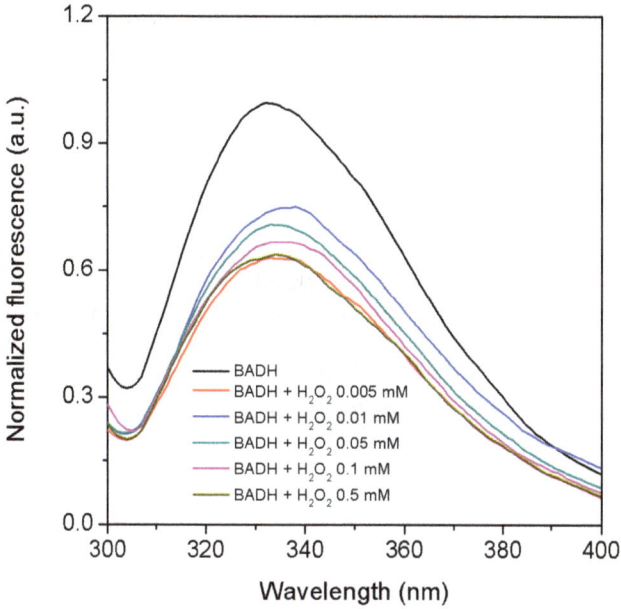

Fig. 2. Fluorescence spectra for skBADH in the presence of hydrogen peroxide.

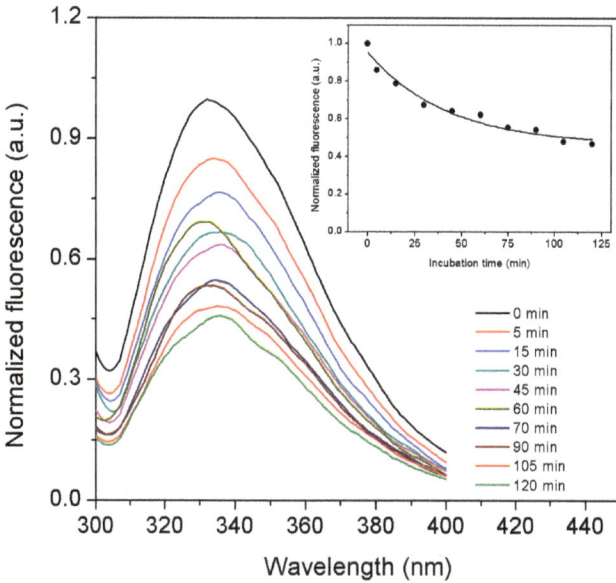

Fig. 3. Fluorescence spectra for skBADH incubated with 100 μM H₂O₂. *Inset:* SkBADH time dependent quenching.

In addition, with respect to the maximum emission wavelength, no blue/red shift could be detected which suggests the direct alteration of the enzyme site rather than a change in protein folding, and which also rules out any enzyme denaturation.

3.3 The effect of ligands on enzyme quenching provoked by hydrogen peroxide

The skBADH fluorescence spectra at both pH 7.0 and pH 8.0 showed that the maximum emission and the wavelength at which the maxima occurred were equal (Fig. 4). Fluorescence spectral centers of mass changed between 1 and 2 nm. At pH 7.0, hydrogen peroxide increased the enzyme's maximum emission but did not cause a blue or red shift (Fig. 4A). The presence of BA produced a red shift indicating a change in solvent exposition of tryptophans, caused by hydrogen peroxide and BA mix (Fig. 4A). However, the enzyme product GB had no effect on the quenching provoked by peroxide. At pH 8.0, as before, skBADH fluorescence was quenched by hydrogen peroxide without any blue or red shift (Fig. 4B). BA and GB did not play a role in the enzyme fluorescence quenching induced by hydrogen peroxide (Fig. 4B).

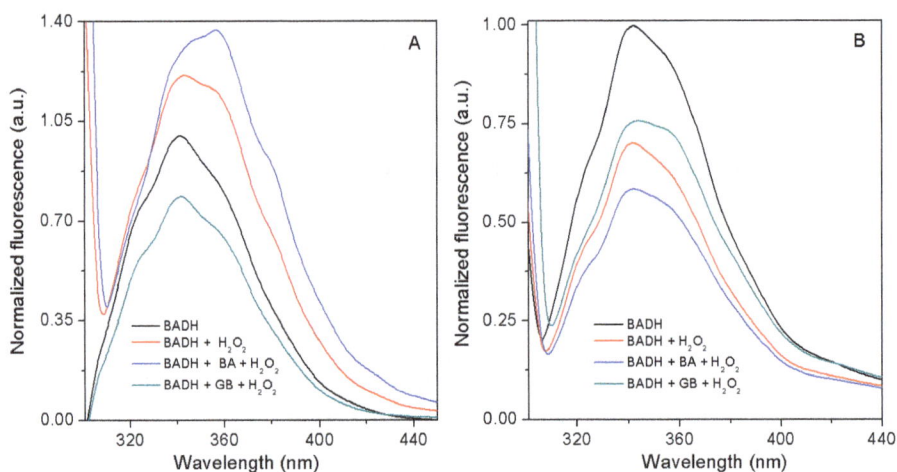

Fig. 4. Fluorescence spectra for skBADH incubated at pH 7.0 (A) and pH 8.0 (B) in the absence and presence of 100 μM H_2O_2, betaine aldehyde (BA) or glycine betaine (GB).

3.4 The effect of copper on enzyme activity

SkBADH incubated with 10 μM $CuCl_2$ at pH 7.0 and 8.0 was inactivated in a time dependent manner (Fig. 5). In assays performed at pH 8.0, enzyme activity decreased 80%, however when KCl was included in the incubation assay, enzyme activity decreased 45% (Fig. 5A). Data analysis revealed single exponential decay kinetics with $k_{obs} = 0.0112 \pm 0.005$ min^{-1} for the enzyme incubated with KCl and $k_{obs} = 0.022 \pm 0.002$ min^{-1} for the enzyme incubated without KCl (Fig. 5A).

The enzyme incubated with 10 μM $CuCl_2$ at pH 7.0 had a different inactivation pattern: skBADH lost 48% of its activity in absence of KCl, whereas in its presence enzyme inactivation reached 59% (Fig. 5B). The inactivation process follows single exponential decay kinetics with k_{obs} = 0.0412 ± 0.0057 min^{-1} for the enzyme incubated with KCl and k_{obs} = 0.0147 ± 0.0048 min^{-1} for the enzyme incubated without KCl (Fig. 5B). Under renal pH and ionic strength conditions skBADH is inactivated by copper concentrations that can occur in kidney.

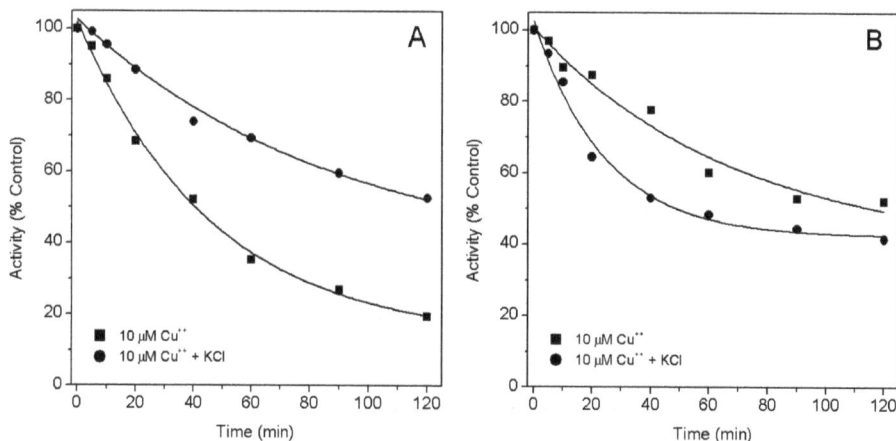

Fig. 5. Time courses of the inactivation of skBADH by copper. (A) Enzyme incubated in Hepes-KOH buffer at pH 8.0; (B) Enzyme incubated in Hepes-KOH buffer at pH 7.0. The lines are the best fit of the inactivation data to a single exponential decay equation.

3.5 Enzyme reactivation kinetics

The fully inactivated enzyme was incubated in the presence of 10 mM DTT, 10 mM GSH or 10 mM 2-mercaptoethanol. SkBADH incubated with DTT at pH 8.0 or pH 7.0 recovered 93% and 96% of its activity, respectively (Fig. 6). When the reducing agent used was 2-mercaptoethanol, under conditions of pH 8.0 or pH 7.0 the enzyme recovered 86% and 84% of its activity (Fig. 6). With GSH at pH 8.0, skBADH recovered 90% of its activity (Fig. 6B), while at pH 7.0 the enzyme recovered 30% of its activity (Fig. 6A). Enzyme recovery activity was faster at pH 8.0 than at pH 7.0, regardless of the reducing agent tested (Fig. 6). SkBADH incubated with reducing agents in the presence of 0.15 M KCl behaved very similarly to the curves in Figure 6 (data not shown). GSH was not able to recover enzyme activity when there was potassium in the assay medium. No significant skBADH activity changes were obtained after long incubation periods.

The failure of GSH to reactivate skBADH and other BADHs inactivated by thiol-specific reagents had been reported (Rosas-Rodríguez & Valenzuela-Soto, 2011; Velasco-García et al., 2003; Vallari & Pietruszko, 1982); however, the reason for this failure is not yet clear and may be related to the larger size of GSH, which prevents it from accessing the active site.

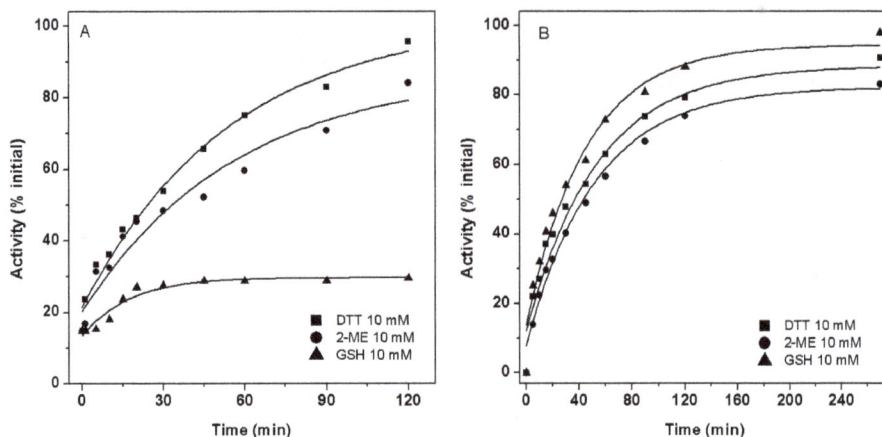

Fig. 6. Reactivation kinetics for skBADH inactivated by copper. Enzyme inactivated by 50 µM CuCl₂ to a residual activity of 15% over a 150 min reaction. Enzyme incubated: at pH 7.0 (A) or pH 8.0 (B). The lines are the best fit of the inactivation data to a single exponential decay equation.

3.6 The effect of copper on enzyme fluorescence

The skBADH fluorescence spectra were obtained from enzyme incubation assays with 10 µM CuCl₂ at pH 8.0 or pH 7.0. Enzyme fluorescence was quenched by copper at both pH values (Fig. 7 and 8). Enzyme maximum emission changed with respect to the time of incubation at both pH values. However, incubation time did not have any effect on the wavelength at which maximum emission occurred.

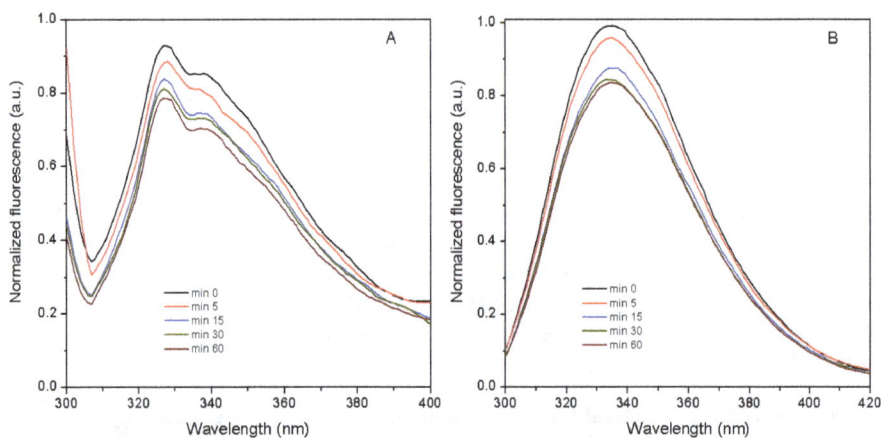

Fig. 7. The effect of copper and incubation time on skBADH fluorescence spectra. Enzyme incubated at pH 7.0 with 10 µM CuCl₂ (A), plus 0.15 M KCl (B).

The drop in the enzyme's maximum emission was similar at both pH values, though the wavelength at which that maximum occurred was different at pH 7.0 and pH 8.0. At pH 7.0 the maximum emission wavelength for the enzyme incubated only with copper was 327 nm (Fig. 7A). Copper caused a blue shift in the enzyme, implying that the tryptophans were less exposed to the solvent. The presence of 0.15 M KCl reversed the blue shift (Fig 7B). The skBADH fluorescence spectra at pH 8.0 with 10 μM CuCl$_2$ did not exhibit a blue or red shift (Fig. 8A) and the presence of KCl did not cause any changes (Fig. 8B).

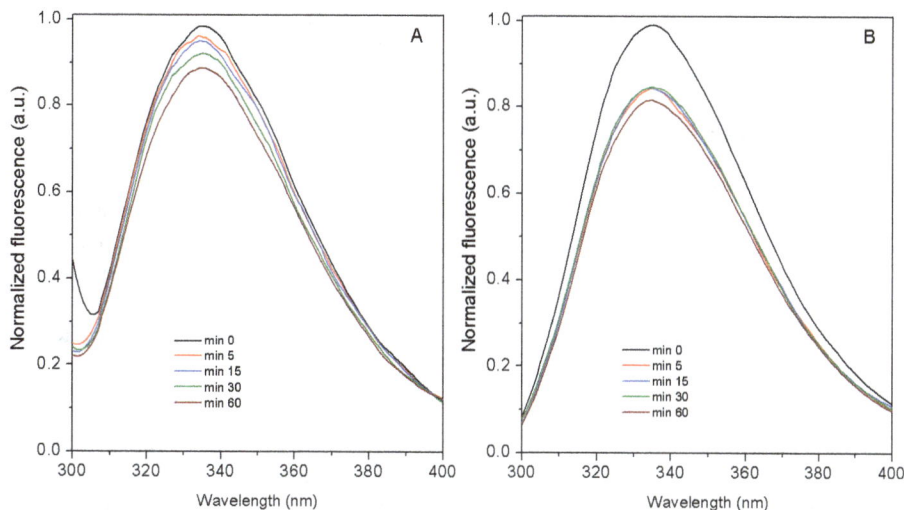

Fig. 8. The effect of copper and incubation time on skBADH fluorescence spectra. Enzyme incubated at pH 8.0 with 10 μM CuCl$_2$ (A), plus 0.15 M KCl (B).

4. Discussion

SkBADH has four tryptophan residues per subunit, which means there are sixteen residues in the enzyme's active conformation (tetramer), making Trp fluorescence a useful technique for structure-function studies. Alvarez et al. (2010) report that the presence of reactive oxygen species could impact the fluorescent properties of a specific protein (Cyan Fluorescent Protein), due to chemical modifications that cause changes in the chromophore pocket (Alvarez et al., 2010). These authors suggest that the alterations could be related to the torsion of amino acid residues, which leads to a quenching process.

The quenching of skBADH fluorescence by hydrogen peroxide is consistent with data reported by Tsourkas et al. (2005) who detected a 50% decrease in fluorescent emission for Red Fluorescent Protein (DsRed) and 95% for Enhanced Yellow Fluorescent Protein (EYFP) in the presence of 50 μM hydrogen peroxide (Tsourkas et al., 2005). The authors attributed the quenching process to intramolecular cross-linking between two cysteines which puts stress on the protein structure (Tsourkas et al., 2005).

A previous study reported that skBADH is inactivated by H_2O_2 in an oxygen independent modification process (Rosas-Rodríguez & Valenzuela-Soto, 2011). Our data for enzyme fluorescence quenching by peroxide indicates that the protein's tertiary structure was not changed (Fig. 2). Together, the inactivation and fluorescence data support the idea that enzyme inactivation is caused by a disulfide bond between vicinal cysteines (C288-C289) at the active site, as previously postulated (Rosas-Rodríguez & Valenzuela-Soto, 2011).

Based on the mechanism of the hydrogen peroxide and cysteine reaction (Luo & Anderson, 2008), we propose a structural model showing the interaction of peroxide with the skBADH active site that leads to enzyme inactivation (Fig. 9). The structural model of skBADH was compared with that of methanol dehydrogenase (pdb 1g72) to look for type VIII folding, because that kind of folding has been found in proteins forming disulfide bridges between vicinal cysteines (Carugo et al., 2003). Both models were very similar, thus we think that skBADH exhibits the type VIII β-turn folding that allows for disulfide bridge formation.

Copper had the strongest inactivation effect on skBADH under physiological pH and ionic strength conditions (Fig. 5). Because the potassium concentrations in kidney medulla cells are those tested in this study, skBADH is a target for copper oxidation which leads to a decrease in glycine betaine synthesis and accumulation. In addition to its function as an osmolyte, glycine betaine functions as an osmoprotector (Rosas-Rodríguez et al., 2010); the insufficient osmoprotection and osmoregulation of renal cells may be related to some kidney pathologies.

The highest levels of BADH activity and protein occur in human liver, the adrenal gland, and kidney (Izaguirre et al, 1991). Besides of the role played by BADH in kidney, mammalian BADH catalyzes the oxidation of different amino aldehydes in addition to betaine aldehyde, so this enzyme is thought to be involved in: (i) the biosynthesis of GB, which may function as a methyl donor for methionine synthesis in liver (duVigneaud, 1946, Muntz, 1950), (ii) polyamine catabolism (Ambroziak & Pietruszko, 1991), (iii) the synthesis of the inhibitory neurotransmitter γ-aminobutyrate (GABA) particularly in adrenal glands where putrescine is a source of γ-aminobutyric acid (Lin et al., 1996), and (iv) carnitine biosynthesis (Vaz et al., 2000). It is possible that all BADHs might be susceptible to copper inactivation, but it is important to conduct studies to test this idea.

SkBADH was inactivated by 10 μM $CuCl_2$ as a consequence of its thiol groups being oxidized (Fig. 5). The oxidation process was reversed by thiol reducing agents, DTT and 2-ME and GSH (at pH 8.0) (Fig. 6B), whereas at pH 7.0 DTT and 2-ME were able to restore enzyme activity which is consistent with the generation of a disulfide bond in the enzyme (Fig. 6A). A disulfide bond could be formed between Cys 288 and Cys 289 at the active site (Fig. 10). GSH was not able to restore enzyme activity at pH 7.0, which suggests that GSH cannot prevent skBADH inactivation *in vivo*.

Inactivation by copper due to thiol oxidation has been observed for other enzymes such as rat and human aldose reductase, where inactivation results from disulfide bridge formation (Cecconi et al., 2002). Similarly, Hadizadeh et al. (2009) observed that the enzyme xanthine oxidase is inactivated by copper in a concentration and time dependent manner (Hadizadeh et al. 2009).

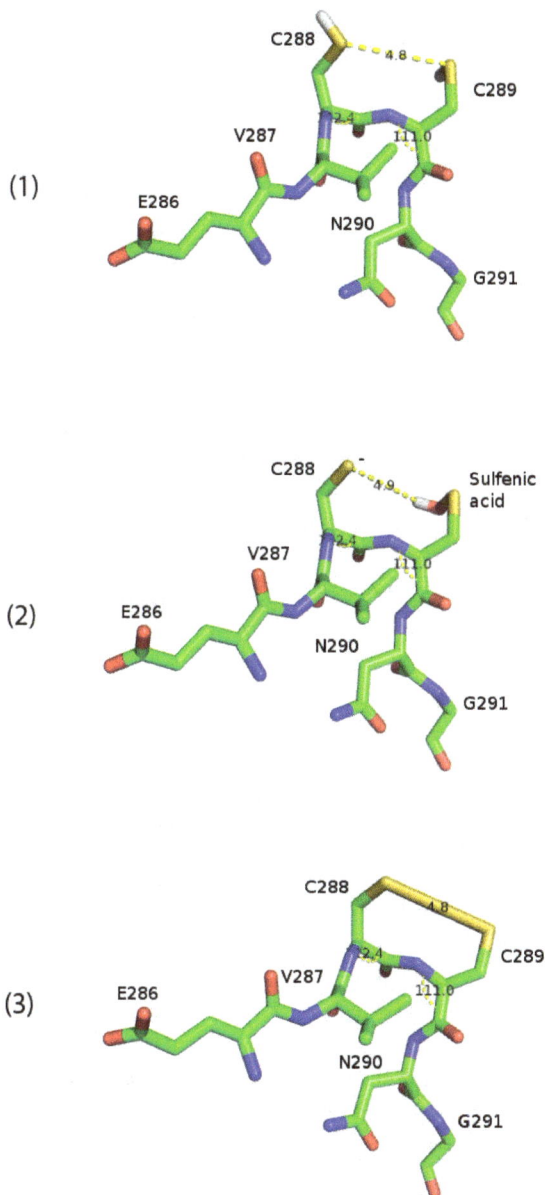

Fig. 9. Proposed mechanism for the reaction of the skBADH active site and H₂O₂. Hydrogen peroxide reacts with the thiol group (CSH) of the catalytic cysteine (C288) or the vicinal cysteine (C289) (1) leading to sulfenic acid formation (RSOH) (2), the proximity of a second thiolate group allows the intermediate sulfenic acid to combine with the thiolate group and form a disulfide bond (RSSR) between cysteines (3).

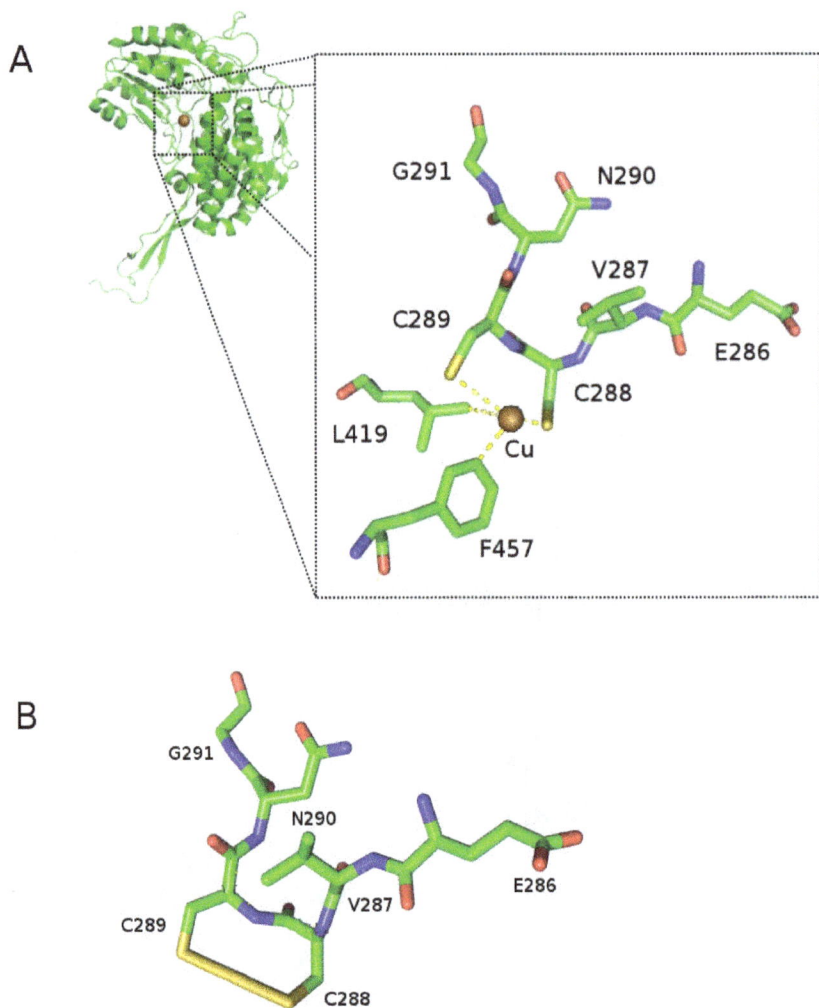

Fig. 10. Proposed mechanism for the reaction of the sKBADH active site and copper. Copper reacts with the thiol group (CSH) of the catalytic cysteine (C288) or the vicinal cysteine (C289) (A), leading to the formation of a disulfide bond (RSSR) between cysteines (B).

SkBADH fluorescence spectra with 10 μM CuCl$_2$ at pH 7.0 and pH 8.0 exhibited differences in the wavelengths at which the maximum emission occurred (Fig. 7 and 8), demonstrating that copper is changing the tryptophans' access to the solvent. In addition, potassium seems to be playing an interesting role in exposing the tryptophans in the enzyme to the solvent. Previous studies have demonstrated that skBADH requires a monovalent cation to maintain its tetrameric conformation (Valenzuela-Soto et al., 2003). SkBADH crystallographic studies are needed to see whether there is a site on the enzyme for potassium or for a different monovalent cation, such as that found in *P. aeruginosa* BADH (González-Segura et al., 2009).

Further studies are underway to determine if skBADH inactivation is dependent on copper concentration, to identify the time course of disulfide generation, and to detect whether there are differences in inactivation when Cu(I) or Cu(II) is used in the treatments.

5. Conclusion

SkBADH was oxidized by hydrogen peroxide and by copper at concentrations that likely reflect physiologically relevant conditions (Atmane et al. 2003; Caselli et al. 1998; Kazi et al., 2008; Meng et al. 2002; Pravodh et al. 2011) in kidney, and the targets of oxidation were thiol groups, mainly those of the active site. Interestingly, the thiol groups were not oxidized to sulfenic or sulfinic acid, but rather a disulfide bond was formed between catalytic cysteine (Cys288) and a vicinal cysteine (C289). Potassium, under physiological conditions, increased the oxidation of the enzyme by copper, possibly because it maintains the enzyme active tetramer or enables the oxidation of the cysteine residues. Since GB synthesis is catalyzed by BADH and it is crucial for the kidney cells to be able to withstand osmotic stress conditions, this type of inhibition can have serious physiological implications. In addition, BADH is an enzyme that is also found in the liver and brain where its inactivation by oxidative stress is associated with other pathologies. In summary, identifying the specific modifications caused by ROS on key enzymes contributes to a greater understanding of the changes that occur during oxidative stress and provides insight into how oxidative stress can be prevented. Further work is underway to determine the mechanisms by which renal BADH can be protected from the oxidative stress that occurs in renal medulla cells.

6. Perspectives

Hydrogen peroxide plays a role as an oxidant and as a secondary messenger, however the molecular basis of these roles and the concentration of H_2O_2 in cells has not been established. In addition, copper can interact with proteins, causing oxidation, or can generate oxidative stress. In light of this it is important to pursue this line of research, also because hydrogen peroxide and copper have been associated with some human pathologies that are characterized by some interaction between hydrogen peroxide and copper and enzymes and proteins. This opens up new areas for developing strategies to contend with some of the diseases directly related to oxidative stress.

7. Acknowledgments

This study was supported by the Consejo Nacional de Ciencia y Tecnología (CONACYT). HFFM received a graduate scholarship from CONACYT. JARR is grateful for support from "Apoyos Complementarios para la Consolidación Institucional de Grupos de Investigación: Retención 149021" awarded by CONACYT. The authors are grateful to Bianca Delfosse, who improved the English.

8. References

Ambroziak, W., & Pietruszko, R. (1991) Human aldehyde dehydrogenase. Activity with aldehyde metabolites of monoamines, diamines, and polyamines. *Journal of Biological Chemistry*, Vol.266, No.20, (July 1991), 13011-13018, ISSN 0021-924.

Alvarez, L., Levin, C.H., Merola, F., Bizouarn, T., Pasquier, H., Baciou, L., Rusconi, F., & Erard, M. (2010). Are the fluorescent properties of the cyan fluorescent protein sensitive to conditions of oxidative stress? *Photochemistry and Photobiology*, Vol.86, No.1, (January-February 2010), 55-61, ISSN 1751-1097.

Arakawa, K., Takabe, T., Sugiyama, T., & Akazawa, T. (1987). Purification of betaine-aldehyde dehydrogenase from spinach leaves and preparation of its antibody. *The Journal of Biochemistry*, Vol.101, No.6, (December 1987), 1485-1488, ISSN 0021-924.

Atmane, N., Dairou, J., Paul, A., Dupret, J-M., & Rodrigues-Lima, F. (2003). Redox Regulation of the Human Xenobiotic Metabolizing Enzyme Arylamine N-Acetyltransferase 1 (NAT1). *Journal of Biological Chemistry*, Vol.278, No.37, (September 2003), 35086-35092, ISSN 0021-9258.

Banday, A.A., & Lokhandwala, M.F. (2011). Oxidative stress causes renal angiotensin II type 1 receptor upregulation, Na^+/H^+ exchanger 3 overstimulation, and hypertension. *Hypertension*, Vol.57, No.2, (March 2011), 452-459, ISSN 0194-911.

Boch, J., Nau-Wagner, G., Kneip, S., & Bremer, E. (1997). Glycine betaine aldehyde dehydrogenase from *Bacillus subtilis:* characterization of an anzyme required for the synthesis of the osmoprotectant glycine betaine. *Archives of Microbiology*, Vol.168, No.4, (September 1997), 282-289, ISSN 1432-072.

Bonda, D.J., Lee, H., Blair, J.A., Zhu, X., Perry, G., & Smith, M.A. (2011) Role of metal dyshomeostasis in Alzheimer's disease. *Metallomics*, Vol.3, No.3, (March 2011), 267-270, ISSN 1756-5901.

Burg, M.B., & Ferraris J.D. (2008). Intracellular organic osmolytes: function and regulation. *Journal of Biological Chemistry*, Vol. 283, No.12, (March 2008), 7309-7313, ISSN 0021-9258.

Burg, M.B., Dmitrieva N.I. & Ferraris J.D. (2007). Cellular response to hyperosmotic stresses. *Physiological Reviews*, Vol.87, No.4 , (October 2007), 1441-1474, ISSN 0031-9333.

Burg, M.B., Kwon, E.D., & Peters, E.M. (1996). Glycerophosphocholine and betaine counteract the effect of urea on pyruvate kinase. *Kidney International*, Vol.50, No.57, (December 1996), S100-104, ISSN 0085-2538.

Burkitt, M.J. (2001) A Critical Overview of the Chemistry of Copper-Dependent Low Density Lipoprotein Oxidation: Roles of Lipid Hydroperoxides, α-Tocopherol, Thiols, and Ceruloplasmin. *Archives of Biochemistry and Biophysics*, Vol.394, No.1, (1 October 2001), 117-135, ISSN 0003-9861.

Butterworth, R.F. (2010). Metal toxicity, liver disease and neurodegeneration. *Neurotoxicity Research*, Vol.18, No.1, (July 2010), 100-105, ISSN 1029-8428.

Carugo, O., Čemažar, M., Zahariev, S., Hudáky, I., Gáspári, Z., Perczel, A., & Pongor, S. (2003) Vicinal disulfide turns. *Protein Engineering*, Vol.16, No.9, (September 2003), 637-639, ISSN 1741-0126.

Caselli, A., Marzocchini, R., Camici, G., Manao, G., Moneti, G., Pieraccini, G., & Ramponi, G. (1998). The inactivation mechanism of low molecular weight phosphotyrosine-protein phosphatase by H_2O_2. *Journal of Biological Chemistry*, Vol.273, No.49, (December 1998), 32554-32560, ISSN 0021-9258.

Cecconi, I., Scalon, A., Rastelli, G., Moroni, M., Vilardo, P., Costantino, L., Cappiello, M., Garland, D., Carper, D., Petrash, J., Del Corso, A., & Mura, U. (2002). Oxidative modification of aldose reductase induced by copper ion. *Journal of Biological Chemistry*, Vol.277, No.44, (November 2002), 42017-42027, ISSN 0021-9258.

Chen, T.H.H., & Murata, N. (2011). Glycinebetaine protects plants against abiotic stress: mechanisms and biotechnological applications. *Plant, Cell and Environment*, Vol. 34, No.1, (January 2011), 1-20, ISSN 1365-3040

Chern, M.K., & Pietruzko R. (1999). Evidence for mitochondrial localization of betaine aldehyde dehydrogenase in rat liver: purification, characterization, and comparison with human cytoplasmic E3 isozyme. *Biochemistry and Cell Biology*, Vol.77, No.3, (July 1999), 179-187. ISSN 0829-8211.

Coughlan, M.T., Thorburn, D.R., Penfold, S.A., Laskowski, A., Harcourt, B.E., Sourris, K.C., Tan, A.L.Y., Fukami, K., Thallas-Bonke, V., Nawroth, P.P., Brownlee, M., Bierhaus, A., Cooper, M.E., & Forbes, J.M. (2009) RAGE-Induced Cytosolic ROS Promote Mitochondrial Superoxide Generation in Diabetes. *Journal of the American Society of Nephrology*, Vol.20, No.4, (April 2009), 742-752, ISSN 1046-6673.

Coyle, C.H., Martinez, L.J., Coleman. M.C., Spitz, D.R., Weintraub, N.L., & Kader, K.N. (2006) Mechanisms of H_2O_2-induced oxidative stress in endothelial cells. *Free Radical Biology & Medicine*, Vol.40, No. 12, (15 June 2006), 2206-2213, ISSN 0891-5849.

Diamant, S., Eliahu, N., Rosenthal, D., & Goloubinoff, P. (2001). Chemical chaperones regulate molecular chaperones in vitro and in cells under combined salt and heat stresses. *Journal of Biological Chemistry*, Vol.276, No.43, (October 2001), 39586-39591, ISSN 0021-9258.

du Vigneaud, V., Simmonds, S., Chandler, J.P., & Cohn, M. (1946). A further investigation of the role of betaine in transmethylation reactions in vivo. *Journal of Biological Chemistry*, Vol.165, No.2, (October 1946), 639-648, ISSN 0021-9258.

Elumalai, M., Antunes, C., & Guilhermino, L. (2002). Effects of single metals and their mixtures on selected enzymes of *Carcinus maenas*. *Water, Air, & Soil Pollution*, Vol.141, No.3, (November 2002), 273–280, ISSN 0049-6979.

Falkenberg, P., & Strom, A.R. (1990). Purification and characterization of osmoregulatory betaine aldehyde deshydrogenase of *Escherichia coli*. *Biochimica et Biophysica Acta*, Vol.1034, No.3, (June 1990), 253-259, ISSN 006-3002.

Figueroa-Soto, C.G., & Valenzuela-Soto, E.M. (2001). Purification of a heterodimeric betaine aldehyde dehydrogenase from wild amaranth plants subjected to water deficit. *Biochemical and Biophysical Research Communications*, Vol.285, No.4, (July 2001), 1052-1058, ISSN 0006-291.

Figueroa-Soto, C.G., & Valenzuela-Soto, E.M. (2000). Kinetic study of porcine kidney betaine aldehyde dehydrogenase. *Biochemical and Biophysical Research Communications*, Vol,269, No.2, (March 2000), 596-603 ISSN 0006-291.

Forbes, J. M., Coughlan M.T., & Cooper M.E. (2008). Oxidative stress as a major culprit in kidney disease in diabetes. *Diabetes*, Vol.57, No.6, (June 2008), 1446-1454, ISSN 0012-1797.

Ganesan, B., Buddhan, S., Anandan R., Sivakumar R., & AnbinEzhilan R. (2010). Antioxidant defense of betaine against isoprenaline-induced myocardial infarction in rats. *Molecular Biology Reports*, Vol.37, No.3, (March 2010), 1319-1327, ISSN 1573-4978.

Giles, N.M., Watts, A.B., Giles, G.I., Fry, F.H., Littlechild, J.A., & Jacob C. (2003). Metal and Redox Modulation Review of Cysteine Protein Function. *Chemistry and Biology*, Vol.10, No.10, (August 2003), 677–693, ISSN 1074-5521.

Go, E.K., Jung, K.J., Kim J.Y., Yu B.P., & Chung, H. (2005). Betaine suppresses proinflammatory signaling during aging: the involvement of nuclear factor-

KappaB via nuclear factor-inducing kinase/IKappaB kinase and mitogen-activated protein kinases. *Journals of Gerontology Series A: Biological Sciences and Medical Sciences*, Vol.60A, No.10, (October 2005), 1252-1264, ISSN 1079-5006.

González-Segura, L., Rudiño-Piñera, E., Muñoz-Clares, R.A., & Horjales, E. (2009). The crystal structure of a ternary complex of betaine aldehyde dehydrogenase from *Pseudomonas aeruginosa* provides new insight into the reaction mechanisms and shows a novel binding mode of the 2´-phosphate of NADP+ and a novel cation binding site. *Journal of Molecular Biology*, Vol.385, No.2, (January 2009), 542-557, ISSN 0022-2836.

González-Segura, L., Velasco-García, R., & Muñoz-Clares, R.A. (2002). Modulation of the reactivity of the essential cysteine residue of betaine aldehyde dehydrogenase from *Pseudomonas aeruginosa*. *Biochemical Journal*, Vol.361, No.3, (February 2002), 577-585, ISSN 1264-6021.

Graf, D., Kurz, A.K., Reinehr, R., Fischer, R., Kircheis, G., & Häussinger, D. (2002). Prevention of bile acid-induced apoptosis by betaine in rat liver. *Hepatology*, Vol.36, No.4, (November 2002), 829-839, ISSN 1665-2681.

Guzman-Partida, A.M., & Valenzuela-Soto, E.M. (1998). Porcine kidney betaine aldehyde dehydrogenase: purification and properties. *Comparative Biochemistry and Physiology, Part B*, Vol.119, No.3, (March 1998), 485-491, ISSN 1096-4959.

Hadizadeh, M., Keyhani, E., Keyhani, J., & Khodadadi, C. (2009). Functional and structural alterations induced by copper in xanthine oxidase. *Acta Biochimica et Biophysica Sinica*, Vol.41, No.7, (July 2009), 603-617, ISSN 1672-9145.

Halliwell, B. (2007) Biochemistry of oxidative stress. *Biochemical Society Transactions*, Vol.35, No.5, (November 2007), 1147-1150, ISSN 0300-5127.

Harper, M., Bevilacqua, L., Hagopian, K., Weindruch, R., & Ramsey, J. (2004). Ageing, oxidative stress, and mitochondrial uncoupling. *Acta Physiologica Scandinavica*, Vol.182, No.4, (August 2004), 321-331, ISSN 001-6772.

Hjelmqvist, L., Norin, A., El-Ahmad, M., Griffiths, W. J., & Jornvall, H. (2003). Distinct but parallel evolutionary patterns between alcohol and aldehyde dehydrogenases: addition of fish/human betaine aldehyde dehydrogenase divergence, *Cellular and Molecular Life Sciences*, Vol.60, No.9, (September 2003), 2009-2016, ISSN 1420-682.

Izaguirre, G., Kikonyogo, A., & Pietruszko, R. (1997). Tissue distribution of human aldehyde dehydrogenase E3 (ALDH9): comparison of enzyme activity with E3 protein and mRNA distribution. *Comparative Biochemistry and Physiology, Part B*, Vol.118, No.1, (February 2003), 59-64, ISSN 1096-4959.

Jomova, K., & Valko, M. (2011). Advances in metal-induced oxidative stress and human disease. *Toxicology*, Vol.283, No.2-3, (May 2011), 65-87, ISSN 0300-483X.

Jomova, K., Vondrakova, D., Lawson, M., & Valko, M. (2010). Metals, oxidative stress and neurodegenerative disorders. *Molecular and Cellular Biochemistry*, Vol.345, No.1, (December 2010), 91-104, ISSN 0300-8177.

Julián-Sánchez, A., Riveros-Rosas, H., Martínez-Castilla, L.P., Velasco-García, R., & Muñoz-Clares, R.A. (2007). Phylogenetic and structural relationships of the betaine aldehyde dehydrogenases. In: *Enzymology and Molecular Biology of Carbonyl Metabolism 13*, H. Weiner, B. Plapp, R. Lindahl, E. Maser, (Eds), pp 64-76, Purdue University Press, ISBN 978-1-55753-447-7.

Kazi, T., Afridi, H., Kazi, N., Jamali, M., Arain, M., Jalbani, N., & Kandhro, G. (2008) Copper, chromium, manganese, iron, nickel, and zinc levels in biological samples of diabetes mellitus patients. *Biological Trace Element Research*, Vol.122, No.1, (April 2008), 1-18, ISSN 0163-4984.

Kurys, G., Ambroziak, W., & Pietruszko, R. (1989). Human aldehyde dehydrogenase. Purification and characterization of a third with low K_m for gamma-aminobutyraldehyde. *Journal of Biological Chemistry*, Vol.264, No.8, (March 1989), 4715-4721, ISSN 0021-9258.

Kwon, M.S., Lim, S.W., & Kwon, H.M. (2009). Hypertonic stress in the kidney: a necessary evil. *Physiology*, Vol.24, No.3, (June 2009), 186-191, ISSN 1548-9213.

Letelier, M.E., Martínez, M., González-Lira, V., Faúndez, M., & Aracena-Parks, P. (2006). Inhibition of cytosolic glutathione S-transferase activity from rat liver by copper. *Chemico-Biological Interactions*, Vol.164, No.1-2, (December 2006), 39-48, ISSN 0009-2797.

Letelier, M.E., Lepe, A. M., Faúndez, M., Salazar, J., Marín, R., Aracena, P., & Speisky, H. (2005). Possible mechanisms underlying copper-induced damage in biological membranes leading to cellular toxicity. *Chemico-Biological Interactions*, Vol.151, No.2, (January 2005), 71-82, ISSN 0009-2797.

Lin, S.W., Chen, J.C., Hsu, L.C. Hsieh, C.L., Yoshida, A. (1996). Human gamma-aminobutyraldehyde dehydrogenase (ALDH9): cDNA sequence, genomic organization, polymorphysm, chromosomal localization, and tissue expression. *Genomics*, Vol.34, No.3, (June 1996), 376-380, ISSN 0888-7543.

Livingstone, J.R., Maruo. T., Yoshida, I. , Tarui, Y., Hirooka, K., Yamamoto, Y., Tsutui, N., & Hirasawa, E. (2003). Purification and properties of betaine aldehyde dehydrogenase from Avena sativa. *Journal of Plant Research*, Vol.116, No.2, (April 2003), 133-140, ISSN 0918-9440.

Luo, D., & Anderson, B.D. (2008). Application of a two-state kinetic model to the heterogeneous kinetics of reaction between cysteine and hydrogen peroxide in amorphous lyophiles. *Journal of Pharmaceutical Sciences*, Vol.97, No.9, (September 2008), 3907-3926, ISSN 0022-3549.

Lushchak, V.I. (2011) Adaptive response to oxidative stress: Bacteria, fungi, plants and animals. *Comparative Biochemistry and Physiology* Part C: Toxicology & Pharmacology, Vol.153, No.2, (March 2011), 175-190, ISSN 1532-0456.

Martin, C.J., & Goeddeke-Merickel, C.M. (2005). Oxidative stress in chronic kidney disease. *Nephrology Nursing Journal*, Vol.32, No.6, (November 2005), 683-685, ISSN 1526-744X.

Mattie, M.D., & Freedman, J.H. (2004). Copper-inducible transcription: regulation by metal- and oxidative stressresponsive pathways. *American Journal Physiology and Cell Physiology*, Vol.286, No.2, (February 2004), C293-C301, ISSN 0363-6143.

Meng, T-C., Fukada, T., & Tonks, N.K. (2002). Reversible Oxidation and Inactivation of Protein Tyrosine Phosphatases In Vivo. *Molecular Cell*, Vol.9, No.2, (February 2002), 387-399, ISSN 1097-2765.

Mori, T., & Cowley, A.W. Jr. (2003). Angiotensin II-NAD(P)H Oxidase-stimulated superoxide modifies tubulovascular nitric oxide cross-talk in renal outer medulla. *Hypertension*, Vol.42, No.4, (October 2003), 588-593, ISSN 0194-911.

Mori, N., Yoshida, N., & Kitamoto, Y. (1992). Purification and properties of betaine aldehyde dehydrogenase from *Xanthomonas translucens*. *Journal of Fermentation and Bioengineering*, Vol.73, No.5, (July 1992), 352-356, ISSN 0922-338X.

Mori, N., Kawakami, B., Hyakutome, K., Tani Y., & Yamada, H. (1980). Characterization of betaine aldehyde dehydrogenase from *Cylindrocarpon didymun* M-1. *Agricultural Biology and Chemistry*, Vol.44, No.12, (December 1980), 3015-3016, ISSN 0002-1369.

Muntz, J.A. (1950). The inability of choline to transfer a methyl group directly to homocysteine for methionine formation, *Journal of Biological Chemistry*, Vol.182, No.2, (February 1950), 489-499, ISSN 0021-9258.

Muñoz-Clares, R.A., Díaz-Sánchez, A.G., González-Segura, L., & Montiel, C. (2010). Kinetic and structural features of betaine aldehyde dehydrogenases: Mechanistic and regulatory implications. *Archives of Biochemistry and Biophysics*, Vol.493, No.1, (January 2010), 71-81, ISSN 003-9861.

Muñoz-Clares, R.A., & Valenzuela-Soto, E.M. (2008). Betaine aldehyde dehydrogenases: evolution, physiological functions, mechanism, kinetics, regulation, structure, and stability. In: *Advances in Protein Physical Chemistry*, E. García-Hernández & D.A. Fernández-Velasco, (Eds), pp 279-302, Research SignPost, ISBN 978-81-7895-324-3, Kerala, India.

Muñoz-Clares, R.A., González-Segura, L., Mújica-Jiménez, C., & Contreras-Díaz, L. (2003). Ligand-induced conformational changes of betaine aldehyde dehydrogenase from Pseudomonas aeruginosa and Amaranthus hypochondriacus L. leaves affecting the reactivity of the catalytic thiol. *Chemico-Biological Interactions*, Vol.143-144, No.3, (Februry 2003), 129-137, ISSN 0009-2797.

Nagasawa, T., Kawabata, Y., Tani, Y., & Ogata, K. (1976). Purification and characterization of betaine aldehyde dehydrogenase from *Pseudomona aeruginosa* A-16. *Agricultural Biology and Chemistry*, Vol.40, No.9, (September 1976), 1743-1749, ISSN 0002-1369.

Neuhofer, W., & Beck, F.X. (2006). Survival in hostile environment: strategies of renal medullary cells. *Physiology*, Vol. 21, No.3, (June 2003), 171-180, ISSN 1548-9213.

Oishi, H., & Ebina, M. (2005). Isolation of cDNA and enzymatic properties of betaine aldehyde dehydrogenase from Zoysia tenuifolia. *Journal of Plant Physiology*, Vol.162, No.10, (October), 1077-1086, ISSN 0176-1617.

Pan, S.M. (1988). Betaine aldehyde dehydrogenase in spinach. *Botanical Bulletin of Academia Sinica*, Vol.29, No.4, (October 1988), 255-263, ISSN 006-8063.

Prabodh, S., Prakash, D., Sudhakar, G., Chowdary, N., Desai, V., & Shekhar, R. (2011) Status of copper and magnesium levels in diabetic nephropathy cases: a case-control study from South India. *Biological Trace Element Research*, Vol.142, No.1, (July 2011), 29-35, ISSN 0163-4989.

Prousek, J. (2007). Fenton chemistry in biology and medicine. *Pure and Applied Chemistry*, Vol.79, No.12, (December 2007), 2325-2338, ISSN 0033-4545.

Prudent, M., & Girault, H.H. (2009) The role of copper in cysteine oxidation: study of intra- and inter-molecular reactions in mass spectrometry. *Metallomics*, Vol.1, No.2, (March 2009), 157-165, ISSN 1756-5901.

Rhee, S.G., Chang, T.-S., Bae, Y.S., Lee, S-R., & Kang, S.W. (2003). Cellular regulation by hydrogen peroxide. *Journal of American Society Nephrology*, Vol.14, No.3, (August 2003), S211-215, ISSN 1533-3450.

Rosas-Rodríguez, J.A., & Valenzuela-Soto, E.M. (2011). Inactivation of porcine kidney betaine aldehyde dehydrogenase by hydrogen peroxide. *Chemico-Biological Interactions*, Vol.191, No.1-3, (May 2011), 159-164, ISSN 0009-2797.

Rosas-Rodríguez, J.A., Figueroa-Soto, C.G., & Valenzuela-Soto, E.M. (2010). Inhibition of porcine kidney betaine aldehyde dehydrogenase by hydrogen peroxide. *Redox Report*, Vol.15, No.6, (December 2010), 282-287, ISSN 1351-0002.

Rothschild, H.A., & Guzman-Barron, E.S. (1954). The oxidation of betaine aldehyde dehydrogenase. *Journal of Biological Chemistry*, Vol.209, (August 1954), 511-523, ISSN 0021-9258.

Salazar-Medina, A.J., García-Rico, L., García-Orozco, K.D., Valenzuela-Soto, E.M., Contreras-Vergara, C.A., Arreola, R., Arvizu-Flores, A., & Sotelo-Mundo, R.R. (2010). Inhibition by Cu2+ and Cd2+ of a mu-class glutathione S-transferase from shrimp *Litopenaeus vannamei*. *Journal of Biochemical and Molecular Toxicology*, Vol.24, No.4, (August 2010), 218-222, ISSN 1099-0461.

Speisky, H., Gómez, M., Burgos-Bravo, F., López-Alarcón, C., Jullian, C., Olea-Azar, C., & Aliaga, M.E. (2009). Generation of superoxide radicals by copper-glutathione complexes: Redox-consequences associated with their interaction with reduced glutathione. *Bioorganic & Medicinal Chemistry*, Vol.17, No.5, (March 2009), 1803-1810, ISSN 0968-0896.

Steinebach, O.M., & Wolterbeek, H.T. (1994). Role of cytosolic copper, metallothionein and glutathione in copper toxicity in rat hepatoma tissue culture cells. *Toxicology*, Vol.92, No.1-3, (September 1994), 75-90, ISSN 0300-483X.

Thengchaisri, N., & Kuo, L. (2003) Hydrogen peroxide induces endothelium-dependent and -independent coronary arteriolar dilation: role of cyclooxygenase and potassium channels. *American Journal of Physiology-Heart and Circulatory Physiology*, Vol.285, No.6, (December 2003), H2255-H2263, ISSN 0363-6135.

Touyz, R.M. (2004). Reactive oxygen species, vascular oxidative stress, and redox signaling in hypertension: what is the clinical significance? *Hypertension*, Vol.44, No.3, (September 2004), 248-252, ISSN 0194-911X.

Touyz, R.M., & Briones, A.M. (2011) Reactive oxygen species and vascular biology: implications in human hypertension. *Hypertension Research*, Vol.34, No.1, (January 2011), 5-14, ISSN 0916-9636.

Tsourkas, A.N.G., Perez, J.M., Basilion, J.P. & Weissleder, R. (2005). Detection of peroxidase/H2O2-mediated oxidation with enhanced yellow fluorescent protein. *Analytical Chemistry*, Vol.77, No.9, (May 2005), 2862-2867, ISSN 0003-2700.

Umenishi, F., Yoshihara, S., Narikiyo, T., & Schrier, R.W. (2005). Modulation of hypertonicity-induced aquaporin-1 by sodium chloride, urea, betaine, and heat shock in murine renal medullary cells. *Journal of the American Society of Nephrology*, Vol.16, No.3, (March 2005), 600-607, ISSN 1533-3450.

Valenzuela-Soto, E. M., Ayala-Castro, H.G., & Muñoz-Clares, R.A., (2005). Effects of monovalent and divalent cations on the thermostability of mammal betaine aldehyde dehydrogenase. In: *Enzymology and Molecular Biology of Carbonyl Metabolism 12*. B. Plapp., E. Maser, R. Lindahl & H. Weiner, (Ed), pp. 104-109, Purdue University Press, ISBN 978-155753-384-5, West Lafayette, IN.

Valenzuela-Soto, E.M., Velasco-García R., Mújica-Jiménez C., Gaviria-González L., & Muñoz-Clares R.A. (2003). Monovalent cations requirements for the stability of betaine aldehyde dehydrogenase from *Pseudomonas aeruginosa*, porcine kidney and amaranth leaves. *Chemico-Biological Interactions*, Vol.143-144, No.3, (February 2003), 139-148, ISSN 0009-2797.

Valenzuela-Soto, E.M., & Muñoz-Clares, R.A. (1994). Purification and properties of betaine aldehyde dehydrogenase extracted from detached leaves of *Amaranthus hypocondriacus* L. Subjected to water deficit. *Journal of Plant Physiology*, Vo.143, No.17, (November 1993), 145-152, ISSN 0176-1617.

Valenzuela-Soto, E.M., & Muñoz-Clares, R.A. (1993) Betaine-aldehyde dehydrogenase from leaves of *Amaranthus hypocondriaus* L. Exhibits an iso ordered Bi Bi steady state mechanism. *Journal of Biological Chemistry*, Vol.268, No.32, (November 1993), 23818-23824, ISSN 0021-9258.

Vasiliou, V., Bairoch, A., Tipton, K.F., & Nebert, D.W. (1999). Eukaryotic aldehyde dehydrogenase (ALDH) genes: human polymorphisms, and recommended nomenclature based on divergent evolution and chromosomal mapping. *Pharmacogenetics*, Vol.9, No.4, (August 1999), 421-434, ISSN 0960-314.

Vallari, R.C., Pietruszko, R. (1982). Human aldehyde dehydrogenase: mechanism of inhibition of disulfiram. *Science*, Vol. 216, No.4546, (May 7 1982), 637-639 ISSN 0036-8075.

Vaz, F.M., Fouchier, S.W., Ofman R., Sommer M., Wanders R.J.A. (2000). Molecular and biochemical characterization of rat γ-trimethylaminobutyraldehyde dehydrogenase and evidence for the involvement of human aldehyde dehydrogenase 9 in carnitine biosíntesis. *Journal of Biological Chemistry*, Vol.275, No.10, (March 2000), 7390-7394, ISSN 0021-9258.

Vaziri, N.D. (2004). Roles of oxidative stress and antioxidant therapy in chronic kidney disease and hypertension. *Current Opinion in Nephrology and Hypertension*, Vol.13, No.1, (January 2004) 93-99, ISSN 1062-4821.

Velasco-García, R., Chacón-Aguilar, V.M., Hervert-Hernández, D., & Muñoz-Clares R.A. (2003). Inactivation of betaine aldehyde dehydrogenase from Pseudomonas aeruginosa and Amaranthus hypochondriacus L. leaves by disulfiram. *Chemico-Biological Interactions*, Vol.143-144, No.3, (February 2003), 149-158 ISSN 0009-2797.

Velasco-García, R., Mújica-Jiménez, C., Mendoza-Hernández, G., & Muñóz-Clares, R.A. (1999). Rapid purification and properties of betaine aldehyde dehydrogenase from *Pseudomonas aeruginosa*. *Journal of Bacteriology*, Vol.181, No.4, (February 1999), 1292-1300, ISSN 0021-9193.

Weretilnyk, E.A., & Hanson, A.D. (1989). Betaine aldehyde dehydrogenase from spinch leaves: purification, *in vitro* translation of the mRNA, and regulation by salinity. *Archives of Biochemistry and Biophysics*, Vol.271, No.1, (May 1989), 56-63, ISSN 003-9861.

Zhang, L., Xiao, N., Pan, Y. Zheng, Y., Pan, Z., Luo, Z., Xu, X., & Liu Z. (2010) Binding and Inhibition of Copper Ions to RecA Inteins from *Mycobacterium tuberculosis*. *Chemistry - A European Journal*, Vol.16, No.14, (April 2010), 4297-4306, ISSN 521-3765.

Zou, A., Li, N., & Cowley, A. (2001). Production and actions of superoxide in the renal medulla. *Hypertension*, Vol.37, No.2, (February 2001), 547-553, ISSN 0194-911.

Section 4

Reactive Species as Signaling Molecules

14

Signalling Oxidative Stress in *Saccharomyces cerevisiae*

Maria Angeles de la Torre-Ruiz[1]*, Luis Serrano[2],
Mima I. Petkova[1] and Nuria Pujol-Carrion[1]
*[1]Dept. Ciències Mèdiques Bàsiques-IRBLleida,
Faculty of Medicine, University of Lleida
[2]Dept. Producció Vegetal & Ciència Forestal,
ETSEA, University of Lleida,
Spain*

1. Introduction

Oxidative stress occurs in the natural environment with exposure to aerobic conditions and UV light. Reactive oxygen species (ROS) are also the consequence of normal cellular metabolism. Aerobic organisms sense redox perturbations and develop several different adaptive mechanisms in order to acquire survival capacity (Zheng & Storz, 2000).

The mitochondrial respiratory chain is the major ROS source of reactive oxygen species. ROS can damage a wide range of molecules, including nucleic acids, proteins and lipids. The accumulation of oxidised proteins, DNA damage and the increased production of ROS, concomitant with a depletion of antioxidant defences, seem to be key factors in aging and cell death.

Mitochondrial oxidation appears to be a major cause of signalling to different pathways; however it is still unclear which one of the inflammatory or the apoptotic signals plays a more relevant role in the mitochondrial generation of ROS. Starvation can increase ROS steady-state concentration and autophagy. Hydrogen peroxide appears to be the major oxidant in these conditions, and would oxidise specific cysteines in autophagyc genes leading to the increase in the autophagosome formation. However it is unknown how the signal is transduced to specific targets (Reviewed in Finkel, 2011).

Mitogen activated protein kinases (MAPK) are required in all the eukaryotic cells to properly activate responses in order to allow cells to respond to the different external stresses. The finality of this is to assure cell survival (Wagner & Nebreda, 2009). Several stimuli, included oxidative stress are signalled to phosphorylate certain MAPK thus activating their kinase activity to phosporylate specific substrates (Shiozaki & Russell, 1995; Nguyen et al., 2000).

The eukaryotic microorganism Saccharomyces cerevisiae serves as a model system to study the signal transduction pathways involved in the response to oxidative stress. Thus, TOR,

---

* Corresponding Author

RAS and CWI pathways are the best characterised routes known to play a relevant role in transducing the oxidative signal in budding yeast.

2. Sensing oxidative stress

Mitochondria are believed to be the major factories of ROS as a byproduct of the respiratory metabolism. However recent studies indicate that ROS produced in mitochondria are signalling molecules capable to activate proteins such as the stress c-Jun N terminal kinase (Finkel, 2011). The activation of the kinase occurs because ROS in fact, inactivate a cysteine dependent phosphatase that regulates c-Jun (Kamata, 2005). The major source of ROS in mitochondrial respiratory chain are complexes I and III. Therefore, mitochondria are ROS generators and consequently act as sensors/transmissors of oxidative stress in eukaryotic cells.

In budding yeast, actin appears to be another possible candidate for sensing oxidative stress (discussed in part 6.2 below)

Transmembrane proteins act also as sensors of a number of stimuli. These proteins can function as detectors of environmental changes and also as a potential transmission molecules, to perhaps downstream signal transduction pathways. Rajavel et al. (1999) identified Mtl1 protein, a Mid2 homologue with a role in cell integrity signalling in vegetative growth. Mtl1 is an element of the cell integrity pathway, overexpression of *MTL1* supress defects in Rho1 function (Sekiya-Kawasaki et al., 2002). In the context of oxidative stress, the most likely candidate for being a transmembrane protein sensor is Mtl1 protein. Vilella et al. (2005) showed that Mtl1 is a cell surface sensor for oxidative stress. Mtl1 functions as oxidatives stress sensor since its function is necessary to survive in response to oxidants and to transmit the oxidative signal to the downstream elements of the cell integrity pathway (Vilella et al., 2005; Petkova et al., 2010a). Mtl1 is a transmembrane protein and localises to the cell periphery in all the stress conditions tested (submitted), supporting the hypothesis that Mtl1 acts as a cell-wall sensor, specifically as an oxidative stress sensor, given that mtl1 mutant cells are only sensitive to oxidative conditions and nutritional starvation (both conditions generate ROS production in the cells). Whether Mtl1 detects extracellular or/and intracellular oxidative stress is still unknown.

3. Transducing oxidative stress

3.1 CWI, TOR and RAS pathways

Signal transduction pathways function as transmissors of environmental estimuli to specific substrates. In budding yeast characteristic routes involved in this processes are CWI, TOR and RAS.

Protein kinase C (PKC) is a protein with a role in oxidative stress response (for a review, Nitti et al., 2008).

In human cells, PKC is involved in the protection against oxidative stress in the heart. The knowledge of this signalling is essential to the development of drugs to treat stroke and cardiac arrithmias (Barnett et al., 2007). Another important feature of the activation of the PKC signal transduction pathway is its role in aging, as reported by Battaini & Pascale

(2005). Importantly, Pascale et al. (2007) exposed in a review how alterations of the PKC cascade may have implications in physiological and pathological brain aging, such as Alzheimer's disease.

The cell-wall integrity pathway in budding yeast involves a protein kinase (MAPKs) cascade which participates in sensing and transmitting several extracellular signals and stresses, including: cell-wall, osmotic, mating and nutritional stress (for a review see Heinisch et al., 1999; Levin, 2005), oxidative (Vilella et al., 2005) and pH (Serrano et al., 2006) stresses. The PKC1-MAPK pathway is integrated by several cell-wall proteins that are putative cell-membrane receptors of different stimuli, they are: the Wsc1-Wsc4 family, Mid2 and Mtl1. They transmit signals to Rom2 which activates the G protein Rho1 which in its turn activates the kinase Pkc1 (this protein has high degree of homology with other isoforms of PKC in eukaryotic cells). Pkc1 activates a MAP kinase module: Bck1 (which is the MAPKKK) phosphorylates the redundant MAPKKs Mkk1 and Mkk2 and together they both activate Slt2, the last kinase member of the pathway. There are two downstream events which correlate to Slt2 activation: transcriptional activity driven by Rlm1 and Swi6 phosphorylation (Heinish et al., 1999; Levin, 2005). Swi6 is one output of Slt2 activity. Swi6 is phosphorylated by Slt2 via the CWI (Sidorova et al., 1995; Madden et al., 1997). However when the external imput is oxidative stress, *SWI6* acts as a sensor through the oxidation of its Cys-404 to a sulfenic residue affecting the cell capability to arrest the cell cycle in G1 (Chiu et al., 2011).

The upper elements of the cell integrity pathway are involved in the organisation of the actin cytoskeleton under different conditions, including cell-wall and nutritional stresses (Helliwell et al., 1998; Delley & Hall, 1999; Torres et al., 2002), oxidative stress (Vilella et al., 2005) and pH (Motizuki et al., 2008) among others.

Exposure of *rom2* to oxidising agent results in diminished Slt2/Mpk1 phosphorylation (Vilella et al., 2005). Pkc1 is also required but the MAP kinase module, downstream of Pkc1, seems to be dispensable for this mechanism. Pkc1 overexpression confers cells with more resistance to oxidising agents. It has been demonstrated that upon oxidative stress Pkc1 translocates to the cell periphery. However, Pkc1 transmits the signal to Slt2/Mpk1 if cells have intact secretory machinery (Vilella et al., 2005)

The Pkc1 pathway is also related to the TOR pathway. Budding yeast have two different TOR genes: *TOR1* and *TOR2*, which share 67% sequence identity and are partly redundant in function (Helliwell et al., 1994). Loewith et al. (2002) purified and identified the components of two distinct TOR complexes, TORC1 and TORC2. TORC1 modulates translation initiation, inhibits protein turnover and represses the transcription of genes related to nutrient starvation. Early studies in *Saccharomyces cerevisiae* indicated that TOR has at least two functions: one regulated by the TORC1 complex which is sensitive to rapamycin and the other which is driven by the TORC2 complex and closely related to the organisation of the actin cytoskeleton and independent of rapamycin inhibition (Loewith et al., 2002; and for reviews Wullschleger et al., 2006; Inoki et al., 2005; Inoki & Guan, 2006). Tor2 functions in both complexes while Tor1 only participates in the TORC1 complex. The unique Tor2 function is related to Pkc1 and the organisation of the actin cytoskeleton (Helliwell et al., 1998). However, the rapamycin-insensitive Tor2-unique function has not been described in other eukaryotic model systems (Crespo and Hall, 2002). Rapamycin also induces depolarisation of the actin cytoskeleton through Sit4 and Tap42, two downstream elements of the TORC1 complex (Torres et al., 2002). TOR function controls a variety of cellular activities. In a global sense TOR inhibits transcription of stress-responsive elements, the nitrogen pathway, starvation-genes, and genes involved in the retrograde response. In this regulation there is a general mechanism: the sequestration of the transcription factors: Msn2/Msn2, Gln3 (Beck & Hall, 1999) and Rtg1/Rtg3 (Crespo et al., 2002; Dilova et al., 2004) in the cytoplasm.

Mitochondrial retrograde signalling (RTG) is a pathway of communication from mitochondria to the nucleus under normal and pathophysiological conditions. The best understood of this pathway is in the budding yeast *Saccharomyces cerevisiae*. It involves multiple factors that sense and transmit mitochondrial signals to induce changes in nuclear gene expression. These changes lead to a reconfiguration of metabolism to accommodate cells to defects in mitochondria that provoke abnormal ROS production. RTG is linked to aging, chronological life span, mitochondrial DNA maintenance, TOR signalling, and nutrient sensing pathways, and is conserved in other fungal species (Liu & Butow, 2006). Lst8 is an integral component of TOR kinase complex. It negatively regulates the RTG pathway at the level of Rtg2. The critical regulatory step of the RTG pathway is the dynamic interaction between Rtg2 and Mks1. The prototypical target of the RTG pathway is *CIT2* (encoding a peroxisomal isoform of citrate synthase, which enables cells to utilize two carbon compounds, such as acetate and ethanol, as sole carbon sources) under the control of Rtg3/Rtg1 heterodimer (Liao et al., 1991).

Tor function also regulates ribosomal protein expression in response to environmental conditions via PKA. This regulation involves the Forkhead factor, *FHL1*, and two cofactors: *IFH1* and *CRF1* (Martin et al., 2004). Tor controls ribosomal gene transcription by maintaining *CRF1* in the cytoplasm, then upon Tor inhibition, *CRF1* translocates to the nucleus and inhibits ribosomal expression, though it is probable that other target transcription factors and different regulatory mechanisms are also involved in this signalling.

The Msn2/Msn4 transcription factor binds and activates genes containing the stress response element (STRE: CCCCT) in response to a wide variety of stresses, including nutritional, osmotic, acidic and oxidative stress (Martínez-Pastor et al., 1996; Schmitt & McEntee, 1996; Beck & Hall, 1999; Hasan et al., 2002). The Ras-cAMP-PKA pathway also negatively regulates Msn2/Msn4 nuclear localisation (Martínez-Pastor et al., 1996; Boy-Marcotte et al., 1998; Görner et al., 1998).

RAS/cAMP pathway is activated when exposed of an optimal carbon source (Broach et al., 1990; Thevelein, 1994). In budding yeast there are two RAS proteins Ras1 and Ras2, both of them are GTPases that signal to the protein kinase PKA and cAMP production (Broach 1991). In optimal growth conditions RAS/cAMP pathway is activated and repress the function of the general stress transcription factor Msn2/Msn4 (Martínez-Pastor et al., 1996; Boy-Marcotte et al., 1998; Görner et al., 2002). On the contrary, nutrient starvation and oxidative stress conditions (Petkova et al., 2010a) are concomitant with RAS/cAMP repression.

SCH9 encodes for a protein kinase involved in life span regulation. Sch9 activates respiratiory metabolism in quiescent phase thus provoking increase in ROS concentration, this effect induces a decrease in life span and increases DNA damage (Madia et al., 2009). Sch9 negatively regulates PKA activity (Zhang et al., 2011). To extend life span it is necessary to reduce TOR activity leading to a decrease of mitochondrial activity, but it is necessary to signal to Sch9 downregulation (Pan & Shadel, 2009).

In recent years several studies there have been published that demonstrate the relationship that exists between the TOR and cAMP-PKA pathways. Schmelzle et al. (2004) suggested that the RAS/cAMP pathway could be a novel TOR effector branch. More recently, Chen & Powers (2006) have demonstrated that the TOR and PKA-cAMP pathways coregulate different biosynthetic pathways which control the expression of genes involved in fermentation and aerobic respiration. Both the TOR and cAMP-PKA pathways regulate the expression of genes needed to overcome the diauxic and stationary phases (Cardenas et al., 1999; Garreau et al., 2000), and in whose regulation Msn2/Msn4 transcriptional activity has been reported to be essential (Powers et al., 2006) (Figure 1). Stationary is a phase prone to

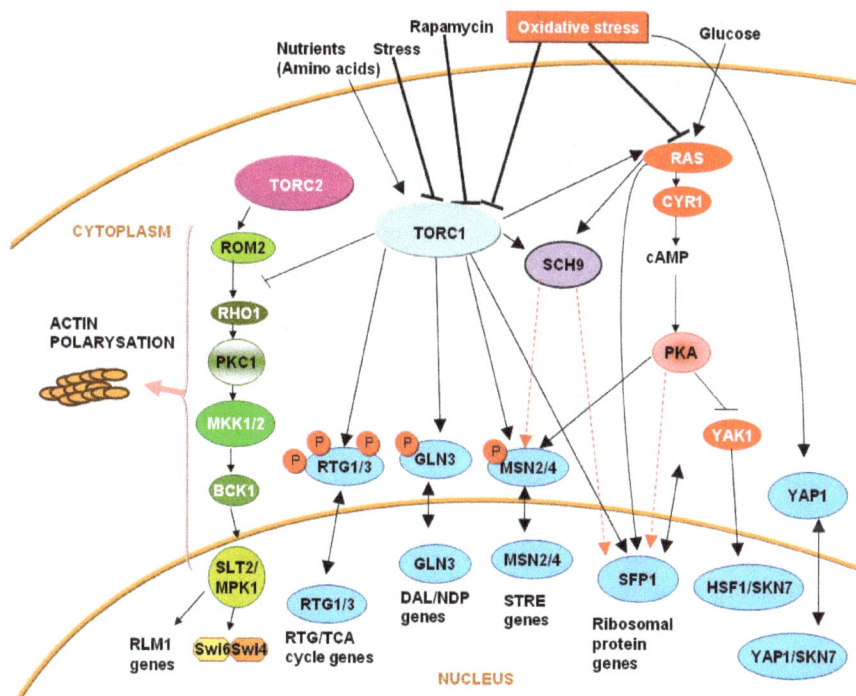

Fig. 1. Oxidative response signalling network in *Saccharomyces cerevisiae*. Oxidative stress negatively regulates TORC1 and RAS activities. TORC1 responds to nutrient availability and is inhibited by rapamycin. This complex when activated promotes the cytoplasm sequestration of specific transcription factors depicted in the figure. TORC1 and RAS/cAMP pathways activate Sfp1 transcription factor inducing ribosomal gene expression. Both TORC1 and RAS converge in *SCH9*. RAS signals to PKA kinase that inhibits both *YAK1* kinase and *MSN2/MSN4* by promoting *MSN2/MSN4* cytoplasm sequestration. *YAK1* in its turn activates *SKN7/HSF1* transcription factor that is required for the oxidative response. For this response TORC1 signals to RAS activation. TORC1 inhibits the CWI activity. However, TORC2 complex signals to cytoplasm elements of the CWI pathway to organise actin cytoskeleton. Red circles containing letters depict phosphorylated amino acid residues. CWI, cell wall integrity; STRE, stress-responsive element; DAL, degradation of urea and allatoin; NDP, nitrogen discrimination pathway; RTG, retrograde pathway.

generate ROS species given that yeast are committed to a respiratory metabolism.The cell integrity pathway is also required for viability in quiescence since Slt2 phosphorylation is necessary for cells to survive in stationary phase and upon rapamycine treatment (Krause & Gray, 2002, Torres et al., 2002). Moreover, rapamycine treatment induces depolarisation of the actin cytoskeleton in a cell-integrity pathway-dependent way (Torres et al., 2002) involving the participation of Wsc1 transmembrane proteins.

3.2 Phospatases as regulatory elements

Although oxidative stress is able to induce MAP kinase cascades, reversal of MAPK activation requires the transcriptional induction of specialised cysteine-based phosphatases that mediate MAPK dephosphorylation. In some occasions, oxidative stress inactivates phosphatases by thiol modification leading to abnormal MAPK upregulation. Recently, Fox et al. (2007) described a mechanism by which the stress inducible MAPK phosphatase Sdp1 acquired enhanced catalytic activity under oxidative conditions. Sdp1 uses an intramolecular disulphide bridge and an invariant histidine side chain to recognise a tyrosine-phosphorylated MAPK substrate in oxidant conditions. The disulphide bridge seems to be essential in order to reach a maximum activity. It is well known that yeast develop several strategies to recognize and adapt to oxidative stress. Reversible formation of disulphide bonds is a major way to regulate oxidative stress response in prokaryotic and eukaryotic microorganisms as well as higher eukaryotes (Lushchak, 2011). This must be one of the most efficient mechanisms in the oxidative context given that it is conserved in regulatory proteins, such as the phosphatase Sdp1. The reverse oxidation seems to be a rapid and effective activating mechanism to regulate stress responsive MAPK proteins.

3.3 ROS detoxifier proteins acting as signalling regulatory molecules

The elements that form part in signalling pathways that are involved in the response to oxidative stress are susceptible to be oxidised and consequently to be impaired in their functions. Therefore, molecules that repair oxidised proteins are likely to be associated to signalling proteins.

In mammals, there are repairing molecules that interplay in the oxidative stress mechanism. These are: thioredoxins, glutaredoxins, peroxiredoxins and other enzymes with an important role on tumourgenesis and oxidative damage resistance. In yeast some peroxiredoxins have been described to play a role in the regulation of the expression of specific stress genes (Ross et al., 2000).

Watson et al. (1999) have demonstrated the interaction between thioredoxins (proteins that reduce oxidised proteins) and PKC. Kahlos et al. (2003) also demonstrated the functional interaction between oxidoreductases and PKC (from endothelial pulmonary cells). In particular, these authors determined that these oxydoreductases were able to reduce disulphide bonds formed in the PKC protein as a consequence of nitric oxide treatment. PICOT protein, is a PKC interacting protein that negatively regulates its function (Witte et al., 2000).

Another form to activate and therefore also regulate MAPK upon oxidative stress has been described in *Schizosaccharomyces pombe* (Day & Veal 2010). These authors demonstrated that redox state regulation of cysteines in Sty1 is needed for the hydrogen-peroxide induced increment of Aft1 mRNA levels. This leads to the transcriptional activation of specific genes and the subsequent augment in the oxidative stress survival (Degols & Russell, 1997).

Also in *S. pombe* Veal et al. (2004) showed the existence of an interesting mechanism by which a peroxiredoxin acts a redox sensor specifically required to activate Sty1. Sty1 is a MAPK responsive to all the stresses in *S. pombe*. High concentrations of hydrogen peroxide promote Sty1 activation by the peroxiredoxine, whereas at low concentrations of the oxidising agent this peroxiredoxin also regulates Pap1 nuclear accumulation (Veal et al., 2007). All these studies show that hydrogen peroxide is capable of oxidising specific kinases, and that certain molecules involved in oxidative repair, such as peroxiredoxins, are required in order to achieve a correct signalling response.

Grx3 and Grx4, two monothiol glutaredoxins of *S. cerevisiae*, regulate Aft1 nuclear localisation and negatively regulate its function. PICOT thioredoxin has a high degree of sequence homology with both Grx3 and Grx4 proteins of *S. cerevisiae*. The absence of both proteins makes the cells sensitive to hydrogen peroxide. In the absence of both Grx3 and Grx4 there is a constitutive oxidative stress induced, in part, by the deregulation of iron homeostasis (Pujol-Carrion et al., 2006). There are some other reports demonstrating similar regulations of other protein kinases. Therefore, there is important to isolate and characterise proteins involved in regulating the redox state of protein kinases in order to maintain an oxidatory equilibrium which allows correct cellular function. It will also be interesting to elucidate the degree of conservation of these proteins and their regulatory functions in evolution.

Actin is a target for oxidative stress. Actin cytoskeleton is of enormous relevance in *S cerevisiae*. It is responsible for all the morphogenetic processes, stress responses, organelles delivery etc. There are two potential cysteines more susceptible for being oxidised. Moreover, other regulatory molecules such as Grx3 and Grx4 regulate actin function in oxidative conditions (Pujol-Carrion et al., 2010). Grx3 and Grx4 are two putative glutaredoxins but there have not been described any enzymatic property related to that putative function. They are involved in the regulation of Aft1 a transcription factor required for the correct iron homeostasis (Pujol-Carrion et al., 2006; Ojeda et al., 2006). Grx4 is required for the maintenance of cable structure. Grx3 plays a redundant role in conjunction with Grx4. Both Grx3 and Grx4 have two theorethical domains, one Trx domain close to the C terminus and a Glutaredoxin domain. Both glutaredoxins through their Trx domains are required to repolarise the actin cytoskeleton in oxidative conditions and also for survival in oxidative stress conditions. Interestingly, Grx4 plays a more direct role in the defence against oxidative stress since Grx4 overproduction increases cell survival when cells are exposed to oxidants (Pujol-Carrion et al., 2010).

4. Transcriptional regulation

Different transcription factors regulate the adaptive response to oxidative stress conditions: the general stress response is mediated by the Msn2/Msn4 transcription factor, whereas specific responses are mediated by Yap1, Skn7 and Hsf1. Msn2/Msn4 nuclear localisation and activity are regulated by both TORC1 and PKA (detailed formerly). For the induction of many antioxidant genes, Skn7 and Yap1 act cooperatively upon oxidative stress (Lee et al.,

1999; Brombacher et al., 2006; He et al., 2005). The contribution of Skn7 to the oxidative stress response does not occur through any of the cysteines of the protein. In addition, SKN7 phosphorylation does not seem to be required in the oxidative stress response (Morgan et al., 1997). It has been proposed a model in which Skn7, when located in the nucleus, cooperates with oxisided Yap1 . The association of Yap1 with Skn7 is a prerequisite for Skn7 phosphorylation and the activation of oxidative stress response genes (He et al., 2009). Skn7 interacts also with Hsf1 and both cooperate to induce heat shock genes specifically in response to oxidative stress (Raitt et al., 2000). Hsf1, like Msn2/Msn4, is negatively regulated by PKA via Yak1 kinase. Sfp1 is a transcriptional factor that induces ribosomal gene expression when located in the nucleus. It is positively regulated by either TOR or RAS activities. In response to oxidative stress and DNA damage Sfp1 translocates to the citoplasm with the consequent adaptive downregulation of ribosomal gene expression (Marion et al., 2004) (Figure 1).

In the pathogen yeast *Candida albicans* there has been characterised a gene *CAP1,* which is homologue to the transcription factor Yap1 and has a role in oxidative stress resistance (Alarco et al., 1999). In *C. albicans* Hog1 pathway is also required for a correct oxidative stress response though through a different pathway than that used by *CAP1* gene (Alonso-Monge et al., 2003).

5. Signal crosstalk events in the oxidative transduction

Mtl1 is a transmembrane cell-wall protein required for cell survival upon oxidative conditions (Vilella et al., 2005). Mtl1 belongs to the CWI pathway and its essential function is to inhibit Tor1 and Ras2 function in conditions of oxidative stress and nutrient depletion. This signal is transduced through Rom2 and Rho1 (both elements of the CWI pathway), however, the rest of the downstream components of the mentioned pathway are dispensable for this essential function (Petkova et al., 2010a) (Figure 2). Consequently, upon oxidative stress both Tor1 and Ras2 functions must be transiently repressed. Downstream outputs of this signal are transcriptional induction mediated by Msn2/Msn4 and ribosomal gene repression, probably due in part to the regulation of Sfp1. In this study the authors propose two possible models: a) the oxidative signal from Mtl1 flows to Tor1 and Ras2 independently and converge in a common pathway; b) this signal flows to Tor1 and then to inhibit Ras2 for the regulation of Msn2/Msn4 activity and the downregulation of ribosomal gene expression. There is another crosstalk involving CWI elements, TOR and RAS, In this case the signal flows from RAS2 and TOR1 inactivation to induce the phosphorylation of Slt2 after treatment with hydrogen hydroxide and upon glucose starvation. Interestingly this backwards signal occurs in the absence of Mtl1 protein (Petkova et al., 2010b).

6. Cellular functions affected by oxidative stress

6.1 Cell cycle oxidation and DNA damage

There exist a wide variety of DNA damaging agents. They provoke different DNA lesions (reviewed in Sage & Harrison, 2011). UV light is a well known DNA damaging agent inducing the formation of thymidin dimmers. DNA damaging agents activate a number of checkpoint genes (Lowndes & Murguia, 2000) in order to transiently block cell cycle progression and simultaneously to activate (transcriptional and/or postraductional) genes that repair possible DNA lesions. UV response is characterised in yeast (Engelberg 1994). Oxidative stress is also known to provoke a wide number of genetic anomalies leading to genome instability cancer and inflammatory diseases. DNA damage checkpoints protect genome integrity (Latif et al., 2001). The DNA damage checkpoint is activated in response to oxidative stress (Leroy et al., 2001).

Since MAPKs are activated in response to several stresses, they are also involved in the UV and oxidative response. In particular JNKs and p38 (Engelberg, 1994; Rouse et al., 1994) are involved in these responses. Both kinases are conserved in yeast. Sty1 in *S pombe* and Hog1 in *S cerevisiae* have a high degree of similarity with JNK and p38 mammalian kinases and are known to be involved in the UV and oxidative stress responses (Haghnazari & Heyer, 2004; Alao & Sunnerhagen, 2008).

Among the possible MAP kinases involved in the UV response, Slt2 and Hog 1 are required for survival whereas Mlp1 and Fus3 are dispensable, (Brian et al, 2004). These authors propose that Slt2 is specifically required for survival in front of UV, based on the observation that *slt2* mutant is not significantly sensitive to MMS (Methyl methanesulfonate) treatment. They demonstrate that addition of sorbitol suppressed the requirement for Slt2, what suggests that probably UV is provoking damage at the level of the cell surface. 8-MOP 8-furocoumarin 8-methoxypsolaren is a chemical agent used in the treatment of psoriasis and other skin diseases. The combination of 8-MOP plus UVA, causes DNA double strand breaks, being Slt2 function required for survival (Dardalhon et al., 2009). At present it is unknown which is the precise role of Slt2 and the other members of the CWI pathway in these responses (Bryan et al., 2004). One possible explanation would be that 8-MOP+UV induces increase in ROS steady-state levels. Another interpretation is that oxygen reacts with 8-MOP and components of the cell membrane leading to lipid peroxidation. This would be the starting point to signal to the CWI (Dall'Acqua & Martelli, 1991; Zarebska, 2000; Dardalhon et al., 2009)

In a recent report, Bandyopadhyay et al. (2010) have used a new methodological approach called differential epistasis mapping. By doing so they have also found a genetic interaction between Slt2 and DNA repair genes. Moreover, MMS treatment induced Slt2 translocation to the nucleus and the transcriptional activation of ribonucleotide reductase genes.

There exist cross-talk between DNA damage and oxidative stress since the UV response in mammals is believed to be induced at the cell membrane through the peroxidation of lipids. On the contrary, alkylating agents might induce the UV response through the oxidation of SH free groups and subsequent glutathione pool depletion (Devary et al., 1993). In addition, the UV transcription response mediated by AP-1 in mammals induces the expression of genes required for the response to oxidative damage (van Dam et al., 1995). In conclusion, there exists some evidence that UV response helps to combat oxidative stress.

In mammals oxidative stress activates all the known MAPK. ERK activation promotes cell survival whereas JNK and p38 suppress apoptosis and induce cellular responses to stress (Runchel et al., 2011).

In mammals a crosstalk between DNA damage checkpoint genes and certain MAPK has been well described (Shafman et al., 1995; Bulavin et al., 2001). In an attempt to describe an equivalent crosstalk in yeast Haghnazari & Heyer (2004) analysed the possible relationship between Hog1, a MAPK sensitive to oxidative stress and Rad53. These two proteins were selected based in the evidence that Hog1 becomes phosphorylated and is also required for cell survival in response to mild oxidative stress. Rad53 is a kinase required for cell cycle arrest upon DNA damage and as a consequence of that, upon an increase in ROS concentration. Haghnazari & Heyer (2004) demonstrated that the oxidative response mediated by Hog1 is independent on that governed by Rad53, consequently there is not such a crosstalk in yeast at least until now. Sublethal oxidative stress signals to Rad53 phosphorylation dependently on Mec1 (a PIK-like kinase whose human homologue is ATR). This checkpoint induces a transient delay in S phase and is also dependent on the Rad17 and Rad24 checkpoint genes (Leroy et al., 2001) (Figure 2). In response to DNA damage, Mec1 also plays a role in the inhibition of mitosis by mediating the phosphorylation of Cdc20, consequently the degradation of Pds1 and Clb2 is abolished. It has also been suggested that Mec1 might phosphorylate PKA in this mechanism (Searle et al., 2004). Queralt & Igual (2005) reported that Pkc1 and Slt2 mutants present synthetic lethality with Rad9 mutants, suggesting a connection between CWI elements and DNA-damage checkpoint genes yet unknown.

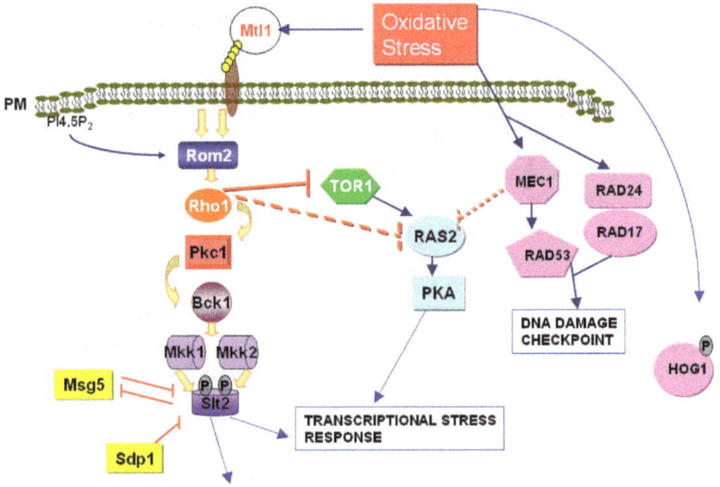

Fig. 2. Schematic diagram of the cross-talk between signal transduction pathways and DNA damage checkpoint genes, in budding yeast. Mtl1 senses oxidative stress and transmit the signal to the downstream elements of the CWI pathway. Rho1 signals to TOR1 and/or RAS2 to induce transcriptional responses that ensure cell survival. Mec1, Rad53, Rad24 and Rad17 are involved in the oxidative response by provoking a DNA-damage checkpoint in S phase. Oxidative stress induces Hog1 phosphorylation independently on Mec1 function. Discontinuous lanes represent signalling events that are very likely to occur.

In mammals there are two central PIK-like kinases whose role is sensing and signalling DNA damage. ATR and ATM (Mec1 and Tel1 in budding yeast, respectively). ATM plays a role in the response to high levels of ROS by repressing mTOR1 expression, although the mechanism by which this occurs is still unknown (Alexander & Walker, 2010). In fission yeast TORC2 complex mediates tolerance to DNA damage and the absence of *tor1* confers cells with more sensitivity to hydroxiurea and MMS, both of them DNA damaging agents. Upon the cell cycle arrest characteristic of the DNA damage checkpoint, *tor1* is required to dephosphorylate and reactivate Cdc2, what elicits the resumption of mitosis (Schonbrun et al., 2009).

6.2 Actin cytoskeleton organisation

The actin molecule is sensitive to oxidative stress (Dalle-Donne et al., 2001). Upon oxidative conditions the actin molecule can be oxidised and a disulphide bond can be formed between cysteins 284 and 373 (Dalle-Donne et al., 2001, 2003).

Cysteine is one amino acid prone to be oxidised when ROS levels increase. In actin Cysteine 374 is the most susceptible to oxidation, according to Takashi (1979). Investigations with erythrocytes from patients suffering sickle cell anemia, demonstrated that these cells posses actin molecules oxidised and forming intramolecular disulphide bonds between C284-C373 (Shartava et al., 1995, 1997; Bencsath et al., 1996). This modification correlates with a decrease in actin polymerisation rates. In yeast these cysteines are homologous to C285 and C374. Recently, Farah et al. (2011) in an elegant study, described the formation of oxidation induced actin bodies (OABs) upon oxidative stress by using budding yeast as a cellular model. These bodies resemble big patches and contain proteins and oxidised actin with intramolecular disulphide bonds between C285 and C374. These authors demonstrated that the formation of C285-374 responds to a protective mechanism against actin oxidation. OABs come from cortical patches. C285-374 are required for the adaptive response and recovery in front to oxidative damage. If actin again is a sensor for oxidative stress, it remains to be elucidated which are all the signalling outputs that govern the cellular responses to oxidative stress once actin oxidation starts the signalling process in the cells. Actin oxidation accelerates cell death in yeast (Dalle-Donne et al., 2001). Studies in eukaryotic model *S. cerevisiae* have allowed the identification of the oxidoreductase *OYE2* (Old Yellow Enzyme 2) that is important to protect actin molecules from being oxidised in Cys285 and Cys374 (Haarer et al., 2004). A deletion in the *OYE2* gene induces an increase in ROS steady-state levels and makes cells more sensitive to oxidation (Farah et al., 2007; Odat et al., 2007). Although Oye enzymes are placed in the signalling network that governs ROS, actin cytoskeleton and survival, it remains unknown at the molecular level which is the connection between any specific signal transduction pathway and Oye2 in response to the redox signal.

Vilella et al. (2005) describe a role for CWI pathway in connecting oxidative stress stimulus with the actin cytoskeleton. This study reveals that oxidative stress depolarises the actin cytoskeleton. None of the CWI elements is required to mediate this depolarisation, however Pkc1 is essential in order to restore the organisation of the actin cytoskeleton in oxidative conditions, concomitantly with an increase in cell viability (Vilella et al., 2005).

In a recent work (Pujol-Carrion et al., 2010) it has been demonstrated that actin polymerisation is a target of hydrogen peroxide. The authors develop an assay based on total protein extracts obtained from different strains of *S. cerevisiae*. These protein extracts are used as polymerisation seeds to study actin assembly. Actin filaments are detected by means of the technique of fluorescence recovery after photobleaching (FRAP). The rationale of this assay is that the association of small amounts of protein extracts with actin monomers could enhance or even inhibit actin nucleation/polymerisation. If the activity of certain protein extracts could promote actin polymerisation, then small oligomers of actin will be created, acting as polymerisation precursors than can accelerate or increase the extent of actin polymerisation (Haarer et al., 1990). By means of this assay the authors demonstrate that Pkc1 plays an important role in promoting actin nucleation both under normal growth conditions and in response to treatment with hydrogen peroxide.

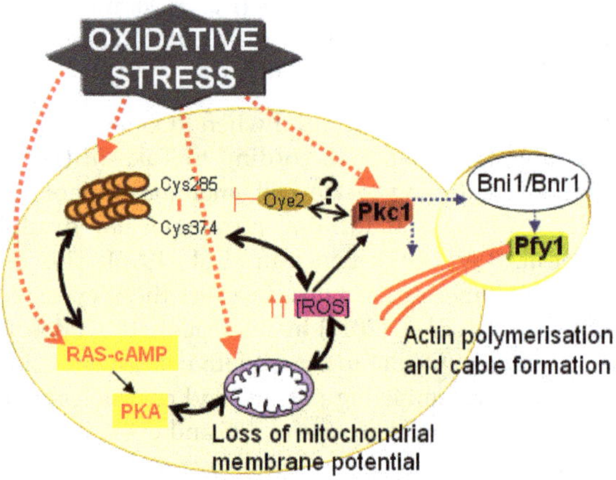

Fig. 3. Diagram of the possible oxidative stress signalling related to mitochondrial function and actin dynamics. Oxidative stress reduces actin dynamics and oxidises actin molecule in the residues depicted in the Figure. Reduced actin dynamics activate RAS/cAMP pathway. Increase of RAS activity activates mitochondrial function releasing high concentrations of ROS to cells. Oye2 repairs actin disulphide bonds and Pkc1 promotes actin dynamics and polymerisation. We have represented this signalling process as a circle of arrows because it is not clear which is the starting point of the cascade. ROS, reactive oxygen species.

Actin is a key element acting as a sensor for oxidative stress and nutritional status and a subsequent linker to ROS dependent mitochondrial release. According to this balance cells will be committed, or not, to cell death (Leadsham et al., 2010). A descent in the actin polymerising activity leads to the formation of actin aggregates associated to the accumulation of ROS in the cytosol (Gourlay & Ayscough, 2005a, 2006; Leadsham et al., 2009). This probably occurs because there is a tight connection between actin cytoskeleton and mitochondria, in addition, the Ras/cAMP pathway plays also an important role in this mechanism. Improper activation of Ras/cAMP leads to higher ROS mitochondrial production. Finally, several studies (Gourlay & Ayscough, 2005a, 2005b; Leadsham et al., 2008; Gourlay et al., 2006) propose the existence of a cross-talk between the actin cytoskeleton dynamics and Ras/cAMP that regulates mitochondrial function and ROS production. Actin is important for the correct distribution of mitochondria between the mother cell and the daughter, but certain proteins required for the remodelling of the cortical actin cytoskeleton induce ROS release from the mitochondria (Gourlay & Ayscough, 2005a, 2005b, 2006). In conclusion, abnormal mitochondrial function increments ROS intracellular levels; high ROS concentration affects actin dynamics and this upregulates the

Ras/cAMP pathway; finally this upregulation signals to the increase in ROS production from the mitochondria (Figure 3). However it not characterised, to date, which is the starting point in this interconnected signalling circle.

7. Conclusion

Oxidative stress provokes different types of damage to each of the components of all the cells. *Saccharomyces cerevisiae* is an optimal eukaryotic model to study signalling events related to this stress, given that the main cascades involved in oxidative stress are highly conserved in the evolution from yeast to men. There exist a number of studies demonstrating that several signal transduction pathways are relevant for this response, PKC, TOR and RAS are central molecules in all the organisms described. The connection between oxidative damage and DNA damage must be very tight. We know that there must be regulatory molecules in cells ensuring a perfect and tight connection between signalling pathways, responding to oxidative stress, and genes involved in DNA-damage checkpoints and DNA repair. We dispose of extended information in the literature regarding this matter, however the precise regulatory pattern that interconnects oxidative sensors, transducers and DNA damage is not totally characterised to date. Another point to be addressed in the future is the characterisation of the different oxidative stress sensors in each of the current cellular models. Actin, mitochondria, transmembrane proteins are good candidates. Future studies will be required to decipher all these questions.

8. References

Alao, J.P. & Sunnerhagen, P. (2008). Rad3 and Sty1 function in *Schizosaccharomyces pombe*: an integrated response to DNA damage and environmental stress?. *Molecular Microbiology*, Vol.68, No.2, (April 2008), pp. 246-254, ISSN 1365-2958

Alarco, A. M. & Raymond, M. (1999). The bZip transcription factor Cap1p is involved in multidrug resistance and oxidative stress response in *Candida albicans*. *Journal of Bacteriology*, Vol.181, No.3, (February 1999), pp. 700–708, ISSN 0021-9193

Alexander, A. & Walker, C.L. (2010). Differential localization of ATM is correlated with activation of distinct downstream signaling pathways. *Cell Cycle*, Vol.9, No.18, (September 2010), pp. 3685-3686, ISSN 1551-4005

Alonso-Monge, R., Navarro-García, F., Román, E., Negredo, A.I., Eisman, B., Nombela, C. & Pla, J. (2003). The Hog1 mitogen-activated protein kinase is essential in the oxidative stress response and chlamydospore formation in *Candida albicans*. *Eukaryotic Cell*, Vol.2, No.2, (April 2003), pp. 351-361, ISSN 1535-9778

Bandyopadhyay, S., Mehta, M., Kuo, D., Sung, M.K., Chuang, R., Jaehnig, E.J., Bodenmiller, B., Licon, K., Copeland, W., Shales, M., Fiedler, D., Dutkowski, J., Guénolé, A., van Attikum, H., Shokat, K.M., Kolodner, R.D., Huh, W.K., Aebersold, R., Keogh, M.C., Krogan, N.J. & Ideker, T. (2010). Rewiring of Genetic Networks in Response to DNA Damage. *Science*, Vol.330, No.6009, (December 2010), pp. 1385–1389 , ISSN 0036-8075

Barnett, M.E., Madgwick, D.K. and Takemoto, D.J. (2007). Protein kinase C as a stress sensor. *Cellular Signalling*, Vol.19, No.9, (September 2007), pp. 1820-1829, ISSN 0898-6568

Battaini, F. & Pascale, A. (2005). Protein kinase C signal transduction regulation in physiological and pathological aging. *Annals of the New York Academy of Sciences*, Vol.1057, (December 2005), pp. 177-192, ISSN 0077-8923

Beck, T. & Hall, M. (1999). The TOR signalling pathway controls nuclear localization of nutrient-regulated transcription factors. *Nature*, Vol.402, No.6762, (December 1999), pp. 689-692, ISSN 0028-0836

Bencsath, F.A., Shartava, A., Monteiro, C.A. & Goodman, S.R. (1996). Identification of the disulfide-linked peptide in irreversibly sickled cell beta-actin. *Biochemistry*, Vol.35, No.14, (April 1996), pp. 4403-4408, ISSN 0006-2960

Boy-Marcotte, E., Perrot, M., Bussereau, F., Boucherie, H. & Jacquet, M. (1998). Msn2p and Msn4p control a large number of genes induced at the diauxic transition which are repressed by cyclic AMP in *Saccharomyces cerevisiae*. *Journal of Bacteriology*, Vol.180, No.5, (March 1998), pp. 1044-1052, ISSN 0021-9193

Broach, J. R. & Deschenes, R. J. (1990). The function of ras genes in *Saccharomyces cerevisiae*. *Advances in Cancer Research*, Vol.54, pp. 79–139, ISSN 0065-230X

Broach, J. R. (1991). RAS genes in *Saccharomyces cerevisiae*: signal transduction in search of a pathway. *Trends in Genetics*, Vol.7, No.1, (January 1991), pp. 28–33, ISSN 0168-9525

Brombacher, K., Fischer, B. B., Rüfenach, K. and Eggen, R. I. L. (2006). The role of Yap1p and Skn7p-mediated oxidative stress response in the defence of *Saccharomyces cerevisiae* against singlet oxygen. *Yeast*, Vol.23, No.10, (July 2006), pp. 741-750, ISSN 0749-503X

Bryan, B.A., Knapp, G.S., Bowen, L.M. & Polymenis, M. (2004). The UV Response in *Saccharomyces cerevisiae* Involves the Mitogen-Activated Protein Kinase Slt2p. *Current Microbiology*, Vol.49, No.1, (July 2004), pp. 32-34, ISSN 0343-8651

Bulavin, D.V., Higashimoto, Y., Popoff, I.J., Gaarde, W.A., Basrur, V., Potapova, O., Appella, E. & Fornace, A.J. Jr. (2001). Initiation of a G2/M checkpoint after ultraviolet radiation requires p38 kinase. *Nature*, Vol.411, No.6833, (May 2001), pp. 102–107, ISSN 0028-0836

Cardenas, M.E., Cutler, N.S., Lorenz, M.C., Di Como, C.J. and Heitman, J. (1999). The TOR signaling cascade regulates gene expression in response to nutrients. *Genes and Development*, Vol.13, No.24, (December 1999), pp. 3271-3279, ISSN 0890-9369

Chen, J.C.Y. and Powers, T. (2006). Coordinate regulation of multiple and distinct biosynthetic pathways by TOR and PKA kinases in *S. cerevisiae*. *Current Genetics*, Vol.49, No.5, (May 2006), pp. 281-293, ISSN 0172-8083

Chiu, J., Tactacan, C.M., Tan, S.X., Lin, R.C., Wouters, M.A. & Dawes, I.W. (2011). Cell cycle sensing of oxidative stress in *Saccharomyces cerevisiae* by oxidation of a specific cysteine residue in the transcription factor Swi6p. *The Journal of Biological Chemistry*, Vol.286, No.7, (February 2011), pp. 5204-5214, ISSN 1083-351X

Crespo, J.L. & Hall, M.N. (2002). Elucidating TOR signaling and rapamycin action: lessons from *Saccharomyces cerevisiae*. *Microbiol. Mol. Biol. Rev.* Vol. 66, No. 44 (December 2002), pp. 579-591, ISSN 0005-3678

Crespo, J.L., Powers, T., Fowler, B. and Hall, M.N. (2002). The TOR-controlled transcription activators *GLN3*, *RTG1*, and *RTG3* are regulated in response to intracellular levels of glutamine. *Proceedings of the National Academy of Sciences of the United States of America*, Vol.99, No.10, (May 2002), pp. 6784-6789, ISSN 0027-8424

Dall'Acqua, F. & Martelli, P. (1991). Photosensitizing action of furocoumarins on membrane components and consequent intracellular events. *Journal of Photochemistry and Photobiology. B, Biology*, Vol.8, No.3, (February 1991), pp. 235–254, ISSN 1011-1344

Dalle-Donne, I., Giustarini, D., Rossi, R., Colombo, R. and Milzani, A. (2003). Reversible S-glutathionylation of Cys 374 regulates actin filament formation by inducing structural changes in the actin molecule. *Free Radical Biology and Medicine*, Vol.34, No.1, (January 2003), pp. 23-32, ISSN 0891-5849

Dalle-Donne, I., Rossi, R., Milzani, A., Di Simplicio, P. and Colombo, R. (2001). The actin cytoskeleton response to oxidants: from small heat shock protein phosphorylation to changes in the redox state of actin itself. *Free Radical Biology and Medicine*, Vol.31, No.12, (December 2001), pp. 1624-1632, ISSN 0891-5849

Dardalhon, M., Agoutin, B., Watzinger, M. & Averbeck, D. (2009) . Slt2 (Mpk1) MAP kinase is involved in the response of *Saccharomyces cerevisiae* to 8-methoxypsoralen plus UVA. *Journal of Photochemistry and Photobiology B: Biology*, Vol.95, No.3, (June 2009), pp. 148–155, ISSN 1873-2682

Day, A.M. and Veal, E.A. (2010). Hydrogen peroxide-sensitive cysteines in the Sty1 MAPK regulate the transcriptional response to oxidative stress. *The Journal of Biological Chemistry*, Vol.285, No.10, (March 2010), pp. 7505-7516, ISSN 1083-351X

Degols, G. & Russell, P. (1997). Discrete roles of the Spc1 kinase and the Atf1 transcription factor in the UV response of *Schizosaccharomyces pombe*. *Molecular and Cellular Biology*, Vol.17, No.6, (June 1997), pp. 3356-3363, ISSN 0270-7306

Delley, P.A. & Hall, M.N. (1999). Cell wall stress depolarizes cell growth via hyperactivation of *RHO1*. *The Journal of Cell Biology*, Vol.147, No.1, (October 1999), pp. 163-174, ISSN 0021-9525

Devary, Y., Rosette, C., DiDonato, J.A. & Karin, M. (1993). NF-kappa B activation by ultraviolet light not dependent on a nuclear signal. *Science*. Vol.261, No.5127, (September 1993), pp. 1442-1445, ISSN 0036-8075

Dilova, I., Aronova, S., Chen, J.C.Y. & Powers, T. (2004). Tor signaling and nutrient-based signals converge on Mks1p phosphorylation to regulate expression of Rtg1.Rtg3p-dependent target genes. *The Journal of Biological Chemistry*, Vol.279, No.45, (November 2004), pp. 46527-46535, ISSN 0021-9258

Engelberg, D., Klein, C., Martinetto, H., Struhl, K. & Karin, M. (1994). The UV response involving the Ras signaling pathway and AP-1 transcription factors is conserved between yeast and mammals. *Cell*, Vol.77, No.3, (May 1994), pp. 381–390, ISSN 0092-8674

Farah, M. E., and Amberg, D. C. (2007). Conserved actin cysteine residues are oxidative stress sensors that can regulate cell death in yeast. *Molecular Biology of the Cell*, Vol.18, No.4, (April 2007), pp. 1359-1365, ISSN 1059-1524

Farah, M.E., Sirotkin, V., Haarer, B., Kakhniashvili, D. & Amberg, D.C. (2011). Diverse protective roles of the actin cytoskeleton during oxidative stress. *Cytoskeleton (Hoboken, N.J.)*, Vol.68, No.6, (June 2011), pp. 340-354, ISSN 1949-3592

Finkel, T. (2011). Signal transduction by mitochondrial oxidants. *The Journal of Biological Chemistry*. (Epub ahead of print)

Fox, G.C., Shafiq, M., Briggs, D.C., Knowles, P.P., Collister, M., Didmon, M.J., Makrantoni, V., Dickinson, R.J., Hanrahan, S., Totty, N., Stark, M.J., Keyse, S.M. & McDonald,

N.Q. (2007). Redox-mediated substrate recognition by Sdp1 defines a new group of tyrosine phosphatases. *Nature*, Vol.447, No.7143, (May 2007), pp. 487-492, ISSN 1476-4687

Garreau, H., Hasan, R.N., Renault, G., Estruch, F., Boy-Marcotte, E. & Jacquet, M. (2000). Hyperphosphorylation of Msn2p and Msn4p in response to heat shock and the diauxic shift is inhibited by cAMP in *Saccharomyces cerevisiae*. *Microbiology*, Vol.146, No.9, (September 2000), pp. 2113-2120, ISSN 1350-0872

Görner, W., Durchschlag, E., Martínez-Pastor, M.T., Estruch, F., Ammerer, G., Halmilton B., Ruis, H. & Schüller, C. (1998). Nuclear localization of the C2H2 zinc finger protein Msn2p is regulated by stress and protein kinase A activity. *Genes and Development*, Vol.12, No.4, (February 1998), pp. 586-597, ISSN 0890-9369

Görner, W., Durchschlag, E., Wolf, J., Brown, E. L., Ammerer, G., Ruis, H. & Schüller, C. (2002). Acute glucose starvation activates the nuclear localization signal of a stress-specific yeast transcription factor. *The EMBO journal*, Vol.21, No. 1-2, (January 2002), pp. 135–144, ISSN 0261-4189

Gourlay, C.W. & Ayscough, K.R. (2005a). Identification of an upstream regulatory pathway controlling actin-mediated apoptosis in yeast. *Journal of Cell Science,*. Vol.118, No. 10, (May 2005), pp. 2119-2132, ISSN 0021-9533

Gourlay, C.W. & Ayscough, K.R. (2005b). A role for actin in aging and apoptosis. *Biochemical Society Transactions*, Vol.33, No. 6, (December 2005), pp. 1260-1264, ISSN 0300-5127

Gourlay, C.W. & Ayscough, K.R. (2006). Actin-induced hyperactivation of the Ras signaling pathway leads to apoptosis in *Saccharomyces cerevisiae*. *Molecular and Cellular Biology*, Vol.26, No. 17, (September 2006), pp. 6487-6501, ISSN 0270-7306

Haarer, B. K. and Amberg, D. C. (2004). Old yellow enzyme protects the actin cytoskeletonfrom oxidative stress. *Molecular Biology of the Cell*, Vol.15, No. 10, (October 2004), pp. 4522-4531, ISSN 1059-1524

Haarer, B. K., Lillie, S. H., Adams, A. E. M., Magdolen, V., Bandlow, W. and Brown, S. S. (1990). Purification of profilin from *Saccharomyces cerevisiae* and analysis of profilin-deficient cells. *The Journal of Cell Biology*, Vol.110, No. 1, (January 1990), pp. 104-114, ISSN 0021-9525

Haghnazari, E. & Heyer, W.D. (2004). The Hog1 MAP kinase pathway and the Mec1 DNA damage checkpoint pathway independently control the cellular responses to hydrogen peroxide. *DNA Repair*, Vol.3, No.7, (July 2004), pp. 769-776, ISSN 1568-7864

Hasan, R., Leroy, C., Isnard, A.D., Labarre, J., Boy-Marcotte, E. & Toledano, M.B. (2002). The control of the yeast H2O2 response by the Msn2/4 transcription factors. *Molecular Microbiology*, Vol.45, No.1, (July 2002), pp. 233-241, ISSN 0950-382X

He, X. J. & Fassier, J. S. (2005). Identification of novel Yap1p and Skn7p binding sites involved in the oxidative stress response of *Saccharomyces cerevisiae*. *Molecular Microbiology*, Vol.58, No.5, (December 2005), pp. 1454-1467, ISSN 0950-382X

He, X.J., Mulford, K.E. & Fassler, J.S. (2009). Oxidative stress function of the *Saccharomyces cerevisiae* Skn7 receiver domain. *Eukaryotic cell*, Vol. 8, No.5, (May 2009), pp. 768-778, ISSN 1535-9786

Heinisch, J.J., Lorberg, A., Schmitz, H.P. & Jacoby, J.J. (1999). The protein kinase C-mediated MAP kinase pathway involved in the maintenance of cellular integrity in

Saccharomyces cerevisiae. Molecular Microbiology, Vol.32, No.4, (May 1999), pp. 671-680, ISSN 0950-382X

Helliwell, S. B., Wagner, P., Kunz, J., Deuter-Reinhard, M., Henriquez, R. & Hall, M.N. (1994). *TOR1* and *TOR2* are structurally and functionally similar but not identical phosphatidylinositol kinase homologues in yeast. *Molecular Biology of the Cell*, Vol.5, No.1, (January 1994), pp. 105-118, ISSN 1059-1524

Helliwell, S.B., Schmidt, A., Ohya, Y. and Hall, M.N. (1998). The Rho1 effector Pkc1, but not Bni1, mediates signalling from Tor2 to the actin cytoskeleton. *Current Biology*, Vol.8, No.22, (November 1998), pp. 1211-1214, ISSN 0960-9822

Inoki, K. and Guan, K.L. (2006). Complexity of the TOR signaling network. *Trends in Cell Biology*, Vol.16, No.4, (April 2006), pp. 206-212, ISSN 0962-8924

Inoki, K., Ouyang, H., Li, Y. and Guan, K.L. (2005). Signaling by target of rapamycin proteins in cell growth control. *Microbiology and Molecular Biology Reviews*, Vol.69, No.1, (March 2005), pp. 79-100, ISSN 1092-2172

Kahlos, K., Zhang, J., Block, E.R. & Patel, J.M. (2003). Thioredoxin restores nitric oxide-induced inhibition of protein kinase C activity in lung endothelial cells. *Molecular and Cellular Biochemistry*, Vol.254, No.1-2, (December 2003), pp. 47-54, ISSN 0300-8177

Kamata, H., Honda, S., Maeda, S., Chang, L., Hirata, H. & Karin, M. (2005). Reactive oxygen species promote TNFalpha-induced death and sustained JNK activation by inhibiting MAP kinase phosphatases. *Cell*, Vol.120, No.5, (March 2005), pp. 649-661, ISSN 0092-8674

Krause, S.A. & Gray, J.V. (2002). The protein kinase C pathway is required for viability in quiescence in *Saccharomyces cerevisiae*. *Current Biology*, Vol.12, No.7, (April 2002), pp. 588-593, ISSN 0960-9822

Latif, C., Harvey, S.H. & O'Connell, M.J. (2001). Ensuring the stability of the genome: DNA damage checkpoints. *TheScientificWorldJournal*, Vol.1, (November 2001), pp. 684-702, ISSN 1537-744X

Leadsham, J.E. & Gourlay, C.W. (2008). Cytoskeletal induced apoptosis in yeast. *Biochimica et Biophysica Acta*, Vol.1783, No.7, (July 2008), pp. 1406-1412, ISSN 0006-3002

Leadsham, J.E., Kotiadis, V.N., Tarrant, D.J. & Gourlay, C.W. (2010). Apoptosis and the yeast actin cytoskeleton. *Cell Death and Differentiation*, Vol.17, No.5, (May 2010), pp. 754-762, ISSN 1476-5403

Leadsham, J.E., Miller, K., Ayscough, K.R., Colombo, S., Martegani, E., Sudbery, P. & Gourlay, C.W. (2009). Whi2p links nutritional sensing to actin-dependent Ras-cAMP-PKA regulation and apoptosis in yeast. *Journal of Cell Science*, Vol.122, No.5, (March 2009), pp. 706-715, ISSN 0021-9533

Lee, J., Godon, C., Lagniel, G., Spector, D., Garin, J., Labarre, J. & Toledano, M.B. (1999). Yap1 and Skn7 control two specialized oxidative stress response regulons in yeast. *The Journal of Biological Chemistry*, Vol.274, No.23, (June 1999), pp. 16040-16046, ISSN 0021-9258

Leroy, C., Mann, C. & Marsolier M.C. (2001). Silent repair accounts for cell cycle specificity in the signaling of oxidative DNA lesions. *The EMBO Journal*, Vol.20, No.11, (June 2001), pp. 2896-2906, ISSN 0261-4189

Levin, D.E. (2005). Cell wall integrity signaling in *Saccharomyces cerevisiae*. *Microbiology and Molecular Biology Reviews*, Vol.69, No.2, (June 2005), pp. 262-291, ISSN 1092-2172

Liao, X., Small, W.C., Srere, P.A. and Butow, R.A. (1991). Intramitochondrial functions regulate nonmitochondrial citrate synthase (CIT2) expression in *Saccharomyces cerevisiae. Molecular and Cellular Biology*, Vol.11, No.1, (January 1991), pp. 38-46, ISSN 0270-7306

Liu, Z. & Butow, R.A. (2006). Mitochondrial retrograde signaling. *Annual Review of Genetics*, Vol.40, pp. 159-185, ISSN 0066-4197

Loewith, R., Jacinto, E., Wullschleger, S., Lorberg, A., Crespo, J.L., Bonenfant, D., Oppliger, W., Jenoe, P. & Hall, M.N. (2002). Two TOR complexes, only one of which is rapamycin sensitive, have distinct roles in cell growth control. *Molecular Cell*, Vol.10, No.3, (September 2002), pp. 457-468, ISSN 1097-2765

Lowndes, N.F. & Murguia, J.R. (2000). Sensing and responding to DNA damage. *Current Opinion in Genetics & Development*, Vol.10, No.1, (February 2000), pp. 17-25, ISSN 0959-437X

Lushchak, V.I. (2011). Adaptive response to oxidative stress: Bacteria, fungi, plants and animals. *Comp. Biochem. Physiol. C. Toxicol. Pharmacol*, Vol. 153, No. 2, (March 2011), pp. 175-190, ISSN 1532-0456

Madden, K., Sheu, Y.J., Baetz, K., Andrews, B. & Snyder, M. (1997). SBF cell cycle regulator as a target of the yeast PKC-MAP kinase pathway. *Science*, Vol.275, No.5307, (March 1997), pp. 1781-1784, ISSN 0036-8075

Madia, F., Wei, M., Yuan, V., Hu, J., Gatazo, C., Pham, P., Goodman, M.F. & Longo, V. (2009). Oncogene homologue Sch9 promotes age-dependent mutations by a superoxide and Rev1/Polzeta-dependent mechanism. *The Journal of Cell Biology*, Vol.186, No.4, (August 2009), pp. 509-523, ISSN 1540-8140

Marion, R.M., Regev, A., Segal, E., Barash, Y., Koller, D., Friedman, N. & O'Shea, E.K. (2004). Sfp1 is a stress- and nutrient-sensitive regulator of ribosomal protein gene expression. *Proceedings of the National Academy of Sciences of the United States of America*, Vol. 101, No.40, (October 2004), pp. 14315-14322, ISSN 0027-8424

Martin, D.E., Soulard, A. & Hall, M.N. (2004). TOR regulates ribosomal protein gene expression via PKA and the Forkhead transcription factor FHL1. *Cell*, Vol.119, No.7, (December 2004), pp. 969-979, ISSN 0092-8674

Martínez-Pastor, M.T., Marchler, G.C.S., Marchler-Bauer, A., Ruis, H. & Estruch, F. (1996). The *Saccharomyces cerevisiae* zinc finger proteins Msn2p and Msn4p are required for transcriptional induction through the stress response element (STRE). *The EMBO Journal*, Vol.15, No.9, (May 1996), pp. 2227-2235, ISSN 0261-4189

Morgan, B.A., Banks, G.R., Toone, W.M., Raitt, D.C., Kuge, S. & Johnston, L.H. (1997). The Skn7 response regulator controls gene expression in the oxidative stress response of the budding yeast *Saccharomyces cerevisiae. The EMBO Journal*, Vol.16, No.5, (March 1997), pp. 1035-1044, ISSN 0261-4189

Motizuki, M., Yokota, S. & Tsurugi, K. (2008). Effect of low pH on organization of the actin cytoskeleton in *Saccharomyces cerevisiae. Biochimica et Biophysica Acta*, Vol.1780, No.2, (February 2008), pp. 179-184, ISSN 0006-3002

Nguyen, A.N., Lee, A., Place, W. & Shiozaki, K. (2000). Multistep phosphorelay proteins transmit oxidative stress signals to the fission yeast stress-activated protein kinase. *Molecular Biology of the Cell*, Vol.11, No.4, (April 2000), pp. 1169-1181, ISSN 1059-1524

Nitti, M., Pronzato, M.A., Marinari, U.M. & Domenicotti, C. (2008). PKC signaling in oxidative hepatic damage. *Mol. Aspects Med.* Vol. 29, No. 1-2, (February-April 2008), pp. 36-42, ISSN 0098-2997

Odat, O., Matta, S., Khalil, H., Kampranis, S. C., Pfau, R., Tsichlis, P. N. & Makris, A. M. (2007). Old yellow enzymes, highly homologous FMN oxidoreductases with modulating roles in oxidative stress and programmed cell death in yeast. *The Journal of Biological Chemistry*, Vol.282, No.49, (December 2007), pp. 36010-36023, ISSN 0021-9258

Ojeda, L., Keller, G., Muhlenhoff, U., Rutherford, J.C., Lill, R. & Winge, D.R. (2006). Role of glutaredoxin-3 and glutaredoxin-4 in the iron regulation of the Aft1 transcriptional activator in *Saccharomyces cerevisiae*. *The Journal of Biological Chemistry*, Vol. 281, No.26, (June 2006), pp. 17661-17669, ISSN 0021-9258

Pan, Y. & Shadel, G.S. (2009). Extension of chronological life span by reduced TOR signaling requires down-regulation of Sch9p and involves increased mitochondrial OXPHOS complex density. *Aging*, Vol.1, No.1, (January 2009), pp. 131-145, ISSN 1945-4589

Pascale, A., Amadio, M., Govoni, S. & Battaini, F. (2007). The aging brain, a key target for the future: the protein kinase C involvement. *Pharmacological research*, Vol.55, No.6, (June 2007), pp. 560-569, ISSN 1043-6618

Petkova, M.I., Pujol-Carrion, N., Arroyo, J., García-Cantalejo, J. & de la Torre-Ruiz, M.A. (2010a). Mtl1 is required to activate general stress response through Tor1 and Ras2 inhibition under conditions of glucose starvation and oxidative stress. *The Journal of Biological Chemistry*, Vol.285, No.25, (June 2010), pp. 19521-19531, ISSN 1083-351X

Petkova, M.I., Pujol-Carrion, N., de la Torre-Ruiz, M.A. (2010b). Signal flow between CWI/TOR and CWI/RAS in budding yeast under conditions of oxidative stress and glucose starvation. *Communicative & Integrative Biology*, Vol.3, No.6, (November 2010), pp. 555-557, ISSN 1942-0889

Pujol-Carrion, N. & de la Torre-Ruiz, M.A. (2010). Glutaredoxins Grx4 and Grx3 of *Saccharomyces cerevisiae* play a role in actin dynamics through their Trx domains, which contributes to oxidative stress resistance. *Applied and Environmental Microbiology*, Vol.76, No.23, (December 2010), pp. 7826-7835, ISSN 1098-5336

Pujol-Carrion, N., Belli, G., Herrero, E., Nogues, A. & de la Torre-Ruiz, M.A. (2006). Glutaredoxins Grx3 and Grx4 regulate nuclear localisation of Aft1 and the oxidative stress response in *Saccharomyces cerevisiae*. *Journal of Cell Science*, Vol.119, No.21, (November 2006), pp. 4554-4564, ISSN 0021-9533

Queralt, E and Igual, J.C. (2005). Functional connection between the Clb5 cyclin, the protein kinase C pathway and the Swi4 transcription factor in *Saccharomyces cerevisiae*. *Genetics*, Vol.171, No.4, (December 2005), pp. 1485–1498, ISSN 0016-6731

Raitt, D.C., Johnson, A.L., Erkine, A.M., Makino, K., Morgan, B., Gross, D.S. & Jonston, L.H. (2000). The Skn7 response regulator of *Saccharomyces cerevisiae* interacts with Hsf1 in vivo and is required for the induction of heat shock genes by oxidative stress. *Molecular Biology of the Cell*, Vol.11, No.7, (December 2010), pp. 2335-2347, ISSN 1059-1524

Rajavel, M., Philip, B., Buehrer, B.M., Errede, B. & Levin, D.E. (1999). Mid2 is a putative sensor for cell integrity signaling in *Saccharomyces cerevisiae*. *Molecular and Cellular Biology*, Vol.19, No.6, (June 1999), pp. 3969-3976, ISSN 0270-7306

Ross, S.J., Findlay, V.J., Malakasi, P. & Morgan, B.A. (2000). Thioredoxin peroxidase is required for the transcriptional response to oxidative stress in budding yeast. *Molecular Biology of the Cell*, Vol.11, No.8, (August 2000), pp. 2631-2642, ISSN 1059-1524

Rouse, J., Cohen, P., Trigon, S., Morange, M., Alonso-Llamazares, A., Zamanillo, D., Hunt, T. & Nebreda, A.R. (1994). A novel kinase cascade triggered by stress and heat shock that stimulates MAPKAP kinase-2 and phosphorylation of the small heat shock proteins. *Cell*, Vol. 78, No.6, (September 1994), pp. 1027-1037, ISSN 0092-8674

Runchel, C., Matsuzawa, A. & Ichijo, H. (2011). Mitogen-activated protein kinases in mammalian oxidative stress responses. *Antioxidants & redox signaling*, Vol.15, No.1, (July 2011), pp. 205-218, ISSN 1557-7716

Sage, E. & Harrison, L. (2011). Clustered DNA lesion repair in eukaryotes: relevance to mutagenesis and cell survival. *Mutation research*, Vol.711, No.1-2, (June 2011), pp. 123-133, ISSN 0027-5107

Schmelzle, T., Beck, T., Martín, D.E. & Hall, M.N. (2004). Activation of the RAS/cyclic AMP pathway suppresses a TOR deficiency in yeast. *Molecular and Cellular Biology*, Vol.24, No.1, (January 2004), pp. 338-351, ISSN 0270-7306

Schmitt, A.P. & McEntee, K. (1996). Msn2p, a zinc finger DNA-binding protein, is the transcriptional activator of the multistress response in *Saccharomyces cerevisiae*. *Proceedings of the National Academy of Sciences of the United States of America*, Vol.93, No.12, (June 1996), pp. 5777-5782, ISSN 0027-8424

Schonbrun, M., Laor, D., López-Maury, L., Bähler, J., Kupiec, M. & Weisman, R. (2009). TOR complex 2 controls gene silencing, telomere length maintenance, and survival under DNA-damaging conditions. *Molecular and Cellular Biology*, Vol.29, No.16, (August 2009), pp. 4584-4594, ISSN 1098-5549

Searle, J.S. and Sanchez, Y. (2004). Stopped for repairs: A new role for nutrient sensing pathways?. *Cell Cycle*, Vol.3, No.7, (July 2004), pp. 865-868, ISSN 1551-4005

Sekiya-Kawasaki, M., Abe, M., Saka, A., Watanabe, D., Kono, K., Minemura-Asakawa, M., Ishihara, S., Watanabe, T. & Ohya, Y. (2002). Dissection of upstream regulatory components of the Rho1p effector, 1,3-beta-glucan synthase, in *Saccharomyces cerevisiae*. *Genetics*, Vol.162, No.2, (October 2002), pp. 663-676, ISSN 0016-6731

Serrano, R., Martín, H., Casamayor, A. & Ariño, J. (2006). Signaling alkaline pH stress in the yeast *Saccharomyces cerevisiae* through the Wsc1 cell surface sensor and the Slt2 MAPK pathway. *The Journal of Biological Chemistry*, Vol.281, No.52, (December 2006), pp. 39785-39795, ISSN 0021-9258

Shafman, T.D., Saleem, A., Kyriakis, J., Weichselbaum, R., Kharbanda, S. & Kufe, D.W. (1995). Defective induction of stress-activated protein kinase activity in ataxia-telangiectasia cells exposed to ionizing radiation. *Cancer Research*, Vol.55, No.15, (August 1995), pp. 3242-3245, ISSN 0008-5472

Shartava, A., Korn, W., Shah, A.K. & Goodman, S.R. (1997). Irreversibly sickled cell beta-actin: defective filament formation. *American Journal of Hematology*, Vol.55, No.2, (June 1997), pp. 97-103, ISSN 0361-8609

Shartava, A., Monteiro, C.A., Bencsath, F.A., Schneider, K., Chait, B.T., Gussio, R., Casoria-Scott, L.A., Shah, A.K., Heuerman, C.A. & Goodman, S.R. (1995). A posttranslational modification of beta-actin contributes to the slow dissociation of the spectrin-protein 4.1-actin complex of irreversibly sickled cells. *The Journal of Cell Biology*, Vol.128, No.5, (March 1995), pp. 805-818, ISSN 0021-9525

Shiozaki, K. & Russell, P. (1995). Cell-cycle control linked to extracellular environment by MAP kinase pathway in fission yeast. *Nature*, Vol.378, No.6558, (December 1995), pp. 739-743, ISSN 0028-0836

Sidorova, J.M., Mikesell, G.E. & Breeden, L.L. (1995). Cell cycle-regulated phosphorylation of Swi6 controls its nuclear localization. *Molecular Biology of the Cell*, Vol.6, No.12, (December 1995), pp. 1641-1658, ISSN 1059-1524

Takashi, R. (1979). Fluorescence energy transfer between subfragment-1 and actin points in the rigor complex of actosubfragment-1. *Biochemistry*, Vol.18, No.23, (November 1979), pp. 5164-5169, ISSN 0006-2960

Thevelein, J. M. (1994). Signal transduction in yeast. *Yeast*, Vol.10, No.13, (December 1994), pp. 1753–1790, ISSN 0749-503X

Torres, J., Di Como, C.J., Herrero, E. & de la Torre-Ruiz, M.A. (2002). Regulation of the cell integrity pathway by rapamycin-sensitive TOR function in budding yeast. *The Journal of biological chemistry*, Vol. 277, No.45, (November 2002), pp. 43495-43504, ISSN 0021-9258

van Dam, H., Wilhelm, D., Herr, I., Steffen, A., Herrlich, P. & Angel, P. (1995). ATF-2 is preferentially activated by stress-activated protein kinases to mediate c-jun induction in response to genotoxic agents. *The EMBO Journal*, Vol. 14, No.8, (April 1995), pp. 1798-1811, ISSN 0261-4189

Veal, E.A., Findlay, V.J., Day, A.M., Bozonet, S.M., Evans, J.M., Quinn, J. & Morgan, B.A. (2004). A 2-Cys peroxiredoxin regulates peroxide-induced oxidation and activation of a stress-activated MAP kinase. *Molecular Cell*, Vol. 15, No.1, (July 2004), pp. 129-139, ISSN 1097-2765

Veal, E.A., Day, A.M. & Morgan, B.A. (2007). Hydrogen peroxide sensing and signaling. Mol. Cell, Vol. 26, No. 1 (April 2007), pp. 1-14, ISSN: 1097-2765

Vilella, F., Herrero, E., Torres, J. & de la Torre-Ruiz, M.A. (2005). Pkc1 and the upstream elements of the cell integrity pathway in *Saccharomyces cerevisiae*, Rom2 and Mtl1, are required for cellular responses to oxidative stress. *The Journal of Biological Chemistry*, Vol. 280, No.10, (March 2005), pp. 9149-9159, ISSN 0021-9258

Wagner, E.F. and Nebreda, A.R. (2009). Signal integration by JNK and p38 MAPK pathways in cancer development. *Nature Reviews. Cancer*, Vol. 9, No.8, (March 2005), pp. 537-549, ISSN 1474-1768

Watson, J.A., Rumsby, M.G. & Wolowacz, R.G. (1999). Phage display identifies thioredoxin and superoxide dismutase as novel protein kinase C-interacting proteins: thioredoxin inhibits protein kinase C-mediated phosphorylation of histone. *The Biochemical journal*, Vol. 343, No.2, (October 1999), pp. 301-305, ISSN 0264-6021

Witte, S., Villalba, M., Bi, K., Liu, Y., Isakov N. & Altman, A. (2000). Inhibition of the c-Jun N-terminal kinase/AP-1 and NF-kappaB pathways by PICOT, a novel protein kinase C-interacting protein with a thioredoxin homology domain. *The Journal of Biological Chemistry*, Vol. 275, No.3, (January 2000), pp. 1902-1909, ISSN 0021-9258

Wullschleger, S., Loewith, R. and Hall, M.N. (2006). TOR signaling in growth and metabolism. *Cell*, Vol. 124, No.3, (February 2006), pp. 471-484, ISSN 0092-8674

Zarebska, Z., Waszkowska, E., Caffieri, S. & Dall'Acqua, F. (2000). PUVA (psoralen + UVA) photochemotherapy: processes triggered in the cells. *Farmaco*, Vol. 55, No.8, (August 2000), pp. 515–520, ISSN 0014-827X

Zhang, A., Shen, Y., Gao, W. & Dong, J. (2011). Role of Sch9 in regulating Ras-cAMP signal pathway in *Saccharomyces cerevisiae*. *FEBS letters*, Vol. 585, No.19, (October 2011), pp. 3026-3032, ISSN 1873-3468

Zheng, M. & Storz, G. (2000). Redox sensing by prokaryotic transcription factors. *Biochemical pharmacology*, Vol. 59, No.1, (January 2000), pp. 1-6, ISSN 0006-2952

15

The Yeast Genes *ROX1, IXR1, SKY1* and Their Effect upon Enzymatic Activities Related to Oxidative Stress

Ana García Leiro, Silvia Rodríguez Lombardero,
Ángel Vizoso Vázquez, M. Isabel González Siso
and M. Esperanza Cerdán
Departamento de Biología Celular y Molecular,
Universidad de A Coruña,
Spain

1. Introduction

Aerobic organisms are characterized by the use of molecular oxygen as the final electron acceptor in the process known as respiration. In the inner membrane of mitochondria the four respiratory complexes transport the electrons and protons from FADH and NAD(P)H to oxygen and produce H_2O. This mechanism is coupled to energy generation, but incomplete reduction of O_2 causes the appearance of reactive oxygen species (ROS) such as hydrogen peroxide (H_2O_2), the hydroxyl radical (OH·) and the superoxide anion ($O_2^{·-}$). ROS are highly reactive in the cell and interact with nucleic acids, proteins and lipids, thus causing a wide spectrum of damages. Increase in steady state ROS level leads to oxidative stress (Lushchak, 2011) and stimulates defence systems. Along evolution of aerobic organisms diverse antioxidant strategies were developed. Many proteins have the function of removing ROS or are able to correct the damage caused by them. Glutathione (GSH), a tripeptide formed by cysteine, glutamic acid and glycine, is the major non-protein thiol-based redox buffer present in the cell (Penninckx, 2002; Perrone et al., 2005). Glutathione is synthesized in its reduced form and transformed to the oxidized form (GSSG) by the formation of one inter-molecular disulfide bond. The principal function of glutathione is to maintain the intracellular redox balance, reducing oxidized molecules and detoxifying ROS, xenobiotics and heavy metals (Grant et al., 1996b; Yu & Zhou, 2007). In fact, for long time the GSH/GSSG ratio was used to describe the redox state of the cell. GSH binds to and directly reduces oxidized molecules, but more often GSH is used as a donor of reducing equivalents to other antioxidant enzymes, like glutaredoxins, glutathione peroxidases and glutathione transferases (Avery & Avery, 2001; Garcera et al., 2006; Lillig et al., 2008).

Saccharomyces cerevisiae, with a predominant fermentative metabolism under aerobic conditions, is considered an eukaryote model for exploring the complex response induced by oxidative stress (Li et al., 2009; Lushchak, 2010, 2011). Besides the use of oxidants, like hydrogen peroxide or menadione, other compounds containing metals also induce the oxidative stress response in yeasts cells (Martins et al., 2008; Thorsen et al., 2009). Sugar oxidation re-routing by different metabolic pathways may also influence the oxidative stress response and, on the contrary, the onset of an oxidative stress response may open previously-blocked metabolic pathways in yeasts (González-Siso et al., 2009).

In *S. cerevisiae*, the Yap family of b-ZIP proteins is involved in a variety of stress-related programs, including the response to DNA damage and oxidative, osmotic and toxic metal stresses. To sum up, functionally, Yap1 is the major regulator of oxidative stress response, Yap2 of cadmium stress, Yap4 and Yap6 of osmotic stress and Yap8 of arsenic stress (Thorsen et al., 2009). We have recently found that in *S. cerevisiae* the transcriptional factor Ixr1 is also related to the oxidative stress response. Moreover, a cross-regulation, affecting transcription, exits between Ixr1 and Rox1, which is the aerobic transcriptional repressor of hypoxic genes (Castro-Prego et al., 2010a). Ixr1 and Sky1 are both related in mediating the cyto-toxicity of the anticancer drug cisplatin (cis-Diaminodichloroplatinum) in yeasts. Cisplatin-induced cell toxicity is also associated with oxidative stress, redox state unbalance, impairment of energetic metabolism and apoptosis (Martins et al., 2008).

As previously described, in aerobic organisms there are multiple connections between the oxidative stress-response, changes in metabolic pathways related to energy production and oxygen utilization and the onset of a cellular response elicited by metals, metalloids or therapeutic compounds containing metals such as cisplatin. In this work we have overtaken a study trying to obtain integrative information about the role of the yeast genes *IXR1*, *ROX1* and *SKY1* in the oxidative stress response induced by As (V), Cd (II) and cisplatin in terms of modulation of four enzymatic activities. The four enzymatic activities tested are glucose-6-phosphate dehydrogenase (G6PDH) that is related to the pentose phosphate pathway (PPP), catalase (CAT) that breaks down H_2O_2 into O_2 and H_2O, glutathione reductase (GLR) and thioredoxin reductase (TRR). The Figure 1 summarizes the metabolic role of the enzymatic activities analyzed and their relationships.

1.1 The relationships between Ixr1, Rox1 and the oxidative stress response

In *S. cerevisiae*, adaptation to environmental signals requires the transcriptional regulation of multiple genes organized in regulons controlled by specific transcriptional regulators. Rox1 and Ixr1 are two yeast transcriptional regulators, which share several structural and functional characteristics in common. Structurally, both contain HMG (high-mobility group) domains, which bind to and bend DNA (Deckert et al., 1995; Deckert et al., 1999; McA'Nulty et al., 1996). Functionally, both control genes that are expressed at higher levels when oxygen is low (hypoxia) or absent (anoxia) than during normoxia (Bourdineaud et al., 2000; Castro-Prego et al., 2010a, 2010b; Kastaniotis & Zitomer, 2000; Klinkenberg et al., 2005; Lambert et al., 1994; Zitomer & Lowry, 1992; Zitomer et al., 1997;). Moreover, a transcriptional cross-regulation between the genes *ROX1* and *IXR1* has been reported (Castro-Prego et al., 2010b). During aerobic growth, low levels of *IXR1* expression are maintained by Rox1 repression through the general co-repressor complex Tup1–Ssn6. Ixr1 is also required for hypoxic repression of *ROX1* and binds to its promoter (Castro-Prego et al., 2010b).

Interestingly, it has been previously reported that low-oxygen levels induce an oxidative stress response accompanied by a rise in ROS levels in *S. cerevisiae* (Dirmeier et al., 2002). Among the evidences of oxidative stress during this transient state are DNA oxidation and

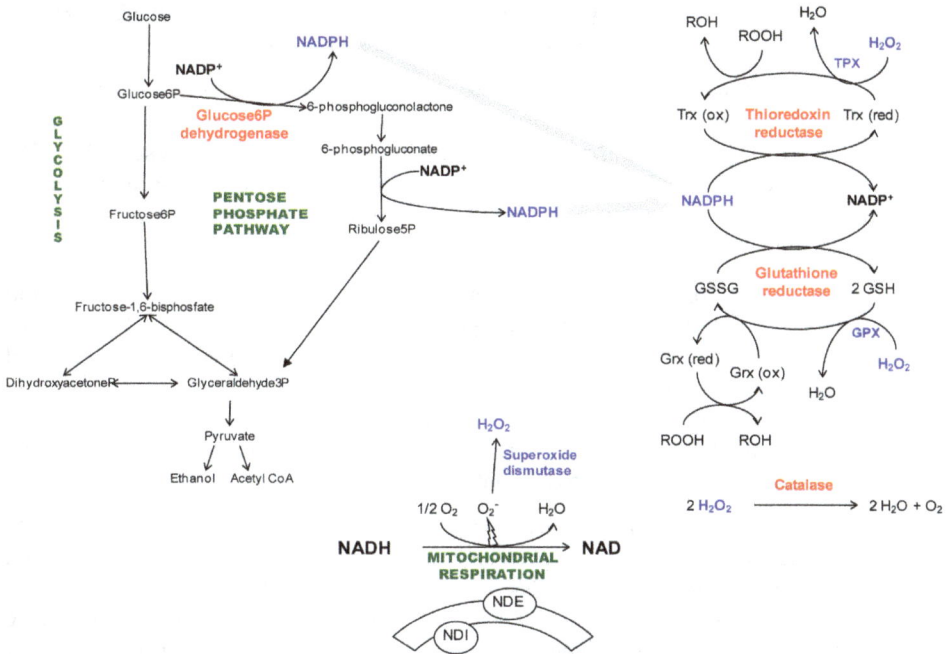

Fig. 1. Metabolic pathways producing and consuming NAD(P)H and connections to the stress response.

selective protein carbonylation. Only certain proteins, such as glyceraldehyde-3-phosphate dehydrogenase, pyruvate decarboxylase, enolase and aconitase, are the targets of oxidants generated during the shift from normoxia to anoxia. The same proteins are also modified during direct exposure of yeast cells to hydrogen peroxide. Besides, *SOD1* (encoding Cu/Zn superoxide dismutase) expression initially declines and then increases during the shift to anoxia, indicating an oxidative stress response (Dirmeier et al., 2002).

Several connections between the transcriptional regulators Rox1, Ixr1 and the yeast response to oxidative stress have been shown. Peroxiredoxins, a family of antioxidant enzymes, play an important role in the cellular defence against oxidative and nitrosative stresses. They have peroxidase and peroxynitrite reductase activities supported by thioredoxin, cyclophilin and glutaredoxin, as well as other electron donors. In *S. cerevisiae*, the transcription of *TSA2*, encoding for peroxiredoxin, is regulated by transcriptional activators, like Yap1 or Skn7, which respond to oxidative signals, but also by Rox1 and the Rox1 transcriptional activator Hap1 (Wong et al., 2003).

In a transtriptome approach comparing wild type and *Δrox1* null strains, several genes involved in mitigating oxidative stress, including *CTT1* (catalase T), *SOD1* and *TSA1* (thioredoxin peroxidase), are up-regulated in absence of Rox1. It is believed that they are not directly repressed by Rox1 because these genes are down-regulated under anoxia, when Rox1 levels diminish; but probably, they change their expression by complex interactions of regulatory networks affected by Rox1 (Lai et al., 2006). Rox1 also appears to play a role in the control of redox balance through the genes *GPM2*, *GMP3* and *CDC19* of the late steps of glycolysis and *ADH1* or *ADH5* of ethanol biosynthesis (Lai et al., 2006).

It has been proposed that caloric restriction extends life span by a process that initially raises ROS levels. But, in turn, it produces protection from acute doses of oxidant, providing adaptation, and Rox1 is active during this adaptive response (Kelley & Ideker, 2009). The mechanisms by which Rox1 is activated after mild pre-treatment with oxidants are unknown, but it has been proposed that a fall in heme levels via degradation induced by hydrogen peroxide may be the signal (Kelley & Ideker, 2009).

During anaerobic growth, *S. cerevisiae* requires both a sterol (at or beyond zymosterol) and unsaturated fatty acids, which must be exogenously supplied. During anaerobiosis the genes required for sterol import and nearly all of the genes involved in the latter portion of sterol biosynthesis (beyond farnesylpyrophosphate) are induced. Many of them are regulated by Rox1. It has been recently shown that oxidative stress triggers repression of *ERG2* and *ERG11* transcription, two genes that are necessary for sterol biosynthesis and this response is partially dependent on Rox1 (Montañés et al., 2011).

About the regulator Ixr1, there are also some reports directly or indirectly related to oxidative stress in yeasts. Hypoxic expression of *SRP1* (*TIR1*) is dependent on Ixr1 and Yap1, the main regulator of the oxidative stress response. Besides, the effect of *Δixr1* is epistatic to *Δyap1* (Bourdineaud et al., 2000). *IXR1* expression is moderately activated by H_2O_2 and this induction is Yap1-dependent (Castro-Prego et al., 2010a). In multi-cellular eukaryotes connexions between the oxidative stress response and *IXR1* homologues also exist. Thus, in surgically resected hepatocellular carcinomas, TRX, a disulfide-reducing intracellular tioredoxin that functions as a cellular defence mechanism against oxidative stress, and HMG proteins type 1, with significant homology to the yeast protein Ixr1, are co-overexpressed when compared to normal tissue (Kawahara et al., 1996). Besides, active transcription of peroxiredoxins is dependent on Ets transcription factors and HMGB1 was shown to function as a coactivator through direct interactions with these Ets transcription factors (Shiota et al., 2008). By other hand, the protein HMGB1 was identified as a substrate of glutaredoxin that reduces the disulfide bond between Cys23 and Cys45. The conformational changes following this event may serve as a basis for redox-dependent control of gene expression, DNA replication, protection and repair (Hoppe et al., 2006).

1.2 The role of Ixr1 and Sky1 in the sensitivity to cisplatin

The yeast *S. cerevisiae* has been used as a simple eukaryotic model to identify genes related to cisplatin-sensitivity or cisplatin-resistance (Fox et al., 1994; Huang et al., 2005; Schenk et al., 2001, 2003). Among the genes that confer cisplatin-resistance are *IXR1* and *SKY1*.

Ixr1 is a yeast HMG-domain protein which binds the major DNA adducts formed with cisplatin (Brown et al., 1993). It has been demonstrated than in the excision repair mutants *Δrad2*, *Δrad4* and *Δrad14*, deletion of *IXR1* does not increase the resistance of *S. cerevisiae* cells to cisplatin (McA´Nulty et al., 1996). This result gives support to the hypothesis that Ixrl and other HMG-domain proteins can block repair of the major cisplatin-DNA adducts *in vivo* (McA´Nulty & Lippard, 1996). Therefore, the cisplatin sensitivity in cells expressing Ixr1 might be caused by an architectural role of this HMG-protein in the chromatin assembles that protects the area from the machinery of DNA repair, thus inducing cell death. The non-histone chromosomal protein high mobility group 1 (HMG1), which is ubiquitously expressed in higher eukaryotic cells, preferentially binds to cisplatin-modified DNA. HMG1 is overexpressed in cisplatin-resistant cell lines from human epidermoid cancer and the specific factor CTF/NF-1 regulates HMG1 gene expression (Nagatani et al., 2001).

Sky1 is a yeast rich serine-arginine (SR) protein-specific kinase and experimental data suggest that its kinase function is essential in the cytotoxicity of cisplatin (Schenk et al., 2001). SR protein-specific kinases and the SR proteins that they phosphorylate are thought to be key regulators of RNA processing and, in mammalian cells, alternative splicing through multiple mechanisms (Siebel et al., 1999). *SKY1* mRNA levels do not change after treatment with cisplatin, which suggests that its expression could be regulated by autophosphorylation or posttranslational modification by upstream components (Schenk et al., 2001). In *Δsky1* cells, lower cisplatin accumulation or DNA platination were not observed, which indicates that the resistance to cisplatin is not related to decreased drug import or increased drug export (Schenk et al., 2002). Besides, *Δsky1* cells display a mutator phenotype, which suggests that Sky1 might play a significant role in specific DNA repair pathways (Schenk et al., 2002). SRPK1, the human homologue of Sky1, is predominantly found in the testis, where it phosphorylates protamine 1 as well as a cytoplasmic pool of other SR proteins (Papoutsopoulou et al., 1999). Protamines are small highly basic proteins that replace histones during spermatogenesis, resulting in extreme chromatin condensation (Oliva & Dixon, 1991). In *S. cerevisiae* Sky1 is a key regulator of inward transport of polyamines such as putrescine, spermine and spermidine (Erez & Kahana, 2001) and it has been suggested that SRPK1 might have a role in spermatogenesis by direct or indirect regulation of intracellular concentrations of polyamines (Schenk et al., 2004). Inactivation of *SRPK1* using antisense oligo-deoxynucleotides directed against the translation initiation site of its mRNA induces cisplatin resistance in a human ovarian carcinoma cell line and *SRPK1* heterologous expression is able to complement the cisplatin-resistant phenotype of a *Δsky1* yeast strain (Schenk et al., 2001).

1.3 Cellular response to cisplatin and oxidative stress

Several connections exist between cellular responses to cisplatin and oxidative stress. Deletion of the yeast *QDR3* gene, encoding for a drug/H+ antiporter, confers sensitivity to cisplatin while its over-expression confers resistance to this drug in yeast (Tenreiro et al., 2005). It has been shown that *QDR3* transcription is up-regulated in response to polyamines by a mechanism dependent on the oxidative stress transcriptional regulator Yap1 (Teixeira et al., 2010). *NPR2* (nitrogen permease regulator 2) is a gene whose disruption confers resistance to cisplatin and hypersensitivity to cadmium chloride (Schenk et al., 2003). In turn, Cd (II) is related to the onset of oxidative stress in yeast cells as summarized in section 1.4.

The clinical use of cisplatin is highly limited by its nephrotoxicity and this effect is caused by cisplatin-induced mitochondrial damage in kidney. It has been proposed that oxidative stress exists in the early stage of cisplatin-induced nephrotoxicity and also in hepatotoxicity (Iraz et al., 2006; Mansour et al., 2006; Pratibha et al., 2006; Satoh et al., 2000). In rats, mitochondrial dysfunction in kidney and liver was evidenced after cisplatin treatment. Impairment of mitochondrial function and structure, depletion of the antioxidant defence system and cellular death by apoptosis were observed (Santos et al., 2008; Martins et al., 2008). In rats, cisplatin increased lactate dehydrogenase and acid phosphatase activities whereas, the activities of malate dehydrogenase, glucose-6-phosphatase, superoxide dismutase and CAT, as well as phosphate transport significantly decreased (Khan et al., 2009).

Consequently, there are reports of different antioxidants, which protect cells from the oxidative damage caused by cisplatin and whose use represents a possible strategy to minimize the nephrotoxicity induced by this antitumor agent. The hydroxyl radical scavenger dimethylthiourea (DMTU) shows a protective effect against cisplatin-induced alterations of renal mitochondrial bioenergetics, redox state and oxidative stress defence (Santos et al., 2008). Green tea consumption increases the activities of the enzymes of carbohydrate metabolism, brush-border membrane, oxidative stress and phosphate transport (Khan et al., 2009). Carvedilol, a beta-blocker with strong antioxidant properties, prevents lipid peroxidation, oxidation of cardiolipin, oxidation of protein sulfhydryls, depletion of the non-enzymatic antioxidant defence and increased activity of caspase-3 (Rodrigues et al., 2011).

ROS production in eukaryotic cells is also characterized by their ability to cause damage to DNA. The cytosolic serine peptidase tripeptidyl-peptidase II (TPPII) translocates into the nucleus of most tumor cell lines in response to gamma-irradiation and ROS production and also after treatment with several types of DNA-damaging drugs including the DNA cross-linker cisplatin (Preta et al., 2010). This demonstrates its participation in mechanisms elicited by both treatments and suggests common connections between ROS production and DNA damage. Antioxidants are also able to prevent DNA damage. Thus, lutein, the second most prevalent carotenoid in human serum and also abundant in green vegetables, reduces the formation of crosslinks and chromosome instability induced by cisplatin (Serpeloni et al., 2010). Lutein also increases GSH levels without affecting CAT activity (Serpeloni et al., 2010).

1.4 Cadmium and arsenate toxicity, the role of oxidative stress

Several metals are toxic for the cell and it has been suggested that one of the mechanisms of metal toxicity might be the induction of oxidative stress (Stohs & Bagchi, 1995). In yeast, Cd (II) has been shown to induce lipid peroxidation and oxidative stress (Brennan & Schiestl, 1996; Howlett & Avery, 1997). As (III) does not produce these effects in a wild type strain, but oxidative stress and lipid peroxidation were detected in $\Delta yap8$ or $\Delta yap1$ mutants (Menezes et al., 2008), suggesting that As (III) also enhances ROS levels in yeast. Since As (V) is reduced to As (III) inside the cell by the action of the arsenate reductase Acr2, using GSH and glutaredoxin as electron donors (Mukhopadhyay & Rosen, 1998; Mukhopadhyay et al., 2000) similar alterations to those produced by As (III) are expected. Different metals have diverse mechanisms to induce oxidative stress (Wysocki & Tamás, 2010) and therefore we have focused on reviewing published data and hypotheses about the mechanisms that induce the oxidative stress response after treatment with Cd (II) or As (V), both with a high capacity to bind thiols. We might consider three ways by which the metals could generate the oxidative stress. First, the metal may stimulate directly or indirectly the generation of ROS; second, it may cause depletion of antioxidant pools; third, it may inhibit specific enzymes necessary to maintain the redox balance in the cell (Beyersmann & Hartwig, 2008; Ercal et al., 2001; Stohs & Bagchi, 1995).

Contrary to As (III or V), Cd (II) is redox-inactive and therefore only indirect mechanisms are possible for the generation of ROS. It has been proposed that redox-inactive metals may alter the Fe metabolism (Kitchin & Wallace, 2008a), increasing the levels of free Fe in the cell, which could be involved in Fenton-type reactions and increase ROS levels. Regarding the question whether As (III), Cd (II) and various oxidants might have similar toxicity profiles in *S. cerevisiae*, a set of genes, which products are responsible for metal tolerance (Thorsen et al., 2009), was compared to other genes previously reported to mediate tolerance to a number of ROS-generating agents including hydrogen peroxide, menadione, cumene hydroperoxide, diamide and linoleic acid 13-hydroperoxide (Thorpe et al., 2004). Some of the genes required for metal tolerance are also necessary for oxidative stress tolerance. However, from this comparison it was not possible to conclude the source and type of ROS that As (III) and Cd (II) generate in the cell and that cause their toxicity (Thorsen et al., 2009).

In relation to metal toxicity and the transcriptional regulators of metal-compounds transport, it has been reported that Rox1 represses *FET4* expression in aerobic conditions causing up-regulation in the *S. cerevisiae Δrox1* mutant and increased Cd (II) toxicity (Jensen & Culotta, 2002). Fet4 is the major importer of Cd (II) into the cell during hypoxic growth (Jensen & Culotta, 2002). GSH is the main antioxidant molecule in yeast cells but it is also used for chelating metals (Thorsen et al., 2009; Wysocki & Tamás, 2010). Cellular mechanisms for Cd (II) or As (III) detoxification depend on their chelation with glutathione, which facilitates their export outside the cell or their sequestration into vacuole. As (III) is exported by the Acr3 transporter (Ghosh et al., 1999; Wysocki et al., 1997). It has been proposed that Yor1 mediates Cd (II) efflux in the form of Cd $(GS)_2$ (Cui et al., 1996; Nagy et al., 2006). Cd (II) and As (III) GSH-conjugates are imported into vacuole by Ycf1. This ABC transporter represents the major pathway for vacuolar sequestration of metals in *S. cerevisiae* (Paumi et al., 2009) although their homologs Bpt1 and Vmr1 might also play a minor role in Cd (II) detoxification (Wysocki & Tamás, 2010). In spite of the fact that metal detoxification requires GSH consumption, it is not probable that always metal treatment causes GSH depletion. The intracellular GSH concentration is in the millimolar range in yeast, whereas Cd (II) is toxic in the micromolar range (Lafaye et al., 2005). However, As (V) is toxic in the millimolar range and therefore it cannot be excluded that some metals could decrease the GSH pool to an extent where GSH-dependent enzyme activities, such as glutathione peroxidases, glutathione S-transferases and glutaredoxins, might be affected. Other argument against this mechanism of Cd (II) or As (V) induction of the oxidative stress response by depletion of the GSH antioxidant pool is the observation that GSH levels strongly increase in response to Cd (II) (Lafaye et al., 2005) and As (III) (Thorsen et al., 2007) exposure.

Regarding Cd (II) or As (IV) inhibition of enzymes, which are necessary for redox balance and protection against oxidative stress, several data have been published. The metals can inhibit these enzymes by different mechanism. They can bind specific thiols that take part of the active site, change the redox state of the protein or diminish enzymatic activity by other complex interactions. Cd (II) inhibits human thiol transferases (GLR, TRR, and thioredoxin) *in vitro*, possibly by binding to vicinal cysteines in their active sites (Chrestensen et al., 2000). Cd (II) may also displace Zn and Ca ions from metalloproteins (Faller et al., 2005; Schutzendubel & Polle, 2002; Stohs & Bagchi, 1995) and zinc-finger proteins (Hartwig, 2001). As (III) has been shown to interact with TRR, pyruvate dehydrogenase and many other proteins (Kitchin & Wallace, 2008b; Menzel et al., 1999; Samikkannu et al., 2003; Wang et al., 2007). Besides, certain proteins may be more susceptible to As (III)-induced protein oxidation than to direct binding of As (III) to critical thiols (Samikkannu et al., 2003).

In this study we have tested the role of the genes *ROX1*, *IXR1* and *SKY1*, as well as their interconnections, in the yeast response to oxidative stress elicited by As (V), Cd (II) and cisplatin and in terms of modulation of glucose-6-phosphate dehydrogenase, catalase, glutathione reductase and thioredoxin reductase enzymatic activities.

2. Materials and methods

2.1 Yeast strains and construction of double knock-outs

Yeast cells from the strain BY4741 (*MATa his3Δ1 leu2Δ0 met15Δ0 ura3Δ0*) and its derivatives *Δixr1* (*MATa his3Δ1 leu2Δ0 met15Δ0 ura3Δ0 YKL032c::kanMX4*), *Δrox1* (MATa *his3Δ1 leu2Δ0 met15Δ0 ura3Δ0 YPR065W::kanMX4*) and *Δsky1* (MATa h*is3Δ1 leu2Δ0 met15Δ0 ura3Δ0 YMR216C::kanMX4*) were obtained from EUROSCARF (http://web.uni-frankfurt.de/fb15/mikro/euroscarf/). The *Δixr1Δrox1* (MATa *his3Δ1 leu2Δ0 met15Δ0 ura3Δ0 YKL032c::kanMX4 YPR065W::URA3*) and *Δixr1Δsky1* (MATa *his3Δ1 leu2Δ0 met15Δ0 ura3Δ0 YKL032c::kanMX4 YMR216C::URA3*) double knock-out strains were obtained by the one-step replacement method and verified by PCR as follows. The plasmid YEplac 195 (Gietz & Sugino, 1998) digested with *EcoRI* was used as template to amplify by PCR a linear fragment containing the *URA3* gene and two flanking regions of homology to the 5´and 3´ends of the selected open reading frame (ORF) as explained in Figure 2A

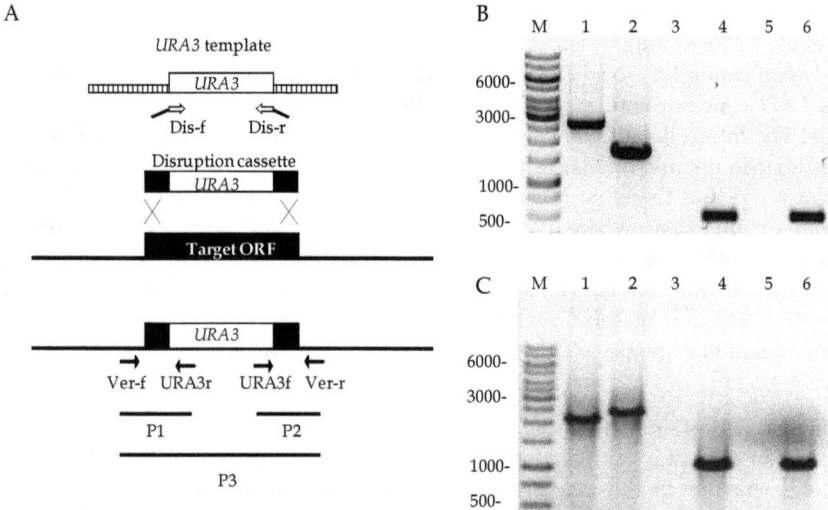

Fig. 2. (A) Strategy for construction of knock-out strains. B) Verification of the BY4741-*Δixr1Δsky1* strain C) verification of the BY4741-*Δixr1Δrox1* strain. In B and C the positions are indicated as follows, M, size marker (bp ladder); 1 and 2, P3; 3 and 4, P1; 5 and 6, P2. The reactions with the starting strain, BY4741-*Δixr1*, are shown in 1, 3, 5 and with the double nulls in 2, 4, 6.

Target ORF: *SKY1*						
Primer name	Code	Sequence	HP SKY1	HP URA3	PCR products	PCR size bp
Dis-f	ECV741	atgggttcatcaattaactatcctgggtttgC CACCTGACGTCTAAGAAACC	+1	-312	P1 SKY1 P1 Δsky1::URA3	- 481
Dis-r	ECV742	tcaatgtctttatgatcgcggacttcttcCCT TTAGCTGTTCTATATGCTGC	+2229	+996	P2 SKY1 P2 Δsky1::URA3	- 483
Ver-f	ECV717SR	GATCACCTGGCGCTGAGAA	-97	-	P3 SKY1 P3 Δsky1::URA3	2562 1701
Ver-r	ECV718SR	CGAGTATGGATTCAAAAACC GC	+2465	-		
URA3-f	ECV716SR	GAGAAGATGCGGCCAGCA	-	+780		
URA3-r	ECV715SR	GGATGAGTAGCAGCACGTTC C	-	+41		
Target ORF: *ROX1*						
Primer name	Code	Sequence	HP ROX1	HP URA3	PCR products	PCR size bp
Dis-f	ECV688AV	atgaatcctaaatcctctacacctaagattCC TTTAGCTGTTCTATATGCTGC	+1	-312	P1 ROX1 P1 Δrox1::URA3	- 1014
Dis-r	ECV689AV	tcatttcggagaaactaggctagttttagcCC ACCTGACGTCTAAGAAACC	+1107	+996	P2 ROX1 P2 Δrox1::URA3	- 1023
Ver-f	ECV698AV	GTGATCTTCGGCTCGGC	-557	-	P3 ROX1 P3 Δrox1::URA3	2091 2352
Ver-r	ECV700AV	TTGTACTTGGCGGATAATGC	+1534	-		
URA3-f	ECV699AV	AAGAGATGAAGGTTACGATT GGT	-	+570		
URA3-r	ECV701AV	ATATCTTGCAGTCCATCCTCG	-	+254		

Table 1. Oligonucleotides used in the construction and verification of knock-out strains. HP, Hybridization position. Oligo position and PCR product designations are as defined in Figure 2A.

After transformation of the *S. cerevisiae* strain BY4741-*Δixr1* with these fragments, cells were selected in complete media without uracyl (CM-Ura) and supplemented with 40 mg/mL geneticin. The correct replacement in the *S. cerevisiae* genome was verified by PCR as previously described (Tizón et al., 1999). Genomic DNAs isolated from the BY4741-*Δixr1* and the null candidates were amplified with two pairs of primers. Internal primers, URA3f and URA3r, were designed for annealing divergently inside *URA3* and external primers, were designed for convergent annealing in the sequences of the *S. cerevisiae* genome, flanking to the knock-out ORFs, but external to the regions of homology used for the recombination event. The strategy and results obtained in the verification of the replacement of the ORFs with the URA3 marker is summarized in Figure 2 and the primers used are shown in Table 1.

2.2 Yeast treatments with arsenate, cadmium and cisplatin

Stress treatments with arsenate, cadmium and cisplatin were performed as follows. Cd (II) was added to the media in the form of cadmium sulphate 8/3-hydrate and As (V) in the form of sodium arsenate dibasic hepta-hydrate. Arsenate and cadmium were added in concentrations of 500 µM and 10 µM respectively to the culture media, and cells were collected when OD_{600nm} reached 1. Cisplatin was added in concentration 150 µM in DMSO when cultures reached an OD_{600nm} of 1. In this case, after addition, cells were incubated during 4 h before protein extraction.

2.3 Determination of enzymatic activities

For the determination of enzymatic activities, protein extracts were prepared as follows. Cultures were grown in Erlenmeyer flask (with a ratio flask-capacity/volume of medium of 5) at 30 °C in YPD medium. The cells from 20 mL of culture were collected by centrifugation at 3000 x g and resuspended in 1 mL of buffer A (0.2 M Tris-HCl (pH=7.0), 0.3 M $(NH_4)_2SO_4$, 10 mM $MgCl_2$, 1 mM EDTA, 10% glycerol) per gram of wet weight. Cells were broken by vortexing with glass beads (45 µm) in 10 seconds pulses. After centrifugation at 8000 x g during 15 minutes, the supernatant was used for enzymatic determinations. For quantification of G6PDH, GLR and TRR activities, proteins were frozen at -80°C until assays were performed. Protein extracts for measurement of CAT activity were immediately used. Protein concentration was measured by the method of Bradford (1976), using bovine serum albumin as a standard.

Enzymatic activities were determined following the methods established by Smith et al., (1988) for GLR; Holmgren & Björnstedt (1995) for TRR; Kuby & Noltmann (1966) for G6PDH and Aebi (1984) for CAT. All protocols, with the exception of CAT, were scaled down to reduce the final volume in order to measure absorbance in a 96-well microplate, using a GENios spectrophotometer (TECAN). CAT activity was assayed using a UV-1700 PharmaSpec spectrophotometer (Shimadzu).

Both methods for measuring GLR and TRR enzymatic activities were based in the reduction of 5-5´-ditio-bis (2-nitrobenzoic acid) or DTNB to 2-nitro-5-tiobenzoic acid or TNB. For measurement of GLR activity, the two coupled reactions were the following: $NADPH+H^++GSSG \rightarrow NADP^++2GSH$; $GSH+DTNB \rightarrow GSTNB+TNB$. The reaction mix contained (final volume = 100 µL) 0.1 M phosphate buffer pH 7.5, 0.5 mM EDTA, 0.75 mM DTNB, 0.1 mM NADPH, 1 mM oxidized glutathione (GSSG) and the protein extracts added in aliquots of 5 and 10 µL in the two respective replicates. The reaction was started by the addition of GSSG (5 µL, 20 mM). The increase of absorbance was recorded at λ = 412 nm, and at 24°C, during 2 minutes. Specific activity, enzymatic units (EU)/mg, was defined as µmol of TNB formed per minute and per mg of protein and it was calculated using the Lambert-Beer law, taking into account that the extinction coefficient of TNB is $13.6*10^3$ $M^{-1}*cm^{-1}$.

Determination of TRR activity was based in the comparison of the reduction of DTNB to TNB in the samples to a standard curve made, by triplicate, with different quantities (0, 10,

20, 30, 40, 50 and 70 μL) of 20 nM TRR from mammals, diluted in the moment of use from a stock of 88.1 UE/mg protein in TE buffer with 100 μg/mL of bovine serum albumin. The reaction mix contained (final volume = 120 μL): 0.3 M HEPES pH 7.6, 0.07 M EDTA, 13.3 mg/mL NADPH, 3.3 mg/mL insulin, 5 μM thioredoxin from *E. coli* and the protein samples added in quantities varying between 10 and 70 μL. Reaction was started by adding thioredoxin (10 μL, 60 μM) and continued for 20 minutes at 37 °C. The reaction was stopped by adding 50 μL of 0.4 mg/mL DTNB in 6 M guanidine hydrochloride 0.2 M Tris-HCl pH 8. Absorbance was read at $\lambda = 412$ nm. The specific activity (EU/mg) corresponded to μmol of TNB formed per minute and per mg of protein.

Determination of G6PDH activity was based in the following reaction: D-glucose-6-phosphate + NADP$^+$ → 6-phosphoglucolactone + NADPH + H$^+$. The reaction mix contained 100 μL of 0.1 M glyclglicine buffer (pH 8.0) 0.03 M glucose-6-phosphate; 0.01 M NADP$^+$ and 0.15 M magnesium sulphate. Aliquots of 5 or 10 μL of the protein sample were added. Increase of absorbance was quantified in the spectrophotometer at $\lambda = 340$ nm and at 30°C during 4 minutes. The specific activity (EU/mg) corresponded to μmol of generated product (NADPH) per minute and per mg of protein, and it was calculated using the Lambert-Beer law, taking into account that the extinction coefficient of NADPH (E_{NADPH}) is $6.22*10^3$ M$^{-1}*$cm^{-1}.

Quantification of CAT activity was based in the following reaction: $2H_2O_2$ → $2H_2O + O_2$. Two reaction mixes were made. The first one (A) contained 30 mM of H_2O_2 in 50 mM phosphate buffer pH 7. The second one (B) contained a dilution of the protein sample in 50 mM phosphate buffer pH 7. Mix B was prepared in several dilutions, and these should be used in a time not longer than 5-10 minutes after their preparation. Reaction started when 0.33 mL of the A mix were added to 0.67 mL of the B mix. Decrease of absorbance was recorded in the spectrophotometer at $\lambda = 240$ nm during 30 seconds and at room temperature. The specific activity (EU/mg) corresponded to 1 μmol of consumed substrate (H_2O_2) per minute and per mg of protein. Concentration of the substrate was determined by applying the Lambert-Beer law, taking into account that $E_{H_2O_2}$ is $3.94*10^3$ M$^{-1}*$cm^{-1}.

2.4 Statistical analyses

Data were expressed as mean ± standard deviation (SD). The statistical significance of differences between means was evaluated by one-way ANOVA with Tukey post-test or by Kruskal-Wallis test with Dunn post-test, both at the 95% confidence level. The program GraphPad Instat was used.

3. Results and discussion

The *S. cerevisiae* response to As (V), Cd (II) and cisplatin was evaluated in terms of modulation of enzymatic activities related to the PPP (G6PDH); break down of H_2O_2 into O_2 and H_2O (CAT); or glutathione and thioredoxin reduction (GLR and TRR). The BY4741 wild type strain and its derivatives *Δixr*, *Δrox1*, *Δsky1*, *Δixr1Δrox1* and *Δixr1Δsky1* strains, which are described in the Materials and Methods section, were used. At least two independent cultures of each case were performed and enzymatic activities were measured in duplicates from each. Multiple statistical comparisons of means were performed classifying the data of the six strains by enzyme activity and treatment. Significant differences found in each case are outlined in Figures 3 to 5 and Tables 2A and 2B.

Treatment: As (V)

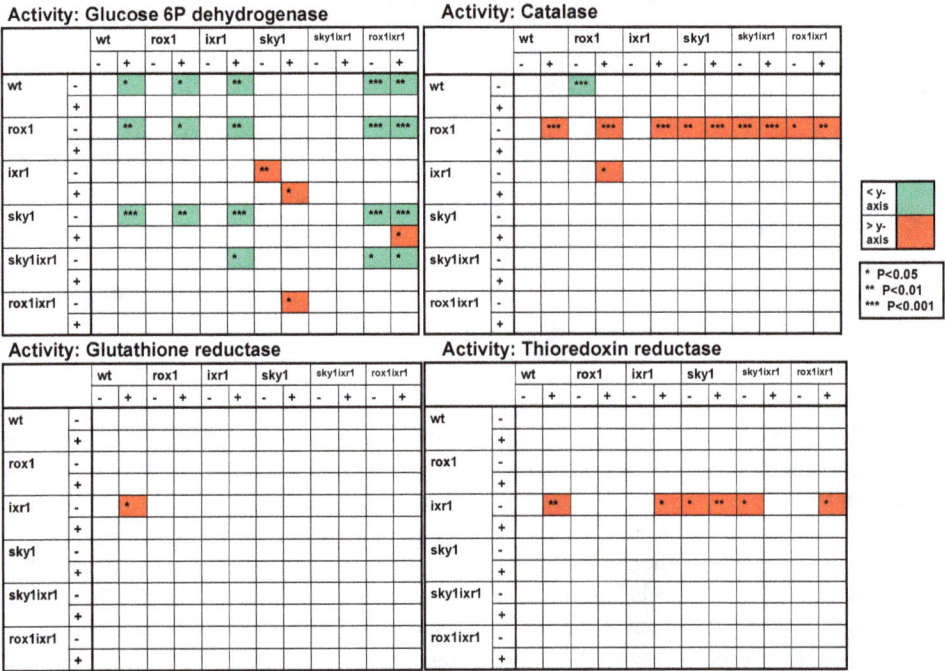

Activity: Glucose 6P dehydrogenase

	wt −	wt +	rox1 −	rox1 +	ixr1 −	ixr1 +	sky1 −	sky1 +	sky1ixr1 −	sky1ixr1 +	rox1ixr1 −	rox1ixr1 +
wt −		*				*		**			***	**
wt +												
rox1 −		**				*		**			***	***
rox1 +												
ixr1 −							**					
ixr1 +								*				
sky1 −		***		**		***					***	***
sky1 +												*
sky1ixr1 −						*					*	*
sky1ixr1 +												
rox1ixr1 −								*				
rox1ixr1 +												

Activity: Catalase

	wt −	wt +	rox1 −	rox1 +	ixr1 −	ixr1 +	sky1 −	sky1 +	sky1ixr1 −	sky1ixr1 +	rox1ixr1 −	rox1ixr1 +
wt −			***									
wt +												
rox1 −	***		***		***	**	***	***	***		*	**
rox1 +												
ixr1 −					*							
ixr1 +												
sky1 −												
sky1 +												
sky1ixr1 −												
sky1ixr1 +												
rox1ixr1 −												
rox1ixr1 +												

Activity: Glutathione reductase

	wt −	wt +	rox1 −	rox1 +	ixr1 −	ixr1 +	sky1 −	sky1 +	sky1ixr1 −	sky1ixr1 +	rox1ixr1 −	rox1ixr1 +
wt −												
wt +												
rox1 −												
rox1 +												
ixr1 −		*										
ixr1 +												
sky1 −												
sky1 +												
sky1ixr1 −												
sky1ixr1 +												
rox1ixr1 −												
rox1ixr1 +												

Activity: Thioredoxin reductase

	wt −	wt +	rox1 −	rox1 +	ixr1 −	ixr1 +	sky1 −	sky1 +	sky1ixr1 −	sky1ixr1 +	rox1ixr1 −	rox1ixr1 +
wt −												
wt +												
rox1 −												
rox1 +												
ixr1 −		**					*	*	**	*		*
ixr1 +												
sky1 −												
sky1 +												
sky1ixr1 −												
sky1ixr1 +												
rox1ixr1 −												
rox1ixr1 +												

Legend:
< y-axis
> y-axis
* P<0.05
** P<0.01
*** P<0.001

Fig. 3. Statistical comparison of the effects of As (V) treatment on the four enzymatic activities and six strains studied in this work. N=4. Only significant differences are marked.

Figure 3 and Table 2B shows the effect of the treatment with As (V) on the four enzymatic activities in the six strains assayed. Treatment with As (V) caused, as most outstanding results, the following activity-dependent and strain-dependent responses: increase of G6PDH activity in wild type and Δrox1 backgrounds; decrease of CAT activity in the Δrox1 background; decrease of TRR activity in the Δixr1 background.

The treatment with Cd (II) only affected significantly the activity of GLR among the four studied enzymes, which increased in the wild type and Δixr1 backgrounds. However this increase was not observed in other single or double mutants (Figure 4 and Table 2B). The effect of Cd (II) on TRR and G6PDH activities was also statistically analyzed using a non-parametric test without finding significant differences (data not shown).

The effect of the treatment with cisplatin is represented in Figure 5 and Table 2B. The most outstanding results were the decrease of GLR activity in the double mutant Δixr1Δsky1 and the decrease of TRR activity in the double mutant Δixr1Δrox1.

Treatment: Cd (II)

Activity: Glutathione reductase

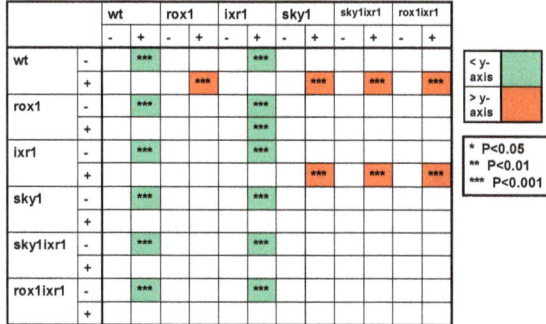

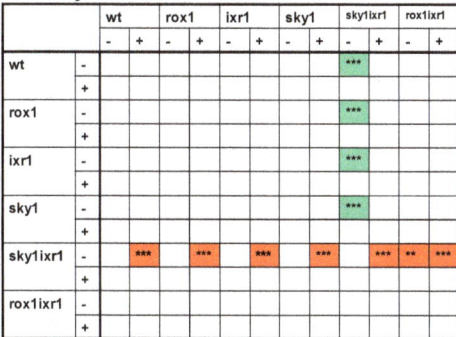

		wt		rox1		ixr1		sky1		sky1ixr1		rox1ixr1	
		-	+	-	+	-	+	-	+	-	+	-	+
wt	-		***				***						
	+				***				***		***		***
rox1	-		***				***						
	+						***						
ixr1	-		***				***						
	+								***		***		***
sky1	-		***				***						
	+												
sky1ixr1	-		***				***						
	+												
rox1ixr1	-		***				***						
	+												

Legend: ◼ < y-axis (green) ◼ > y-axis (orange) — * P<0.05 ** P<0.01 *** P<0.001

Fig. 4. Statistical comparison of the effects of Cd (II) treatment on GLR activity in the six strains studied in this work. N=4. Only significant differences are marked.

Treatment: cisPt

Activity: Catalase

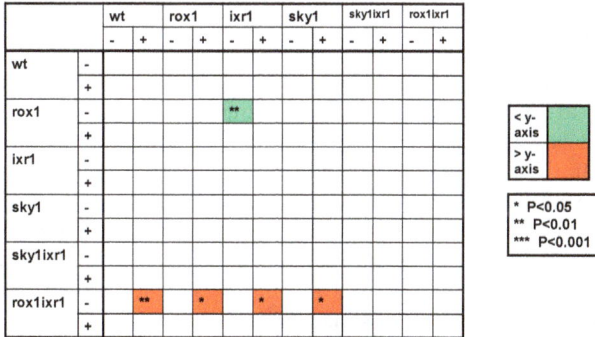

		wt		rox1		ixr1		sky1		sky1ixr1		rox1ixr1	
		-	+	-	+	-	+	-	+	-	+	-	+
wt	-												
	+												
rox1	-						**						
	+												
ixr1	-												
	+												
sky1	-												
	+												
sky1ixr1	-												
	+												
rox1ixr1	-		**				*		*		*		
	+												

Legend: ◼ < y-axis (green) ◼ > y-axis (orange) — * P<0.05 ** P<0.01 *** P<0.001

Activity: Glutathione reductase

		wt		rox1		ixr1		sky1		sky1ixr1		rox1ixr1	
		-	+	-	+	-	+	-	+	-	+	-	+
wt	-									***			
	+												
rox1	-									***			
	+												
ixr1	-									***			
	+												
sky1	-									***			
	+												
sky1ixr1	-		***		***		***		***		***	**	***
	+												
rox1ixr1	-												
	+												

Activity: Thioredoxin reductase

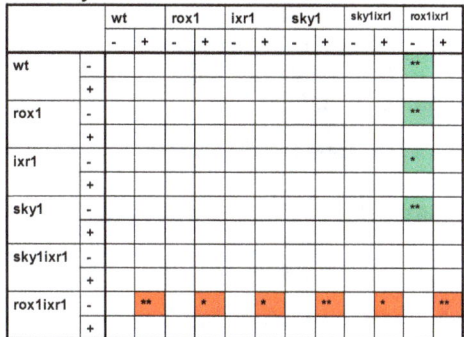

		wt		rox1		ixr1		sky1		sky1ixr1		rox1ixr1	
		-	+	-	+	-	+	-	+	-	+	-	+
wt	-									**			
	+												
rox1	-									**			
	+												
ixr1	-									*			
	+												
sky1	-									**			
	+												
sky1ixr1	-												
	+												
rox1ixr1	-		**		*		*		**		*		**
	+												

Fig. 5. Statistical comparison of the effects of cisplatin treatment on the enzymatic activities in the six strains studied in this work. N=4. Only significant differences are marked.

3.1 Changes in glucose-6-phosphate dehydrogenase activity

Glucose-6-phosphate dehydrogenase is the protein that catalyzes the first step in the oxidative branch of the pentose phosphate pathway (PPP), the conversion of glucose-6-phosphate into ribulose-5-phosphate. Nevertheless, the protein does not seem to be essential, since mutants in its coding gene, ZWF1, can still grow in both respiratory and fermentative carbon sources (Nogae & Johston 1990; Saliola et al., 2007). The enzyme uses $NADP^+$ as a coenzyme, thus converting it to the reduced form NADPH, which is used by proteins with antioxidant functions. In fact, the PPP is the major source of this coenzyme during situations of oxidative stress (Minard et al., 2005). Mutants in ZWF1 also show methionine auxotrophy (Thomas et al., 1991) probably caused by the interconnections between methionine biosynthesis and glutathione biosynthesis.

If we analyzed the results of this study, shown in Figures 3, 4 and 5, by enzyme activity (Table 2) we observed that G6PDH was significantly affected only by As (V) treatment (increase of activity in wild type and $\Delta rox1$ backgrounds). A previous work (Godon et al., 1998) showed that S. cerevisiae treated with H_2O_2 is able to oxidize more glucose through the PPP than through glycolysis in order to obtain NADPH necessary in the oxidative defence reactions. In fact it has been proved that G6PDH mutants are more sensitive to oxidative stress caused by H_2O_2 (Izawa et al., 1998; Junhke et al., 1996).

Why As (V) affects G6PDH activity and probably increases the glucose utilization via PPP, while treatment with Cd (II) or cisplatin does not produce a similar effect is striking since it has been reported that all these treatments stimulate intracellular ROS generation. A possible explanation could be related to the different ways that these metals and derivative compounds use to enter into the cells. Cd (II) enters yeasts cells through proteins involved in the uptake of other bivalent cations, which are essential for cell survival. Zn (II) enters through Zrt1, Mn (II) though Smf1 or Smf2, Fe (II) though Fet4 and Ca (II) though Mid1; all these proteins are also Cd (II) importers (reviewed in Wysocki & Tamas, 2010). In S. cerevisiae the import of cisplatin inside the cell is mediated by the copper transporter Ctr1 and the N-terminal methionine-rich motifs that are dispensable for copper transport play a critical role for cisplatin uptake (Adle et al., 2007). The arsenate As (V) oxyanion is a structural analogue of inorganic phosphate and is taken up through phosphate transporters. Phosphate import into S. cerevisiae is mediated by two high-affinity permeases, Pho84p and Pho89p, and two low-affinity permeases, Pho87p and Pho90p (Persson et al., 1999; Wykoff & O'Shea, 2001). Perhaps in presence of As (V) the cellular homeostasis of phosphate change, this might affect the energetic balance and indirectly cause a redistribution of sugar utilization by different metabolic pathways.

About the target regulators investigated in the five mutants analyzed in this work we may conclude that Ixr1 and Sky1 are necessary to directly or indirectly mediate the observed increase in G6PDH activity, while Rox1 is dispensable.

3.2 Changes in catalase activity

Catalase catalyzes the breakdown of H_2O_2 into oxygen and water. In S. cerevisiae, there are two genes for CAT, CTA1 that encodes the peroxisomal and mitochondrial isoforms, and CTT1, that encodes the protein in the cytosol (Jamieson, 1998). CAT is one of the principal members that conforms the H_2O_2 stimulon, the set of proteins induced in S. cerevisiae in response to H_2O_2 (Godon et al., 1998). Also, *Schyzosaccharomyces pombe* (Vivancos et al., 2006) and *Kluyveromyces lactis* (Becerra et al., 2004; Tarrío et al., 2008), other two yeasts used frequently in this area of research, show increased levels of CAT in response of oxidative stress conditions. It has been shown that CAT function, even though it is an important antioxidant protein, can be partially substituted by other enzymatic activities. For example, in *S. cerevisiae*, when CAT activity is inhibited with 3-aminotriazol (3-AT), a compound commonly used as herbicide, simultaneous decrease in G6PDH and increase of GLR activity is observed (Bayliak et al., 2008).

Catalase activity did not increase after the treatments assayed in this work (Table 2). Among the hypothetical mechanism by which metals produce the onset of oxidative stress defence the production of ROS has been proposed (Stohs & Bagchi, 1995; Ercal et al., 2001; Beyersmann & Hartwig, 2008). However, from our data we might assume that the production of hydrogen peroxide is not significant in these conditions and perhaps other ROS are predominant after exposure to these metal compounds. Contrary to our results, previous studies (Muthukumar & Nachiappan, 2010) found that GSH levels were increased in cells exposed to Cd (II), as well as CAT and glutathione peroxidase activities. However, these changes were observed when yeast cells were exposed to 100 μM Cd (II), ten folds higher than the concentration used in our study.

	As (V)				Cd (II)				cisPt			
	G6PDH	CAT	GLR	TRR	G6PDH	CAT	GLR	TRR	G6PDH	CAT	GLR	TRR
wt	↑						↑					
Δrox1	↑	↓										
Δixr1				↓			↑					
Δsky1												
Δixr1Δsky1											↓	
Δixr1Δrox1										↓		↓

Table 2A. Most significant effects of As (V), Cd (II) and cisplatin (cisPt) treatment on the four enzymatic activities and six strains under study in this work. Green arrow: increase, red arrow: decrease.

Activity	Strain	Treatment	-	+	Ratio
G6PDH	wt	As	0.20±0.06	0.34±0.04	+1.7
	Δrox1	As	0.19±0.03	0.32±0.04	+1.7
CAT	*Δrox1*	As	0.82±0.02	0.02±0.02	-41.0
	Δixr1Δrox1	cisPt	0.32±0.15	0.06±0.05	-5.3
GLR	wt	Cd	0.06±0.04	0.33±0.04	+5.5
	Δixr1	Cd	0.06±0.03	0.28±0.04	+4.7
	Δixr1Δsky1	cisPt	0.07±0.04	0.01±0.01	-7
TRR	*Δixr1*	As	1.35±0.35	0.66±0.16	-2.0
	Δixr1Δrox1	cisPt	0.11±0.06	0.03±0.00	-3.7

Table 2B. Numerical data corresponding to significant effects reported in Table 2A. Media ± standard deviation (0.00 = < 0.005). Enzymatic units are defined in the text (- without and + with treatment). Ratio +, fold increase; ratio -, fold decrease.

About the control exerted by the selected regulators on CAT activity, it is interesting to say that Rox1 is necessary to maintain wild type activity levels after treatment with As (V) and cisplatin (Table 2), since the enzymatic activity diminished in the mutant background.

3.3 Changes in glutathione reductase activity

Once GSH is oxidized to GSSG, it becomes toxic to the cell and it cannot be accumulated for long. The principal protein in charge of catalyzing the transformation of one molecule of GSSG into two of GSH is glutathione reductase (GLR). Glutathione reductase is an enzyme belonging to the family of flavoproteins with oxidoreductase activity (Mustacich & Powis, 2000), and together with GSH and glutaredoxins, it constitutes the glutathione/glutaredoxin system. GLR has a double function, since it uses NADPH as a coenzyme and reduces GSSG. Besides it also produces $NADP^+$ that can be reduced by other enzymes, such as G6PDH. S. cerevisiae has only one gene to codify for GLR, and it contains two in-frame start codons (Collinson & Dawes, 1995). Translation from AUG1 or AUG2 generates the mitochondrial or cytosolic isoforms of the protein, respectively (Outten & Culotta, 2004). GLR is not essential to cell survival (Collinson & Dawes, 1995) but it is required for defence against oxidative stress (Grant et al., 1996ab).

GLR was significantly increased by Cd (II) treatment in the wild type strain (Table 2), which is in agreement with previous data that indicate that the oxidative stress caused by Cd (II) and the processes related to metal detoxification are highly dependent on the GSH/GRX system (Paumi et al., 2009; Wysocki & Tamás, 2010). Comparing the effects in the wild type and the five mutants analyzed in this work we may conclude that Rox1 and Sky1 are necessary to directly or indirectly mediate the observed increase in GLR activity, while Ixr1 is dispensable.

3.4 Changes in thioredoxin reductase activity

Thioredoxin reductase is an enzyme that belongs to the flavoprotein family of pyridine nucleotide-disulphide oxidoreductases. Its primary function is to reduce oxidized thioredoxins (TRXs). TRXs are small peptides between 10 and 12 kDa, which can supply reducing equivalents to enzymes such as ribonucleotide reductase (Laurent et al., 1964) and thioredoxin peroxidase (Chae et al., 1994). They also produce thiol-disulphide exchange and may reduce key Cys residues in certain transcription factors, which increases their ability to bind to DNA and regulate gene transcription. TRR and TRXs form the called "thioredoxin system" and TRRs from S. cerevisiae are induced by H_2O_2 (Godon et al., 1998).

In S. cerevisiae there are two genes encoding TRR: TRR1 and TRR2, coding TRR1 the cytosolic and TRR2 the mitochondrial isoform. There are also three TRXs genes. TRX1 and TRX2 encode cytosolic forms, whereas TRX3 the mitochondrial form. In S. cerevisiae, deletion of both TRR1 and TRR2 genes inhibits vacuole inheritance, decreases the rate of DNA synthesis, increases the cell size and the generation time and makes the cells auxotrophic for methionine/cysteine (Pedrajas et al., 1999). In a long term, cells lacking TRR1 are unviable (Pedrajas et al., 1999).

TRR activity was not increased after the treatments applied in this work (Table 2). This observation might indicate that oxidative stress caused by these metals is counteracted principally by the glutathione/glutaredoxin system instead of the thioredoxin system. However, the thioredoxin system is important for Cd (II) tolerance, since deletion of TRR1 or both *TRX1* and *TRX2* results in cadmium-hypersensitivity (Vido et al., 2001). Also, it has been proved that organic arsenicals can inhibit the thioredoxin and glutathione reductases leading to the increase of ROS steady state levels in the cell (Lin et al., 1999; Styblo et al., 1997).

3.5 Interactions between Ixr1, Rox1 and Sky1 in the maintenance of CAT, GLR and TRR activities during response to cisplatin

Data from Table 2 clearly show that enzymatic activities in the double mutants were affected by cisplatin treatment in reference to the wild type. Glutathione reductase activity decreased in *Δixr1Δsky1* strain, while CAT and TRR activities decreased in the *Δixr1Δrox1* strain. These data indicate that both Ixr1-Rox1 and Ixr1-Sky1 interactions are necessary in the response to cisplatin, since none of the single mutants analyzed demonstrated significant effect. However, probably the interconnections of Ixr1 with Rox1 and Sky1 play specialized functions in this change, as deduced from the different enzymatic activities that change in each case. The nature of these interactions, physic or genetic, has not been yet explored and constitutes and interesting subject for further studies.

4. Conclusions

Summarizing the results from this work (Figures 3-5 and Tables 2A and 2B), we observed in the wild type strain an increase of G6PDH activity upon As (V) treatment and increase of GLR activity upon Cd (II) treatment. In the *Δrox1* mutant, treatment with As (V) caused increase of G6PDH activity and decreased CAT activity. The *Δixr1* mutant showed decrease of TRR activity upon As (V) treatment and increased GLR activity upon Cd (II) treatment. The double mutants were affected by cisplatin; GLR activity decreased in the *Δixr1Δsky1* strain and TRR activity decreased in *Δixr1Δrox1* strain.

We may conclude that changes caused by As (V), Cd (II) and cisplatin treatments could not be considered as only general oxidative stress response. On the contrary, each treatment induced changes on specific enzymatic activities without affecting the others. G6PDH was enhanced by As (V) while GLR activity was increased by Cd (II). Besides, the increase of these activities depended on different regulatory factors; G6PDH seems to be regulated by Ixr1 and Sky1, while GLR by Rox1 and Sky. After treatment with cisplatin, maintenance of enzymatic activities in the levels observed in the wild type strain was also under the control of complex interaction between Rox1, Ixr1 and Sky1. Further studies will be necessary to understand the nature of these interactions.

5. Acknowledgements

This research was supported by grant BFU2009-08854 from MICINN (Spain), co financed by FEDER (CEE). General support to the laboratory during 2008-11 was funded by Xunta de Galicia (Consolidación C.E.O.U.2008/008), co-financed by FEDER.

6. References

Adle, D. J., Sinani, D., Kim, H. & Lee, J. (2007). A cadmium-transporting P1B-type ATPase in yeast *Saccharomyces cerevisiae*. *The Journal of Biological Chemistry*, Vol. 282, No. 2, (Jan 12), pp. (947-955), ISSN 0021-9258.

Aebi, H. (1984). Catalase in Vitro. *Methods in Enzymology*, Vol. 105, pp (121-126.), ISSN 0076-6879.

Avery, A. M. & Avery, S. V. (2001). *Saccharomyces cerevisiae* expresses three phospholipid hydroperoxide glutathione peroxidases. *The Journal of Biological Chemistry*, Vol. 276, No. 36, (Sep 7), pp. (33730-33735), ISSN 0021-9258; 0021-9258.

Bayliak, M., Gospodaryov, D., Semchyshyn, H. & Lushchak, V. (2008). Inhibition of catalase by aminotriazole *in vivo* results in reduction of glucose-6-phosphate dehydrogenase activity in *Saccharomyces cerevisiae* cells. *Biochemistry.Biokhimiia*, Vol. 73, No. 4, (Apr), pp. (420-426), ISSN 0006-2979.

Becerra, M., Tarrío, N., González-Siso, M. I. & Cerdán, M. E. (2004). Genome-wide analysis of *Kluyveromyces lactis* in wild-type and *rag2* mutant strains. *Genome* Vol. 47, No. 5, (Oct), pp. (970-978), ISSN 0831-2796.

Beyersmann, D. & Hartwig, A. (2008). Carcinogenic metal compounds: recent insight into molecular and cellular mechanisms. *Archives of Toxicology*, Vol. 82, No. 8, (Aug), pp. (493-512), ISSN 0340-5761.

Bourdineaud, J. P. (2001). At acidic pH, the GPA2-cAMP pathway is necessary to counteract the *ORD1*-mediated repression of the hypoxic *SRP1/TIR1* yeast gene. *Yeast* (Chichester, England), Vol. 18, No. 9, (Jun 30), pp. (841-848), ISSN 0749-503X.

Bourdineaud, J. P., De Sampaio, G. & Lauquin, G. J. (2000). A Rox1-independent hypoxic pathway in yeast. Antagonistic action of the repressor Ord1 and activator Yap1 for hypoxic expression of the *SRP1/TIR1* gene. *Molecular Microbiology*, Vol. 38, No. 4, (Nov), pp. (879-890), ISSN 0950-382X.

Bradford, M. M. (1976). A rapid and sensitive method for the quantitation of microgram quantities of protein utilizing the principle of protein-dye binding. *Analytical Biochemistry*, Vol. 72, (May 7), pp. (248-254), ISSN 0003-2697.

Brennan, R. J. & Schiestl, R. H. (1996). Cadmium is an inducer of oxidative stress in yeast. *Mutation Research*, Vol. 356, No. 2, (Sep 23), pp. (171-178), ISSN 0027-5107.

Brown, S. J., Kellett, P. J. & Lippard, S. J. (1993). Ixr1, a yeast protein that binds to platinated DNA and confers sensitivity to cisplatin. *Science* (New York, N.Y.), Vol. 261, No. 5121, (Jul 30), pp. (603-605), ISSN 0036-8075.

Castro-Prego, R., Lamas-Maceiras, M., Soengas, P., Carneiro, I., González-Siso, I. & Cerdán, M. E. (2010a). Regulatory factors controlling transcription of *Saccharomyces cerevisiae* *IXR1* by oxygen levels: a model of transcriptional adaptation from aerobiosis to hypoxia implicating *ROX1* and *IXR1* cross-regulation. *The Biochemical Journal*, Vol. 425, No. 1, (Dec 14), pp. (235-243), ISSN 1470-8728; 0264-6021.

Castro-Prego, R., Lamas-Maceiras, M., Soengas, P., Fernández-Leiro, R., Carneiro, I., Becerra, M., González-Siso, M. I. & Cerdán, M. E. (2010b). Ixr1p regulates oxygen-dependent *HEM13* transcription. *FEMS Yeast Research*, Vol. 10, No. 3, (May), pp. (309-321), ISSN 1567-1364; 1567-1356.

Chae, H. Z., Chung, S. J. & Rhee, S. G. (1994). Thioredoxin-dependent peroxide reductase from yeast. *The Journal of Biological Chemistry*, Vol. 269, No. 44, (Nov 4), pp. (27670-27678), ISSN 0021-9258.

Chow, C. S., Whitehead, J. P. & Lippard, S. J. (1994). HMG domain proteins induce sharp bends in cisplatin-modified DNA. *Biochemistry*, Vol. 33, No. 50, (Dec 20), pp. (15124-15130), ISSN 0006-2960.

Chrestensen, C. A., Starke, D. W. & Mieyal, J. J. (2000). Acute cadmium exposure inactivates thioltransferase (Glutaredoxin), inhibits intracellular reduction of protein-glutathionyl-mixed disulfides and initiates apoptosis. *The Journal of Biological Chemistry*, Vol. 275, No. 34, (Aug 25), pp. (26556-26565), ISSN 0021-9258.

Collinson, L. P. & Dawes, I. W. (1995). Isolation, characterization and overexpression of the yeast gene, *GLR1*, encoding glutathione reductase. *Gene*, Vol. 156, No. 1, (Apr 14), pp. (123-127), ISSN 0378-1119.

Cui, Z., Hirata, D., Tsuchiya, E., Osada, H. & Miyakawa, T. (1996). The multidrug resistance-associated protein (MRP) subfamily (Yrs1/Yor1) of *Saccharomyces cerevisiae* is important for the tolerance to a broad range of organic anions. *The Journal of Biological Chemistry*, Vol. 271, No. 25, (Jun 21), pp. (14712-14716), ISSN 0021-9258.

Deckert, J., Khalaf, R. A., Hwang, S. M. & Zitomer, R. S. (1999). Characterization of the DNA binding and bending HMG domain of the yeast hypoxic repressor Rox1. *Nucleic Acids Research*, Vol. 27, No. 17, (Sep 1), pp. (3518-3526), ISSN 1362-4962.

Deckert, J., Rodríguez Torres, A. M., Simon, J. T. & Zitomer, R. S. (1995). Mutational analysis of Rox1, a DNA-bending repressor of hypoxic genes in *Saccharomyces cerevisiae*. *Molecular and Cellular Biology*, Vol. 15, No. 11, (Nov), pp. (6109-6117), ISSN 0270-7306.

Dirmeier, R., O'Brien, K. M., Engle, M., Dodd, A., Spears, E. & Poyton, R. O. (2002). Exposure of yeast cells to anoxia induces transient oxidative stress. Implications for the induction of hypoxic genes. *The Journal of Biological Chemistry*, Vol. 277, No. 38, (Sep 20), pp. (34773-34784), ISSN 0021-9258.

Ercal, N., Gurer-Orhan, H. & Aykin-Burns, N. (2001). Toxic metals and oxidative stress part I: mechanisms involved in metal-induced oxidative damage. *Current Topics in Medicinal Chemistry*, Vol. 1, No. 6, (Dec), pp. (529-539), ISSN 1568-0266.

Erez, O. & Kahana, C. (2001). Screening for modulators of spermine tolerance identifies Sky1, the SR protein kinase of *Saccharomyces cerevisiae*, as a regulator of polyamine transport and ion homeostasis. *Molecular and Cellular Biology*, Vol. 21, No. 1, (Jan), pp. (175-184), ISSN 0270-7306.

Faller, P., Kienzler, K. & Krieger-Liszkay, A. (2005). Mechanism of Cd^{2+} toxicity: Cd^{2+} inhibits photoactivation of Photosystem II by competitive binding to the essential Ca^{2+} site. *Biochimica et Biophysica Acta*, Vol. 1706, No. 1-2, (Jan 7), pp. (158-164), ISSN 0006-3002.

Fox, M. E., Feldman, B. J. & Chu, G. (1994). A novel role for DNA photolyase: binding to DNA damaged by drugs is associated with enhanced cytotoxicity in *Saccharomyces cerevisiae*. *Molecular and Cellular Biology*, Vol. 14, No. 12, (Dec), pp. (8071-8077), ISSN 0270-7306.

Garcera, A., Barreto, L., Piedrafita, L., Tamarit, J. & Herrero, E. (2006). *Saccharomyces cerevisiae* cells have three Omega class glutathione S-transferases acting as 1-Cys thiol transferases. *The Biochemical Journal*, Vol. 398, No. 2, (Sep 1), pp. (187-196), ISSN 1470-8728; 0264-6021.

Ghosh, M., Shen, J. & Rosen, B. P. (1999). Pathways of As (III) detoxification in *Saccharomyces cerevisiae*. *Proceedings of the National Academy of Sciences of the United States of America*, Vol. 96, No. 9, (Apr 27), pp. (5001-5006), ISSN 0027-8424.

Gietz, R. D. & Sugino, A. (1988). New yeast-*Escherichia coli* shuttle vectors constructed with in vitro mutagenized yeast genes lacking six-base pair restriction sites. *Gene*, Vol. 74, No. 2, (Dec 30), pp. (527-534), ISSN 0378-1119.

Godon, C., Lagniel, G., Lee, J., Buhler, J. M., Kieffer, S., Perrot, M., Boucherie, H., Toledano, M. B. & Labarre, J. (1998). The H$_2$O$_2$ stimulon in *Saccharomyces cerevisiae. The Journal of Biological Chemistry*, Vol. 273, No. 35, (Aug 28), pp. (22480-22489), ISSN 0021-9258.

González-Siso, M. I., García-Leiro, A., Tarrío, N. & Cerdán, M. E. (2009). Sugar metabolism, redox balance and oxidative stress response in the respiratory yeast *Kluyveromyces lactis. Microbial Cell Factories*, Vol. 8, (Aug 30), pp. (46), ISSN 1475-2859.

Grant, C. M., Collinson, L. P., Roe, J. H. & Dawes, I. W. (1996a). Yeast glutathione reductase is required for protection against oxidative stress and is a target gene for yAP-1 transcriptional regulation. *Molecular Microbiology*, Vol. 21, No. 1, (Jul), pp. (171-179), ISSN 0950-382X.

Grant, C. M., MacIver, F. H. & Dawes, I. W. (1996b). Glutathione is an essential metabolite required for resistance to oxidative stress in the yeast *Saccharomyces cerevisiae. Current Genetics*, Vol. 29, No. 6, (May), pp. (511-515), ISSN 0172-8083.

Grant, C. M., Perrone, G. & Dawes, I. W. (1998). Glutathione and catalase provide overlapping defenses for protection against hydrogen peroxide in the yeast *Saccharomyces cerevisiae. Biochemical and Biophysical Research Communications*, Vol. 253, No. 3, (Dec 30), pp. (893-898), ISSN 0006-291X.

Hartwig, A. (2001). Zinc finger proteins as potential targets for toxic metal ions: differential effects on structure and function. *Antioxidants and Redox Signaling*, Vol. 3, No. 4, (Aug), pp. (625-634), ISSN 1523-0864.

Holmgren, A. & Bjornstedt, M. (1995). Thioredoxin and thioredoxin reductase. *Methods in Enzymology*, Vol. 252, pp. (199-208), ISSN 0076-6879.

Hoppe, G., Talcott, K. E., Bhattacharya, S. K., Crabb, J. W. & Sears, J. E. (2006). Molecular basis for the redox control of nuclear transport of the structural chromatin protein Hmgb1. *Experimental Cell Research*, Vol. 312, No. 18, (Nov 1), pp. (3526-3538), ISSN 0014-4827.

Howlett, N. G. & Avery, S. V. (1997). Induction of lipid peroxidation during heavy metal stress in *Saccharomyces cerevisiae* and influence of plasma membrane fatty acid unsaturation. *Applied and Environmental Microbiology*, Vol. 63, No. 8, (Aug), pp. (2971-2976), ISSN 0099-2240.

Huang, R. Y., Eddy, M., Vujcic, M. & Kowalski, D. (2005). Genome-wide screen identifies genes whose inactivation confer resistance to cisplatin in *Saccharomyces cerevisiae. Cancer Research*, Vol. 65, No. 13, (Jul 1), pp. (5890-5897), ISSN 0008-5472.

Iraz, M., Ozerol, E., Gulec, M., Tasdemir, S., Idiz, N., Fadillioglu, E., Naziroglu, M. & Akyol, O. (2006). Protective effect of caffeic acid phenethyl ester (CAPE) administration on cisplatin-induced oxidative damage to liver in rat. *Cell Biochemistry and Function*, Vol. 24, No. 4, (Jul-Aug), pp. (357-361), ISSN 0263-6484.

Izawa, S., Maeda, K., Miki, T., Mano, J., Inoue, Y. & Kimura, A. (1998). Importance of glucose-6-phosphate dehydrogenase in the adaptive response to hydrogen peroxide in *Saccharomyces cerevisiae. The Biochemical Journal*, Vol. 330, No. Pt 2, (Mar 1), pp. (811-817), ISSN 0264-6021.

Jamieson, D. J. (1998). Oxidative stress responses of the yeast *Saccharomyces cerevisiae. Yeast* (Chichester, England), Vol. 14, No. 16, (Dec), pp. (1511-1527), ISSN 0749-503X.

Jensen, L. T. & Culotta, V. C. (2002). Regulation of *Saccharomyces cerevisiae* FET4 by oxygen and iron. *Journal of Molecular Biology*, Vol. 318, No. 2, (Apr 26), pp. (251-260), ISSN 0022-2836.

Juhnke, H., Krems, B., Kotter, P. & Entian, K. D. (1996). Mutants that show increased sensitivity to hydrogen peroxide reveal an important role for the pentose phosphate pathway in protection of yeast against oxidative stress. *Molecular and General Genetics : MGG*, Vol. 252, No. 4, (Sep 25), pp. (456-464), ISSN 0026-8925.

Kastaniotis, A. J. & Zitomer, R. S. (2000). Rox1 mediated repression. Oxygen dependent repression in yeast. *Advances in Experimental Medicine and Biology*, Vol. 475, pp. (185-195), ISSN 0065-2598.

Kawahara, N., Tanaka, T., Yokomizo, A., Nanri, H., Ono, M., Wada, M., Kohno, K., Takenaka, K., Sugimachi, K. & Kuwano, M. (1996). Enhanced coexpression of thioredoxin and high mobility group protein 1 genes in human hepatocellular carcinoma and the possible association with decreased sensitivity to cisplatin. *Cancer Research*, Vol. 56, No. 23, (Dec 1), pp. (5330-5333), ISSN 0008-5472.

Kelley, R. & Ideker, T. (2009). Genome-wide fitness and expression profiling implicate Mga2 in adaptation to hydrogen peroxide. *Plos Genetics*, Vol. 5, No. 5, (May), pp. (e1000488), ISSN 1553-7404; 1553-7390.

Khan, S. A., Priyamvada, S., Khan, W., Khan, S., Farooq, N. & Yusufi, A. N. (2009). Studies on the protective effect of green tea against cisplatin induced nephrotoxicity. *Pharmacological Research*, the official journal of the Italian Pharmacological Society, Vol. 60, No. 5, (Nov), pp. (382-391), ISSN 1096-1186; 1043-6618.

Kitchin, K. T. & Wallace, K. (2008a). Evidence against the nuclear *in situ* binding of arsenicals -oxidative stress theory of arsenic carcinogenesis. *Toxicology and Applied Pharmacology*, Vol. 232, No. 2, (Oct 15), pp. (252-257), ISSN 1096-0333; 0041-008X.

Kitchin, K. T. & Wallace, K. (2008b). The role of protein binding of trivalent arsenicals in arsenic carcinogenesis and toxicity. *Journal of Inorganic Biochemistry*, Vol. 102, No. 3, (Mar), pp. (532-539), ISSN 0162-0134.

Klinkenberg, L. G., Mennella, T. A., Luetkenhaus, K. & Zitomer, R. S. (2005). Combinatorial repression of the hypoxic genes of *Saccharomyces cerevisiae* by DNA binding proteins Rox1 and Mot3. *Eukaryotic Cell*, Vol. 4, No. 4, (Apr), pp. (649-660), ISSN 1535-9778.

Kuby, S. A. & Noltmann, E. A. (1966). Glucose 6-phosphate dehydrogenase crystalline from Brewers´ yeast. *Methods in Enzymology*, Vol. 9, pp (116-125), ISSN 0076-6879.

Lafaye, A., Junot, C., Pereira, Y., Lagniel, G., Tabet, J. C., Ezan, E. & Labarre, J. (2005). Combined proteome and metabolite-profiling analyses reveal surprising insights into yeast sulfur metabolism. *The Journal of Biological Chemistry*, Vol. 280, No. 26, (Jul 1), pp. (24723-24730), ISSN 0021-9258.

Lai, L. C., Kosorukoff, A. L., Burke, P. V. & Kwast, K. E. (2006). Metabolic-state-dependent remodeling of the transcriptome in response to anoxia and subsequent reoxygenation in *Saccharomyces cerevisiae*. *Eukaryotic Cell*, Vol. 5, No. 9, (Sep), pp. (1468-1489), ISSN 1535-9778.

Lambert, J. R., Bilanchone, V. W. & Cumsky, M. G. (1994). The *ORD1* gene encodes a transcription factor involved in oxygen regulation and is identical to *IXR1*, a gene that confers cisplatin sensitivity to *Saccharomyces cerevisiae*. *Proceedings of the National Academy of Sciences of the United States of America*, Vol. 91, No. 15, (Jul 19), pp. (7345-7349), ISSN 0027-8424.

Laurent, T. C., Moore, E. C. & Reichard, P. (1964). Enzymatic Synthesis of Deoxyribonucleotides. Iv. Isolation and Characterization of Thioredoxin, the

Hydrogen Donor from *Escherichia coli* B. *The Journal of Biological Chemistry*, Vol. 239, (Oct), pp. (3436-3444), ISSN 0021-9258.

Lillig, C. H., Berndt, C. & Holmgren, A. (2008). Glutaredoxin systems. *Biochimica et Biophysica Acta*, Vol. 1780, No. 11, (Nov), pp. (1304-1317), ISSN 0006-3002.

Li, Q., Harvey, L. M. & McNeil, B. (2009). Oxidative stress in industrial fungi. Critical *Reviews inBiotechnology*, Vol. 29, No. 3, pp. (199-213), ISSN 1549-7801; 0738-8551.

Lin, S., Cullen, W. R. & Thomas, D. J. (1999). Methylarsenicals and arsinothiols are potent inhibitors of mouse liver thioredoxin reductase. *Chemical Research in Toxicology*, Vol. 12, No. 10, (Oct), pp. (924-930), ISSN 0893-228X.

Lushchak, V. I. (2010). Oxidative stress in yeast. *Biochemistry.Biokhimiia*, Vol. 75, No. 3, (Mar), pp. (281-296), ISSN 1608-3040; 0006-2979.

Lushchak, V. I. (2011). Adaptive response to oxidative stress: Bacteria, fungi, plants and animals. *Comparative Biochemistry and Physiology. Toxicology and Pharmacology : CBP*, Vol. 153, No. 2, (Mar), pp. (175-190), ISSN 1532-0456.

Mansour, H. H., Hafez, H. F. & Fahmy, N. M. (2006). Silymarin modulates cisplatin-induced oxidative stress and hepatotoxicity in rats. *Journal of Biochemistry and Molecular Biology*, Vol. 39, No. 6, (Nov 30), pp. (656-661), ISSN 1225-8687.

Martins, N. M., Santos, N. A., Curti, C., Bianchi, M. L. & Santos, A. C. (2008). Cisplatin induces mitochondrial oxidative stress with resultant energetic metabolism impairment, membrane rigidification and apoptosis in rat liver. *Journal of Applied Toxicology : JAT*, Vol. 28, No. 3, (Apr), pp. (337-344), ISSN 0260-437X.

McA'Nulty, M. M. & Lippard, S. J. (1996). The HMG-domain protein Ixr1 blocks excision repair of cisplatin-DNA adducts in yeast. *Mutation Research*, Vol. 362, No. 1, (Jan 2), pp. (75-86), ISSN 0027-5107.

McA'Nulty, M. M., Whitehead, J. P. & Lippard, S. J. (1996). Binding of Ixr1, a yeast HMG-domain protein, to cisplatin-DNA adducts in vitro and in vivo. *Biochemistry*, Vol. 35, No. 19, (May 14), pp. (6089-6099), ISSN 0006-2960.

Menezes, R. A., Amaral, C., Batista-Nascimento, L., Santos, C., Ferreira, R. B., Devaux, F., Eleutherio, E. C. & Rodrigues-Pousada, C. (2008). Contribution of Yap1 towards *Saccharomyces cerevisiae* adaptation to arsenic-mediated oxidative stress. *The Biochemical Journal*, Vol. 414, No. 2, (Sep 1), pp. (301-311), ISSN 1470-8728; 0264-6021.

Menzel, D. B., Hamadeh, H. K., Lee, E., Meacher, D. M., Said, V., Rasmussen, R. E., Greene, H. & Roth, R. N. (1999). Arsenic binding proteins from human lymphoblastoid cells. *Toxicology Letters*, Vol. 105, No. 2, (Mar 29), pp. (89-101), ISSN 0378-4274.

Minard, K. I. & McAlister-Henn, L. (2005). Sources of NADPH in yeast vary with carbon source. *The Journal of Biological Chemistry*, Vol. 280, No. 48, (Dec 2), pp. (39890-39896), ISSN 0021-9258.

Montañés, F. M., Pascual-Ahuir, A. & Proft, M. (2011). Repression of ergosterol biosynthesis is essential for stress resistance and is mediated by the Hog1 MAP kinase and the Mot3 and Rox1 transcription factors. *Molecular Microbiology*, Vol. 79, No. 4, (Feb), pp. (1008-1023), ISSN 1365-2958; 0950-382X.

Mukhopadhyay, R. & Rosen, B. P. (1998). *Saccharomyces cerevisiae ACR2* gene encodes an arsenate reductase. *FEMS Microbiology Letters*, Vol. 168, No. 1, (Nov 1), pp. (127-136), ISSN 0378-1097.

Mukhopadhyay, R., Shi, J. & Rosen, B. P. (2000). Purification and characterization of ACR2p, the *Saccharomyces cerevisiae* arsenate reductase. *The Journal of Biological Chemistry*, Vol. 275, No. 28, (Jul 14), pp. (21149-21157), ISSN 0021-9258.

Muthukumar, K. & Nachiappan, V. (2010). Cadmium-induced oxidative stress in *Saccharomyces cerevisiae*. *Indian Journal of Biochemistry and Biophysics*, Vol. 47, No. 6, (Dec), pp. (383-387), ISSN 0301-1208.

Mustacich, D. & Powis, G. (2000). Thioredoxin reductase. *The Biochemical Journal*, Vol. 346 Pt 1, (Feb 15), pp. (1-8), ISSN 0264-6021.

Nagatani, G., Nomoto, M., Takano, H., Ise, T., Kato, K., Imamura, T., Izumi, H., Makishima, K. & Kohno, K. (2001). Transcriptional activation of the human *HMG1* gene in cisplatin-resistant human cancer cells. *Cancer Research*, Vol. 61, No. 4, (Feb 15), pp. (1592-1597), ISSN 0008-5472.

Nagy, Z., Montigny, C., Leverrier, P., Yeh, S., Goffeau, A., Garrigos, M. & Falson, P. (2006). Role of the yeast ABC transporter Yor1p in cadmium detoxification. *Biochimie*, Vol. 88, No. 11, (Nov), pp. (1665-1671), ISSN 0300-9084.

Nogae, I. & Johnston, M. (1990). Isolation and characterization of the *ZWF1* gene of *Saccharomyces cerevisiae*, encoding glucose-6-phosphate dehydrogenase. *Gene*, Vol. 96, No. 2, (Dec 15), pp. (161-169), ISSN 0378-1119.

Oliva, R. & Dixon, G. H. (1991). Vertebrate protamine genes and the histone-to-protamine replacement reaction. *Progress in Nucleic Acid Research and Molecular Biology*, Vol. 40, pp. (25-94), ISSN 0079-6603.

Outten, C. E. & Culotta, V. C. (2004). Alternative start sites in the *Saccharomyces cerevisiae* *GLR1* gene are responsible for mitochondrial and cytosolic isoforms of glutathione reductase. *The Journal of Biological Chemistry*, Vol. 279, No. 9, (Feb 27), pp. (7785-7791), ISSN 0021-9258.

Papoutsopoulou, S., Nikolakaki, E., Chalepakis, G., Kruft, V., Chevaillier, P. & Giannakouros, T. (1999). SR protein-specific kinase 1 is highly expressed in testis and phosphorylates protamine 1. *Nucleic Acids Research*, Vol. 27, No. 14, (Jul 15), pp. (2972-2980), ISSN 0305-1048.

Paumi, C. M., Chuk, M., Snider, J., Stagljar, I. & Michaelis, S. (2009). ABC transporters in *Saccharomyces cerevisiae* and their interactors: new technology advances the biology of the ABCC (MRP) subfamily. *Microbiology and Molecular Biology Reviews : MMBR*, Vol. 73, No. 4, (Dec), pp. (577-593), ISSN 1098-5557; 1092-2172.

Pedrajas, J. R., Kosmidou, E., Miranda-Vizuete, A., Gustafsson, J. A., Wright, A. P. & Spyrou, G. (1999). Identification and functional characterization of a novel mitochondrial thioredoxin system in *Saccharomyces cerevisiae*. *The Journal of Biological Chemistry*, Vol. 274, No. 10, (Mar 5), pp. (6366-6373), ISSN 0021-9258.

Penninckx, M. J. (2002). An overview on glutathione in *Saccharomyces* versus non-conventional yeasts. *FEMS Yeast Research*, Vol. 2, No. 3, (Aug), pp. (295-305), ISSN 1567-1356.

Perrone, G. G., Grant, C. M. & Dawes, I. W. (2005). Genetic and environmental factors influencing glutathione homeostasis in *Saccharomyces cerevisiae*. *Molecular Biology of the Cell*, Vol. 16, No. 1, (Jan), pp. (218-230), ISSN 1059-1524.

Persson, B. L., Petersson, J., Fristedt, U., Weinander, R., Berhe, A. & Pattison, J. (1999). Phosphate permeases of *Saccharomyces cerevisiae*: structure, function and regulation. *Biochimica et Biophysica Acta*, Vol. 1422, No. 3, (Nov 16), pp. (255-272), ISSN 0006-3002.

Pratibha, R., Sameer, R., Rataboli, P. V., Bhiwgade, D. A. & Dhume, C. Y. (2006). Enzymatic studies of cisplatin induced oxidative stress in hepatic tissue of rats. *European Journal of Pharmacology*, Vol. 532, No. 3, (Feb 27), pp. (290-293), ISSN 0014-2999.

Preta, G., de Klark, R., Chakraborti, S. & Glas, R. (2010). MAP kinase-signaling controls nuclear translocation of tripeptidyl-peptidase II in response to DNA damage and oxidative stress. *Biochemical and Biophysical Research Communications*, Vol. 399, No. 3, (Aug 27), pp. (324-330), ISSN 1090-2104; 0006-291X.

Rodrigues, M. A., Rodrigues, J. L., Martins, N. M., Barbosa, F., Curti, C., Santos, N. A. & Santos, A. C. (2011). Carvedilol protects against cisplatin-induced oxidative stress, redox state unbalance and apoptosis in rat kidney mitochondria. *Chemico-Biological Interactions*, Vol. 189, No. 1-2, (Jan 15), pp. (45-51), ISSN 1872-7786; 0009-2797.

Saliola, M., Scappucci, G., De Maria, I., Lodi, T., Mancini, P. & Falcone, C. (2007). Deletion of the glucose-6-phosphate dehydrogenase gene *KlZWF1* affects both fermentative and respiratory metabolism in *Kluyveromyces lactis*. *Eukaryotic Cell*, Vol. 6, No. 1, (Jan), pp. (19-27), ISSN 1535-9778.

Samikkannu, T., Chen, C. H., Yih, L. H., Wang, A. S., Lin, S. Y., Chen, T. C. & Jan, K. Y. (2003). Reactive oxygen species are involved in arsenic trioxide inhibition of pyruvate dehydrogenase activity. *Chemical Research in Toxicology*, Vol. 16, No. 3, (Mar), pp. (409-414), ISSN 0893-228X.

Santos, N. A., Bezerra, C. S., Martins, N. M., Curti, C., Bianchi, M. L. & Santos, A. C. (2008). Hydroxyl radical scavenger ameliorates cisplatin-induced nephrotoxicity by preventing oxidative stress, redox state unbalance, impairment of energetic metabolism and apoptosis in rat kidney mitochondria. *Cancer Chemotherapy and Pharmacology*, Vol. 61, No. 1, (Jan), pp. (145-155), ISSN 0344-5704.

Satoh, M., Shimada, A., Zhang, B. & Tohyama, C. (2000). Renal toxicity caused by cisplatinum in glutathione-depleted metallothionein-null mice. *Biochemical Pharmacology*, Vol. 60, No. 11, (Dec 1), pp. (1729-1734), ISSN 0006-2952; 0006-2952.

Schenk, P. W., Boersma, A. W., Brandsma, J. A., den Dulk, H., Burger, H., Stoter, G., Brouwer, J. & Nooter, K. (2001). *SKY1* is involved in cisplatin-induced cell kill in *Saccharomyces cerevisiae*, and inactivation of its human homologue, *SRPK1*, induces cisplatin resistance in a human ovarian carcinoma cell line. *Cancer Research*, Vol. 61, No. 19, (Oct 1), pp. (6982-6986), ISSN 0008-5472.

Schenk, P. W., Boersma, A. W., Brok, M., Burger, H., Stoter, G. & Nooter, K. (2002). Inactivation of the *Saccharomyces cerevisiae SKY1* gene induces a specific modification of the yeast anticancer drug sensitivity profile accompanied by a mutator phenotype. *Molecular Pharmacology*, Vol. 61, No. 3, (Mar), pp. (659-666), ISSN 0026-895X.

Schenk, P. W., Brok, M., Boersma, A. W., Brandsma, J. A., Den Dulk, H., Burger, H., Stoter, G., Brouwer, J. & Nooter, K. (2003). Anticancer drug resistance induced by disruption of the *Saccharomyces cerevisiae NPR2* gene: a novel component involved in cisplatin- and doxorubicin-provoked cell kill. *Molecular Pharmacology*, Vol. 64, No. 2, (Aug), pp. (259-268), ISSN 0026-895X.

Schenk, P. W., Stoop, H., Bokemeyer, C., Mayer, F., Stoter, G., Oosterhuis, J. W., Wiemer, E., Looijenga, L. H. & Nooter, K. (2004). Resistance to platinum-containing chemotherapy in testicular germ cell tumors is associated with downregulation of the protein kinase SRPK1. *Neoplasia* (New York, N.Y.), Vol. 6, No. 4, (Jul-Aug), pp. (297-301), ISSN 1522-8002; 1476-5586.

Schutzendubel, A. & Polle, A. (2002). Plant responses to abiotic stresses: heavy metal-induced oxidative stress and protection by mycorrhization. *Journal of Experimental Botany*, Vol. 53, No. 372, (May), pp. (1351-1365), ISSN 0022-0957; 0022-0957.

Serpeloni, J. M., Grotto, D., Mercadante, A. Z., de Lourdes Pires Bianchi, M. & Antunes, L. M. (2010). Lutein improves antioxidant defence in vivo and protects against DNA damage and chromosome instability induced by cisplatin. *Archives of Toxicology*, Vol. 84, No. 10, (Oct), pp. (811-822), ISSN 1432-0738.

Shiota, M., Izumi, H., Miyamoto, N., Onitsuka, T., Kashiwagi, E., Kidani, A., et al. (2008). Ets regulates peroxiredoxin1 and 5 expressions through their interaction with the high-mobility group protein B1. *Cancer Science*, Vol. 99, No. 10, (Oct), pp. (1950-1959), ISSN 1349-7006; 1347-9032.

Siebel, C. W., Feng, L., Guthrie, C. & Fu, X. D. (1999). Conservation in budding yeast of a kinase specific for SR splicing factors. *Proceedings of the National Academy of Sciences of the United States of America*, Vol. 96, No. 10, (May 11), pp. (5440-5445), ISSN 0027-8424.

Smith, I. K., Vierheller, T. L. & Thorne, C. A. (1988). Assay of glutathione reductase in crude tissue homogenates using 5,5'-dithio bis(2-nitrobenzoic acid). *Analytical Biochemistry*, Vol. 175, No. 2, (Dec), pp. (408-413), ISSN 0003-2697.

Stohs, S. J. & Bagchi, D. (1995). Oxidative mechanisms in the toxicity of metal ions. *Free Radical Biology & Medicine*, Vol. 18, No. 2, (Feb), pp. (321-336), ISSN 0891-5849.

Styblo, M., Serves, S. V., Cullen, W. R. & Thomas, D. J. (1997). Comparative inhibition of yeast glutathione reductase by arsenicals and arsenothiols. *Chemical Research in Toxicology*, Vol. 10, No. 1, (Jan), pp. (27-33), ISSN 0893-228X.

Tarrío, N., García-Leiro, A., Cerdán, M. E. & González-Siso, M. I. (2008). The role of glutathione reductase in the interplay between oxidative stress response and turnover of cytosolic NADPH in *Kluyveromyces lactis*. *FEMS Yeast Research*, Vol. 8, No. 4, (Jun), pp. (597-606), ISSN 1567-1356.

Teixeira, M. C., Cabrito, T. R., Hanif, Z. M., Vargas, R. C., Tenreiro, S. & Sa-Correia, I. (2010). Yeast response and tolerance to polyamine toxicity involving the drug:H^+ antiporter Qdr3 and the transcription factors Yap1 and Gcn4. *Microbiology* (Reading, England), (Dec 9), ISSN 1465-2080; 1350-0872.

Tenreiro, S., Vargas, R. C., Teixeira, M. C., Magnani, C. & Sa-Correia, I. (2005). The yeast multidrug transporter Qdr3 (Ybr043c): localization and role as a determinant of resistance to quinidine, barban, cisplatin, and bleomycin. *Biochemical and Biophysical Research Communications*, Vol. 327, No. 3, (Feb 18), pp. (952-959), ISSN 0006-291X.

Thomas, D., Cherest, H. & Surdin-Kerjan, Y. (1991). Identification of the structural gene for glucose-6-phosphate dehydrogenase in yeast. Inactivation leads to a nutritional requirement for organic sulfur. *The EMBO Journal*, Vol. 10, No. 3, (Mar), pp. (547-553), ISSN 0261-4189.

Thorpe, G. W., Fong, C. S., Alic, N., Higgins, V. J. & Dawes, I. W. (2004). Cells have distinct mechanisms to maintain protection against different reactive oxygen species: oxidative-stress-response genes. *Proceedings of the National Academy of Sciences of the United States of America*, Vol. 101, No. 17, (Apr 27), pp. (6564-6569), ISSN 0027-8424.

Thorsen, M., Lagniel, G., Kristiansson, E., Junot, C., Nerman, O., Labarre, J. & Tamas, M. J. (2007). Quantitative transcriptome, proteome, and sulfur metabolite profiling of the

Saccharomyces cerevisiae response to arsenite. *Physiological Genomics*, Vol. 30, No. 1, (Jun 19), pp. (35-43), ISSN 1531-2267; 1094-8341.

Thorsen, M., Perrone, G. G., Kristiansson, E., Traini, M., Ye, T., Dawes, I. W., Nerman, O. & Tamas, M. J. (2009). Genetic basis of arsenite and cadmium tolerance in *Saccharomyces cerevisiae*. *BMC Genomics*, Vol. 10, (Mar 12), pp. (105), ISSN 1471-2164.

Tizón, B., Rodríguez-Torres, A. M. & Cerdán, M. E. (1999). Disruption of six novel *Saccharomyces cerevisiae* genes reveals that YGL129c is necessary for growth in non-fermenTable carbon sources, YGL128c for growth at low or high temperatures and YGL125w is implicated in the biosynthesis of methionine. *Yeast* (Chichester, England), Vol. 15, No. 2, (Jan 30), pp. (145-154), ISSN 0749-503X.

Vido, K., Spector, D., Lagniel, G., Lopez, S., Toledano, M. B. & Labarre, J. (2001). A proteome analysis of the cadmium response in *Saccharomyces cerevisiae*. *The Journal of Biological Chemistry*, Vol. 276, No. 11, (Mar 16), pp. (8469-8474), ISSN 0021-9258.

Vivancos, A. P., Jara, M., Zuin, A., Sanso, M. & Hidalgo, E. (2006). Oxidative stress in *Schizosaccharomyces pombe*: different H_2O_2 levels, different response pathways. *Molecular Genetics and Genomics : MGG*, Vol. 276, No. 6, (Dec), pp. (495-502), ISSN 1617-4615.

Wang, Z., Zhang, H., Li, X. F. & Le, X. C. (2007). Study of interactions between arsenicals and thioredoxins (human and *E. coli*) using mass spectrometry. *Rapid Communications in Mass Spectrometry : RCM*, Vol. 21, No. 22, pp. (3658-3666), ISSN 0951-4198.

Wong, C. M., Ching, Y. P., Zhou, Y., Kung, H. F. & Jin, D. Y. (2003). Transcriptional regulation of yeast peroxiredoxin gene *TSA2* through Hap1p, Rox1p, and Hap2/3/5p. *Free Radical Biology & Medicine*, Vol. 34, No. 5, (Mar 1), pp. (585-597), ISSN 0891-5849.

Wykoff, D. D. & O'Shea, E. K. (2001). Phosphate transport and sensing in *Saccharomyces cerevisiae*. *Genetics*, Vol. 159, No. 4, (Dec), pp. (1491-1499), ISSN 0016-6731.

Wysocki, R. & Tamás, M. J. (2010). How *Saccharomyces cerevisiae* copes with toxic metals and metalloids. *FEMS Microbiology Reviews*, Vol. 34, No. 6, (Nov), pp. (925-951), ISSN 1574-6976; 0168-6445.

Wysocki, R., Bobrowicz, P. & Ulaszewski, S. (1997). The *Saccharomyces cerevisiae* ACR3 gene encodes a putative membrane protein involved in arsenite transport. *The Journal of Biological Chemistry*, Vol. 272, No. 48, (Nov 28), pp. (30061-30066), ISSN 0021-9258.

Yu, J. & Zhou, C. Z. (2007). Crystal structure of glutathione reductase Glr1 from the yeast *Saccharomyces cerevisiae*. *Proteins*, Vol. 68, No. 4, (Sep 1), pp. (972-979), ISSN 1097-0134.

Zitomer, R. S. & Lowry, C. V. (1992). Regulation of gene expression by oxygen in *Saccharomyces cerevisiae*. *Microbiological Reviews*, Vol. 56, No. 1, (Mar), pp. (1-11), ISSN 0146-0749.

Zitomer, R. S., Carrico, P. & Deckert, J. (1997). Regulation of hypoxic gene expression in yeast. *Kidney International*, Vol. 51, No. 2, (Feb), pp. (507-513), ISSN 0085-2538.

16

Complex Regulatory Interplay Between Multidrug Resistance and Oxidative Stress Response in Yeast: The *FLR1* Regulatory Network as a Systems Biology Case-Study

Miguel C. Teixeira[1,2]
[1]IBB - Institute for Biotechnology and Bioengineering,
Centre for Biological and Chemical Engineering,
Instituto Superior Técnico, Lisboa,
[2]Department of Bioengineering,
Instituto Superior Técnico,
Technical University of Lisbon, Lisboa,
Portugal

1. Introduction

Multidrug resistance (MDR), the intrinsic or acquired ability to tolerate toxic concentrations of structurally and functionally diverse chemicals, is a widespread phenomenon that can be found in all living organisms, from bacteria to man (Hayes and Wolf 1997). Its negative consequences include the failure of many therapeutic, antimicrobial and crop protection actions. At the same time, the ability to tolerate multiple stresses is a highly desirable phenotype in organisms used as cell factories that have to cope with fermentation-related stresses (Teixeira et al. 2011b). It is, thus, crucial to understand the molecular basis underlying this phenomenon to be able to circumvent it or to explore it to design more robust industrial strains.

Oxidative stress, on the other hand, is usually considered the result from an imbalance between the generation or influx of reactive oxygen species and the cell ability to readily neutralize these molecules (Ikner and Shiozaki 2005; Lushchak 2011). Increased ROS concentration may lead, in term, to the modification of susceptible biomolecules, such as [Fe-S]-clusters-containing enzymes, proteins exhibiting reactive thiol groups, DNA and lipids, which may undergo peroxidation. These corrupted molecules may, in a small number of cases, be regenerated, but in most cases are degraded or accumulated in cells. The steady-state accumulation of ROS and associated ROS-damaged biomolecules has been linked to ageing and to the development of certain pathologies, such as diabetes mellitus, atherosclerosis and cardiovascular and neurodegenerative diseases (Lushchak 2011). Oxidative stress is usually linked to cell exposure to reactive oxygen species, including hydrogen peroxide, or to redox-cycling agents such as menadione, which leads to superoxide radical generation. However, mounting evidence appears to suggest that many chemical compounds, including widely used pesticides and pharmaceuticals, can induce oxidative stress indirectly, acting as pro-oxidant agents, at the same time that drug resistance mechanisms are activated. The comprehension of the underlying molecular mechanisms is, thus crucial to evaluate the toxicity of these xenobiotics and to design strategies to deal with the arising of multidrug resistance.

Being now clear that these cellular protection programs, crucial to prevent or delay disease progression and ageing, are highly interconnected, it is pivotal to fully understand the underlying cross-mechanisms. Thus, this chapter integrates current knowledge of the link between oxidative stress and multidrug resistance transcriptional control in *S. cerevisiae*, extending it to pathogenic yeasts. The particular case of the regulation of the multidrug resistance transporter Flr1 is further explored as an example of the use of systems biology approaches, including the combination of experimental and computational techniques, to increase our understanding of complex regulatory networks, shedding light into the cross-talk between the MDR phenomenon and oxidative stress response.

2. The multidrug resistance network in yeast

Multidrug resistance is often acquired through the activation of multidrug efflux pumps, belonging to the ATP-Binding Cassette (ABC) or Major Facilitator Superfamilies (MFS), this activation occurring, many times, at the transcriptional level. This fact has led to years of research aiming the definition of the transcription regulatory networks that control the expression of multidrug transporters under stress. The first finding in this field was the discovery that the *PDR1* gene (Saunders and Rank 1982), latter characterized as a transcription factor (Balzi et al. 1987), confers multidrug resistance in the model eukaryote *S. cerevisiae*. Soon after, the so-called PDR (Pleiotropic Drug Resistance) network was first described (Balzi and Goffeau 1995) as a very simple network in which Pdr1, and its homologous transcription factor Pdr3, were found to control the transcription of the *PDR5* gene (Balzi et al. 1994), encoding an ABC drug efflux pump. This network was rapidly extended to include other ABC multidrug transporters, such as Snq2 (Decottignies et al. 1995), but also members of a new family of multidrug transporters of the MFS (Sá-Correia et al. 2009), predicted to function as Drug:H$^+$ Antiporters (DHA) and uncovered mostly upon the release of the *S. cerevisiae* genome sequence (Goffeau et al. 1996), including Flr1 (Brôco et al. 1999; Tenreiro et al. 2001) and Tpo1 (do Valle Matta et al. 2001; Teixeira and Sá-Correia 2002). The use of genome-wide expression analysis tools helped to enlarge this network, while the genome-wide targets of Pdr1 and Pdr3 were uncovered (DeRisi et al. 2000). Apparently, several unrelated drugs and xenobiotics are able to bind to the so-called xenobiotic-binding domain of Pdr1p family members in budding yeast and in the human pathogen *Candida glabrata*, resulting in the over-expression of drug efflux pumps, this finding, providing new clues for the development of novel targets for antifungal drugs (Thakur et al. 2008). Additionally, new transcription factors were also found to belong to the PDR network, based on their homology to Pdr1 and Pdr3. These include Yrr1 (Le Crom et al. 2002), Pdr8 (Hikkel et al. 2003) and Yrm1 (Lucau-Danila et al. 2003), their target-genes also being identified through microarray analysis and, more directly, through ChIP(Chromatine ImmunoPrecipitation)-on-chip analysis. Considering only the canonical PDR transcription factors and the genes encoding predicted multidrug transporters of the ABC and MFS superfamilies, we get a relatively small, but intricate network controlling multidrug resistance in *S. cerevisiae*, as depicted in Fig. 1. However, if we consider all the targets of the same five transcription factors, the PDR network is found to include nearly 500 target genes with a broad scope of biological functions.

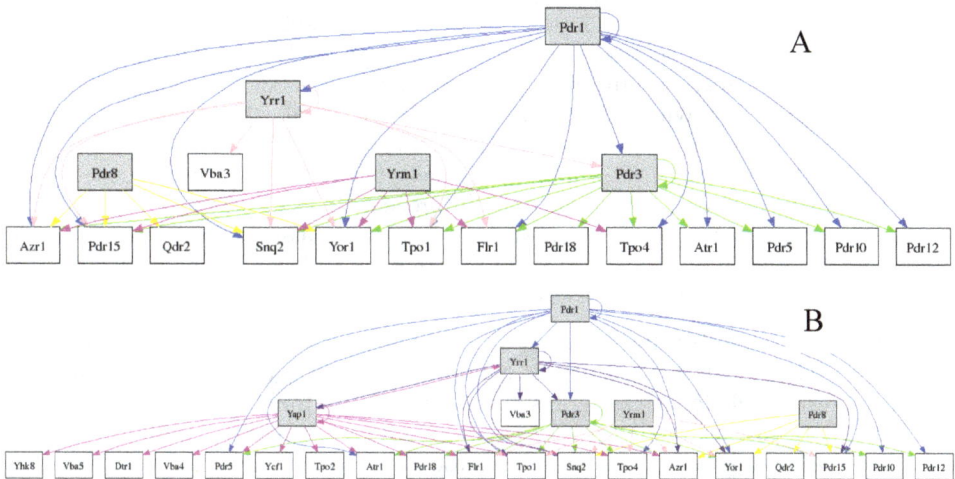

Fig. 1. A: The PDR network, considering only the canonical PDR transcription factors and the genes encoding predicted multidrug transporters of the ABC and MFS superfamilies. B: The PDR network, considering the canonical PDR transcription factors, the oxidative stress response regulator Yap1, and the genes encoding predicted multidrug transporters of the ABC and MFS superfamilies. Both networks were built based on the information gathered in the YEASTRACT database (www.yeastract.com), considering as evidence for transcriptional association between transcription factor and target genes either expression and/or DNA-binding evidence.

Further extending this network, it became clear that other transcription factors, whose function is not primarily linked to multidrug resistance, are also involved in the transcriptional control of drug efflux pumps. The first non-PDR transcription factor to join this network was Yap1, the major regulator of oxidative stress response in yeast (Rodrigues-Pousada et al. 2010), found to confer resistance to the drug diazaborine via the Pdr3 and, less significantly, Pdr1 transcription factors (Jungwirth et al. 2000; Wendler et al. 1997). Two other transcriptions factors found to relate to the PDR network are Rpn4 (Owsianik et al. 2002; Teixeira et al. 2008) and Hsf1 (Hahn et al. 2006), regulators of proteasomal genes and of the heat shock response, respectively.

3. The role of yap1 in multidrug resistance in yeast

Recent studies in this field focusing the model eukaryote *Saccharomyces cerevisiae* have shown that there seems to be a close cross-talk between the multidrug resistance regulatory network and the oxidative stress response transcription factor Yap1. Indeed, Yap1 was demonstrated to confer resistance against a wide variety of drugs, including quinine, rapamycin, trenimon and diazaborine, but also to antifungal agents, such as cerulenin, benomyl, cycloheximide, fenpropimorph, mancozeb, to herbicides, including sulfometuron methyl, 2,4-dichlorophenoxyacetic acid (2,4-D) and paraquat and to the food preservative acetic acid. Although some of these compounds have been described as pro-oxidants molecules (Dias et al. 2010; Semchyshyn et al.; Teixeira et al. 2004), the role of Yap1 in drug resistance seems to rely not only in the control of antioxidant defenses, but also on the control of multidrug resistance transporters. In fact, Yap1 has been shown to underlie the stress-induced up-regulation of the multidrug ABC transporters Pdr5, Pdr18, Snq2, and Ycf1 (Cabrito et al. 2011; Jungwirth and Kuchler 2006; Teixeira et al. 2006) and of the drug:H^+ antiporters Atr1, Azr1, Dtr1, Flr1, Qdr3, Tpo1, Tpo2, Tpo4, and Yhk8 (Sá-Correia et al. 2009; Teixeira et al. 2011a) (Table 1). If we take a global look, it becomes clear that the role of Yap1 in the regulation of the PDR network is even broader. Indeed, using the YEASTRACT database, a repository of all demonstrated regulatory associations in *S. cerevisiae* (Abdulrehman et al. 2011; Monteiro et al. 2008; Teixeira et al. 2006), it is possible to see that Yap1 co-regulates around 18% of the Pdr1-target genes (Fig. 2A).

Interestingly, Yap1 displays two different activation mechanisms depending on the nature of the imposed stress. In both cases, the molecular events triggering Yap1 activation are apparently responsible for releasing this transcription factor from the interaction with the exportin Crm1, thus leading to its nuclear accumulation (Yan et al. 1998). One of the activation mechanisms occurs due to the increase in intracellular ROS concentration, due, for example, to cellular exposure to H_2O_2. Hydrogen peroxide appears to, indirectly, lead to the formation of an intramolecular disulfide bond between Cys303 and Cys598 of Yap1 (Delaunay et al. 2000). On the other hand, a second redox centre was later found in this transcription factor, and suggested to involve the direct binding of electrophiles such as N-ethylmaleimide (Azevedo et al. 2003) to Cys598, Cys620 and Cys629, thus inducing a conformational change that also prevents Yap1-Crm1 binding and, thus, leads to Yap1 accumulation in the nucleus. Given this differential activation mechanism, the question of whether Yap1 could regulate distinct target-gene sets under different stress conditions arose.

Microarray analysis was recently used to compare the Yap1-dependent transcriptional response to hydrogen peroxide and to the thiol-reactive compounds N-ethylmaleimide (NEM) and acrolein (Ouyang et al. 2011). The obtained results showed that 56 genes are exclusive of the response to H_2O_2, while 327 are exclusive of the response to NEM or acrolein. Although both responses were primarily under the control of the same transcription factor, in each case the elicited response resulted in the expression of protective genes specific for each of the imposed stresses (Ouyang et al. 2011). This specificity appears to result from the differential mechanisms of Yap1 activation imposed by the analyzed stress agents. The global analysis of the role of Yap1 in yeast response to benomyl induced stress had also highlighted the differences between the gene-sets up-regulated by Yap1 in response to ROS or to thiol-reactive compounds (Lucau-Danila et al. 2005). Genes required for the maintenance of redox balance were shown to be up-regulated in both cases, while specific genes such as *SOD1* and *CTT1*, encoding the cytosolic superoxide dismutase and catalase, respectively, are only responsive to ROS. An interesting discovery from this study was that the promoter occupancy by Yap1, when activated by benomyl, increases in all the promoters of Yap1 targets genes, including highly up-regulated genes such as *FLR1*, but also non-responsive genes such as *CTT1* and *SOD1* (Lucau-Danila et al. 2005). This finding

Gene	Binding evidence	Expression evidence (under stress)	References
Drug:H+ Antiporters			
ATR1	+	Arsenic, Arsenite, Hydrogen peroxide, Nitric oxide, Selenite	(Coleman et al. 1997; Harbison et al. 2004; Haugen et al. 2004; Horan et al. 2006; Kelley and Ideker 2009; Lucau-Danila et al. 2005; Salin et al. 2008; Thorsen et al. 2007; Workman et al. 2006)
AZR1	+	Arsenic	(Haugen et al. 2004; Salin et al. 2008)
DTR1	+	-	(Salin et al. 2008)
FLR1	+	Arsenic, Arsenite, Benomyl, Diamide, Diazaborine, Diethylmaleate, Hydrogen peroxide, Hydroxyurea, Methylmethane sulfonate, Tert-butyl hydroperoxide, Selenite	(Alarco et al. 1997; Brôco et al. 1999; Dubacq et al. 2006; Haugen et al. 2004; Jungwirth et al. 2000; Kelley and Ideker 2009; Lucau-Danila et al. 2005; Nguyen et al. 2001; Salin et al. 2008; Teixeira et al. 2010; Teixeira et al. 2008; Tenreiro et al. 2001; Thorsen et al. 2007; Workman et al. 2006)
QDR3	-	Spermine, Spermidine	(Teixeira et al. 2011a)
TPO1	-	Benomyl	(Lucau-Danila et al. 2005)
TPO2	-	Arsenite	(Thorsen et al. 2007)
TPO4	-	Arsenic	(Haugen et al. 2004)
YHK8	+	Nitric oxide	(Harbison et al. 2004; Horan et al. 2006; Lee et al. 2002)
Pleiotropic Drug Resistance ABC transporters			
PDR5	+	Arsenite, Benomyl, Hydrogen peroxide	(Kelley and Ideker 2009; Lucau-Danila et al. 2005; Salin et al. 2008; Thorsen et al. 2007)
PDR18	+	2,4-D	(Cabrito et al. 2011; Salin et al. 2008)
SNQ2	+	Arsenite, Benomyl	(Harbison et al. 2004; Lee et al. 2002; Lucau-Danila et al. 2005; Salin et al. 2008; Thorsen et al. 2007; Workman et al. 2006)
YCF1	+	Arsenic, Cadmium, Diazaborine	(Haugen et al. 2004; Jungwirth et al. 2000; Lucau-Danila et al. 2005; Salin et al. 2008; Wemmie et al. 1994)

Table 1. *S. cerevisiae* multidrug resistance transporter encoding genes under the control of Yap1, according to the YEASTRACT database (www.yeastract.com). Whether there is evidence (+) or not (-) for Yap1 binding to the promoter regions of the selected genes is indicated. The stress conditions leading to target gene up-regulation under Yap1 control are also highlighted. Supporting references are provided.

reinforces the possibility that the different mechanisms of Yap1 activation lead to diverse conformational changes which do not deeply affect Yap1 binding ability, but rather its action as a transcriptional activator, allowing this transcription factor to discriminate, among its target genes, those that should be up-regulated in each condition.

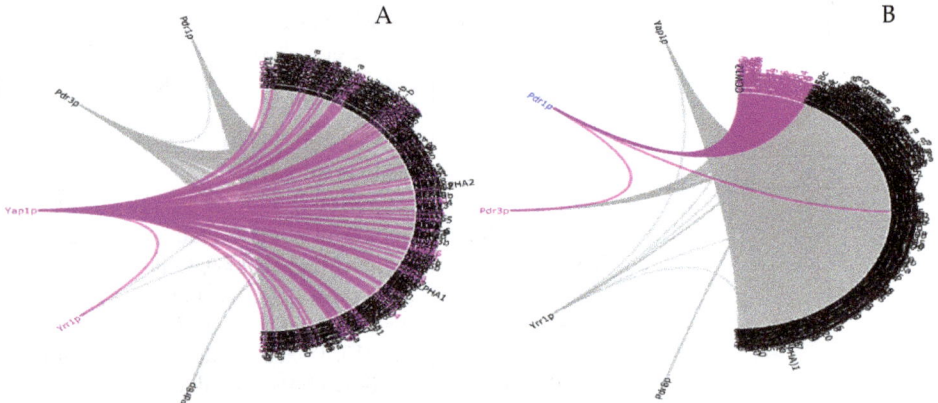

Fig. 2. Global overlapping between the Yap1 and the PDR regulatory networks. On the left, the participation of the Yap1 transcription factor in the regulation of around 18% of the Pdr1 targets is highlighted, while the right panel emphasizes the role of Pdr1 in the regulation of about 15% of the Yap1 regulon. Both networks were built based on the information gathered in the YEASTRACT database (www.yeastract.com), considering only DNA-binding evidence for the establishment of transcriptional association between transcription factors and target genes.

4. The role of the pdr network in oxidative stress response

Since there is a clear role of Yap1 in MDR, the hypothesis that the PDR network may also play a role in oxidative stress seems logical. Although, to date there is no evidence supporting that the PDR network plays a clear role in the response to oxidative stress induced by ROS, several studies have highlighted the role of this multidrug resistance network in the response to pro-oxidant drugs and xenobiotics (Lelandais and Devaux). These include the agricultural fungicides mancozeb (Teixeira et al. 2010; Teixeira et al. 2008) and benomyl (Lucau-Danila et al. 2005), the herbicide 2,4-dichlorophenoxyacetic acid (Teixeira et al. 2007), the redox-cycling agent menadione and selenite (Salin et al. 2008). Indeed, in yeast cells exposed to pro-oxidants and metalloids, a cooperation between the transcription factors Pdr1/Pdr3 and Yap1, Rpn4 and Hsf1 in the modulation of oxidative stress response appears to exist.

In the particular case of the seletine stress response, microarray analysis was used to check the transcriptome-wide effect of the deletion of the transcription factor encoding genes *YAP1*, *RPN4*, *PDR1* and *PDR3*. It was found that the absence of Pdr1 or Pdr3 affected the expression of around 20% of the Yap1 targets genes induced under selenite stress. These shared genes were found to include chemical stress response genes such as *FLR1*, as expected, but also a sub-group of oxidative stress responsive genes. When taking a global look at the information gathered in the YEASTRACT database (Abdulrehman et al. 2011; Monteiro et al. 2008; Teixeira et al. 2006), it becomes clear that around 15% of the Yap1 target genes are also Pdr1 targets (Fig. 2B). Among these shared targets a small, but significant set of genes encoding direct antioxidant enzymes have been shown to be direct targets of the PDR transcription factors, including the cytosolic catalase Ctt1 (Devaux et al. 2002; Hikkel et al. 2003), and the alkyl hydroperoxide reductase Ahp1 (Larochelle et al. 2006) (Fig. 3).

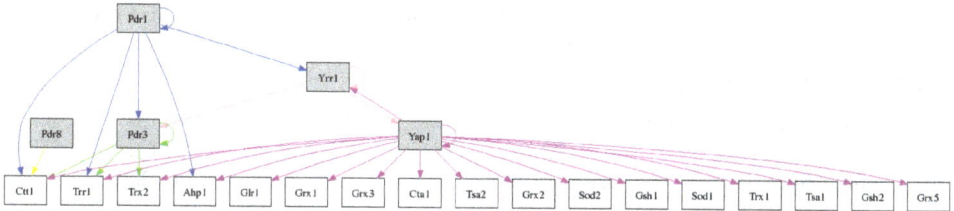

Fig. 3. The core oxidative stress response regulatory control, focused on the transcription factor Yap1 and on the yeast genes encoding antioxidant enzymes or proteins required for the maintenance of cellular redox balance. The role of the PDR network transcription factors in the regulation of these genes is displayed, based on the information gathered in the YEASTRACT database (www.yeastract.com).

In this context, it is also important to point out the specific role of Pdr3, but not of Pdr1, in the response to mitochondrial dysfunction (Devaux et al. 2002; Hallstrom and Moye-Rowley 2000), one of the main natural sources of oxidative imbalance. Indeed, upon the deletion of the mitochondrial genome, Pdr3 was seen to play a role in the activation of 14, out of 54, genes whose expression changes in these circumstances, placing Pdr3 as one of the transcription factors responsible for the retrograde response pathway (Devaux et al. 2002). Differently from what happens under chemical stress, upon mitochondrial dysfunction Pdr3 was seen to be post-translationally modified, but the exact nature of this modification was not clarified to date. The proposed role of this transcription factor in the response to mitochondrial dysfunction, as a controller of plasma membrane properties, still remains to be elucidated.

5. Cross-talk between multidrug resistance control and oxidative stress response in pathogenic yeasts

The study of multidrug resistance in pathogenic yeast species has been guided, to some extent, by the knowledge gathered for *S. cerevisiae*, as a model organism. A particularly close degree of similarity can be found when comparing this model organism with *Candida glabrata*, while the observations made for *C. albicans* and other *Candida* species reveal a lower conservation in terms of MDR regulation.

Clinical multiple antifungal drug resistance in *C. albicans* is mostly found to be based on the over-expression of the ABC multidrug efflux pumps encoded by *CDR1* (Prasad et al. 1995) and *CDR2* (Sanglard et al. 1997) genes, which share a high degree of homology with *S. cerevisiae PDR5*, and of the MFS drug:H+ antiporter *MDR1* (Goldway et al. 1995), a close homologue to *S. cerevisiae FLR1* gene. *FLU1*, another *C. albicans* drug:H+ antiporter encoding gene, was also found to confer fluconazole resistance, but to a lesser extent (Calabrese et al.

2000). Interestingly, the regulation of these multidrug transporters shares some similarity to that of their homologues in budding yeast. The *C. albicans* transcription factor Tac1, belonging to the *S. cerevisiae* Pdr1/Pdr3 protein family, was found to be required for *CDR1* and *CDR2* up-regulation induced by the drug fluphenazine (Coste et al. 2004). Another Pdr1/Pdr3 homologous transcription factor, Mrr1, was found to control the expression of *MDR1*, in both fluconazole-resistant clinical isolates and in benomyl-or hydrogen peroxide-challenged cells (Morschhauser et al. 2007). Of particular interest, in the context of this review, is the fact that the Yap1 homologue from *C. albicans*, named Cap1, is also involved in multidrug resistance (Alarco et al. 1997; Alarco and Raymond 1999), controlling *MDR1* expression, directly binding to its promoter region (Znaidi et al. 2009).

In *C. glabrata*, multidrug resistance relies mostly on the ABC drug efflux pumps *CgCDR1* (a *ScPDR5* homologue), *CgPDH1/CgCDR2* (a *ScPDR15* homologue) and *CgSNQ2* (a *ScSNQ2* homologue), but also on the drug-H+ antiporter *CgFLR1* (Chen et al. 2007). In this pathogenic yeast a single homologue of the budding yeast transcription factors Pdr1/Pdr3, CgPdr1, appears to control antifungal drug resistance through its action as an activator of all of the above mentioned ABC transporter encoding genes. This role is not only seen in the response of laboratory strains to suddenly imposed stress, but also in azole-resistant clinical isolates (Torelli et al. 2008; Tsai et al. 2006). The role of other *C. glabrata* CgPdr1 homologues, such as CgYrm1, has not been inspected so far. The *C. glabrata* Yap1 homologue, Cgap1, has also been related to the control of multidrug resistance transporters. Specifically, it was found to be required for *CgFLR1* up-regulation in response to benomyl-induced stress (Chen et al. 2007). Although this transcription factor was not seen to confer antifungal drug resistance its expression does increase *C. glabrata* tolerance to toxic concentrations of various oxidants and other xenobiotics (Chen et al. 2007). It is expectable that the understanding of the complex transcriptional regulation of MDR in this less well-studied organism will increase in the near future, guided by the huge amount of information that is being provided through genome-wide approaches. For example, microarray analysis revealed that ORF *CAGL0G08624g*, encoding a close homolog to the *S. cerevisiae* MFS-MDR transporter Qdr2 [T4, (Vargas et al. 2004)], is transcriptionally activated in response to fluconazole induced stress, in the dependency of the CgPdr1 transcription factor (Vermitsky and Edlind 2004). A more recent transcriptomics study showed that the expression of *CgFLR1* and ORF *CAGL0G03927g* (a *ScTPO1* homologue) genes is up-regulated in cells challenged with benomyl, under the control of the Cgap1 transcription factor (Lelandais et al. 2008). Although this subject has only now began to be unraveled in *C. glabrata*, current results already allow us to build a relatively small PDR network (Caudle et al.; Ferrari et al.; Lelandais et al. 2008; Tsai et al. 2006), including Cgap1, as depicted in Fig. 4.

Altogether, these results reinforce the notion that the transcriptional control of multidrug resistance and oxidative stress response are highly interconnected processes in yeasts and suggest that this crosstalk may be extended to other more complex eurakyotes.

6. The combinatorial regulation of the multidrug transporter Flr1: A systems biology case-study

The *FLR1* gene, encoding a plasma membrane drug:H+ antiporter, was one of the first of its family to be characterized. Although it derives its name from FLuconazol Resistance (Alarco et al. 1997), Flr1 has been shown to confer resistance to a large number of chemically and

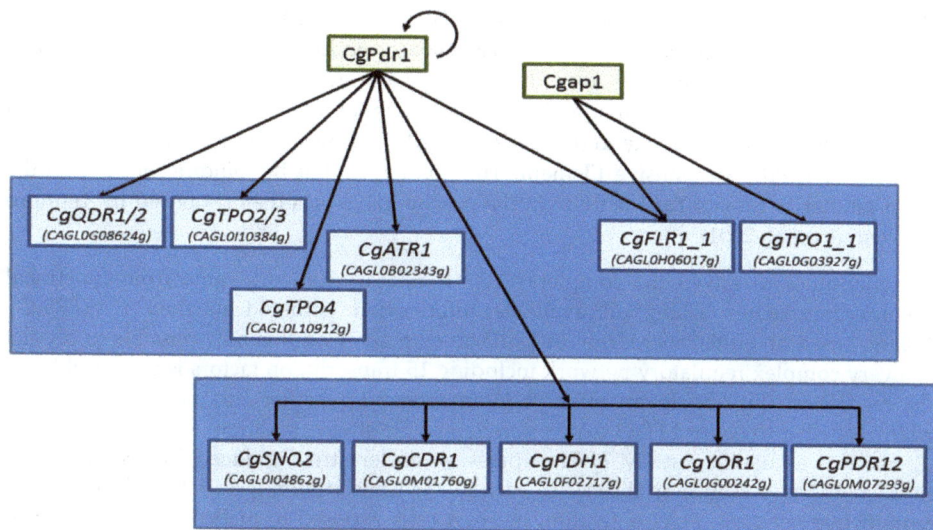

Fig. 4. The PDR network in the human pathogen *Candida glabrata*, based on the results of the few global studies carried so far on the subject (Caudle et al.; Ferrari et al.; Lelandais et al. 2008; Tsai et al. 2006). The role of the transcription factors CgPdr1 and Cgap1 in the regulation of the genes encoding predicted drug:H+ antiporters of the MFS (upper box) or multidrug resistance ABC efflux pumps (lower box) is highlighted.

structurally unrelated xenobiotics and drugs, including cycloheximide, 4-nitroquinoline-1-oxide (4-NQO), benomyl, methotrexate, diazaborine, cerulenin, diamide, diethylmaleate, menadione, paracetamol and mancozeb (Alarco et al. 1997; Brôco et al. 1999; Jungwirth et al. 2000; Srikanth et al. 2005; Teixeira et al. 2008). Unlike many of the DHA encoding genes (Sá-Correia et al. 2009), *FLR1* is highly induced at the transcriptional level when yeast cells are exposed to the stresses this gene confers resistance to. This high responsiveness to stress, made *FLR1* transcriptional control an attractive working model to study complex transcriptional regulation mechanisms. The first indication on *FLR1* transcriptional control, came very early on, when *FLR1* was identified as a Yap1 target in *S. cerevisiae* (Alarco et al. 1997). Upon Yap1 deletion, the up-regulation of *FLR1*, found to occur in yeast cells exposed to the fungicide benomyl, was seen to be completely abrogated (Brôco et al. 1999; Tenreiro et al. 2001). At the same time, maximal activation of *FLR1* under benomyl stress was found to be dependent on the presence of an additional transcription factor, Pdr3 (Brôco et al. 1999; Tenreiro et al. 2001). Interestingly, *FLR1* promoter region was found to include three putative Yap1-Responsive Elements (YRE1-3) (Fig. 6). In a detailed study of the role of each of these predicted binding sites, Nguyen and co-workers found that the three binding sites were functional, but that their relative importance depends on the imposed stress (Nguyen et al. 2001). Indeed, using site-directed mutagenesis to remove each of the Yap1-binding sites, YRE3 (-364) was found to be the major player in *FLR1* activation under stress imposed by benomyl and diethylmaleate. However, YRE2 (-167) becomes the most significant YRE in *FLR1* up-regulation induced by hydrogen peroxide, diamide and *tert*-butyl hydroperoxide. Finally, all three YREs are equally important to assure full activation of *FLR1* in response to methylmethane sulfonate (Nguyen et al. 2001). This finding may relate to the fact that at least some of these stresses lead to different Yap1 conformations, which in term may change the transcription factor's affinity towards the possible variations of the Yap1-Responsive Element. Furthermore, the fact that YRE3 is responsible for 90% of the *FLR1* up-regulation induced by benomyl, may relate to the fact this Yap1-binding site is in the proximity of the predicted Pdr1/Pdr3-Responsive Element. The possibility that the binding of Pdr3 to the *FLR1* promoter facilitates Yap1 activity in the nearby YRE3 was then proposed by Tenreiro et al (Tenreiro et al. 2001).

Additional clues to unveil the complete *FLR1* regulatory network came from functional genomics approaches. Using all available information gathered in the YEASTRACT database, based on either microarray or ChIP-on-chip analysis, it is now possible to predict that a very complex regulatory network including 15 transcription factors is responsible for *FLR1* regulation (Fig. 5). In an attempt to understand whether this network could be working together to control *FLR1* expression in a single stress condition, an *FLR1* promoter-lacZ fusion was used to study *FLR1* expression under stress induced by the fungicide mancozeb, in the presence or absence of each of the transcription factors found to occur in the predictive network. *FLR1* activation in yeast cells exposed to mancozeb was found to depend upon Yap1 and Pdr3, as previously registered under benomyl stress. However, two additional transcription factors were found to be required for mancozeb-induced *FLR1* maximal activation: Yrr1, one of the transcription factors controlling the PDR network, and Rpn4, a proteasomal gene regulator (Teixeira et al. 2008).

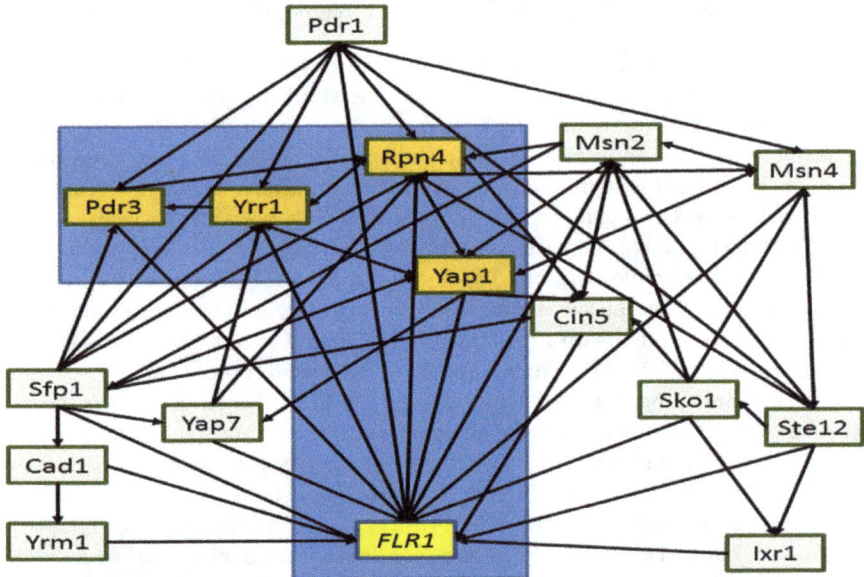

Fig. 5. Network of transcription factors documented as affecting *FLR1* expression, as retrieved from the YEASTRACT database (www.yeastract.com). The L-shaped box indicates the sub-network found to affect *FLR1* up-regulation occurring in yeast cells challenged with the fungicide mancozeb (Teixeira et al. 2008).

Given the fact that it is becoming increasingly clear that transcriptional regulation is far more complex that initially foreseen, the use of systems biology approaches seems to be the only way to pursue the goal of understanding such biological processes in all their depth. Indeed, Systems Biology aims at understanding living processes as systems of multiple interacting components, preferably at a genome-wide scale, and through the combination of experimental and computational approaches. Such an approach was undertaken to further analyze the *FLR1* regulatory network. The profiles of *FLR1*, *YAP1*, *PDR3*, *YRR1* and *RPN4* transcript levels were registered during the period of adaptation of a yeast cell population to stress imposed by the fungicide mancozeb, in wild-type cells and in mutants devoid of each of the four transcription factors (Teixeira et al. 2008). This information was used to build a mathematical description of the *FLR1* network (Teixeira et al. 2010), taking advantage of the freely available GNA software (de Jong et al. 2003). This modeling approach allowed the testing of new hypothesis in silico (Monteiro et al. 2011), providing guidance for the design of further experimental work (Teixeira et al. 2010). The comparison between simulated and experimentally obtained results led to a refined understanding of the network, including the realization that a fifth still unidentified transcription factor, denominated FactorX, has to be included in the network to fully explain the observed transcriptional profiles (Fig. 6). Furthermore, combined results suggested that Yap1 and Yrr1 may function together, eventually working as a heterodimer, in the co-regulation of their shared target genes, which include the multidrug transporter encoding genes *FLR1*, *AZR1* and *SNQ2* and the transcription factor encoding genes *PDR3*, *YRR1* and *RPN4* (Teixeira et al. 2010).

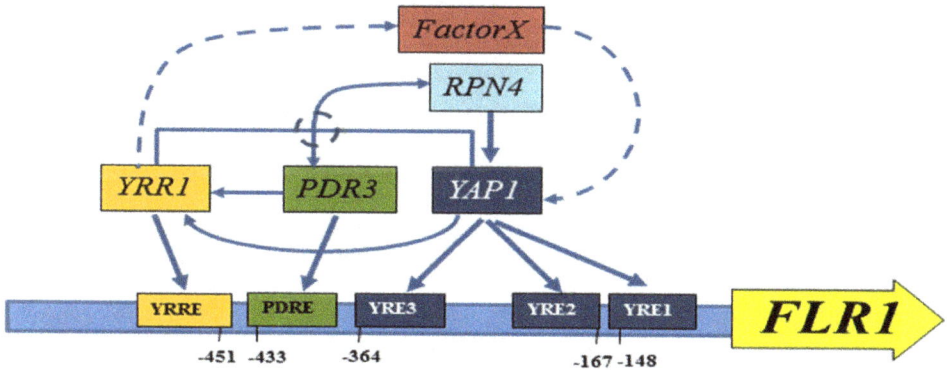

Fig. 6. *FLR1* regulatory network structure, found to be functional in yeast cells exposed to mancozeb stress, as obtained from the combination of computational and experimental approaches described by Teixeira et al (Monteiro et al. 2011; Teixeira et al. 2010). Dashed arrows indicate the new aspects of the network suggested by the used systems biology approach that are still to be validated.

Studies in *C. glabrata* and *C. albicans* showed that the *ScFLR1* homologues *CgFLR1* and *CaMDR1* are regulated by Yap1p homologs in either species (Chen et al. 2007; Znaidi et al. 2009). Whether the *C. glabrata* homologs of the *S. cerevisiae* Yrr1 and Rpn4 transcription factors also play a role in *CgFLR1* regulation is still an open question. An interesting clue comes from the fact that the *C. albicans* homologue of ScPdr3, CaMrr1, is also required for full *CaMDR1* activation (Morschhauser et al. 2007). These results strongly suggest that there may be a significant degree of conservation between multidrug resistance control in these related yeast species.

The current version of the *FLR1* regulatory network is small but intricate. We believe that it is a good example of how complex transcription control in eukaryotes may be. Furthermore, it provides a good platform for further studies on the connection between oxidative stress response and two additional biological processes: multidrug resistance and protein degradation through the proteasome.

7. Conclusion and perspectives

Multidrug resistance control and stress response, in particular oxidative stress response, are now recognized as two complex cellular processes that coexist and interplay to allow cells to thrive in harsh environments. This chapter highlights current knowledge on the cross-talk between oxidative stress and MDR regulation in the model eukaryote *S. cerevisiae*. It emphasizes that, although being more focused on their canonical roles, both Yap1 and the PDR regulators play important functions in the regulation of multidrug efflux pumps and antioxidant enzymes, respectively.

At the moment, the logical explanation for this cross-talk appears to lie on the fact that many drugs and other xenobiotics may exert pro-oxidant effects, thus activating both multidrug resistance and oxidative stress response pathways. A question that still remains to be answered relates to why multidrug transporters should be controlled by Yap1 and more specifically what might be their role, if any, in oxidative stress response. For the single case of the vacuolar membrane ABC transporter Ycf1, there seems to be a possible connection. Ycf1, a close homologue to the human MRP1 multidrug transporter, confers resistance to chemical stress, including metal ions, antimonite, arsenite, 1-chloro-2,4-dinitrobenzene and diazaborine. The role of Ycf1 in metal ion resistance was further explored and this transporter was found to mediate the vacuolar compartmentalization of S-glutathione conjugates (Li et al. 1996). Glutathione, whose concentration is controlled by Yap1, through the regulation of the glutathione synthetase encoding gene *GSH1*, plays a crucial role in the maintenance of the intracellular redox potential. It is thus reasonable to think that the oxidative stress response regulator Yap1 may coordinately control the expression of *GSH1* and *YCF1*, to assure the maintenance of the physiological concentration of free cytosolic reduced glutathione. As for the remaining multidrug transporters controlled by Yap1, such a close link between drug detoxification and oxidative stress response remains to be established. Nonetheless, it is reasonable to think that, since many of the natural chemical stress inducers are also capable of unbalancing the cellular redox state, the oxidative stress signaling would also control the expression of membrane transporters capable of relieving the cell from the exogenous source of oxidative stress. Interestingly, in bacterial systems a rather similar coordination of the response to xenobiotics and oxidants can also be found, under the control of the SoxRS regulon. Indeed, the *E. coli* SoxRS transcription factor controls the expression of both antioxidant enzymes and also of, at least, the outer membrane protein (porin) F, OmpF, suggested to play a role in reducing cell membrane permeability towards ROS or ROS-generating compounds (reviewed in (Lushchak 2011)).

Altogether, the results reviewed herein also highlight the fact that transcriptional control is much more complex than initially foreseen. Indeed, we come to realize that it is not possible to study individual phenomena, as if there was no influence from the surrounding, and even interconnected, cellular processes. The compilation of all the regulatory associations identified, so far, in *S. cerevisiae*, deposited in the YEASTRACT database (Abdulrehman et al. 2011; Monteiro et al. 2008; Teixeira et al. 2006), reveals that there are more than 48,000 regulatory associations between transcription factors and target genes. This number, which indicates that, on average, each yeast gene is controlled by at least 8 different transcription factors, rises up to nearly 375,000, when making in silico predictions based on the occurrence of transcription factor recognition sequences in the yeast promoter regions. Given this high degree of complexity, the study of biological networks using the new interdisciplinary approaches of Systems Biology seems to be the most suitable way to tackle this issue. The small, but intricate case-study explored herein, focused on the *S. cerevisiae FLR1* regulatory network, suggests that the use of computer modeling, as a systems biology tool, will be crucial to increase our understanding of the cross-talk between regulatory networks.

8. Acknowledgements

Work on transcriptional control of yeast multidrug resistance transporters in our laboratory is being financially supported by FEDER and Fundação para a Ciência e Tecnologia (FCT) (contracts PTDC/BIO/72063/2006, PTDC/BIA-MIC/72577/2006 and PTDC/EBB-BIO/119356/2010).

9. References

Abdulrehman D, Monteiro PT, Teixeira MC, Mira NP, Lourenco AB, Dos Santos SC, Cabrito TR, Francisco AP, Madeira SC, Aires RS, Oliveira AL, Sá-Correia I, Freitas AT (2011) YEASTRACT: providing a programmatic access to curated transcriptional regulatory associations in *Saccharomyces cerevisiae* through a web services interface. *Nucleic Acids Res,* 39, pp. D136-140

Alarco AM, Balan I, Talibi D, Mainville N, Raymond M (1997) AP1-mediated multidrug resistance in *Saccharomyces cerevisiae* requires *FLR1* encoding a transporter of the major facilitator superfamily. *J Biol Chem,* 272, pp. 19304-19313

Alarco AM, Raymond M (1999) The bZip transcription factor Cap1p is involved in multidrug resistance and oxidative stress response in *Candida albicans. J Bacteriol,* 181, pp. 700-708

Azevedo D, Tacnet F, Delaunay A, Rodrigues-Pousada C, Toledano MB (2003) Two redox centers within Yap1 for H_2O_2 and thiol-reactive chemicals signaling. *Free Radic Biol Med,* 35, pp. 889-900

Balzi E, Chen W, Ulaszewski S, Capieaux E, Goffeau A (1987) The multidrug resistance gene *PDR1* from *Saccharomyces cerevisiae. J Biol Chem,* 262, pp. 16871-16879

Balzi E, Goffeau A (1995) Yeast multidrug resistance: the PDR network. *J Bioenerg Biomembr,* 27, pp. 71-76

Balzi E, Wang M, Leterme S, Van Dyck L, Goffeau A (1994) PDR5, a novel yeast multidrug resistance conferring transporter controlled by the transcription regulator PDR1. *J Biol Chem,* 269, pp. 2206-2214

Brôco N, Tenreiro S, Viegas CA, Sá-Correia I (1999) *FLR1* gene (ORF *YBR008c*) is required for benomyl and methotrexate resistance in *Saccharomyces cerevisiae* and its benomyl-induced expression is dependent on Pdr3 transcriptional regulator. *Yeast*, 15, pp. 1595-1608

Cabrito TR, Teixeira MC, Singh A, Prasad R, Sá-Correia I (2011) The yeast ABC transporter Pdr18 (ORF *YNR070w*) controls plasma membrane sterol composition, playing a role in multidrug resistance. *Biochem J*, 440, 195-202

Calabrese D, Bille J, Sanglard D (2000) A novel multidrug efflux transporter gene of the major facilitator superfamily from *Candida albicans* (*FLU1*) conferring resistance to fluconazole. *Microbiology*, 146 (Pt 11), pp. 2743-2754

Caudle KE, Barker KS, Wiederhold NP, Xu L, Homayouni R, Rogers PD Genomewide expression profile analysis of the *Candida glabrata* Pdr1 regulon. *Eukaryot Cell*, 10, pp. 373-383

Chen KH, Miyazaki T, Tsai HF, Bennett JE (2007) The bZip transcription factor Cgap1p is involved in multidrug resistance and required for activation of multidrug transporter gene *CgFLR1* in *Candida glabrata*. *Gene*, 386, pp. 63-72

Coleman ST, Tseng E, Moye-Rowley WS (1997) *Saccharomyces cerevisiae* basic region-leucine zipper protein regulatory networks converge at the *ATR1* structural gene. *J Biol Chem*, 272, pp. 23224-23230

Coste AT, Karababa M, Ischer F, Bille J, Sanglard D (2004) *TAC1*, transcriptional activator of CDR genes, is a new transcription factor involved in the regulation of *Candida albicans* ABC transporters *CDR1* and *CDR2*. *Eukaryot Cell*, 3, pp. 1639-1652

de Jong H, Geiselmann J, Hernandez C, Page M (2003) Genetic Network Analyzer: qualitative simulation of genetic regulatory networks. *Bioinformatics*, 19, pp. 336-344

Decottignies A, Lambert L, Catty P, Degand H, Epping EA, Moye-Rowley WS, Balzi E, Goffeau A (1995) Identification and characterization of Snq2, a new multidrug ATP binding cassette transporter of the yeast plasma membrane. *J Biol Chem*, 270, pp. 18150-18157

Delaunay A, Isnard AD, Toledano MB (2000) H2O2 sensing through oxidation of the Yap1 transcription factor. *EMBO J*, 19, pp. 5157-5166

DeRisi J, van den Hazel B, Marc P, Balzi E, Brown P, Jacq C, Goffeau A (2000) Genome microarray analysis of transcriptional activation in multidrug resistance yeast mutants. *FEBS Lett*, 470, pp. 156-160

Devaux F, Carvajal E, Moye-Rowley S, Jacq C (2002) Genome-wide studies on the nuclear PDR3-controlled response to mitochondrial dysfunction in yeast. *FEBS Lett*, 515, pp. 25-28

Dias PJ, Teixeira MC, Telo JP, Sá-Correia I (2010) Insights into the mechanisms of toxicity and tolerance to the agricultural fungicide mancozeb in yeast, as suggested by a chemogenomic approach. *OMICS*, 14, pp. 211-227

do Valle Matta MA, Jonniaux JL, Balzi E, Goffeau A, van den Hazel B (2001) Novel target genes of the yeast regulator Pdr1p: a contribution of the *TPO1* gene in resistance to quinidine and other drugs. *Gene*, 272, pp. 111-119

Dubacq C, Chevalier A, Courbeyrette R, Petat C, Gidrol X, Mann C (2006) Role of the iron mobilization and oxidative stress regulons in the genomic response of yeast to hydroxyurea. *Mol Genet Genomics*, 275, pp. 114-124

Ferrari S, Sanguinetti M, Torelli R, Posteraro B, Sanglard D Contribution of CgPDR1-regulated genes in enhanced virulence of azole-resistant *Candida glabrata*. *PLoS One*, 6, pp. e17589

Goffeau A, Barrell BG, Bussey H, Davis RW, Dujon B, Feldmann H, Galibert F, Hoheisel JD, Jacq C, Johnston M, Louis EJ, Mewes HW, Murakami Y, Philippsen P, Tettelin H, Oliver SG (1996) Life with 6000 genes. *Science*, 274, pp. 546, 563-547

Goldway M, Teff D, Schmidt R, Oppenheim AB, Koltin Y (1995) Multidrug resistance in *Candida albicans*: disruption of the *BENr* gene. *Antimicrob Agents Chemother*, 39, pp. 422-426

Hahn JS, Neef DW, Thiele DJ (2006) A stress regulatory network for co-ordinated activation of proteasome expression mediated by yeast heat shock transcription factor. *Mol Microbiol*, 60, pp. 240-251

Hallstrom TC, Moye-Rowley WS (2000) Multiple signals from dysfunctional mitochondria activate the pleiotropic drug resistance pathway in *Saccharomyces cerevisiae*. *J Biol Chem*, 275, pp. 37347-37356

Harbison CT, Gordon DB, Lee TI, Rinaldi NJ, Macisaac KD, Danford TW, Hannett NM, Tagne JB, Reynolds DB, Yoo J, Jennings EG, Zeitlinger J, Pokholok DK, Kellis M, Rolfe PA, Takusagawa KT, Lander ES, Gifford DK, Fraenkel E, Young RA (2004) Transcriptional regulatory code of a eukaryotic genome. *Nature*, 431, pp. 99-104

Haugen AC, Kelley R, Collins JB, Tucker CJ, Deng C, Afshari CA, Brown JM, Ideker T, Van Houten B (2004) Integrating phenotypic and expression profiles to map arsenic-response networks. *Genome Biol*, 5, pp. R95

Hayes JD, Wolf CR (1997) Molecular genetics of drug resistance. In: *Modern Genetics* v. 3:Amsterdam, the Netherlands: Harwood Academic.

Hikkel I, Lucau-Danila A, Delaveau T, Marc P, Devaux F, Jacq C (2003) A general strategy to uncover transcription factor properties identifies a new regulator of drug resistance in yeast. *J Biol Chem*, 278, pp. 11427-11432

Horan S, Bourges I, Meunier B (2006) Transcriptional response to nitrosative stress in *Saccharomyces cerevisiae*. *Yeast*, 23, pp. 519-535

Ikner A, Shiozaki K (2005) Yeast signaling pathways in the oxidative stress response. *Mutat Res*, 569, pp. 13-27

Jungwirth H, Kuchler K (2006) Yeast ABC transporters-- a tale of sex, stress, drugs and aging. *FEBS Lett*, 580, pp. 1131-1138

Jungwirth H, Wendler F, Platzer B, Bergler H, Hogenauer G (2000) Diazaborine resistance in yeast involves the efflux pumps Ycf1p and Flr1p and is enhanced by a gain-of-function allele of gene *YAP1*. *Eur J Biochem*, 267, pp. 4809-4816

Kelley R, Ideker T (2009) Genome-wide fitness and expression profiling implicate Mga2 in adaptation to hydrogen peroxide. *PLoS Genet*, 5, pp. e1000488

Larochelle M, Drouin S, Robert F, Turcotte B (2006) Oxidative stress-activated zinc cluster protein Stb5 has dual activator/repressor functions required for pentose phosphate pathway regulation and NADPH production. *Mol Cell Biol*, 26, pp. 6690-6701

Le Crom S, Devaux F, Marc P, Zhang X, Moye-Rowley WS, Jacq C (2002) New insights into the pleiotropic drug resistance network from genome-wide characterization of the *YRR1* transcription factor regulation system. *Mol Cell Biol*, 22, pp. 2642-2649

Lee TI, Rinaldi NJ, Robert F, Odom DT, Bar-Joseph Z, Gerber GK, Hannett NM, Harbison CT, Thompson CM, Simon I, Zeitlinger J, Jennings EG, Murray HL, Gordon DB, Ren B, Wyrick JJ, Tagne JB, Volkert TL, Fraenkel E, Gifford DK, Young RA (2002) Transcriptional regulatory networks in *Saccharomyces cerevisiae*. *Science*, 298, pp. 799-804

Lelandais G, Devaux F (2010) Comparative functional genomics of stress responses in yeasts. *OMICS*, 14, pp. 501-515

Lelandais G, Tanty V, Geneix C, Etchebest C, Jacq C, Devaux F (2008) Genome adaptation to chemical stress: clues from comparative transcriptomics in *Saccharomyces cerevisiae* and *Candida glabrata*. *Genome Biol*, 9, pp. R164

Li ZS, Szczypka M, Lu YP, Thiele DJ, Rea PA (1996) The yeast cadmium factor protein (YCF1) is a vacuolar glutathione S-conjugate pump. *J Biol Chem*, 271, pp. 6509-6517

Lucau-Danila A, Delaveau T, Lelandais G, Devaux F, Jacq C (2003) Competitive promoter occupancy by two yeast paralogous transcription factors controlling the multidrug resistance phenomenon. *J Biol Chem*, 278, pp. 52641-52650

Lucau-Danila A, Lelandais G, Kozovska Z, Tanty V, Delaveau T, Devaux F, Jacq C (2005) Early expression of yeast genes affected by chemical stress. *Mol Cell Biol*, 25, pp. 1860-1868

Lushchak VI (2011) Adaptive response to oxidative stress: Bacteria, fungi, plants and animals. *Comp Biochem Physiol C Toxicol Pharmacol*, 153, pp. 175-190

Monteiro PT, Dias PJ, Ropers D, Oliveira AL, Sá-Correia I, Teixeira MC, Freitas AT (2011) Qualitative modeling and formal verification of the *FLR1* gen mancozeb response in *Saccharomyces cerevisiae*. *IET Systems Biology*, 5, pp.308–316

Monteiro PT, Mendes ND, Teixeira MC, d'Orey S, Tenreiro S, Mira NP, Pais H, Francisco AP, Carvalho AM, Lourenco AB, Sá-Correia I, Oliveira AL, Freitas AT (2008) YEASTRACT-DISCOVERER: new tools to improve the analysis of transcriptional regulatory associations in *Saccharomyces cerevisiae*. *Nucleic Acids Res*, 36, pp. D132-136

Morschhauser J, Barker KS, Liu TT, Bla BWJ, Homayouni R, Rogers PD (2007) The transcription factor Mrr1p controls expression of the *MDR1* efflux pump and mediates multidrug resistance in *Candida albicans*. *PLoS Pathog*, 3, pp. e164

Nguyen DT, Alarco AM, Raymond M (2001) Multiple Yap1p-binding sites mediate induction of the yeast major facilitator *FLR1* gene in response to drugs, oxidants, and alkylating agents. *J Biol Chem*, 276, pp. 1138-1145

Ouyang X, Tran QT, Goodwin S, Wible RS, Sutter CH, Sutter TR (2011) Yap1 activation by H_2O_2 or thiol-reactive chemicals elicits distinct adaptive gene responses. *Free Radic Biol Med*, 50, pp. 1-13

Owsianik G, Balzi l L, Ghislain M (2002) Control of 26S proteasome expression by transcription factors regulating multidrug resistance in *Saccharomyces cerevisiae*. *Mol Microbiol*, 43, pp. 1295-1308

Prasad R, De Wergifosse P, Goffeau A, Balzi E (1995) Molecular cloning and characterization of a novel gene of *Candida albicans*, CDR1, conferring multiple resistance to drugs and antifungals. *Curr Genet*, 27, pp. 320-329

Rodrigues-Pousada C, Menezes RA, Pimentel C (2010) The Yap family and its role in stress response. *Yeast*, 27, pp. 245-258

Sá-Correia I, Santos S, Teixeira M, Cabrito T, Mira N (2009) Drug:H^+ antiporters in chemical stress response in yeast. *Trends Microbiol*, 17, pp. 22-31

Salin H, Fardeau V, Piccini E, Lelandais G, Tanty V, Lemoine S, Jacq C, Devaux F (2008) Structure and properties of transcriptional networks driving selenite stress response in yeasts. *BMC Genomics*, 9, pp. 333

Sanglard D, Ischer F, Monod M, Bille J (1997) Cloning of *Candida albicans* genes conferring resistance to azole antifungal agents: characterization of *CDR2*, a new multidrug ABC transporter gene. *Microbiology*, 143 (Pt 2), pp. 405-416

Saunders GW, Rank GH (1982) Allelism of pleiotropic drug resistance in *Saccharomyces cerevisiae. Can J Genet Cytol*, 24, pp. 493-503

Semchyshyn HM, Abrat OB, Miedzobrodzki J, Inoue Y, Lushchak VI (2011) Acetate but not propionate induces oxidative stress in bakers' yeast *Saccharomyces cerevisiae. Redox Rep*, 16, pp. 15-23

Srikanth CV, Chakraborti AK, Bachhawat AK (2005) Acetaminophen toxicity and resistance in the yeast *Saccharomyces cerevisiae. Microbiology*, 151, pp. 99-111

Teixeira MC, Cabrito TR, Hanif ZM, Vargas RC, Tenreiro S, Sá-Correia I (2011a) Yeast response and tolerance to polyamine toxicity involving the drug:H+ antiporter Qdr3 and the transcription factors Yap1 and Gcn4. *Microbiology*, 157, pp. 945-956

Teixeira MC, Dias PJ, Monteiro PT, Sala A, Oliveira AL, Freitas AT, Sá-Correia I (2010) Refining current knowledge on the yeast *FLR1* regulatory network by combined experimental and computational approaches. *Mol Biosyst*, 6, pp. 2471-2481

Teixeira MC, Dias PJ, Simoes T, Sá-Correia I (2008) Yeast adaptation to mancozeb involves the up-regulation of *FLR1* under the coordinate control of Yap1, Rpn4, Pdr3, and Yrr1. *Biochem Biophys Res Commun*, 367, pp. 249-255

Teixeira MC, Duque P, Sá-Correia I (2007) Environmental genomics: mechanistic insights into toxicity of and resistance to the herbicide 2,4-D. *Trends Biotechnol*, 25, pp. 363-370

Teixeira MC, Mira NP, Sá-Correia I (2011b) A genome-wide perspective on the response and tolerance to food-relevant stresses in *Saccharomyces cerevisiae. Curr Opin Biotechnol*, 22, pp. 150-156

Teixeira MC, Monteiro P, Jain P, Tenreiro S, Fernandes AR, Mira NP, Alenquer M, Freitas AT, Oliveira AL, Sá-Correia I (2006) The YEASTRACT database: a tool for the analysis of transcription regulatory associations in *Saccharomyces cerevisiae. Nucleic Acids Res*, 34, pp. D446-451

Teixeira MC, Sá-Correia I (2002) *Saccharomyces cerevisiae* resistance to chlorinated phenoxyacetic acid herbicides involves Pdr1p-mediated transcriptional activation of *TPO1* and *PDR5* genes. *Biochem Biophys Res Commun*, 292, pp. 530-537

Teixeira MC, Telo JP, Duarte NF, Sá-Correia I (2004) The herbicide 2,4-dichlorophenoxyacetic acid induces the generation of free-radicals and associated oxidative stress responses in yeast. *Biochem Biophys Res Commun*, 324, pp. 1101-1107

Tenreiro S, Fernandes AR, Sá-Correia I (2001) Transcriptional activation of *FLR1* gene during *Saccharomyces cerevisiae* adaptation to growth with benomyl: role of Yap1p and Pdr3p. *Biochem Biophys Res Commun*, 280, pp. 216-222

Thakur JK, Arthanari H, Yang F, Pan SJ, Fan X, Breger J, Frueh DP, Gulshan K, Li DK, Mylonakis E, Struhl K, Moye-Rowley WS, Cormack BP, Wagner G, Naar AM (2008) A nuclear receptor-like pathway regulating multidrug resistance in fungi. *Nature*, 452, pp. 604-609

Thorsen M, Lagniel G, Kristiansson E, Junot C, Nerman O, Labarre J, Tamas MJ (2007) Quantitative transcriptome, proteome, and sulfur metabolite profiling of the *Saccharomyces cerevisiae* response to arsenite. *Physiol Genomics*, 30, pp. 35-43

Torelli R, Posteraro B, Ferrari S, La Sorda M, Fadda G, Sanglard D, Sanguinetti M (2008) The ATP-binding cassette transporter-encoding gene *CgSNQ2* is contributing to the *CgPDR1*-dependent azole resistance of *Candida glabrata. Mol Microbiol*, 68, pp. 186-201

Tsai HF, Krol AA, Sarti KE, Bennett JE (2006) *Candida glabrata PDR1*, a transcriptional regulator of a pleiotropic drug resistance network, mediates azole resistance in clinical isolates and petite mutants. *Antimicrob Agents Chemother*, 50, pp. 1384-1392

Vargas RC, Tenreiro S, Teixeira MC, Fernandes AR, Sá-Correia I (2004) *Saccharomyces cerevisiae* multidrug transporter Qdr2p (Yil121wp): localization and function as a quinidine resistance determinant. *Antimicrob Agents Chemother*, 48, pp. 2531-2537

Vermitsky JP, Edlind TD (2004) Azole resistance in *Candida glabrata*: coordinate upregulation of multidrug transporters and evidence for a Pdr1-like transcription factor. *Antimicrob Agents Chemother*, 48, pp. 3773-3781

Wemmie JA, Szczypka MS, Thiele DJ, Moye-Rowley WS (1994) Cadmium tolerance mediated by the yeast AP-1 protein requires the presence of an ATP-binding cassette transporter-encoding gene, *YCF1. J Biol Chem*, 269, pp. 32592-32597

Wendler F, Bergler H, Prutej K, Jungwirth H, Zisser G, Kuchler K, Hogenauer G (1997) Diazaborine resistance in the yeast *Saccharomyces cerevisiae* reveals a link between *YAP1* and the pleiotropic drug resistance genes *PDR1* and *PDR3. J Biol Chem*, 272, pp. 27091-27098

Workman CT, Mak HC, McCuine S, Tagne JB, Agarwal M, Ozier O, Begley TJ, Samson LD, Ideker T (2006) A systems approach to mapping DNA damage response pathways. *Science*, 312, pp. 1054-1059

Yan C, Lee LH, Davis LI (1998) Crm1p mediates regulated nuclear export of a yeast AP-1-like transcription factor. *EMBO J*, 17, pp. 7416-7429

Znaidi S, Barker KS, Weber S, Alarco AM, Liu TT, Boucher G, Rogers PD, Raymond M (2009) Identification of the *Candida albicans* Cap1p regulon. *Eukaryot Cell*, 8, pp. 806-820

17

Role of the Yap Family in the Transcriptional Response to Oxidative Stress in Yeasts

Christel Goudot[1,2,3], Frédéric Devaux[4] and Gaëlle Lelandais[1,2,3]
[1]Dynamique des Structures et Interactions des
Macromolécules Biologiques (DSIMB),
[2]Univ Paris Diderot, Sorbonne Paris Cité, Paris,
[3]INTS, 75015, Paris,
[4]Laboratoire de Génomique des Microorganismes,
Univ Pierre et Marie Curie, Paris,
France

1. Introduction

Oxidative stress can be defined as physiological changes that arise in a living organism, in reaction to an abnormal level of cytotoxic oxidants and free radicals in the environment. Because they have unpaired valence shell electrons, free radicals are very unstable and constantly seek to bond to other molecules in order to increase their stability. Within the cell, free radicals can therefore cause considerable damages in different components such as DNA (Storz et al., 1987), lipid membranes (Davies, 1985) or proteins (Smith et al., 1984). Free radicals are implicated in the ageing process (Wickens, 2001), in some autoimmune diseases (Blake et al., 1987) and in the development of cancer (Trush and Kensler, 1991). Notably, the generation of free radicals is a natural process in cell functioning and for instance, reactive oxygen species (ROS) are continuously produced as side products of aerobic metabolic pathways. To neutralize them, living systems have developed specific strategies. In normal conditions, equilibrium thus exists between the generation and the degradation of free radicals, whereas in case of oxidative stress conditions a persistent imbalance is observed.

In this chapter, we will focus on the role of the Yap proteins in the regulation of the transcriptional response to oxidative stress in yeasts. As other aerobically growing organisms, yeasts are constantly exposed to ROS molecules and have acquired sophisticated mechanisms to control modifications of its redox status due to impaired metabolism or to an excess of oxidative molecules in its environment. Yeasts thus represent an interesting model to understand how eukaryotes can cope with different levels of oxidative stress. In that respect, the species *Saccharomyces cerevisiae* has been extensively studied and in the literature several reviews present in details the antioxidant defence systems of this model organism (Costa and Moradas-Ferreira, 2001; Herrero et al., 2008; Ikner and Shiozaki, 2005; Jamieson, 1998; Lushchak, 2006, 2010; Moye-Rowley, 2002). Together these reviews draw one of the most complete pictures of the genomic strategy used by a cell to protect their components against ROS molecules. Yeasts possess both enzymatic and non-enzymatic defence strategies. Enzymatic system comprise enzymes including catalases, superoxide dismutases

(SOD), glutathione reductases or glutathione peroxidases, which are able to remove partially reduced forms of molecular oxygen and/or to repair the cellular damages caused by oxidative stress. Non-enzymatic defence system involves small molecules that act as radical scavengers, *i.e.* they can be oxidized by ROS and thereby allow to neutralize cytotoxic oxidants. One of the best-known examples of a non-enzymatic defence system is glutathione (GSH), a tripeptide γ-L-glutamyl-L-cystinylglycine. In this context, to correctly produce the numerous enzymes and components required for its oxidative defence, the expression of the corresponding genes must be precisely synchronized. A major role is played by specific transcription factors (TFs), *i.e.* proteins that bridge sensors of cytotoxic oxidants and the transcriptional activation of particular sets of genes. In yeast *S. cerevisiae*, the TFs Msn2p/Msn4p, Snk7p, Hsf1p or Yap1p are known to be involved in the response to oxidant-induced stress. The focus of this review will be on the Yap protein family that includes the protein Yap1p, *i.e.* the main transcriptional regulator of the genomic response to oxidative stress (Lelandais et al., 2008; Lucau-Danila et al., 2005; Rodrigues-Pousada et al., 2010). Note that other reviews provide complete descriptions of the other TF roles (Ikner and Shiozaki, 2005; Lushchak, 2010; Moye-Rowley, 2002).

2. Yap proteins and the response to oxidative stress

Yap proteins belong to the b-ZIP super family of TFs that is widely conserved from yeast to human (Rodrigues-Pousada et al., 2010). In yeast *S. cerevisiae*, the role of Yap1p in oxidative stress response was initially suggested by the analysis of *YAP1* mutant strains, in which increased sensitivity to cadmium and hydrogen peroxide (H_2O_2) was observed (Delaunay et al., 2000; Hirata et al., 1994; Kuge and Jones, 1994; Wu et al., 1993). Functional studies next demonstrated that Yap1p is able to activate the transcription of genes that encode key proteins in oxidative stress response such as *TRX2* (thioredoxin) and *GSH1* (γ-glutamylcysteine synthase) (Kuge and Jones, 1994), *GSH2* (glutathione synthase) (Wu and Moye-Rowley, 1994), *TRR1* (thioredoxin reductase) (Sugiyama et al., 2000), *GPX2* (glutathione peroxidase) (Sugiyama et al., 2000), *TSA1* (thioredoxin peroxidase) (Grant et al., 1996; Inoue et al., 1999) or *AHP1* (alkylhydroperoxide reductase) (Lee et al., 1999a; Lee et al., 1999b) (see (Lushchak, 2010) for review). Others analyses found that Yap1p controlled expression of the ABC transporter gene *YCF1* (Wemmie et al., 1994) and of the multidrug resistance (MDR) transporter genes *FLR1* and *ATR1* (Alarco et al., 1997; Coleman et al., 1999). The activation of the Yap1p protein in case of stress relates to a post-translational modification of the protein (Wemmie et al., 1997). In normal condition the protein Yap1p is constantly produced and exported from the nucleus to the cytoplasm, whereas in case of oxidant exposure the protein is recruited in the nucleus (Kuge et al., 1997), therefore allowing the transcriptional activation of its target genes. A key modulator of the sub cellular distribution of Yap1p is the nuclear export regulator Crm1p (Kuge et al., 1998; Yan et al., 1998) and it was demonstrated that the protein Gpx3p is involved in the activation of Yap1p in case of stress induced by H_2O_2 (Delaunay et al., 2000). Notably, the response of Yap1p varies according to the nature of the oxidative stress (induced by H_2O_2 or diamide for instance) (Wemmie et al., 1997) and it was proposed that Yap1p has two distinct molecular redox centers, one triggered by ROS (hydroperoxides and the superoxide anion) and the other triggered by chemicals with thiol reactivity (electrophiles and divalent heavy metal cations) (Azevedo et al., 2003). This last class of chemical Yap1p activators does not require the presence of the protein Gpx3p to initiate the oxidative stress response. In addition to these post-translational controls, Yap1p is transcriptionnally activated by strong and long-standing oxidative stresses, possibly through positive autoregulatory loop, as suggested by the observation that Yap1p binds to its own promoter (Salin et al., 2008). Analyses of promoter sequences of genes whose transcription is modulated by Yap1p allowed the identification of specific regulatory DNA motifs. In particular, different sequences were experimentally characterized with a clear preference for the TTACTAA motif (Nguyen et al., 2001). Yap1p binding sites are referred to as Yap Response Element (YRE). Recently, it was shown that the Yap1p protein cover a larger DNA fragment than strictly the TTA•TAA half sites, with a conserved adenine (and to a less extend a cytosine) located in 5′ of the canonical YREs (Goudot et al., 2011).

Interestingly, functional homologous proteins of Yap1p were identified in the pathogenic yeast species *Candida glabrata* and *Candida albicans*. They are respectively named Cgap1p and Cap1p and are also involved in the response to oxidative stress (Chen et al., 2007; Lelandais et al., 2008; Znaidi et al., 2009). Infections caused by fungal pathogens have become major life-threatening diseases, especially in patients with defects in their immune system. The yeast species *C. albicans* and *C. glabrata* respectively rank as the first and the second causes of invasive infections, including systemic candidiasis and candidemia (Pfaller and Diekema, 2007). They are both human pathogens that can colonize numerous sites within their human host and are thus continuously exposed to rapid and drastic changes in their external milieu. Azole antifungals, especially fluconazole, were widely used to treat these fungal diseases, but rapidly an important number of clinical drug resistance cases was reported (White et al., 1998). Recently, it was proposed that drug resistance and the response to oxidative stress are interconnected processes in *C. albicans* (Znaidi et al., 2009). This connection exists through the activity of several TFs like Cap1p, Tac1p, Mrr1p or Upc2p for which (*i*) gain-of-function mutations were observed as being related to clinical azole resistance and (*ii*) genome-wide studies identified several of their target genes as being involved in the general response to oxidative stress (Liu et al., 2007; Morschhauser et al., 2007; Znaidi et al., 2009; Znaidi et al., 2008). Also, it is interesting to remind that, in *S. cerevisiae*, *YAP1* was initially described as a gene involved in pleiotropic drug resistance (Wu et al., 1993). Moreover, as a defence strategy against fungal infections for the animal host consists in using oxidative killing carried out by macrophages, an essential feature for the virulence of pathogenic yeasts relates to their ability to tolerate ROS molecules (Abegg et al., 2010). In this context, understanding the role of Yap TFs in the genomic response to oxidant-induced stress in fungi provide valuable information to better understand fungal pathogenesis.

Our text is organized in the following manner. We will first present genome-wide strategies to decipher the genomic response related to the activity of Yap1p, Cgap1p and Cap1p TFs in case of oxidative stress. Data coming from high-throughput experimental technologies will be presented. Next, three different yeast species (the pathogen species *C. glabrata* and *C. albicans* and the model yeast *S. cerevisiae*) will be compared. As a large amount of genomic information is available for *S. cerevisiae* (genomic sequence, transcriptome data, gene functional annotations, TF DNA binding sequences, etc.), this species will serve as a reference to describe the genomic oxidative stress response in *Candida* species. Finally, we will discuss oxidative stress response in the light of the other Yap TFs. In *S. cerevisiae*, this family comprises eight members (Yap1p to Yap8p) that are TFs carrying both overlapping and distinct biological functions (Fernandes et al., 1997; Rodrigues-Pousada et al., 2010). In yeasts *C. glabrata* and *C. albicans*, we will present a description of the Yap families, comprising respectively seven and four members. We will discuss how modifications in the number of paralogous TFs that belong to the Yap family in each species imply similarities and differences in the regulatory control of their target genes (genes being transferred from one factor to another).

3. Integration of multiple data sources for the characterisation of AP-1 transcriptional modules

3.1 General principle

In a recent study, we proposed a strategy to identify the sets of genes for which transcription was activated by the regulatory proteins Yap1p (in *S. cerevisiae*), Cgap1p (in *C. glabrata*) and Cap1p (in *C. albicans*), in response to a specific stress induced by benomyl (Goudot et al., 2011). These sets of genes are referred to as AP- 1 transcriptional modules (TMs). In each yeast species, benomyl is known to activate an oxidative stress, for which genomic response of the cell is primarily dependant on the TFs Yap1p (Lucau-Danila et al., 2005), Cgap1p (Lelandais et al., 2008) and Cap1p (Znaidi et al., 2009). To summarize, our strategy consisted in considering three different layers of information based on the analysis of genome-wide datasets. First we used expression patterns of genes to identify those that are significantly up regulated in response to benomyl induced-stress (see below Step 1). Second we analyzed the transcriptome alterations in yeast strains deleted for the genes coding the yeast AP-1 TFs (respectively Yap1p, Cgap1p and Cap1p) (see Step 2), and third we searched for genomic locations of the TF binding DNA sequences using ChIP-chip experiments (see Step 3). A global overview of the computational framework is presented Figure 1. Note that each experimental dataset was carefully chosen in order to ensure both intra and inter-species comparisons of the obtained results, *i.e.* comparable benomyl concentrations in each species and similar time point measurements for transcriptome analyses.

3.2 Step 1: Genome-wide expression data to measure the transcriptional response induced by benomyl

In a previous study (Lelandais et al., 2008), we carried out microarray analyses of the transcriptome responses of yeasts *S. cerevisiae* and *C. glabrata*, following identical treatments with the antifungal agent benomyl (Gupta et al., 2004) (20 µg/ml benomyl concentration and time measurements at 2, 4, 10, 20 and 40 minutes). More recently, Znaidi *et al.* (2009) performed similar microarray experiments in *C. albicans*, analyzing the transcriptome modifications after 30 minutes of benomyl treatment with a 30 µg/ml benomyl concentration. As each dataset was originally examined using different bioinformatics methodologies, we collected the initial raw data from the GEO database (Barrett et al., 2009) and applied in each species the same procedure for identifying genes whose transcription was significantly modified after benomyl addition. Besides data pre-processing, we used a combination of three different algorithms: SAM (Tusher et al., 2001), LIMMA (Wettenhall and Smyth, 2004) and SMVar (Jaffrezic et al., 2007). These algorithms were chosen because they are representative of different variance modelling strategies in gene expression data (Jeanmougin et al., 2010). In total, 786 genes were identified as being significantly up regulated in *S. cerevisiae*, 327 genes in *C. glabrata* and 337 genes in *C. albicans* (Figure 1, step 1).

3.3 Step 2: Mutant analyses to measure the deletion impact of genes coding for yeast AP-1 transcription factors

Previous studies shown that even if Yap1p (in *S. cerevisiae*), Cgap1p (in *C. glabrata*) and Cap1p (in *C. albicans*) are important coordinators of the early transcriptional response to benomyl stress (Lelandais et al., 2008; Znaidi et al., 2009), other TFs are also involved in the system (Lelandais et al., 2008). The lists of genes identified in Step 1 therefore comprised in one hand, target genes for Yap1p, Cgap1p and Cap1p TFs, but also on the other hand, target genes for additional regulators (for instance Msn2p/Msn4p (Lelandais et al., 2008)). To specifically highlight the yeast AP-1 responsive genes, we analyzed in each species transcriptome data comparing gene expression between wild-type strains and strains deleted for genes coding yeast AP-1 TFs: *ΔYAP1* (in *S. cerevisiae*), *ΔCgAP1* (in *C. glabrata*) and *ΔCAP1* (in *C. albicans*). The yeast AP-1 TF target genes are genes whose expression is significantly altered in the deleted strains. Mutant datasets were therefore collected from the studies of Lucau-Danila *et al.* (2005) (*ΔYAP1* analyses with a 20 µg/ml benomyl concentration), Lelandais *et al.* (2008) (*ΔCgAP1* analyses with a 20 µg/ml benomyl concentration) and Znaidi *et al.* (2009) (*ΔCAP1* analyses with a 30 µg/ml benomyl concentration). Besides data pre-processing, we used the combination of the three algorithms SAM, LIMMA and SMVar (Goudot et al., 2011). As a result, 33 genes were identified as Yap1p-dependent genes in *S. cerevisiae*, 134 genes as Cgap1p-dependent genes in *C. glabrata* and 168 genes as Cap1p-dependent genes in *C. albicans* (Figure 1, step 2).

3.4 Step 3: ChIP-chip experiments to identify yeast AP-1 transcription factor binding sites *in vivo*

As the data sources analyzed in Step 1 and 2 provided only indirect information concerning the transcriptional regulation between yeast AP-1 TFs and their target genes, the goal for this additional step was to discern dependencies between the gene expression patterns observed in Step 1 and 2 and the physical interactions revealed by genome-wide location profiling experiments. We thus analyzed ChIP-chip experiments performed for TFs Yap1p, Cgap1p and Cap1p. Note that experiments in benomyl condition were available only in *S. cerevisiae* for Yap1p (20 µg/ml benomyl concentration) (Salin et al., 2008). Other experiments in *C. glabrata* and *C. albicans* were performed in non-stress induced conditions (respectively (Kuo et al., 2010) and (Znaidi et al., 2009)). Besides data pre-processing, we identified promoter sequences that were significantly bond by the yeast AP-1 TFs by a combination of the results obtained with the SAM, LIMMA and SMVar methods, and those obtained using the ChIPmix algorithm (Martin-Magniette et al., 2008). Unlike SAM, LIMMA and SMVar methods that work on log ratio, ChIPmix has the originality to directly analyze the signals of IP (DNA fragments cross-linked to TF protein) and INPUT (genomic DNA) by modelling the distribution of the IP signal conditional to the INPUT signal (Martin-Magniette et al., 2008). Therefore, 260 genes were found to be bond by Yap1p, 327 genes by Cgap1p and finally 373 genes by Cap1p (Figure 1, step 3).

3.5 Final step: Data integration

Results obtained in Step 1, 2 and 3 were finally combined as described in Figure 1. Genes selected in "Step 1 and Step 2", or in "Step 1 and Step 3" were conserved in the final AP-1 TMs. In *S. cerevisiae* the Yap1p TM therefore comprised 67 genes, in *C. glabrata* the Cgap1p

TM comprised 98 genes, and finally in *C. albicans* the Cap1p TM comprised 130 genes. Complete list of genes in each TM together with their corresponding functional annotations can be found in (Goudot et al., 2011). Note that the number of genes in each AP-1 TM is underestimated compared to information available in database like YEASTRACT (Abdulrehman et al., 2010) (in which more than 400 genes are annotated as potential targets for Yap1p). However, one must keep in mind that the TMs described here have the particularity *(i)* to be focused on the AP-1 responsive genes in benomyl stress-induced conditions (genes regulated by AP-1 TFs in other conditions are not considered), and *(ii)* to include only genes for which at least two types of experimental evidences were available for interactions with Yap1p, Cgap1p or Cap1p. All together, we integrated experimental results arising from more than 80 individual microarray experiments applying four different bioinformatics methodologies. The predictive strength of the strategy is based on the combined constraints that arise from the use of multiple biological and bioinformatics data sources.

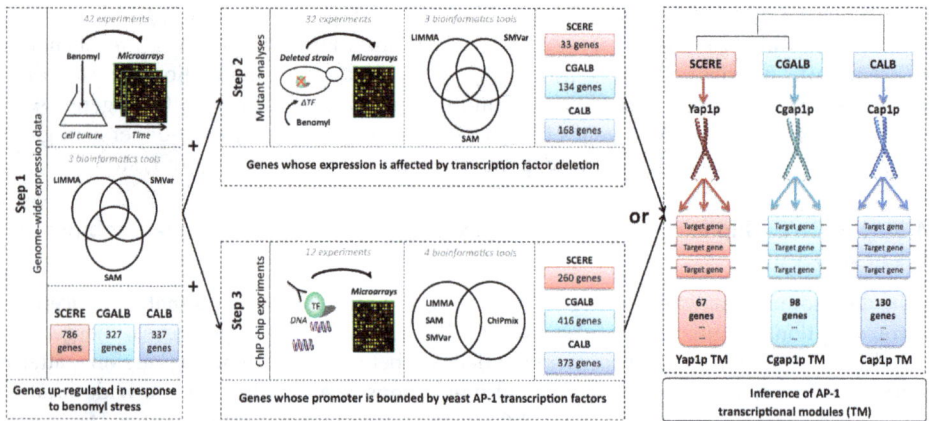

Fig. 1. General strategy to characterize the yeast AP-1 transcriptional modules (TMs). Characterizing a TM consists in indentifying all genes whose transcription is modulated by a particular transcription factor (TF). Applied to the analysis of AP-1 TMs in yeasts during benomyl induced-stress, the general strategy shown here proceeds in several successive steps, using datasets which focus on different levels of transcriptional regulation: (Step 1) identification of differentially expressed genes during benomyl stress, (Step 2) identification of genes whose benomyl induction is affected by the deletion of the gene encoding the AP-1 TFs, and (Step 3) genome-wide location of the AP-1 TFs, as determined by ChIP-chip analyses (see the main text). Genes selected in "Step 1 and Step 2", or in "Step 1 and Step 3" were conserved in the final AP-1 TMs. In *S. cerevisiae* the Yap1 TM therefore comprised 67 genes, in *C. glabrata* the Cgap1p TM comprised 98 genes, and finally in *C. albicans* the Cap1p TM comprised 130 genes.

4. Comparative analysis of yeast AP-1 transcriptional modules based on sequence orthology

One of the main objective for genomics studies is to transfer functional annotations from well-studied organisms to the newly sequence species. For that, the most widely used approach for function prediction is based on orthology assignments. Orthology defines the relationship between genes in different species that originate from a single gene in the last ancestor of these species (Fitch, 2000; Sonnhammer and Koonin, 2002). Therefore, orthologous genes are most likely to share the same function and should, in principle, have conserved their expression patterns together with their regulatory control by TFs. To test this hypothesis, the three AP-1 TMs defined separately in *S. cerevisiae*, *C. glabrata*, and *C. albicans* represent valuable information. Indeed, as they were characterized using in each yeast species independent experimental datasets (see previous Section), they represent an interesting reference model. We applied the INPARANOID algorithm (O'Brien et al., 2005) comparing all protein sequences of the three yeast species (Goudot et al., 2011). Orthologous links were thus inferred for 80% of genes between *S. cerevisiae* and *C. glabrata*, 61% of genes between *S. cerevisiae* and *C. albicans* and 63% of genes between *C. glabrata* and *C. albicans*. These results were coherent with the phylogeny of the yeast species analyzed here, *i.e.* *C. glabrata* being more closely related to *S. cerevisiae* than *C. albicans* is. Then, we determined whether orthologous genes were present in each of the three AP-1 TMs. Strikingly we could only distinguish 11 orthologous genes between the *S. cerevisiae* and *C. glabrata* AP-1 TMs (16%), 7 between the *S. cerevisiae* and *C. albicans* AP-1 TMs (10%) and 14 between the *C. glabrata* and the *C. albicans* AP-1 TMs (14%) (Goudot et al., 2011). Considering the global amount of orthologous genes between the three species (more than 60%), these values were surprisingly low and demonstrated that in yeasts, the functioning of Yap1p, Cgap1p and Cap1p TFs during the transcriptional response to benomyl stress has been significantly rewired. Gene duplication and multigenic protein families are parameters that can explain changes in the yeast AP-1 TMs. For instance, we observed that OYE paralogous genes, which encode NADPH oxydoreductases involved in sterol metabolism and oxidative stress response, were present in AP-1 TMs in each species, but with a different number of copies. In *S. cerevisiae*, only two OYE genes (*OYE2* and its paralogue *OYE3*) belong to the Yap1p TM, whereas 3 and 4 OYE paralogues were identified respectively in Cgap1p and Cap1p TMs (in *C. glabrata* and *C. albicans*) (see (Goudot et al., 2011) for a detailed description of these genes). In the three yeasts, the general function mediated by the OYE genes is therefore conserved, but because of several duplication events, direct orthologous relationships between genes were lost. Additionally, AP-1 proteins belong to Yap families that comprised 3 to 8 paralogous genes in the *Hemiascomycetes*. In the model yeast *S. cerevisiae*, the Yap family is composed of eight proteins (Yap1p to Yap8p) that are TFs carrying both overlapping and distinct biological functions (Rodrigues-Pousada et al., 2010), and recognize similar DNA consensus (Tan et al., 2008). Structural features of Yap proteins in *S. cerevisiae* are presented in Figure 2. Considering that these factors can interact functionally, they certainly cross-influenced the evolution of their respective TM. To clarify this point, we searched for potential roles in the response to oxidative stress for other members of the Yap families, using genomic sequence analyses and functional information stored in the databases.

5. Characterisation of the Yap families in *Candida* species

In contrast to *S. cerevisiae*, in which an important number of functional information concerning Yap proteins is available (see (Rodrigues-Pousada et al., 2010) for review), very few studies have yet been carried out in *Candida* species to analyze the redundant Yap families (Chen et al., 2007; Singh et al., 2011). Most information is currently based on sequence similarity with Yap genes of *S. cerevisiae*. We searched for all potential Yap

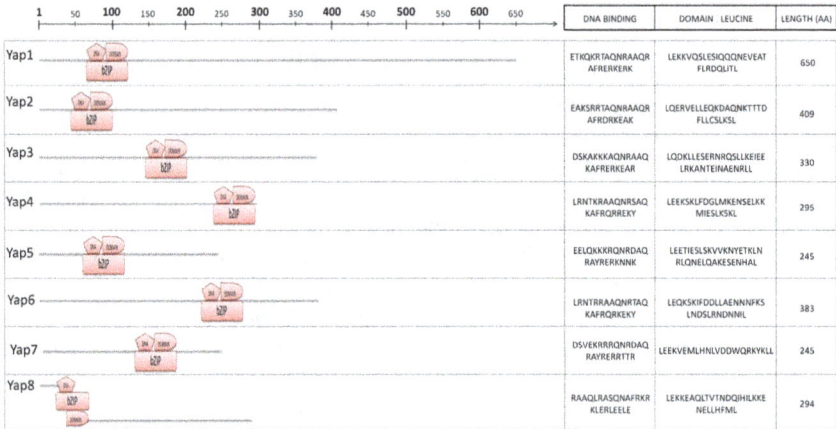

	DNA BINDING	DOMAIN LEUCINE	LENGTH (AA)
Yap1	ETKQKRTAQNRAAQR AFRERKERK	LEKKVQSLESIQQQNEVEAT FLRDQLITL	650
Yap2	EAKSRRTAQNRAAQR AFRDRKEAK	LQERVELLEQKDAQNKTTTD FLLCSLKSL	409
Yap3	DSKAKKKAQNRAAQ KAFRERKEAR	LQDKLLESERNRQSLLKEIEE LRKANTEINAENRLL	330
Yap4	LRNTKRAAQNRSAQ KAFRQRREKY	LEEKSKLFDGLMKENSELKK MIESLKSKL	295
Yap5	EELQKKKRQNRDAQ RAYRERKNNK	LEETIESLSKVVKNYETKLN RLQNELQAKESENHAL	245
Yap6	LRNTRRAAQNRTAQ KAFRQRKEKY	LEQKSKIFDQLLAENNNFKS LNDSLRNDNNIL	383
Yap7	DSVEKRRRQNRDAQ RAYRERRTTR	LEEKVEMLHNLVDDWQRKYKLL	245
Yap8	RAAQLRASQNAFRKR KLERLEELE	LEKKEAQLTVTNDQIHILKKE NELLHFML	294

Fig. 2. Structural features of the Yap proteins in the yeast *S. cerevisiae*. Yap family comprises 8 members in yeast *S. cerevisiae*, named Yap1p to Yap8p. Structural information shown here was obtained from the database UniProtKB http://www.uniprot.org/uniprot/. This database provides functional and structural information on proteins as well as links to other databases. Here only information concerning the DNA binding and the leucine-zipper domains are shown, together with the sequence length. One bZIP thus consists of a leucine-zipper domain and a DNA binding domain. The left representation represents the position of the bZIP in the sequence of each Yap protein. As an illustration, the coding sequence for the Yap1p protein comprises 650 amino acids with a bZIP region of around 100 amino acids. The sequence of the DNA binding domain is ETKQKRTAQNRAAQRAFRERKERK and the sequence of the leucine-zipper domain is LEKKVQSLESIQQQNEVEATFLRDQLITL.

proteins in complete genomes of *C. glabrata* and *C. albicans* yeast species, applying a methodology derived from the strategy used in *S. cerevisiae* to identify the eight Yap family members (Fernandes et al., 1997). Starting from the protein sequences of Cgap1p and Cap1p in *Candida* species (referred to as "query sequences"), our approach consists in comparing these AP-1 sequences with all other protein sequences available in *Candida* genomes (referred to as "target sequences"). For each comparison between query and target sequences, a score is calculated (see Equation 1). This score (between 0 and 1) as the advantage to take into account *(i)* the amino acid similarity observed after a *global* alignment between the query and the target sequences, *(ii)* the amino acid similarity observed after a *local* alignment and *(iii)* the amino acid similarity observed by restricting the alignment to the DNA binding region of the queried protein sequences (Cgap1p or Cap1p). The higher the score, the more the probability is for the target sequence to belong to the Yap families in one of the *Candida* species. Note that we could observe that a score value higher than 0.4 represents a "highly confident" homology relationship, a score value between 0.3 and 0.4 a "confident" homology relationship and a score below 0.3 a "poorly confident" homology relationship. Global and local alignments were performed using respectively the Needleman-Wunsch and the Smith-Waterman algorithms implemented in EMBOSS (Lamprecht et al., 2011). After careful manual inspection of the obtained results, we identified 7 members in the *C. glabrata* Yap family and 4 members in *C. albicans* (Table 1). Note that this lower number of Yap proteins in *C. albicans* is connected to the whole genome duplication that arose in the common history of *S. cerevisiae* and *C. glabrata*, but not in the *C. albicans* ancestor.

Systematic name (C. glabrata)	Standard name	Description (Génolevures database)
CAGL0H04631g	CgAP1	Similar to uniprot|P19880 Saccharomyces cerevisiae YML007w YAP1 (ohnolog of YDR423C) Basic leucine zipper (bZIP) transcription factor required for oxidative stress tolerance
CAGL0F03069g	-	Some similarities with uniprot|P24813 Saccharomyces cerevisiae YDR423c CAD1 (ohnolog of YML007W) AP-1-like basic leucine zipper (bZIP) transcriptional activator involved in stress responses, iron metabolism, and pleiotropic drug resistance
CAGL0K02585g	-	Some similarities with uniprot|P38749 Saccharomyces cerevisiae YHL009c YAP3 Basic leucine zipper (bZIP) transcription factor
CAGL0M10087g	-	Some similarities with uniprot|P38749 Saccharomyces cerevisiae YHL009c YAP3 Basic leucine zipper (bZIP) transcription factor
CAGL0M08800g	-	Weakly similar to uniprot|Q03935 Saccharomyces cerevisiae YDR259c YAP6 (ohnolog of YOR028C) Putative basic leucine zipper (bZIP) transcription factor
CAGL0K08756g	-	Weakly similar to uniprot|P40574 Saccharomyces cerevisiae YIR018w YAP5 (ohnolog of YOL028C) Basic leucine zipper (bZIP) transcription factor
CAGL0M08800g	-	Weakly similar to uniprot|Q03935 Saccharomyces cerevisiae YDR259c YAP6 (ohnolog of YOR028C) Putative basic leucine zipper (bZIP) transcription factor
CAGL0F01265g	-	Some similarities with uniprot|Q08182 Saccharomyces cerevisiae YOL028c YAP7 (ohnolog of YIR018W) Putative basic leucine zipper (bZIP) transcription factor
Systematic name (C. albicans)	Standard name	Description (CGD database)
orf19.1623	CAP1	Transcription factor, AP-1 bZIP family; role in oxidative stress response and resistance, multidrug resistance; oxidative stress regulates nuclear localization; partially complements S. cerevisiae yap1 mutation; human neutrophil-ind
orf19.3193	FCR3	Transcriptional regulator of the bZIP family; partially functionally complements the fluconazole sensitivity of an S. cerevisiae pdr1 pdr3 double mutant; probable ortholog of S. cerevisiae Yap3p; Hap43p-induced
orf19.681	HAP43	CCAAT-binding factor-dependent transcriptional repressor required for low iron response; similar to bZIP transcription factor AP-1; HAP4L-bZIP bipartite domain; gene negatively regulated by Sfu1p; ciclopirox olamine induced
orf19.861	CAP4	Predicted transcriptional regulator with bZip domain; possibly an essential gene, disruptants not obtained by UAU1 method

Table 1. **Characterisation of the Yap families in *Candida* species.** Systematic protein sequence comparisons were performed to identify Yap family respectively in *C. glabrata* and *C. albicans* yeast species (see the main text). After manual inspection of the obtained results, 7 genes were identified as members of the Yap family members in *C. glabrata* genome and 4 in the *C. albicans* genome. Systematic and standard gene names are presented here, with the associated descriptions found in the Génolevures http://www.genolevures.org/ (for *C. glabrata*) or Candida Genome Database (CGD) http://www.candidagenome.org/ (for *C. albicans*).

$$Score(Sq,St) = \frac{NW_{global}(Sq,St) + SW_{local}(Sq,St) + SW_{DNAbinding}(Sq,St)}{3}$$

Equation 1. Score used to identify Yap family members in *Candida* genomes. Sq and St respectively means "Query sequence" and "Target sequence". NW_{global} corresponds to the score value returned by the Needleman-Wunsch algorithm after a global alignment of the query and target sequences. SW_{local} corresponds to the score value returned by the Smith-Waterman algorithm after a local alignment of the query and target sequences. Finally, $SW_{DNAbinding}$ corresponds to the score value returned by the Smith-Waterman algorithm restricting the alignment to the DNA binding region.

In a second step, we performed cross-species comparisons of the Yap sequences identified in *Candida* genomes with these described in the model yeast *S. cerevisiae*, *i.e.* proteins Yap1p to Yap8p. The highest values of the comparative score based on protein sequence alignments (see previous section) are presented in Table 2. As expected, we could observe important score values between Yap1p (in *S. cerevisiae*) and Cgap1p (in *C. glabrata*) (score value = 0.48),

Yap family in S. cerevisiae	Yap family in C. glabrata	Score value (between protein sequences)
Yap1p	Cgap1p	0.4767
Yap2p	CAGL0F03069g	0.4298
Yap3p	CAGL0K02585g	0.4788
	CAGL0M10087g	0.4309
Yap4p	CAGL0M08800g	0.3408
Yap5p	CAGL0K08756g	0.3976
Yap6p	CAGL0M08800g	0.3352
Yap7p	CAGL0F01265g	0.3536
Yap8p	CAGL0H04631g	0.2558
Yap family in S. cerevisiae	**Yap family in C. albicans**	**Score value (between protein sequences)**
Yap1p	Cap1p	0.4158
Yap2p		0.3713
Yap3p		0.3957
Yap4p	Fcr3p	0.2499
Yap5p		0.2757
Yap6p		0.2317
Yap7p	Hap43p	0.2970
Yap8p		0.2680
-	Cap4p	< 0.25

Table 2. Cross-species comparisons of the Yap families in *Candida* species with the Yap family in *S. cerevisiae*. Protein sequences of each member of the Yap families described in *Candida* species were compared with the protein sequences of each member of the Yap family in *S. cerevisiae*. Highest scores values are shown here and allow to assign homologous relationships between proteins. A score value higher than 0.4 represents a "highly confident" homology relationship, a score value between 0.3 and 0.4 a "confident" homology relationship and a score below 0.3 a "poorly confident" homology relationship.

and Yap1p and Cap1p (in *C. albicans*) (score value = 0.41). For Yap2p protein, we found a homologous protein in *C. glabrata* named CAGL0F03069g (score value = 0.43). This result is in agreement with the annotation stored in the Génolevures database (Sherman et al., 2006). In *C. albicans*, the best homologous sequence was Cap1p (score value = 0.37). Interestingly in *S. cerevisiae* Yap2p protein is the Yap member that shares highest sequence similarity and functional redundancy with Yap1p (Figure 2). Considering that the whole genome duplication arose only in *S. cerevisiae* and *C. glabrata* genomes, the protein Cap1p certainly retains functional properties in *C. albicans* that are split between Yap1p and Yap2p in *S. cerevisiae*, and Cgap1p and CAGL0F03069g in *C. glabrata*. In the case of the protein Yap3p, we could identify two proteins with important scores in *C. glabrata*: CAGL0K02585g (score value = 0.48) and CAGL0M10087g (score value = 0.43). On the other hand Yap4p and Yap6p proteins in *S. cerevisiae* appeared to have a unique homologous protein CAGL0M08800g in *C. glabrata* (scores values are respectively 0.34 and 0.33). Finally, Yap5p and Yap7p proteins exhibited unique homologous proteins in *C. glabrata*, which are respectively CAGL0K08756g (score value = 0.39) and CAGL0F01265g (score value = 0.35). In *C. albicans*, the protein Fcr3p (orf19.3193) shared an important score value with Yap3p (score value = 0.39) and lower values with Yap4p (score value = 0.25), Yap5p (score value = 0.27) and Yap6p (score value = 0.24). Yap7p and Yap8p proteins also had a single homologous protein named Hap43p (orf19.681, score values = 0.30 and 0.27). Finally, Yap8p protein in *S. cerevisiae* appeared to have the lowest score values with the identified Yap proteins in both *Candida* species (lower than 0.3) meaning that the homology relationships are "poorly confident". Such an observation suggest an important evolution of Yap8p protein in yeast species with a potential functional transfer of the Yap8p regulated genes to other Yap factors in *Candida* species. No confident homologous protein for Cap4p (orf19.861, in *C. albicans*) could also be identified in *S. cerevisiae*. Again, this indicates that in yeast species, significant rewiring occurred in the Yap families during evolution.

6. Potential roles in oxidative stress response for others members of the Yap family

Yeast species possess very flexible and complex gene expression programs to respond quickly and efficiently to several environmental challenges. The complexity of the underlying transcriptional networks is only become to light. Since the pioneer study of Harbison *et al.* (2004), an important challenge consists in understanding combinatorial regulations (Bhardwaj et al., 2010a; Bhardwaj et al., 2010b), *i.e.* how TFs partner with each other to control the expression of common target genes. This allows a limited number of regulators to govern many genes and obtain more plastic and specific functions in the cell. The observation that multiple homologous links exist between the Yap proteins identified in the three different species (see Table 2), together with the observation that only 10% of genes share orthologous relationships between AP-1 TMs (see Section 3) raise important questions concerning the inter- and intra-species functional redundancies that exist between Yap proteins. In the model yeast *S. cerevisiae*, several functional studies were performed in order to characterize the role of the different Yap family members (see (Rodrigues-Pousada et al., 2010) for review). It has been shown that Yap2p (also names Cad1p) plays a role in the response to toxic compounds such as cadmium (Azevedo et al., 2007). Yap4p and Yap6p were identified as being involved in the osmotic stress response (Posas et al., 2000). Yap5p has been shown to play a role in the metabolism and storage of iron and Yap8p, the last Yap family member, has a central role in the detoxification of the cell and more specifically in arsenic stress response (Ilina et al., 2008). Only the roles of TFs Yap3p and Yap7p remain to be established.

In order to consider the potential roles of the different Yap family members in the specific response to oxidative stress, we inferred a transcriptional regulatory network based on information stored in databases. Our approach consisted, in a first step, to select all the genes in *S. cerevisiae*, annotated in the Gene Ontology database (2010) as being involved in oxidative stress (functional category GO:0006979). 88 genes were thus collected and were searched in a second step, for potential regulatory interactions with Yap TFs. For that, we used information deposited in the YEASTRACT database (Abdulrehman et al., 2011). YEASTRACT was developed to support the analysis of transcription regulatory associations in *S. cerevisiae* and stores regulatory associations described in the literature. Among the 88 genes selected in the first step, more than 50% (41 genes) were annotated as target genes for one of the Yap TFs (only using YEASTRACT direct evidences). Interactions between Yap TFs and their target genes are represented Figure 3A. As expected, the transcription factor Yap1p appears to have the predominant role in the transcriptional control of genes associated with oxidative stress. Interactions with 24 analyzed genes (27%) were identified. Interestingly our results reveal that Yap4p, Yap7p and Yap2p potentially exert significant transcriptional controls on a significant number of oxidative responsive genes. They are the regulators of respectively 14 genes (16%), 13 genes (15%) and 11 genes (12.5%). Yap5p, Yap6p and Yap3p present only a small number of target genes in the oxidative stress response GO category (respectively 8, 6 and 2 genes). Therefore, the Yap transcriptional network involved in response to oxidative stress appeared to be larger than this generally stated in the literature (Moye-Rowley, 2002). Despite its incontestable and crucial role, Yap1p is certainly not the unique Yap regulator necessary to obtain an optimal oxidative response. It belongs to a redundant network, with a majority of genes being potentially regulated by several Yap transcription factors (only 17 genes are target genes of a single Yap protein).

Our observations suggest a potential important role for Yap2p, Yap4p and Yap7p proteins in the transcriptional control of oxidative stress response. Interestingly, Yap2p and Yap4p share common target genes with Yap1p. These genes are known to play a key role in the response of oxidative stress. For example, the gene *TRX2* is a highly conserved oxydoreductase, required to maintain the redox homeostasis of the cell. In *S. cerevisiae*, thioredoxin *TRX2* and thioredoxin reductase *TRR1* were shown to be necessary to protect the cell against oxidative stress induced by the presence of ROS compounds in condition H_2O_2 (Boisnard et al., 2009) or exposure to dithiothreitol (DTT) (Garrido and Grant, 2002). This agrees with the review of Rodrigues-Pousada *et al.* (2010), in which Yap2p and Yap4p were described as involved in several types of stress responses, including oxidative stress (Cohen et al., 2002; Nevitt et al., 2004). Yap5p and Yap6p proteins also seem to be involved in the oxidative stress response, but in a more general context. Indeed, whereas Yap1p, Yap2p, Yap4p (and probably Yap7p) are involved in the core response to oxidative stress by activating target genes (catalases, reductases, etc.) specialized in response to the presence ROS compounds in the cell or its environment, Yap5p and Yap6p rather activate the transcription of genes involved in cell protection against damages caused by oxidative stress (for instance *GRX4*, *GRX5*, *CTT1*), in the biosynthetic pathway of ergosterol (for instance *MCR1*), in the regulation of apoptosis (for instance *MCA1*) or in the maintain of membrane

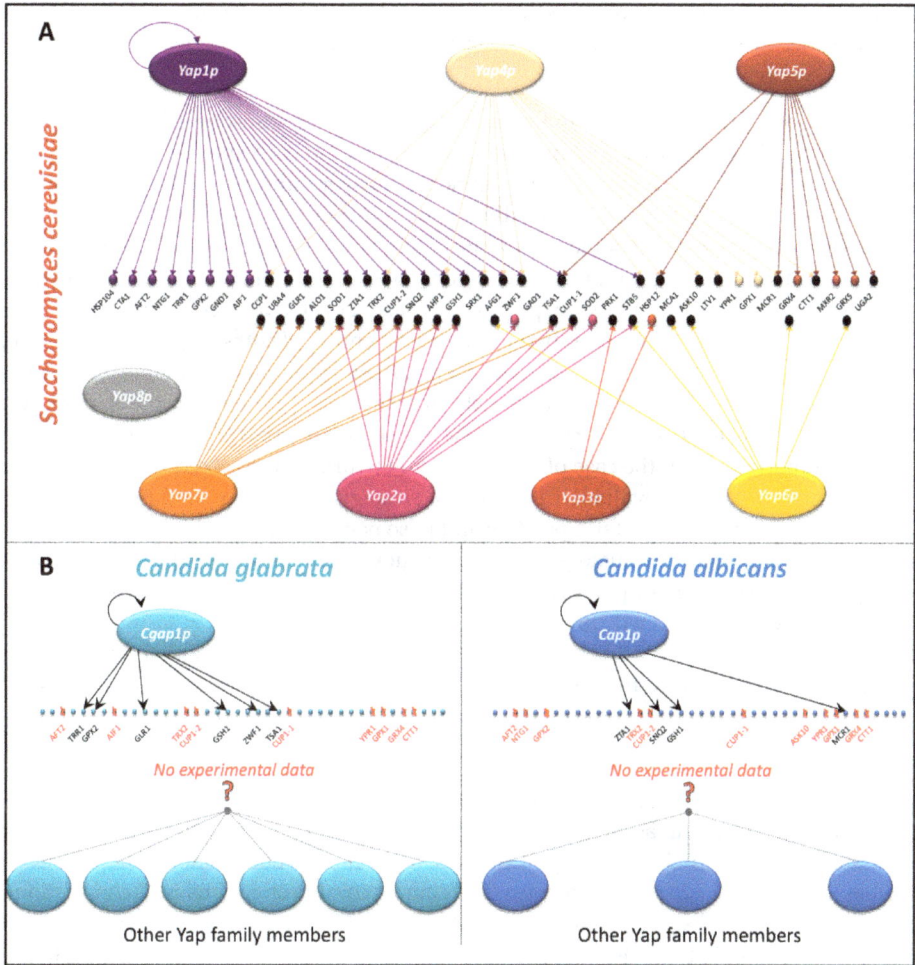

Fig. 3. Potential transcriptional networks involved in yeast oxidative stress response. (A) Using information stored in the GO (2010) and YEASTRACT (Abdulrehman et al., 2011) databases, we inferred a Yap transcriptional network in yeast *S. cerevisiae* (see the main text). Regulatory interactions are color-coded according to their dependant YAP TF: purple for Yap1p, pink for Yap2p, red for Yap3p, beige for Yap4p, brown for Yap5p, yellow for Yap6p, orange for Yap7p. No interaction with Yap8p TF was identified. (B) Representation of the transcriptional networks obtained in *Candida* species using orthologous relationships with *S. cerevisiae* genes and regulatory interactions described in (Goudot et al., 2011). Genes for which no orthologue was found are crossed out with red crosses. These networks are incomplete compared to this obtained in *S. cerevisiae* and highlight the need for more experimental data in *C. glabrata* and *C. albicans* species.

organization (for instance *HSP12*). Interestingly, potential roles of Yap7p in response to stress in general, has not been defined (Rodrigues-Pousada et al., 2010). Even if additional experimental validations are required to confirm the role of Yap7p in the stress response in *S. cerevisiae*, this review gives interesting information and represents a good starting point.

To go further, we tried to extend our representation of the Yap transcriptional network to *C. glabrata* and *C. albicans* species. In a first step, we searched in each species for a list of genes involved in oxidative stress response. GO annotations were unfortunately incomplete and compelled us to restrict our search to orthologous *S. cerevisiae* genes. Only 32 genes were thus selected in *C. glabrata* genome and 31 in *C. albicans* genome. They are represented in Figure 3B. Concerning the regulatory interactions, literature information was just available for Cgap1p (in *S. cerevisiae*) and Cap1p (in *C. albicans*) TFs (Goudot et al., 2011). Therefore only 7 and 5 interactions were identified in *C. glabrata* and *C. albicans* respectively. It is clear that the obtained networks in *Candida* species are largely incomplete compared to this of *S. cerevisiae*. The classical approach that consists in transferring functional annotations from well-studied organisms (like *S. cerevisiae*) to newly sequence species (like *Candida* species) cannot be properly used in the case of a regulatory network that involved highly redundant TFs. In a recent study, we evaluated that such a strategy would have led (in case of AP-1 TMs, see Section 2) to a rate of false positive and false negatives predictions higher than 70% (Goudot et al., 2011). Fortunately, the recent development of costless and efficient multispecies transcriptomic platforms should lead to a rapid accumulation of new experimental datasets obtained directly in *Candida* species. It will allow to reproduced the analysis performed with AP-1 TFs to other factors and hence will greatly enhance our comprehension of the role of the Yap family in the response to oxidative stress in yeasts.

7. Conclusion

Comparative functional analyses have been made possible by the accumulation of large-scale gene expression datasets for a large number of organisms, due directly to the exponential increase in the number of species for which whole genome sequences are available (Liolios et al., 2010). To increase the accuracy of investigations in the evolution of transcriptional networks, we would like in an ideal case, compare species with different lifestyle and physiological properties. In this respect, yeasts are ideal organisms for comparative functional genomic studies, since they have evolved in niches with constantly varying nutrient availability and growth conditions. Also, the sequencing of genomes of a dozen of different species open new possibility for studying the adaptation of transcriptional networks to environmental stresses (see (Lelandais and Devaux, 2010; Lelandais et al., 2011) for reviews). Analysis of the genomic events that underlie the response to oxidative stress demonstrated that although the gene expression patterns characterizing the response are relatively well conserved between yeast species (Lelandais et al., 2008; Znaidi et al., 2009), part of the underlying regulatory networks differed. In particular, the roles of the oxidative stress response TFs Yap1p (in *S. cerevisiae*), Cgap1p (in *C. glabrata*) and Cap1p (in *C. albicans*) appeared to have diverged (Goudot et al., 2011; Lelandais et al., 2008). In *C. glabrata*, the preferred DNA binding sites of Cgap1p protein is different from the *S. cerevisiae* and *C. albicans* canonical YRE sequence (Goudot et al., 2011; Kuo et al., 2010). The functioning of Yap1p TM is therefore more similar to the functioning of Cap1p TM than to the Cgap1p TM. Considering that *C. glabrata* is phylogenetically more related to *S. cerevisiae* than *C. albicans* is (Dujon, 2010), this observation illustrates that evolution of regulatory modules may be different from the phylogeny established from comparative genomics. Our review of Yap families suggests that the associated TMs have been importantly shuffled during evolution. Constrains exists to maintain key cellular functions as oxidative defence, but the genomic strategies to obtain these function evolve more rapidly than originally expected. Deciphering the regulatory interactions, as well as the dynamics of these processes, represent important challenges to understand the genomic control of stress response, a highly conserved process in eukaryotes.

8. References

(2010). The Gene Ontology in 2010: extensions and refinements. Nucleic Acids Res *38*, D331-335.

Abdulrehman, D., Monteiro, P.T., Teixeira, M.C., Mira, N.P., Lourenco, A.B., Dos Santos, S.C., Cabrito, T.R., Francisco, A.P., Madeira, S.C., Aires, R.S., *et al.* (2010). YEASTRACT: providing a programmatic access to curated transcriptional regulatory associations in Saccharomyces cerevisiae through a web services interface. Nucleic Acids Res.

Abdulrehman, D., Monteiro, P.T., Teixeira, M.C., Mira, N.P., Lourenco, A.B., dos Santos, S.C., Cabrito, T.R., Francisco, A.P., Madeira, S.C., Aires, R.S., *et al.* (2011). YEASTRACT: providing a programmatic access to curated transcriptional regulatory associations in Saccharomyces cerevisiae through a web services interface. Nucleic Acids Res *39*, D136-140.

Abegg, M.A., Alabarse, P.V., Casanova, A., Hoscheid, J., Salomon, T.B., Hackenhaar, F.S., Medeiros, T.M., and Benfato, M.S. (2010). Response to oxidative stress in eight pathogenic yeast species of the genus Candida. Mycopathologia *170*, 11-20.

Alarco, A.M., Balan, I., Talibi, D., Mainville, N., and Raymond, M. (1997). AP1-mediated multidrug resistance in Saccharomyces cerevisiae requires FLR1 encoding a transporter of the major facilitator superfamily. J Biol Chem *272*, 19304-19313.

Azevedo, D., Nascimento, L., Labarre, J., Toledano, M.B., and Rodrigues-Pousada, C. (2007). The S. cerevisiae Yap1 and Yap2 transcription factors share a common cadmium-sensing domain. FEBS Lett *581*, 187-195.

Azevedo, D., Tacnet, F., Delaunay, A., Rodrigues-Pousada, C., and Toledano, M.B. (2003). Two redox centers within Yap1 for H2O2 and thiol-reactive chemicals signaling. Free Radic Biol Med *35*, 889-900.

Barrett, T., Troup, D.B., Wilhite, S.E., Ledoux, P., Rudnev, D., Evangelista, C., Kim, I.F., Soboleva, A., Tomashevsky, M., Marshall, K.A., *et al.* (2009). NCBI GEO: archive for high-throughput functional genomic data. Nucleic Acids Res *37*, D885-890.

Bhardwaj, N., Carson, M.B., Abyzov, A., Yan, K.K., Lu, H., and Gerstein, M.B. (2010a). Analysis of combinatorial regulation: scaling of partnerships between regulators with the number of governed targets. PLoS Comput Biol *6*, e1000755.

Bhardwaj, N., Kim, P.M., and Gerstein, M.B. (2010b). Rewiring of transcriptional regulatory networks: hierarchy, rather than connectivity, better reflects the importance of regulators. Sci Signal *3*, ra79.

Blake, D.R., Allen, R.E., and Lunec, J. (1987). Free radicals in biological systems--a review orientated to inflammatory processes. Br Med Bull 43, 371-385.

Boisnard, S., Lagniel, G., Garmendia-Torres, C., Molin, M., Boy-Marcotte, E., Jacquet, M., Toledano, M.B., Labarre, J., and Chedin, S. (2009). H2O2 activates the nuclear localization of Msn2 and Maf1 through thioredoxins in Saccharomyces cerevisiae. Eukaryot Cell 8, 1429-1438.

Chen, K.H., Miyazaki, T., Tsai, H.F., and Bennett, J.E. (2007). The bZip transcription factor Cgap1p is involved in multidrug resistance and required for activation of multidrug transporter gene CgFLR1 in Candida glabrata. Gene 386, 63-72.

Cohen, B.A., Pilpel, Y., Mitra, R.D., and Church, G.M. (2002). Discrimination between paralogs using microarray analysis: application to the Yap1p and Yap2p transcriptional networks. Mol Biol Cell 13, 1608-1614.

Coleman, S.T., Epping, E.A., Steggerda, S.M., and Moye-Rowley, W.S. (1999). Yap1p activates gene transcription in an oxidant-specific fashion. Mol Cell Biol 19, 8302-8313.

Costa, V., and Moradas-Ferreira, P. (2001). Oxidative stress and signal transduction in Saccharomyces cerevisiae: insights into ageing, apoptosis and diseases. Mol Aspects Med 22, 217-246.

Davies, K.J. (1985). Free radicals and protein degradation in human red blood cells. Prog Clin Biol Res 195, 15-27.

Delaunay, A., Isnard, A.D., and Toledano, M.B. (2000). H2O2 sensing through oxidation of the Yap1 transcription factor. EMBO J 19, 5157-5166.

Dujon, B. (2010). Yeast evolutionary genomics. Nat Rev Genet 11, 512-524.

Fernandes, L., Rodrigues-Pousada, C., and Struhl, K. (1997). Yap, a novel family of eight bZIP proteins in Saccharomyces cerevisiae with distinct biological functions. Mol Cell Biol 17, 6982-6993.

Fitch, W.M. (2000). Homology a personal view on some of the problems. Trends Genet 16, 227-231.

Garrido, E.O., and Grant, C.M. (2002). Role of thioredoxins in the response of Saccharomyces cerevisiae to oxidative stress induced by hydroperoxides. Mol Microbiol 43, 993-1003.

Goudot, C., Etchebest, C., Devaux, F., and Lelandais, G. (2011). The reconstruction of condition-specific transcriptional modules provides new insights in the evolution of yeast AP-1 proteins. PLoS One 6, e20924.

Grant, C.M., Collinson, L.P., Roe, J.H., and Dawes, I.W. (1996). Yeast glutathione reductase is required for protection against oxidative stress and is a target gene for yAP-1 transcriptional regulation. Mol Microbiol 21, 171-179.

Gupta, K., Bishop, J., Peck, A., Brown, J., Wilson, L., and Panda, D. (2004). Antimitotic antifungal compound benomyl inhibits brain microtubule polymerization and dynamics and cancer cell proliferation at mitosis, by binding to a novel site in tubulin. Biochemistry 43, 6645-6655.

Harbison, C.T., Gordon, D.B., Lee, T.I., Rinaldi, N.J., Macisaac, K.D., Danford, T.W., Hannett, N.M., Tagne, J.B., Reynolds, D.B., Yoo, J., et al. (2004). Transcriptional regulatory code of a eukaryotic genome. Nature 431, 99-104.

Herrero, E., Ros, J., Belli, G., and Cabiscol, E. (2008). Redox control and oxidative stress in yeast cells. Biochim Biophys Acta *1780*, 1217-1235.

Hirata, D., Yano, K., and Miyakawa, T. (1994). Stress-induced transcriptional activation mediated by YAP1 and YAP2 genes that encode the Jun family of transcriptional activators in Saccharomyces cerevisiae. Mol Gen Genet *242*, 250-256.

Ikner, A., and Shiozaki, K. (2005). Yeast signaling pathways in the oxidative stress response. Mutat Res *569*, 13-27.

Ilina, Y., Sloma, E., Maciaszczyk-Dziubinska, E., Novotny, M., Thorsen, M., Wysocki, R., and Tamas, M.J. (2008). Characterization of the DNA-binding motif of the arsenic-responsive transcription factor Yap8p. Biochem J *415*, 467-475.

Inoue, Y., Matsuda, T., Sugiyama, K., Izawa, S., and Kimura, A. (1999). Genetic analysis of glutathione peroxidase in oxidative stress response of Saccharomyces cerevisiae. J Biol Chem *274*, 27002-27009.

Jaffrezic, F., Marot, G., Degrelle, S., Hue, I., and Foulley, J.L. (2007). A structural mixed model for variances in differential gene expression studies. Genet Res *89*, 19-25.

Jamieson, D.J. (1998). Oxidative stress responses of the yeast Saccharomyces cerevisiae. Yeast *14*, 1511-1527.

Jeanmougin, M., de Reynies, A., Marisa, L., Paccard, C., Nuel, G., and Guedj, M. (2010). Should we abandon the t-test in the analysis of gene expression microarray data: a comparison of variance modeling strategies. PLoS One *5*, e12336.

Kuge, S., and Jones, N. (1994). YAP1 dependent activation of TRX2 is essential for the response of Saccharomyces cerevisiae to oxidative stress by hydroperoxides. EMBO J *13*, 655-664.

Kuge, S., Jones, N., and Nomoto, A. (1997). Regulation of yAP-1 nuclear localization in response to oxidative stress. EMBO J *16*, 1710-1720.

Kuge, S., Toda, T., Iizuka, N., and Nomoto, A. (1998). Crm1 (Xpo1) dependent nuclear export of the budding yeast transcription factor yAP-1 is sensitive to oxidative stress. Genes Cells *3*, 521-532.

Kuo, D., Licon, K., Bandyopadhyay, S., Chuang, R., Luo, C., Catalana, J., Ravasi, T., Tan, K., and Ideker, T. (2010). Coevolution within a transcriptional network by compensatory trans and cis mutations. Genome Res *20*, 1672-1678.

Lamprecht, A.L., Naujokat, S., Margaria, T., and Steffen, B. (2011). Semantics-based composition of EMBOSS services. J Biomed Semantics *2 Suppl 1*, S5.

Lee, J., Godon, C., Lagniel, G., Spector, D., Garin, J., Labarre, J., and Toledano, M.B. (1999a). Yap1 and Skn7 control two specialized oxidative stress response regulons in yeast. J Biol Chem *274*, 16040-16046.

Lee, J., Spector, D., Godon, C., Labarre, J., and Toledano, M.B. (1999b). A new antioxidant with alkyl hydroperoxide defense properties in yeast. J Biol Chem *274*, 4537-4544.

Lelandais, G., and Devaux, F. (2010). Comparative functional genomics of stress responses in yeasts. OMICS *14*, 501-515.

Lelandais, G., Goudot, C., and Devaux, F. (2011). The evolution of gene expression regulatory networks in yeasts. C R Biol *334*, 655-661.

Lelandais, G., Tanty, V., Geneix, C., Etchebest, C., Jacq, C., and Devaux, F. (2008). Genome adaptation to chemical stress: clues from comparative transcriptomics in Saccharomyces cerevisiae and Candida glabrata. Genome Biol 9, R164.

Liolios, K., Chen, I.M., Mavromatis, K., Tavernarakis, N., Hugenholtz, P., Markowitz, V.M., and Kyrpides, N.C. (2010). The Genomes On Line Database (GOLD) in 2009: status of genomic and metagenomic projects and their associated metadata. Nucleic Acids Res 38, D346-354.

Liu, T.T., Znaidi, S., Barker, K.S., Xu, L., Homayouni, R., Saidane, S., Morschhauser, J., Nantel, A., Raymond, M., and Rogers, P.D. (2007). Genome-wide expression and location analyses of the Candida albicans Tac1p regulon. Eukaryot Cell 6, 2122-2138.

Lucau-Danila, A., Lelandais, G., Kozovska, Z., Tanty, V., Delaveau, T., Devaux, F., and Jacq, C. (2005). Early expression of yeast genes affected by chemical stress. Mol Cell Biol 25, 1860-1868.

Lushchak, V.I. (2006). Budding yeast Saccharomyces cerevisiae as a model to study oxidative modification of proteins in eukaryotes. Acta Biochim Pol 53, 679-684.

Lushchak, V.I. (2010). Oxidative stress in yeast. Biochemistry (Mosc) 75, 281-296.

Martin-Magniette, M.L., Aubert, J., Bar-Hen, A., Elftieh, S., Magniette, F., Renou, J.P., and Daudin, J.J. (2008). Normalization for triple-target microarray experiments. BMC Bioinformatics 9, 216.

Morschhauser, J., Barker, K.S., Liu, T.T., Bla, B.W.J., Homayouni, R., and Rogers, P.D. (2007). The transcription factor Mrr1p controls expression of the MDR1 efflux pump and mediates multidrug resistance in Candida albicans. PLoS Pathog 3, e164.

Moye-Rowley, W.S. (2002). Transcription factors regulating the response to oxidative stress in yeast. Antioxid Redox Signal 4, 123-140.

Nevitt, T., Pereira, J., and Rodrigues-Pousada, C. (2004). YAP4 gene expression is induced in response to several forms of stress in Saccharomyces cerevisiae. Yeast 21, 1365-1374.

Nguyen, D.T., Alarco, A.M., and Raymond, M. (2001). Multiple Yap1p-binding sites mediate induction of the yeast major facilitator FLR1 gene in response to drugs, oxidants, and alkylating agents. J Biol Chem 276, 1138-1145.

O'Brien, K.P., Remm, M., and Sonnhammer, E.L. (2005). Inparanoid: a comprehensive database of eukaryotic orthologs. Nucleic Acids Res 33, D476-480.

Pfaller, M.A., and Diekema, D.J. (2007). Epidemiology of invasive candidiasis: a persistent public health problem. Clin Microbiol Rev 20, 133-163.

Posas, F., Chambers, J.R., Heyman, J.A., Hoeffler, J.P., de Nadal, E., and Arino, J. (2000). The transcriptional response of yeast to saline stress. J Biol Chem 275, 17249-17255.

Rodrigues-Pousada, C., Menezes, R.A., and Pimentel, C. (2010). The Yap family and its role in stress response. Yeast 27, 245-258.

Salin, H., Fardeau, V., Piccini, E., Lelandais, G., Tanty, V., Lemoine, S., Jacq, C., and Devaux, F. (2008). Structure and properties of transcriptional networks driving selenite stress response in yeasts. BMC Genomics 9, 333.

Sherman, D., Durrens, P., Iragne, F., Beyne, E., Nikolski, M., and Souciet, J.L. (2006). Genolevures complete genomes provide data and tools for comparative genomics of hemiascomycetous yeasts. Nucleic Acids Res *34*, D432-435.

Singh, R.P., Prasad, H.K., Sinha, I., Agarwal, N., and Natarajan, K. (2011). Cap2-HAP complex is a critical transcriptional regulator that has dual but contrasting roles in regulation of iron homeostasis in Candida albicans. J Biol Chem *286*, 25154-25170.

Smith, C.V., Hughes, H., and Mitchell, J.R. (1984). Free radicals in vivo. Covalent binding to lipids. Mol Pharmacol *26*, 112-116.

Sonnhammer, E.L., and Koonin, E.V. (2002). Orthology, paralogy and proposed classification for paralog subtypes. Trends Genet *18*, 619-620.

Storz, G., Christman, M.F., Sies, H., and Ames, B.N. (1987). Spontaneous mutagenesis and oxidative damage to DNA in Salmonella typhimurium. Proc Natl Acad Sci U S A *84*, 8917-8921.

Sugiyama, K., Izawa, S., and Inoue, Y. (2000). The Yap1p-dependent induction of glutathione synthesis in heat shock response of Saccharomyces cerevisiae. J Biol Chem *275*, 15535-15540.

Tan, K., Feizi, H., Luo, C., Fan, S.H., Ravasi, T., and Ideker, T.G. (2008). A systems approach to delineate functions of paralogous transcription factors: role of the Yap family in the DNA damage response. Proc Natl Acad Sci U S A *105*, 2934-2939.

Trush, M.A., and Kensler, T.W. (1991). An overview of the relationship between oxidative stress and chemical carcinogenesis. Free Radic Biol Med *10*, 201-209.

Tusher, V.G., Tibshirani, R., and Chu, G. (2001). Significance analysis of microarrays applied to the ionizing radiation response. Proc Natl Acad Sci U S A *98*, 5116-5121.

Wemmie, J.A., Steggerda, S.M., and Moye-Rowley, W.S. (1997). The Saccharomyces cerevisiae AP-1 protein discriminates between oxidative stress elicited by the oxidants H2O2 and diamide. J Biol Chem *272*, 7908-7914.

Wemmie, J.A., Szczypka, M.S., Thiele, D.J., and Moye-Rowley, W.S. (1994). Cadmium tolerance mediated by the yeast AP-1 protein requires the presence of an ATP-binding cassette transporter-encoding gene, YCF1. J Biol Chem *269*, 32592-32597.

Wettenhall, J.M., and Smyth, G.K. (2004). limmaGUI: a graphical user interface for linear modeling of microarray data. Bioinformatics *20*, 3705-3706.

White, T.C., Marr, K.A., and Bowden, R.A. (1998). Clinical, cellular, and molecular factors that contribute to antifungal drug resistance. Clin Microbiol Rev *11*, 382-402.

Wickens, A.P. (2001). Ageing and the free radical theory. Respir Physiol *128*, 379-391.

Wu, A., Wemmie, J.A., Edgington, N.P., Goebl, M., Guevara, J.L., and Moye-Rowley, W.S. (1993). Yeast bZip proteins mediate pleiotropic drug and metal resistance. J Biol Chem *268*, 18850-18858.

Wu, A.L., and Moye-Rowley, W.S. (1994). GSH1, which encodes gamma-glutamylcysteine synthetase, is a target gene for yAP-1 transcriptional regulation. Mol Cell Biol *14*, 5832-5839.

Yan, C., Lee, L.H., and Davis, L.I. (1998). Crm1p mediates regulated nuclear export of a yeast AP-1-like transcription factor. EMBO J *17*, 7416-7429.

Znaidi, S., Barker, K.S., Weber, S., Alarco, A.M., Liu, T.T., Boucher, G., Rogers, P.D., and Raymond, M. (2009). Identification of the Candida albicans Cap1p regulon. Eukaryot Cell 8, 806-820.

Znaidi, S., Weber, S., Al-Abdin, O.Z., Bomme, P., Saidane, S., Drouin, S., Lemieux, S., De Deken, X., Robert, F., and Raymond, M. (2008). Genomewide location analysis of Candida albicans Upc2p, a regulator of sterol metabolism and azole drug resistance. Eukaryot Cell 7, 836-847.

18

ROS as Signaling Molecules and Enzymes of Plant Response to Unfavorable Environmental Conditions

Dominika Boguszewska[1] and Barbara Zagdańska[2]
[1]Plant Breeding and Acclimatization Institute-
National Research Institute Division Jadwisin,
[2]Warsaw University of Life Sciences
Poland

1. Introduction

Research concern plant response to unfavourable environmental conditions is becoming increasingly important and most climate-change scenarios suggest an increase in aridity in many areas of the globe (Petit et al. 1999, Blum 2011). On a global principle, drought (taking into account soil and/or atmospheric water deficit) in combination with coincident high temperature and radiation, determines the most important environmental constraints to plant survival and to crop productivity (Boyer 1982). Agriculture is a major user of water resources and abiotic stress is the principal cause of decreasing the average yield of major crops by more than 50%, which causes losses worth hundreds of millions of dollars each year. Understanding mechanism of abiotic stress tolerance and defense is important for crop improvement. Many works concerning this problem were developed over last decades, discussing subjects from plant strategies to control water status under drought (Schulze 1986) to the physiological and biochemical processes underlying plant response to water shortage (Chaves 1991, Cornin and Massacci 1996) and oxidative stress (Smirnoff 1998). In this chapter various aspects of reactive oxygen species and enzymes in plant response to drought (abiotic stress) are discussed.

2. Environmental stresses

Stress in physical terms is defined as mechanical force per unit area applied to an object. In biological systems stress can be defined as an adverse force, effect, or influence that trends to inhabit normal systems from functioning.

A wide range of unfavourable environmental conditions may induce stresses in plants that alter plant growth, development and metabolism, and even may lead to plant death. These stresses include mechanical damage, herbicides, UV radiation, salt, low/high temperature, soil drought, flooding, high speed wind, nutrient loss and anaerobic conditions are very important stress factors limiting crop productivity (Lawlor, 2002). Among them drought is a major abiotic factor that limits agricultural crop production.

To sense these environmental signals plants have evolved a complex signalling network, which may also cross-talk. Stress signal transduction pathway start with signal perception by receptors like phytochromes, histidine kinases, receptor like kinases, G-protein-coupled receptors (GPCR), hormone receptors etc., which generate secondary signalling molecules like inositol phosphatase, reactive oxygen species (ROS), abscisic acid (ABA), etc. Mentioned secondary molecules can modulate the intracellular Ca^{2+} level and initiate protein phosphorylation cascades i.e. mitogen-activated protein kinase MAPK, calcium dependent protein kinase CDPK, protein phosphatase, SOS3/protein kinase S etc. Drought and salinity occur their influence on cell mainly by disrupting the ionic and osmotic equilibrium leading to a cascade of events, which can be grouped under ionic and osmotic signalling pathway. These stresses are marked by symptoms of stress injury. Stress injury may occur through denaturation of cellular proteins/enzymes or through production of ROS. In response to injury stress plants start detoxification process which include changes in expression of LEA/dehydrin-type gene synthesis of proteinases, enzymes for scavenging ROS and other detoxification proteins.

3. Plant defence against reactive oxygen species

Reactive oxygen species are continuously produced during aerobic metabolism as byproducts of different metabolic pathways which are localized in some cellular compartments such as chloroplast, mitochondria and peroxisomes – organelles with a highly oxidizing metabolic activity or with intense rate of electron flow. Therefore, plants are well equipped with antioxidants and enzymes scavenging ROS to keep their levels low under favourable conditions of growth. However, under unfavourable environmental conditions production of reactive oxygen species may increase and lead to oxidative stress in many plant species (Smirnoff 1993, Mullineaux & Karpinski 2002, Miller et al., 2010).

Reactive oxygen species inactivate enzymes and damage important cellular components. ROS are responsible for protein, lipid and nucleic acid modification and are thought to play a major role in ageing and cell death (Jacobson 1996). To avoid the accumulation of these compounds to toxic level, animals and plants possess several detoxifying enzymatic systems that comprise a variety of antioxidant molecules and enzymes. Two main classes of **plant defenses** have been described and can be classified as non-enzymatic and enzymatic systems. The first class - non-enzymatic constituents, including lipid-soluble and membrane-associated tocopherols, water soluble reductants, ascorbic acid and glutathione, and enzymatic constituents, including superoxide dismutase (EC 1.15.1.1), catalase (EC 1.11.1.6), peroxidase (EC 1.11.1.7), ascorbate peroxidase (EC 1.11.1.11) and associated antioxidant enzyme, glutathione reductase (EC 1.1.4.2).

Under steady state conditions, the ROS molecules are scavenged by antioxidant mechanisms (AOS) mentioned earlier (Fig. 1). The equilibrium between the production and the scavenging of ROS may be disturbed by stress factors. However, it is widely discussed that the relationship between metabolism and redox state is complex and subtle (Hare et al., 1998; Bohnert & Jensen 1996). Oxidation involves different signals that plants use to make appropriate adjustments of gene expression or cell structure in response to environmental factors. Increased level of intercellular ROS due to increased production and/or insufficient antioxidant protection can cause significant modification of cell structure or even lead to cell death. In many cases, the situation of enhancing oxidation, which itself is negative may be a signal for adjusting plant metabolism to the new conditions of growth.

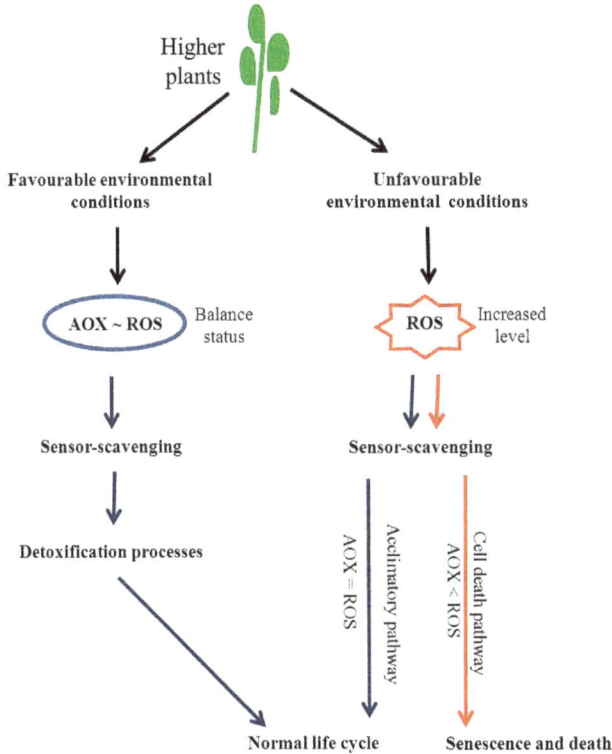

Fig. 1. Antioxidants and redox signaling in plants

4. Reactive oxygen species involving in plant defence

Stress of any kind, biotic or abiotic leads to an increased level of ROS production and/or to inactivation of antioxidants, particularly enzymes. The oxidative stress that accompanies unfavorable environmental conditions should not be viewed as symptom of cellular dysfunction but it can represent a perquisite signal for a plant to induce proper acclimation mechanisms (Jaspers & Kandasjarvi, 2010; Miller et al. 2010). Reactive oxygen species can also play a role of "oxidation signalling molecules" (Foyer & Noctor 2009). Signals of ROS originating at different organelles have been shown to induce large transcriptional changes and cellular reprogramming that can either protect the plant cell or induce programmed cell death (Foyer & Noctor 2005).

It is important to remember that whether ROS will act as damaging, protective or signalling factor depends on sensible equilibrium between ROS production and scavenging at the proper site and time. It is known that ROS can damage cells and initiate different responses like expression of new genes. Evoking of cell response highly depends on several factors. One of them is subcellular localization for formation of ROS. Stress-induced ROS accumulation is counteracted by enzymatic and nonenzymatic antioxidant. Thus, plant stress tolerance may be improved by increasing of *in vivo* levels of antioxidants enzymes.

Antioxidants found in almost all cellular compartments, demonstrating the importance of ROS detoxification for cellular survival. It has been presented that ROS influence expression of a number of genes and signal transduction pathways witch suggest that cells have evolved strategies to use ROS as biological stimuli and signals that activate and control various genetic stress-response programs.

Reactive oxygen species signalling is also highly integrated with hormonal signalling networks processing and transmitting environmental inputs in order to induce plant appropriate responses to environmental constraints (Mittler et al., 2011). Involvement of hormons such as auxins, cytokinins, ethylene, ABA, jasmonic (JA) and salicylic (SA) acids in signalling together with ROS signaling allow plants to regulate developmental processes and adaptive response to environmental cues (Fig. 2). A protective signaling role of plant hormones may lead to activation of acclimation responses such as stomatal closure, regulation of hydraulic conductivity and developmental processes that affect senescence and abscission (Boursiac et al. 2008; Sakamoto et al., 2008; Miller et al. 2010). Salicylic acid, similarly as ROS, is involved in both, defence and cell death responses e.g. increased level of ROS can cause SA accumulation which in turn is involved in SA-induced stomatal closure. There is also cross-talk between ROS and ABA. It has been proved that gibberellin (GA) signalling is connected with ROS by stimulating the destruction of DELLA proteins that regulate transcript levels of antioxidant enzymes. The integration of ROS with auxin signalling networks, caused by recognition of environmental factors as the stress-induced morphogenic response, lead to ROS and auxin metabolism interaction, and in effect to morphological changes that help to avoid deleterious effect of environmental stresses.

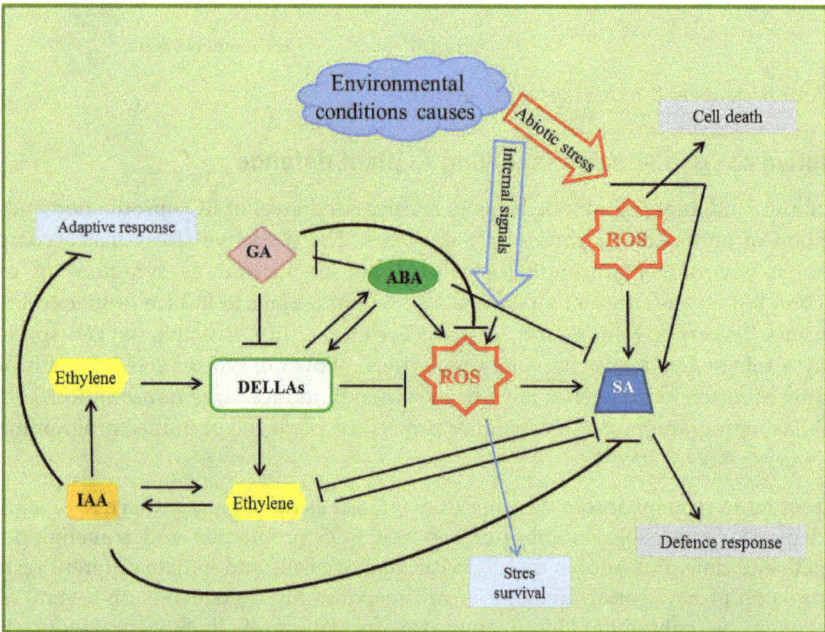

Fig. 2. A schematic model showing interactions between ROS and hormonal signalling pathways

The most common ROS are hydrogen peroxide (H_2O_2), superoxide ($O_2\cdot^-$), the hydroxyl radical (HO·) and singlet oxygen ($^1O_2$).

The major site of **superoxide radical** ($O_2\cdot^-$)production is the reaction centers of photosystem I (PSI) and a photosystem II (PSII) in chloroplast thylakoids. In mitochondria, complex I, II and complex III in the electron transport chain contribute to superoxide radical production. The terminal oxidases-cytochrome c oxidase and the alternative oxidase react with O_2, four electrons are transferred and H_2O is released. There is situation when O_2 can react with other ETC components and there in only one electron transferred with the result of $O_2\cdot^-$ release. It has been shown that in plants 1-2% of O_2 consumption leads to $O_2\cdot^-$ production (Puntarulo., et al. 1988).

Singlet oxygen is the first excited electronic state of O_2. Insufficient energy dissipation during photosynthesis can lead to formation of chlorophyll (Chl) triplet state. And the Chl triplet state can react with $3O_2$ to give up very reactive singlet oxygen. It has been proved that singlet oxygen formation during photosynthesis can have damaging effect on PSI and PSII and on whole machinery of photosynthesis.

Hydroxyl radicals (HO·) are the highest reactive ROS. It can be produced from $O_2\cdot^-$ and H_2O_2 at neutral pH and ambient temperature by iron-catalyzed.

Hydrogen peroxide (H_2O_2) is produced by univalent reduction of $O_2\cdot^-$. H_2O_2 is moderately reactive. It has been proved that excess of hydrogen peroxide leads to oxidative stress. This molecule may also inactivate enzymes by oxidizing their thiol groups. Moreover, H_2O_2 play dual role in plants. At low concentration it can act as a signal molecule involved in acclimatory signaling triggering tolerance to different biotic and abiotic stresses. At high concentration it leads to programmed cell death (Quan et al., 2008). It has been proved that H_2O_2 act as a key regulator of in a wide range of physiological processes like photorespiration and photosynthesis (Noctor & Foyer 1998), stomatal movement (Bright et al., 2006), cell cycle (Mittler et al., 2006) and growth and development (Foreman 2003). H_2O_2 is taking as a second messenger for signals generated by means of ROS due to its relatively long life and high permeability across membranes. Many of the general stress genes are regulated by a signaling pathways using H_2O_2 as the messenger (Möller & Sweetlove 2010).

5. Major cellular sources of reactive oxygen species

ROS are produced continuously as byproducts of various metabolic pathways that are localized in different cellular compartments such as chloroplasts, mitochondria and peroxisomes.

Plant **mitochondria** called "energy factories" are known, apart from chloroplasts, as a main place of ROS production like H_2O_2 and also the ROS target (Rasmusson et al., 2004). Plant mitochondria have specific electron transfer chain (ETC) components and functions in processes like photorespiration. The mitochondrial ETC harbors electrons with sufficient free energy to directly reduced O_2 which is the unavoidable primary source of mitochondrial ROS generation in aerobic respiration (Rhoads et al., 2006). Nevertheless, ROS production in mitochondria takes place also under normal respiratory condition but can be enhanced due to various biotic and abiotic stress factors. Complex I and II is well known as a place of $O_2\cdot^-$ production. In aqueous solution $O_2\cdot^-$ is moderately reactive but this $O_2\cdot^-$ can be reduced by SOD dismutation to H_2O_2 (Quan, 2008; Möller 2001; Grene 2002). Next H_2O_2 can react with Fe^{2+} and Cu^+ to produce highly toxic HO·, and these uncharged HO· can penetrate membranes and leave the mitochondrion (Rhoads et al., 2006). Abstraction of hydrogen atom by ROS, especially by HO·, starts peroxidation of mitochondrial membrane PUFA (polyunsaturated fatty acid). The consequence of this is formation of cytotoxic lipid aldehydes, alkenals and hydroxyalkenals etc. However, plant mitochondria may control ROS generation by means of energy-dissipating systems. Therefore mitochondria may play a central role in plant adaptation to abiotic stress.

ROS formation is possible also in **chloroplasts**, where photosynthesis takes place, which contain a highly organized thylakoid membrane system that harbours all components of light-capturing photosynthetic apparatus. Oxygen generated in chloroplast during photosynthesis is able to accept electrons passing through the photosystem resulting on $O_2\cdot^-$ formation. The presence of ROS producing centers like triplet Chl, electron transfer chains in PSI and PSII make chloroplast a site of ROS ($O_2\cdot^-$, $^1O_2$, H_2O_2) production.

Small, usually spherical microbodies bounded by a single lipid bilayer membrane called **peroxisomes** – organelles with an essentially oxidative type of metabolism are sites of intracellular ROS production. Similar to mitochondria and chloroplasts, peroxisomes produce $O_2\cdot^-$ radicals in their normal metabolism. There are two sites of $O_2\cdot^-$ production in peroxisomes (Rio et al., 2002). One of them is in the organelle matrix, where xanthine oxidase (XOD) catalyzes the oxidation of xanthine and hypoxanthine to uric acid. The second site takes place in peroxisome membranes.

Other important sources of ROS production in plants takes places in cytoplasm, endoplasmic reticulum and in appoplast at plasma membrane level or extracellular (Gill & Tuteja 2010).

6. Reactive oxygen species-mediated damage to macromolecules

It has long been know that he stress-induced formation of reactive oxygen species have been associated with non-specific damage to DNA, proteins and lipids which potentially can result in death of the cell and even the organism.

Oxidative damage to lipids or lipid peroxidation (LPO). The peroxidation of lipids is one of the most damaging processes occurs every living organism. During LPO, products are formed from polyunsaturated precursors that include small hydrocarbon fragments like ketons, MDA etc. Lipid peroxidation takes place when in both cellular and organelle membranes above-threshold levels of ROS are reached. The process of LPO involved three stages: initiation, progression and termination. (Fig.3). The overall effects of LPO are to decrease membrane fluidity, increase the leakiness of the membrane to substances the not normally cross it other than through specific channels and damage membrane proteins, inactivating receptors, enzymes and ion channels (Gill & Tuteja 2010).

Oxidative damage to protein. Proteins are the most abundant cellular component oxidized by ROS constituting up to 68% of the oxidized molecules in the cell (Rinalducci et al., 2008). Protein oxidation is a covalent modification induced by ROS or by products of oxidative stress. Protein oxidation mostly is irreversible, however, a few involving sulfur-containing amino acid are reversible (Ghezi & Bonetto 2003). The most susceptible residues to oxidation

Initiation step

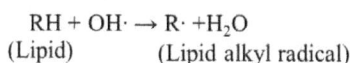

$$RH + OH\cdot \rightarrow R\cdot + H_2O$$
(Lipid)   (Lipid alkyl radical)

Propagationn step

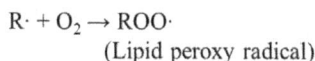

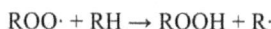

$$R\cdot + O_2 \rightarrow ROO\cdot$$
(Lipid peroxy radical)

$$ROO\cdot + RH \rightarrow ROOH + R\cdot$$

$$ROOH \rightarrow RO\cdot \ \text{Epoxides, hydroperoxides, glycol, aldehydes}$$

Termination step

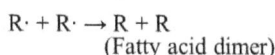

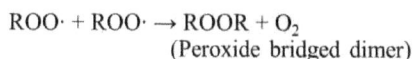

$$R\cdot + R\cdot \rightarrow R + R$$
(Fatty acid dimer)

$$R\cdot + ROO\cdot \rightarrow ROOR$$
(Peroxide bridged dimer)

$$ROO\cdot + ROO\cdot \rightarrow ROOR + O_2$$
(Peroxide bridged dimer)

Fig. 3. The main steps in lipid peroxidation according to Gill & Tuteja (2010).

are the sulphur containing cysteine and methionine. The thiol of cysteine may be oxidized by hydroxyl radicals, superoxide and hydrogen peroxide to a disulfide that can be readily reversible. Oxidation of methionine in many proteins has little effect on protein structure and function. An example of the reversible oxidation of methionine is the inactivation of the small heat shock protein in chloroplasts that is reactivated by thioredoxin in reaction catalysed by methionine sulphoxide reductase (Gustavsson et al., 2002). Tyrosine oxidation can alter residue hydrophobicity with consequent effect on protein structure. Tryptophan oxidation is an irreversible protein modification (Rinalducci et al., 2008). The most commonly occurring oxidative modification of proteins, after oxidation of sulphur-containing amino acids, is protein carbonylation. The oxidation of number of protein amino acid side-chains particularly Arg, His, Lys, Pro, Thr and Trp give free carbonyl groups which may inactivated, cross-linking or breakdown of proteins (Möller et al., 2007; Foyer & Noctor 2009). In leaves of wheat (*Triticum aestivum* L.) mitochondria contained more oxidatively modified proteins than chloroplasts and peroxisomes (Bartoli et al., 2004).

Oxidative damage to DNA. Due to biotic and abiotic stresses DNA is exposed to damage. Endogenously generated damage to DNA is known as "spontaneous DNA damage", which is produced by reactive metabolites ($HO\cdot$, O_2^- and $\cdot NO$). High level of ROS can influence on damage to cell structures, nucleic acids, lipids and proteins. It has been considered that one of the most reactive is $HO\cdot$ causing damage to all components of DNA molecules. This molecule damages purine, pyrimidine and deoxyribose backbone. $^1O_2$ damages guanine, and H_2O_2 and O_2^- do not react at all. Result of DNA damage can be various physiological effects like reduced protein synthesis, cell membrane destruction, damage to photosynthetic proteins what consequently leads to growth and development disorders (Britt 1999).

7. Enzymatic antioxidants

A wide range of unfavorable environmental conditions like mentioned drought, extreme temperatures, salt stress etc. can induce stresses that alter seriously plant metabolism and may increase production of ROS (H_2O_2, $O_2^{·-}$, $^1O_2$, HO·) inducing an oxidative stress in organelles. Plants are unable to escape exposure to these environmental constraints and evolved mechanisms in order to survive. To prevent appearance of these toxic compounds and their consequences plants have a variety of constitutively expressed antioxidant defense mechanisms to scavenge the ROS generated. A lot of researches have been done to emphasize the importance of the cellular antioxidant machinery in protection against various stresses (Dalton, et. al., 1999, Tuteja 2007, Tuteja 2009). ROS-scavenging enzymes such as superoxide dismutase (SOD), catalase (CAT), ascorbate peroxidase (APX) and associated antioxidant enzymes, glutathione reductase (GR) and antioxidants such as "big three" antioxidants (ascorbic acid, glutathione and the pyridine nucleotides) and many redox-active phenolics, carotenoids and tocopherols are essential for ROS detoxification. The components of cellular "antioxidant machinery" and their role in plant protection against various abiotic stresses have been summarized in Fig.4.

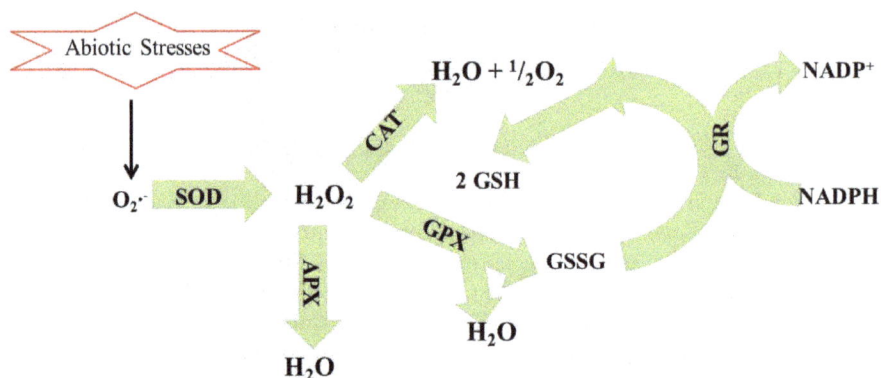

Fig. 4. ROS and antioxidant defense mechanism.

Enzymatic antioxidants include SOD, CAT, APX, GPX (guaiacol peroxidase), MDHAR (monodehydroascorbate reductase), DHAR (dehydroacorbate reductase) and GR. Reactions catalyzed by enzymatic antioxidants are presented in Table 1.

Enzymatic antioxidant	Enzyme code	Reaction catalyzed
Superoxide dismutase (SOD)	EC 1.15.1.1	$O_2^{·-} + O_2^{·-} + 2H^+ \rightarrow 2H_2O_2 + O_2$
Catalase (CAT)	EC 1.11.1.6	$H_2O_2 \rightarrow H_2O + \frac{1}{2} O_2$
Ascorbate peroxidase (APX)	EC 1.11.1.11	$H_2O_2 + AA \rightarrow 2 H_2O + DHA$
Guaiacol peroxidase (GPX)	EC 1.11.1.7	$H_2O_2 + GSH \rightarrow H_2O + GSSG$
Monodehydroascorbate reductase (MDHAR)	EC 1.6.5.4	$MDHA + NAD(P)H \rightarrow AA + NAD(P)^+$
Dehydroascorbate reductase (DHAR)	EC 1.8.5.1	$DHA + 2GSH \rightarrow AA + GSSG$

Table 1. Reaction catalyzed by major ROS-scavenging antioxidant enzymes

Superoxide dismutase (EC 1.15.1.1) is the primary scavenger in the detoxification of active oxygen species in plants discovered by Irwin Fridovich and Joe McCord (1969). SOD constitutes the first line of defense against ROS. Specialization of function among SODs may be due to combination of the influence of subcellular localization of the enzyme and upstream sequences in genomic sequence. SODS remove $O_2^{\cdot-}$ by catalyzing its dismutation, one $O_2^{\cdot-}$ is reduced to H_2O_2 and another to O_2 (Table 1). SODs are metalloproteins and based on their metal cofactor they are classified into three known types: the copper/zinc (Cu/Zn-SOD), the manganese (Mn-SOD) and the iron (Fe-SOD) that are localized in different cellular compartment (Mittler 2002). The activity of SOD isoenzymes can be detected by negative staining and identified on the base of their sensitivity to KCN and H_2O_2. Cu/Zn-SOD is sensitive to both inhibitors; the Mn-SOD is resistant on both inhibitors, whereas Fe-SOD is resistant to KCN and sensitive to H_2O_2. The distribution of SOD isoenzymes is also distinctive. The Cu/Zn-SOD is found in the cytosolic fraction and also in chloroplasts in higher plants. Mn-SOD is found in the mitochondria of eukaryotic cells and in peroxisomes. And the Fe-SOD is usually present in chloroplasts, but they are not often found in plants. The up regulation of SODs has been observed in plants subjected to both abiotic (Boguszewska et al. 2010) and biotic stresses (Torres 2010, Świątek et al., submitted for publication). Overexpression of SODs in transgenic plants resulted in higher salt or drought tolerance (Badawi et al., 2004). Thus, SODS have a critical role in the survival of plants under environmental stresses.

Catalase (EC 1.11.1.6) was the first antioxidant enzyme to be discovered and characterized (Mhamdi et al. 2010). Catalase is a heme-containing enzyme that catalyzes the dismutation of two molecules of hydrogen peroxide to water and oxygen (Table 1). All forms of the enzyme are tetramers with each monomer of 50-70 kDa. CAT has one of the highest turnover rates for all enzymes: one molecule of CAT can convert about 6 million molecules of H_2O_2 to H_2O and O_2 per minute. Multiple forms of catalase have been described in many plants. Monocots and dicots contain three catalase genes. The CAT1 gene is mainly expressed in pollen and seeds, CAT2 in photosynthetic tissues but also in roots and seeds and CAT3 in vascular tissues and in leaves. CAT isozymes, CAT1 and CAT2 are localized in peroxisomes and the cytosol, whereas CAT3 is mitochondrial isozyme. Catalase is a light-sensitive protein that has a high rate of turnover and environmental stresses which reduce the rate of protein turnover, such as salinity, heat shock or cold, cause the depletion of catalase activity. This may have significance in the plant's ability to tolerate the oxidative components of these environmental stresses (Boguszewska et al., 2010, Mhamdi et al., 2010). However, the response of CAT2 to soil drought differed from that of mite-infestation. Mite feeding decreased the CAT2 activity band whereas dehydration of leaves increased it only slightly and induced the CAT3 activity band (Świątek et al., submitted for publication). Catalase activity increased in response to low temperatures in germinating embryos of maize lines and the increase in total catalase activity was due to accumulation of the CAT1 and CAT2 isozymes whereas the CAT3 activity decreased (Auh & Scandalios 1997). Kukreja et al. (2005) reported increase of CAT activity in *C. arietinum* roots under salinity stress whereas, in the other study, Sharma & Dubey (2005) reported a decrease in CAT activity in rice seedling under drought stress. It remains unclear whether variability in catalase response to different unfavourable conditions may be of importance in plant stress tolerance level.

Ascorbate peroxidase (EC 1.11.1.11) exists as isoenzymes and plays an important role in the metabolism of H_2O_2 in higher plants. It is clear that a high level of endogenous ascorbate is essential to maintain effectively the antioxidant system that protects plants from oxidative damage due to biotic and abiotic stresses. APX is involved in scavenging of H_2O_2 into water-water and ascorbate-glutathione cycles and utilizes ascorbate as an electron donor (Table 1). There are five different isoforms of APX base on the localization: thylakoid tAPX, glyoxysome membrane APX (gmAPX), chloroplast stromal soluble form (sAPX) and cytosolic form of APX (cAPX). It has been shown enhanced expression of APX in plants growing under unfavourable environmental conditions.

Guaiacol peroxidase (EC 1.11.1.7). Guaiacol peroxidase represent an important peroxidase group which oxidise a large number of organic compounds such as phenols, aromatic amines, hydroquinones etc. but the commonly used reducing substrates are guaiacol or pyrogallol. In most plants, about 90% of the peroxidase activity is referred to as guaiacol ('anionic') peroxidase (Foyer et al., 1994). This haeme-containig protein decomposes indole-3-acetic acid (IAA) and has a role in the biosynthesis of lignin and defense against biotic stresses consuming H_2O_2. It is found in cytoplasm and apoplast. The activity of GPOX depends on plant species and stress condition.

Glutathione reductase (EC 1.6.4.2) is a flavo-protein oxidoreductase. It is an enzyme that is thought to play an essential role in defence system against ROS (Gill & Tuteja 2010, Noctor et al. 2010). Reducing glutathione disulfide (GSSG) to the sulfhydryl form (GSH), which is an important cellular antioxidant in defense against ROS, it sustains the reduced status of GSH. Glutathione disulfide contains of two GSH linked by a disulphide bridge which can be converted back to GSH by GR (Reddy 2006). Glutathione reductase is localized mainly in chloroplasts and small amount of this enzyme has been found in mitochondria and cytosol. By catalyzing the reduction of GSH, GR is an enzyme involved in regulation of cell energy metabolism. Glutathione reductase catalyzes the NADPH-dependent reduction of disulfide bond of GSSG what is important in the maintaining of GSH pool. Increased level of GR has been observed in plants subjected to metal, drought and salt stresses.

Glutathione peroxidase (EC 1.11.1.9) is the name of an enzyme family with peroxidase activity. The biological function of glutathione peroxidase is to reduce lipid hydroperoxides to their corresponding alcohols and to reduce free hydrogen peroxide to water. There are several isozymes encoded by different genes, which vary in cellular location and substrate specificity. Millar at al., (2003) identified a family of seven related proteins in the cytosol, chloroplasts, mitochondria and endoplasmatic reticulum in *Arabidopsis*. Activity of GPX decreased in roots and did not change significantly in the leaves of *Pisum sativum* exposed on Cd-stress (Dixit et al., 2001), whereas activity of this enzyme increased in cultivars of *Capsicum annuum* under Cd-stress (Leon et al., 2002).

Monodehydroascorbate reductase (EC 1.6.5.4). In plants, the monodehydroascorbate reductase (MDHAR) is an enzymatic component of the glutathione-ascorbate cycle that is one of the major antioxidant systems of plant cells for the protection against the damages by reactive oxygen species (ROS). The MDHAR activity has been described in several cell compartments, such as chloroplasts, cytosol, mitochondria, glyoxysomes, and leaf peroxisomes.

Dehydroascorbate reductase (EC 1.8.5.1) contributes to the regulation of the symplastic and apoplastic ascorbate pool size and redox state of the cell. Dehydroascorbate is reduced to ascorbate by DHAR in a reaction requiring glutathione. Thus, dehydroascorbate reductase catalyzes the regeneration of ascorbate from its oxidized state and serves as an important regulator of ascorbate recycling. It has been shown that guard cells in DHAR-overexpressing plants exhibited a reduced level of H_2O_2, decreased responsiveness to H_2O_2 or abscisic acid signaling, and greater stomatal opening (Chen & Gallie 2005). On the contrary, suppression of DHAR expression resulted in higher levels of H_2O_2 in guard cells and enhanced stomatal closure under normal growth conditions or following water deficit. Increasing DHAR expression also provided enhanced tolerance to ozone. Plants suppressed in DHAR expression exhibited a reduced rate of CO_2 assimilation with slower growth and reduced biomass accumulation (Chen & Gallie 2005).

8. Characterization of non-enzymatic antioxidants

The non-enzymatic antioxidants refer to the biological activity of numerous vitamins, secondary metabolites and other phytochemicals aimed to protect plants against ROS activity. Among the most important non-enzymatic antioxidants are ascorbic acid (AA), glutathione (GSH), proline, α-tocopherols, carotenoids and flavonoids.

Ascorbic acid (vitamin C) is the most abundant, powerful and water soluble antioxidant which minimizes or prevents damage caused by ROS in plants. Ascorbic acid (AA) is one of the most studied one and has been detected in majority of plant cell types, organelles and appoplast (Smirnoff & Wheeler 2000). Ascorbic acid reacts not only with H_2O_2, but also with O_2, OH and lipid hydroperoxidases (Wu 2007). In turf grass AA concentration significantly increases during water deficiency (Shao et al., 2006). Ascorbic acid can also directly scavenge $^1O_2$, O_2^- and HO· and regenerate tocopherol from tocopheroxyl radicals providing membrane protection. Moreover, antioxidants like AA and glutathione are involved in neutralization of secondary products of ROS reaction. Fundamental role of AA in the plant defense system is to protect metabolic processes against H_2O_2. Summing up AA reacts non-enzymatically with superoxide, hydrogen peroxide and singlet oxygen (Smirnoff & Wheeler 2000).

Glutathione (GSH) is a tripeptide (α-glutamyl-cysteinyl-glycine), which is considered as the most important intracellular defense against ROS-induced oxidative damage. Glutathione has been detected in all cell compartments such as cytosol, chloroplasts and endoplasmatic reticulum (Foyer & Noctor 2003). Glutathione is the major source of non-protein thiol groups. The nucleophilic nature of the thiol group is important in the formation of mercaptide bonds with metals for reacting with selected electrophiles. Glutathione is involved in control of H_2O_2 levels. The change in the ratio of its reduced (GSH) to oxidized (GSSG) form during the degradation of H_2O_2 is very important in certain signaling pathway. It has been considered that GSH/GSSG ratio, indicative of the cellular redox balance, may be involved in ROS perception (Li & Jin 2007). Glutathione is important in plant chloroplasts because it helps to protect the photosynthetic apparatus from oxidative damage.

Tocopherols, lipid soluble antioxidants are known as potential scavengers of ROS and lipid radicals. Tocopherols are major antioxidant in biomembranes, where they play both antioxidant and non-antioxidant functions. The main role of tocopherols is protection of membrane stability, including quenching or scavenging ROS like $^1O_2$. Tocopherols are localized in plants in the thylakoid membrane of chloroplasts. Among four tocopherols isomers α-tocopherol (vitamin E) has the highest antioxidative activity because of presence of three methyl groups in its molecular structure (Kamal-Eldin & Appelqvist 1996).

Carotenoids are pigments that are found in plants and microorganisms. There are over 600 carotenoids in nature. Carotenoids are lipid soluble antioxidants that plays multitude of function in plant metabolism including oxidative stress tolerance. Carotenoids take part in three different functions in plants. First one, they absorb the light at wavelength between 400 and 550 nm and transfer it to the Chl. Secondly, they protect photosynthetic apparatus by quenching a triplet sensitizer (Chl3), $^1O_2$ and other harmful free radicals which are naturally formed during photosynthesis (an antioxidant function). Thirdly, they are important for the PSI assembly and the stability of light of light harvesting complex protein as well as thylakoid membrane stabilization (structural function) (Sieferman-Harms 1987)

Flavonoids are widely distributed in plants leaves, floral part and pollens. They often accumulate in the plant vacuole as glycosides or as exudates on the leaves surface and other aerial part of the plant. There are four flavonoid classes depending on their structure: flavonols, flavones, isoflavones and anthocyanines. Flavonoides belong to one of the most reactive secondary metabolites of plants (Olsen et al., 2010). Flavonoids play important role as ROS scavenger by locating and neutralizing radicals before they damage cell structure. Flavonoides have function as flowers, fruits and seed pigmentation, they play protective role before UV light, drought and cold and defense against pathogens. Flavonoids play an important role in plant fertility and germination of pollen. They are involved in plant signaling with interaction with plant microbes (Olsen et al., 2010, Gill & Tuteja 2010). It has been proved that they are involved in plant responses to both, biotic or abiotic stresses such as wounding, drought and metal toxicity (Cle et al., 2008).

Proline, α-amino acid is an antioxidant and potential inhibitor of programmed cell death. It has been suggested that free proline act as osmoprotectant, a protein stabilizer, a metal chelator, an inhibitor of lipid peroxidation and OH· and $^1O_2$ scavenger. Increased proline accumulation appears especially during salt, drought and metal stresses (Trovato et al., 2008). Therefore proline is not only an important signaling molecule, but also an effective ROS quencher. It has been found that the important role of proline is in potentiating pentose-phosphatase pathway activity as important component of antioxidative defense mechanism (Hare & Cress 1997).

9. Drought-responsive antioxidant enzymes in potato

The involvement of antioxidant enzymes in limitation of ROS production has been examined in potato cultivars differing in the drought tolerance (Boguszewska et al., 2010). Additionally, it has been interested whether the ROS-itscavenging antioxidant enzymes may be critical for protecting plants against water deficiency in the soil and/or whether they may be responsible for the timely activation of the accliamatory response. In our experiment three weeks after tuberization, half of potato plants grown in pots were subjected to soil drought by cessation of watering during two weeks. The other parts of plants were still watering (control plants). After this dry treatment the plants were rewatered and grown under the same conditions as control plants until maturity. To assess relative water content (RWC) in leaves of control and dehydrated plants at the end of dry period, mature leaves from the third level from the top of the plant, comparable in size, were sampled. Leaves were weighted immediately (fresh weight, FW), floated in dark for 24 h to achieve turgidity (saturated weight, SW), then oven dried (105°C) for 24 h and weighted again (dry weight, DW). Relative water content of leaves was calculated according to Weatherly (1951): RWC (%) = [(fresh weight–dry weight)/(saturated weight–dry weight)] x 100%

Electrophoretic separation was performed using 4% stacking gel and 10% polyacrylamide resolving gel as described by Laemmli (1970). Samples (30 μg) were diluted in loading buffer in relation 1: 1 (50mM Tris-HCl pH 6.8; 0.1 % bromophenol blue, 10% glycine). Gel electrophoresis was run at 4°C for about 1.5h with constant current of 30 mA.

Peroxidases were visualised by incubating the gel in 50 mM potassium phosphate buffer, pH 5.0 containing 2 mM benzidine and 3 mM H_2O_2 until appearance of orange bands.

Superoxide dismutase activity was detected following the method of Beauchamp & Fridovich (1997). Gels were soaked in 50 mM sodium phosphate buffer, pH 7.8 containing 0.098 mM nitroblue tetrazolium and 0.03 mM riboflavin. After 20 min in the dark, gels were immersing in 50 mM sodium phosphate buffer pH 7.8 containing 28 mM TEMED and exposed to a light source at room temperature. In a separate experiments 3 mM KCN and 5 mM H_2O_2 were included during activity staining steps to distinguish amongst Cu/Zn-SOD (inhibited by KCN), Fe-SOD (inactivated by H_2O_2) and Mn-SOD (resistant to both inhibitors).

Catalase (CAT, E.C. 1.11.1.6) was stained according to the method of Racchi & Terragona (1993). After native PAGE gels were washed in deionized water, incubated in 0.003% (w/v) H_2O_2 for 10 min, and stained in 1% ferric chloride and 1% potassium ferricyanide solution.

The obtained results clearly indicated that the investigated cultivars differed in tolerance to applied two-week soil drought (Table 2). According to physiological criterion of dehydration tolerance (RWC expressed in %) the more tolerant cultivar was Owacja, whereas susceptible cultivar was Jutrzenka However, the more intensive wilting and lower regeneration ability of cv Owacja resulted in 56% yield decrease of potato tubers of cv Owacja and significantly lower yield decrease (34%) of cv. Jutrzenka. Therefore, from agricultural point of view, cv. Jutrzenka was more tolerant than that of cv. Owacja.

Specification	cv. Owacja	cv. Jutrzenka
Yield decrease (%)	56	34
Wilting during drought (0,5 – 9 scale)	0,5	2
Regeneration after drought	1	3
RWC Control – RWCDrought (%)	Δ 26	Δ 31

Table 2. Characterization of drought resistance of potato cultivars.

To keep ROS at constant level and maintain redox homeostasis within the plant the activities of diverse antioxidant enzymes are up-regulated in response to abiotic stresses such as drought, salinity (Miller et al., 2010) and many others (Gill & Tuteja, 2010; Jaspers & Kangasjärvi 2010) as well as in response to biotic stresses (Maserti et al., 2011, Torres 2010). Among them, plant peroxidases have also been shown to be upregulated under various abiotic treatments (Foyer & Noctor 2009). Peroxidase activity bands in leaves of droughted plants increased in both cultivars (Fig. 5). However, the increase being higher in drought susceptible cultivar (Owacja with 59% yield decrease) than in the drought resistant (Jutrzenka with 34% yield decrease) cultivar. Soil drought did not induce the new isoforms of peroxidase.

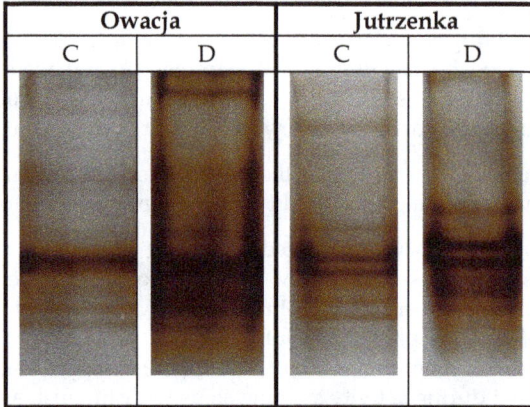

Fig. 5. Peroxidase activity in control, fully turgid leaves (C) and in dehydrated leaves (D) of two potato cultivars

In contrast to the dehydrated leaves, the response of peroxidase from potato tubers to soil drought was negative or stable (Fig. 6). The activity of peroxidase bands remained at the same level in Jutrzenka tubers and decreased in Owacja, but with one band being more active.

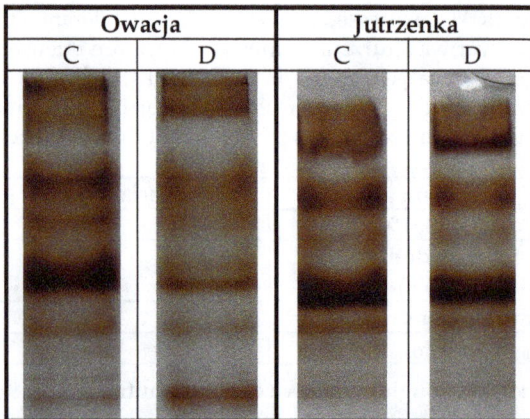

Fig. 6. Peroxidase activity in potato tubers of control, fully turgid plants (C) and in tubers of plants subjected to dehydration (D) of two potato cultivars.

Increased activity of peroxidase band activity more in leaves of drought susceptible cultivar than in resistant cultivar may suggest their involvement, at least in part, in defence against dehydration-induced injuries of plant tissues. One can speculate that this enzyme may locally detoxify over-abundance of H_2O_2 by cell wall lignification and chlorotic lesion formation. Higher responsiveness of leaf and tuber peroxidase of sensitive potato plants exposed to soil drought seems to suggest the involvement of this enzyme in limitation of H_2O_2 amount.

The only enzyme able to dismutate O_2^- to H_2O_2 and O_2 is SOD. An increase in SOD activity under diverse abiotic stresses has been shown in several plants (Gill & Tuteja 2010). In extracts obtained from potato cultivars, three activity bands of superoxide dismutase (SOD) were observed (Boguszewska et al., 2010). In the case of two other potato cultivars, the similar pattern of SOD activity bands was observed (Fig. 7). In response to drought, a stable amount of activity bands was observed in tubers of cv. Jutrzenka, whereas in tubers of cv. Owacja the increased amount of activity bands of SOD was clearly visible. Thus, the response of potato to soil water shortage referring to SOD activity bands was only visible in sensitive cultivar. Substantial increase in the activity of SOD at tuberisation stage of potato plants under soil drought may suggest that they play an important role in O_2^- scavenging and H_2O_2 formation within cells and are involved in an adjustment of plants to this environmental stress as it was suggested earlier (Matters & Scandalios 1986). In contrast to the sensitive plants, practically unchanged activity of SOD bands in plants of drought tolerant cultivars seems to suggest that that under water deficiency cell compartments most probably differ in H_2O_2 concentration.

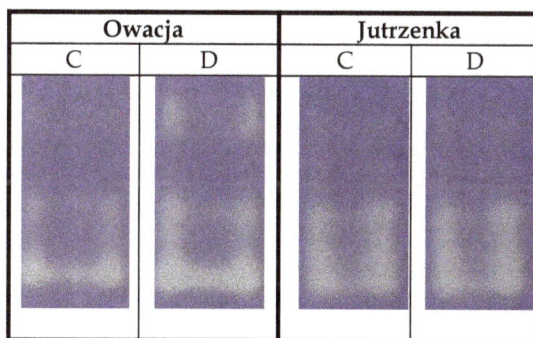

Owacja		Jutrzenka	
C	D	C	D

Fig. 7. SOD activity in potato tubers of control, fully turgid plants (C) and in plants subjected to dehydration (D) of two potato cultivars

The last enzyme investigated was catalase. Catalase is a key enzyme catalysing H_2O_2 decomposition (Apel & Hirt 2004, Mittler et al., 2004). In tubers of both potato cultivars two forms of catalase, namely CAT-2 and CAT-3, fairly the same in activity were detectable (Fig. 8). In tubers of drought treated plants the activity of two catalase activity band decreased, more in cv Jutrzenka than in cv. Owacja. Such observation may indicate that the concentration of H_2O_2 has been limited to the level that was putatively not able to activate catalase efficiently. Although, the reason of CAT decrease is not clear yet, our results are consistent with the previous suggestion that in plants subjected to abiotic stress the diminished activity of catalase and peroxidase in tubers of drought sensitive cultivar resulted in localized increases of ROS and mediates in drought injuries of cells.

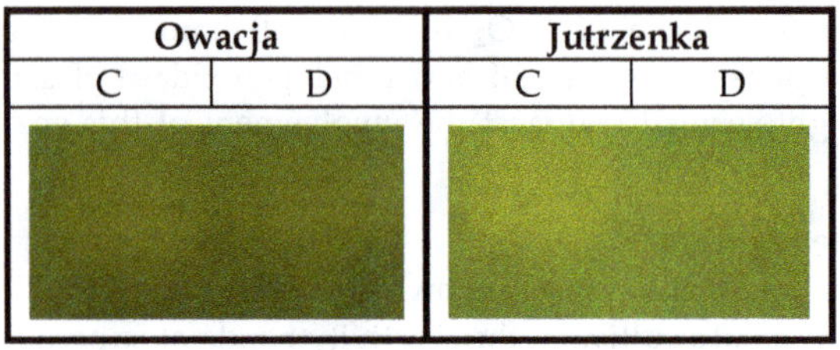

Owacja		Jutrzenka	
C	D	C	D

Fig. 8. CAT activity in potato tubers of ccontrol, fully turgid plants (C) and in tubers of plants subjected into dehydration (D) of two potato cultivars

In conclusion, ROS-scavenging enzymes investigated in potato tubers of two cultivars differing in drought tolerance respond differently to a two-week soil drought. A strong increase in the activity of peroxidase in leaves of potato of sensitive cultivar exposed to drought suggests that this cultivar suffers from oxidative stress and high amounts of ROS may be limited insufficiently as indicated higher wilting intensity and weak regeneration ability of this cultivar. On the contrary, a strong increase of SOD together with a decrease of peroxidase and catalase activities in tubers of sensitive cultivar may indicate on changed conversion of O_2^- to H_2O_2. This findings show that SOD rather than peroxidase and catalase are a target antioxidant protein protecting tuber tissues against oxidative damage.

The present experiments and those made earlier (Boguszewska et al., 2010) showed that the yield of potato tubers (agricultural yield) dependend more on the regeneration ability of rewatered plants after soil drought treatments than on the water loss from the leaves. Although leaf relative water content (RWC) is considered a reliable and widely-used indicator of the plant sensitivity to dehydration (Rampino et al., 2006, Sanchez-Rodriguez et al., 2010), the correlation between RWC in leaves of ten potato cultivars investigated and yield decrease was poor. Moreover, the observed differences in RWC in leaves of investigated potato cultivars indicated that neither the time course of dehydration nor the attained leaf RWC values related to tuber yield. The observed differences in leaf RWC attained at the same drought period indicated only that the investigated cultivars differed in characteristics responsible for dehydration avoidance.

However, the RWC reflected guaiacol peroxidase activity in droughted potato leaves i.e. a high correlation was observed between the relative increase in guaiacol peroxidase activity (i.e. POX activity in droughted leaves minus POX activity in turgid leaves) and RWC in potato leaves. These results are in full agreement with a substantial increase in guaiacol POX activity in maize seedlings acclimated to suboptimal growth temperature (Prasad 1996). It may indicate that a high activity of guaiacol POX induced by such diverse stresses as soil drought and suboptimal growth temperature is a common response of plants to unfavourable environmental conditions. The activity bands of guaiacol peroxidase confirmed the spectrophotometric measurement of POX activity: the bands with the highest activity and an increase in number of activity bands were observed in in dehydrated leaves of cultivar with a yield decrease of 49% whereas in leaf extracts from cultivar with a yield decrease of 26%, only two new activity bands appeared. In contrast, the response of POX in potato tubers was cultivar dependent i.e. it was either negative (activity of peroxidase bands decreased) or remained at the level of control, turgid plants. Increased activity of POX and

SOD in leaves and tubers of dehydrated cultivars tolerant to soil drought seemed to counteract the accumulation of ROS and in effect protected plants against loss of yield.

10. Perspectives

As discussed above, significant progress has been made in understanding the biological role of reactive oxygen species in plant growth and development. Moreover, a growing number of data that have been identified indicate on participation of ROS in plant responses to both, abiotic and biotic stresses. It is becoming increasingly evident that many different stresses activate endogenous production of ROS and that these are not only a deleterious effect of oxidative metabolism but are necessary for plant intracellular communication system. ROS signaling and integration into many others signaling networks together with antioxidants has been shown to be involved in acclimation ability of plants to unfavourable environmental conditions and their responses to pathogen attack. It seems to be of importance to decipher ROS signaling mechanisms because it may lead to the development of agricultural important plants more tolerant to suboptimal conditions of growth. Thus, it should have a significant impact on enhanced and stable yield of plants under less favourable conditions of growth to provide plants for fibbers, improving human health, human food and feed for animals.

It is hoped that the next decade will address many of the unanswered questions pertinent to plant functioning under adverse environmental conditions. We think that one of the most important questions still unanswered is mechanism(s) of ROS sensing and signal transduction. Up to date, sensors of ROS are mostly unknown. We think that the next important question concerns the way in which antioxidants provide the most important information on plant redox state and how they affect gene expression associated with plant responses to both, biotic and abiotic stresses to maximize plant defence systems. It is hoped also that the coming years will shade more light on involvement of photosynthetic and respiratory electron transport chains in ROS signaling pathways, post-transcriptional regulation of gene expression and modification of proteins important for plant survival.

11. References

Apel, H.; Hirt, H. (2004). Reactive oxygen species: metabolism, oxidative stress, and signal transduction. *Annu. Rev. Plant Biol.* 55, 373-399.

Auh, C-K.; Scandalios, J.G. (1997). Spatial and temporal responses of the maize catalases to low temperature. *Physiologia Plantarum* 101, pp. 143-156.

Badawi, G.H.; Yamauchi, E.; Shimada, R.; Sasaki, N.; Kawano, K.; Tanaka, K.; Tanaka, K. (2004). Enhanced tolerance to salt stress and water deficit by overexpressing superoxide dismutase in tabacco (Nicotina tabacum) chloroplast, *Plant Science*, 166, pp. 919-928.

Bartoli, C.; Gòmez, F.; Martinez, D.E.; Guiamet, J.J. (2004). Mitochondria are the main target for oxidative damage in leaves of wheat (Triticum aestivum L). *Journal of Experimental Botany*, 55, pp. 1663-1669.

Beauchamp, C; Fridovich, I. (1971). Superoxide dismutase: improved assay and assay applicable to acrylamide gels, *Analytical Biochemistry*, 44, pp. 276-287.

Blum, A. (2011). Drought resistance –is it really a complex trait? *Functional Plant Biology*, 38, 753-757.

Boguszewska, D.; Grudkowska,M.; Zagdańska, B. (2010). Drought responsive antioxidant enzymes in potato (Solanum tuberosum L.), *Potato Research*, 53, pp. 373-382.

Bohnert, H.J.; Jensen, R.G. (1996). Strategies for engineering water-stress tolerance in plants, *Trends Biotechnology*, 14, pp.89-97.

Boursiac, Y. et al. (2008). Stomata-induced downregulation of root water transport involves reactive oxygen species –activated cell signalling and plasma membrane intrinisic protein internalization. *The Plant Journal*, 56. pp. 207-218.

Boyer, J.S. (1982). Plant productivity and environment, *Science*, 218, pp. 443-448.

Bright et al. (2006). ABA-induced NO generation and stomatal closure in Arabidopsis are dependent on H_2O_2 synthesis. *The Plant Journal*, 45, pp.113–122.

Britt, A.B. (1999). Molecular genetics of DNA repair in higher plants, *Trends in Plant Science*, 4, pp. 337-397.

Chaves, M.M. (1991). Effect of water deficits on carbon assimilation, *Journal of Experimental Botany*, 42, pp. 1-16.

Chen Z.; Galli, D.R. (2008). Dehydroascorbate reductase affects non-photochemical quenching and photosynthetic performance, *J. Biol. Chem*, 283, pp. 21347-21361.

CordMc, J.M.; Fridovich, I. (1969). Superoxide dismutase: an enzymic function for erythrocuprein (hemocuprein). J. Biol. Chem. 244:6049-55.

Chen, Z.; Gallie, D.R. (2005). Increasing tolerance to ozone by elevating folia ascorbic acid confers greater protection against ozone than increasing avoidance, *Plant Physiol.*, 138, pp. 1673-1689.

Cle, D. at al. (2008). Modulation of chlorogenic acid biosynthesis in Solanum lycopersicum; consequences for phenolic accumulation and UV-tolerance, *Phytochemistry*, 69, pp. 2149-2156.

Cornin, G.; Massacci, A. (1996). Leaf photosynthesis under drought stress, In: Baker N.R (Eds), *Photosynthesis and the environment*, New York, Kluwer Academic Publisher, pp. 347-366.

Dalton, T.H.; Shertzer, A.; Puga, A.(1999). Regulation of gene expression by reactive oxygen, Annual Reviev Pharmacol. *Toxicol*, 39, pp. 67-101.

Dixit, V.; Panday, V.; Shyam, R. (2001). Differential oxidative responses to cadmium in roots and leaves of pea (Pisum sativum L cv. Azad*), J.Exp. Bot.*, 52, pp. 465-475.

Foreman, J. et al. (2003). Reactive oxygen species produced by NADPH oxidase regulate plant cell growth, *Nature*, 422, pp. 422-446.

Foyer, C.H.; Lelandais, M.; Kunert, K.J. (1994). Oxidative stress in plants, *Physiol Plant.*, 92, pp. 696–717.

Foyer, C.H.; Noctor, G. (2003). Redox sensing and signalling associated with reactive oxygen in chloroplasts, peroxisomes and mitochondria, *Physiol. Plantarum*, 119, pp. 355-364.

Foyer, C.H.; Noctor, G. (2005). Redox homeostasis and antioxidant signalling: A metabolic interface between stress perception and physiological responses, *Plant Cell* 177, pp. 1866-1875.

Foyer, C.; Noctor, G. (2009). Redox regulation in photosynthetic organisms: signaling, acclimation, and practical implications. *Antioxid. Redox Signal.*, 11, pp. 861-905.

Ghezi, P.; Bonetto. (2003). Redox proteomics: identification of oxidatively modified proteins, *Proteomics* 3, pp. 1145-1153.

Gill, S.S.; Tuteja, N. (2010). Reactive oxygen species and antioxidant machinery in abiotic stress tolerance in crop plants, *Plant Physiology and Biochemistry*, 48, pp. 909-930.

Grene, R. (2002). Oxidative stress and acclimation mechanisms in plants. In: Somerville, C.R; Myerowitz, E.M (Eds), *The Arabidopsis Book*, American Society of Plant Biologist, Rockville, MD http://www.aspb.org/publications/arabidopsis/.

Gustavsson, N.; Kokke, B.P.; Härndahl, U.; Silow, M.; Bechtold, U.; Poghosyan, Z.; Murphy, D.; Boelens, W.C.; Sundby, C. (2002). A peptide methionine sulfoxide reductase highly expressed in photosynthetic tissue in Arabidopsis thaliana can protect the chaperone-like activity of a chloroplast-localized small heat shock protein. *The Plant Journal*. 29, pp. 545-553.

Hancock, J et all. (2006). Doing the unexpected: proteins involved in hydrogen peroxide perception, *Journal of Experimental Botany*, 57, pp. 1711-1718.

Hare, P.D.; Cress, W.A. (1998). Metabolic implications of stress-induced proline accumulation in plants, *Plant Growth Regul.*, 21, pp.79–102.

Hare, P.D; Cress, W.A. (1997). Metabolic implications of stress-induced proline accumulation in plants, *Plant Growth Regul*, 21, pp. 79-102.

Hare, P.D.; Cress, W.A.; Staden, J. (1998). Dissecting the roles of osmolytes accumulation during stress. *Plant Cell Environment*, 21, pp.535-553.

Hong-bo, S. et all. (2008). Higher plant antioxidants and redox signalling under environmental stresses. *C.R. Biologies*, pp. 331, 433-441.

Jacobson, M.D.; Weil, M.; & Raff, M.C. (1996). Role of Ced-3/I family proteases in stuarosporine-induced programmed cell death, *J. Cell Biol.* 133, pp.1041–1051.

Jaleel, C.A. at all. (2009). Antioxidant defense responses: physiological plasticity in higher plants under abiotic constraints, *Act Physiol. Plant.* 31, pp. 427-436.

Jaspers, P.; Kandasjarvi, J. (2010). Reactive oxygen species in abiotic stress signalling, *Physiol. Plant*, 138, pp. 405-413.

Kamal-Eldin, A; Appelqvist, L.A. (1996). The chemistry and antioxidant properties of tocopherols and tocotrienols, *Lipids*, 31,671-701.

Kukreja et al. (2005). Plant water status, H2O2 scavenging enzymes, ethylene evolution and membrane integrity of Cicer arietinum roots as affected by salinity, *Biol. Plant.*, 49, pp. 305-308.

Laemmli, U.K. (1970). Cleavage of structural proteins during the assembly of the head of bacteriophage T4, *Nature*, 227, pp. 680-685.

Lawlor, D.W. (2002). Limitation to photosynthesis in water-stressed leaves: stomata vs. metabolism and the role of ATP. *Annals of Botany*, 89, pp. 871–885.

Leon, A.M. et al. (2002). Antioxidant enzymes in cultivars of pepper plants with different sensitivity to cadmium, *Plant Physiol. Biochem.* 40, pp.813-820.

Li, J.M.; Jin, H. (2007). Regulation of brassinosteroid signalling, *Trends Plant Science*, 12, pp.37-41.

Masserti, B.E et al. (2011). Comparative analysis of proteome changes induced by the two spotted spider mite Tatranychus urticae and methyl jasmonate in citrus leaves. *J. Plant Physiol.* 168, 392-402.

Matters, G.L.; Scandalios, J.G. (1986). Effect of the free radical-generating herbicide paraquat on the expression of the superoxide dismutase (Sod) genes in maize. *Biochim. Biophys. Acta.* 882, 29-38.

Mhamdi, A. et al. (2010). Catalase function in plants: a focus on Arabidopsis mutants as stress-mimic models. *Journal of Experimental Botany*, 61, pp. 4107-4320.

Millar at all. (2003). Control of ascorbate synthesis by respiration and its implication for stress responses, *Plant Physiol.*, 133, pp. 443-447.

Miller, G.; Suzuki, N.; Ciftci-Yilmaz, S.; Mittler, R. (2010). Reactive oxygen species homeostasis and signalling during drought and salinity stresses, *Plant Cell Environ*, 33, pp. 453-467.

Mittler, R. (2002). Oxidative stress, antioxidants and stress tolerance, *Trends in Plant Science*, Vol. 7 no.9, pp. 405-410.

Mittler, R et al. (2004). Reactive oxygen gene network of plants, *Trends in Plant Science*, Vol. 9, No.10, pp. 490-498.

Mittler, R. (2006). Abiotic stress, the field environment and stress combination, *Trends in Plant Science*, Vol.11, No.1, pp.15-19.

Mittler, R.; Blumwald, E. (2010). Genetic engineering for modern agriculture: Challenges and perspectives. *Ann. Rev. Plant Biol.* 61:443-62.

Mittler, R. et al. (2011). ROS signaling: the new wave? *Trends in Plant Science*, Vol. 16, No. 6, pp. 300-309.

Möller, I.M. (2001). Plant mitochondria and oxidative stress: electron transport NADPH turnover, and metabolism of reactive oxygen species, *Annu. Review Plant Physiol.* Mol. Biol. 52, pp. 561-591.

Möller, I. M.; Jensen, P. E.; Hansson, A. (2007). Oxidative modifications to cellular components in plants, *Annu. Rev. Plant Biol.* 51, pp. 459-481.

Möller, I. M.; Sweetlove, L J. (2010). ROS signaling – specificity is required, *Trends in Plant Science*, 15, pp. 370-374,

Mullineaux, P.; Karpinski, S. (2002). Signal transduction in response to excess light: getting out of the chloroplast. *Curr Opin Plant Biol*, 5, pp. 43–48.

Noctor, G.; Foyer, C.H. (1998). A re-evolution of the ATP: NADPH budget during C3 photosynthesis. A contribution from nitrate assimilation and its associated respiratory activity? *Journal of Experimental Botany*, 49, pp. 1895-1908.

Noctor, G. et al. (2010). Peroxide processing in photosynthesis: antioxidant coupling and redox signalling. *Proc. Biol. Sci*, 355, pp. 1465-1475.

Olsen, K.M. et al. (2010). Identification and characterization of CYP75A31, a new flavonoid 3'5'-hydrolase, isolated from Solanum lycopersicum, *BMC Plant Biol*, doi:10.1186/1471-2229-10-21.

Puntarulo, S., Sanchez, R.A., Boveris, A. (1988). Hydrogen peroxide metabolism in soybean embryonic axes at the onset of germination, *Plant Physiol.* 86, pp. 626---630.

Petit et al. (1999). Climate and atmospheric history of the past 420,000 years from the Vostok ice core, Antarctica, *Nature*, 399, pp. 429-436.

Prasad, T.G. (1996) Mechanisms of chilling-induced oxidative stress injury and tolerance in developing maize seedling: changes in antioxidant system, oxidation of proteins and lipids, and protease activities. *Plant J* , 10, pp. 1017–1026

Quan, L. J et al. (2008). Hydrogen peroxide in plants a versatile molecule of reective oxygen specie network, *Journal Integrat. Plant Biol*, 50, pp. 2-18.

Racchi, M.L; Terragona, C. (1993). Catalaze isoenzymes are useful markers of differentiation in maize tissue-cultures, *Plant Science*, 93, pp. 195-202.

Rampino, P.; Patalo, S. et al. (2006). Drought stress response in wheat: physiological and molecular analysis of resistant and sensitive genotypes. *Plant Cell Environ.*, 29, pp. 2143–2152

Rasmusson, A.G.; Soole, K.L.; Elthon, T.E. (2004). Alternative NAD(P)H dehydrogenases of plant mitochondria, *Annu. Rev. Plant Biol.* 55, pp.23-39.

Reddy, A. R.; Chaitanya, K. V.; Vivekanandan, M. (2004). Drought-induced responses of photosynthesis and antioxidant metabolism in higher plants, *Journal of Plant Physiology*, 161, pp. 1189-1202.

Reddy, A.R.; Raghavendra, A.S. (2006). Photooxidative stress. In: Madhava Rao,K.V; Raghavendra, A.S; Reddy, A.R. (Eds), *Physiology and molecular biology of stress Tolerance in Plants*, Springer, The Netherlands, pp.157-186.

Rhoads, D.M. et al. (2006). Mitochondrial reactive oxygen species. Contribution to oxidative stress and interorganellar signaling, *Plant Physiol*, 141, pp. 125-134.

Rio, I.A. et al. (2002). Reactive oxygen species, antioxidants systems and nitric oxide in peroxisomes, *J.Exp. Biol.*, 53, pp. 1255-1272.

Rinalducci, S; Murgiano, L; Zolla, L. (2008). Redox proteomics: basic principles and future perspectives for the detection of protein oxidation in plants. *J Exp Bot* 59:3781–3801.

Sakamoto, M. (2008). Involvement of hyrogen peroxide in leaf abscission signaling revealed by analysis with an in vitro abscission system in *Capsicum* plants. *The Plant Journal*, 56, 13-27.

Sanchez-Rodriguez et al. (2010). Genotypic differences in some physiological parameters symptomatic for oxidative stress under moderate drought in tomato plants. *Plant Sci* 178, pp. 30–40

Schulze, E.D. (1986). Whole-plant responses to drought, *Australian Journal of Plant Physiology*, 13 pp. 127-141.

Shao, H.B.; Chen, X.Y.; Chu, L.Y, et all. (2006). Investigation on the relationship of Proline with wheat antidrought under soil water deficits, *Biointerfaces*, 53, pp. 113-119.

Sharma, P.; Dubey, R.S. (2005). Modulation of nitrate reductase activity in rice seedlings under aluminum toxicity and water stress: role of osmolytes as enzyme protectant, *J.Plant Physiol*, 162, pp. 854-864.

Sieferman-Harms, D. (1987). The light harvesting function of carotenoids in photosynthetic membrane, *Plant Physiology*, 69, pp. 561-568.

Smallwood, M.F et al. (1999). *Plant response to Environmental Stress*, BIOS Scientific Publishers Limited, Guildford, UK.

Smirnoff, N. (1993). The role of active oxygen in the response of plants to water deficit and desiccation. *New Phytol.*, 125, pp.27–58.

Smirnoff, N., (1998). Plant resistance to environmental stress. *Curr. Opin. Biotechnol.*, 9, pp. 214-219.

Smirnoff, N.; Wheeler, G.L. (2000). AA: metabolism and functions of a multi faceted molecule, *Current Opinion in Plant Biology*, 3, pp. 229-235.

Świątek, M.; Kiełkiewicz M.; Zagdańska, B. Differential response of maize antioxidant enzymes induced by the two-spotted spider mite and soil drought (submitted).

Torres, M.A. (2010) ROS in biotic interactions, *Physiol. Plant.* 138, pp. 414-429.

Trovato, M.; Mattioli, R.; Costantino, P. (2008). Multiple roles of proline in plant stress tolerance and development, *Rendiconti Lincei*, 19, pp. 325-346.

Tuteja, N. (2007). Mechanisms of high salinity tolerance in in plants,. *Methods of Enzymology*, 428, pp. 419-438.

Tuteja, N. (2009). Cold, Salinity, and Drought Stress, In: *Plant Stress Biology*, Hirt, H. WILEY-VCH Verlag GmbH &Co. KGaA, Weinheim, pp. 137-159.

Weatherley, P.E . (1951). Studies of the water relations of the cotton plant. II. Diurnal and seasonal variations in relative turgidity and environmental factors. *New Phytol.* 50:36-51.

Wu, G.; Wei, Z.K.; Shao, H.B. (2007). The mutual responses of higher plants to environment: physiological and microbiological aspects, *Biointerferences*, 59, pp.113-119.

Permissions

The contributors of this book come from diverse backgrounds, making this book a truly international effort. This book will bring forth new frontiers with its revolutionizing research information and detailed analysis of the nascent developments around the world.

We would like to thank Prof. Dr. Volodymyr I. Lushchak and Assoc. Prof. Dr. Halyna Semchyshyn, for lending their expertise to make the book truly unique. They have played a crucial role in the development of this book. Without their invaluable contribution this book wouldn't have been possible. They have made vital efforts to compile up to date information on the varied aspects of this subject to make this book a valuable addition to the collection of many professionals and students.

This book was conceptualized with the vision of imparting up-to-date information and advanced data in this field. To ensure the same, a matchless editorial board was set up. Every individual on the board went through rigorous rounds of assessment to prove their worth. After which they invested a large part of their time researching and compiling the most relevant data for our readers. Conferences and sessions were held from time to time between the editorial board and the contributing authors to present the data in the most comprehensible form. The editorial team has worked tirelessly to provide valuable and valid information to help people across the globe.

Every chapter published in this book has been scrutinized by our experts. Their significance has been extensively debated. The topics covered herein carry significant findings which will fuel the growth of the discipline. They may even be implemented as practical applications or may be referred to as a beginning point for another development. Chapters in this book were first published by InTech; hereby published with permission under the Creative Commons Attribution License or equivalent.

The editorial board has been involved in producing this book since its inception. They have spent rigorous hours researching and exploring the diverse topics which have resulted in the successful publishing of this book. They have passed on their knowledge of decades through this book. To expedite this challenging task, the publisher supported the team at every step. A small team of assistant editors was also appointed to further simplify the editing procedure and attain best results for the readers.

Our editorial team has been hand-picked from every corner of the world. Their multi-ethnicity adds dynamic inputs to the discussions which result in innovative outcomes. These outcomes are then further discussed with the researchers and contributors who give their valuable feedback and opinion regarding the same. The feedback is then collaborated with the researches and they are edited in a comprehensive manner to aid the understanding of the subject.

Apart from the editorial board, the designing team has also invested a significant amount of their time in understanding the subject and creating the most relevant covers. They scrutinized every image to scout for the most suitable representation of the subject and create an appropriate cover for the book.

The publishing team has been involved in this book since its early stages. They were actively engaged in every process, be it collecting the data, connecting with the contributors or procuring relevant information. The team has been an ardent support to the editorial, designing and production team. Their endless efforts to recruit the best for this project, has resulted in the accomplishment of this book. They are a veteran in the field of academics and their pool of knowledge is as vast as their experience in printing. Their expertise and guidance has proved useful at every step. Their uncompromising quality standards have made this book an exceptional effort. Their encouragement from time to time has been an inspiration for everyone.

The publisher and the editorial board hope that this book will prove to be a valuable piece of knowledge for researchers, students, practitioners and scholars across the globe.

List of Contributors

Volodymyr I. Lushchak and Halyna M. Semchyshyn
Vassyl Stefanyk Precarpathian National University, Ukraine

Shobha Udipi and Padmini Ghugre
Department of Food Science & Nutrition, India

Chanda Gokhale
S.P.N. Doshi Women's College & Dr. Nanavati BM College of Home Science, S.N.D.T. Women's University, India

Hillar Klandorf and Knox Van Dyke
West Virginia University, USA

Ehab M. M. Ali and Tarek M. Mohamed
Biochemistry Division, Chemistry Department, Faculty of Science, Tanta University, Tanta, Egypt

Soha M. Hamdy
Biochemistry Division, Chemistry Department, Faculty of Science, Fayoum University, Fayoum, Egypt

Sandro Argüelles, Mario F. Muñoz-Pinto, Rafael Ayala, Afrah Ismaiel and Antonio Ayala
Department of Biochemistry and Molecular Biology, Faculty of Pharmacy, Spain

Mercedes Cano
Department of Physiology, University of Seville, Spain

Hiroshi Morimatsu, Toru Takahashi, Hiroko Shimizu, Junya Matsumi, Junko Kosaka and Kiyoshi Morita
Department of Anesthesiology and Resuscitology, Okayama University Hospital, Japan

Tünay Kontaş Aşkar
Department of Biochemistry, University of Aksaray, Turkey

Olga Büyükleblebici
Vocational School, Faculty of Science, University of Çankırı Karatekin, Turkey

Shulin Liu, Xuejun Sun and Hengyi Tao
Second Military Medical University, China

Elisabete Silva and Patrício Soares-da-Silva
Department of Pharmacology and Therapeutics, Faculty of Medicine, Porto University, Portugal

Larry F. Lemanski
Department of Biological and Environmental Sciences, Texas A&M University-Commerce, Commerce, Texas, USA
Department of Anatomy and Cell Biology and The Cardiovascular Research Center, School of Medicine, Temple University, Philadelphia, Pennsylvania, USA

Andrei Kochegarov, Ashley Moses, William Lian, Jessica Meyer and Michael Hanna
Department of Biological and Environmental Sciences, Texas A&M University-Commerce, Commerce, Texas, USA

Sharon L. Lemanski
Department of Anatomy and Cell Biology and The Cardiovascular Research Center, School of Medicine, Temple University, Philadelphia, Pennsylvania, USA

Chi Zhang
Department of Radiation Oncology, New York Presbyterian Hospital, Columbia University Medical Center, New York, New York, USA

Pingping Jia
Miami Project to Cure Paralysis, University of Miami Miller School of Medicine, Miami, Florida, USA

Keith A. Webster and Yuanyuan Jia
Department of Molecular and Cellular Pharmacology, University of Miami, Miller School of Medicine, Miami, Florida, USA

Xupei Huang
Department of Biomedical Science, Florida Atlantic University; Boca Raton, Florida, USA

Mohan P. Achary
Department of Radiation Oncology, School of Medicine, Temple University, Philadelphia, Pennsylvania, USA

Herbert Weissbach
Center for Molecular Biology and Biotechnology, Charles E. Schmidt College of Science, Florida Atlantic University, Jupiter, Florida, USA

Yuejin Li
Department of Biomedical Science, Florida Atlantic University; Boca Raton, Florida, USA
Department of Pediatrics, Division of Cardiology, Johns Hopkins School of Medicine, Baltimore, Maryland, USA

Ana Remesal, Laura San Feliciano and Dolores Ludeña
Department of Paediatrics (A.R., L.SF.), Department of Cellular Biology and Pathology (D.L.), School of Medicine, University of Salamanca, Salamanca University Hospital, Spain

Jesús A. Rosas-Rodríguez and Edgar F. Morán-Palacio
Departamento de Ciencias Químico Biológicas y Agropecuarias, Universidad de Sonora Unidad Regional Sur, Navojoa, Sonora, Mexico

Hilda F. Flores-Mendoza, Ciria G. Figueroa-Soto and Elisa M. Valenzuela-Soto
Centro de Investigación en Alimentación y Desarrollo A.C., Hermosillo, Sonora, México

Maria Angeles de la Torre-Ruiz, Mima I. Petkova and Nuria Pujol-Carrion
Dept. Ciències Mèdiques Bàsiques-IRBLleida, Faculty of Medicine, University of Lleida, Spain

Luis Serrano
Dept. Producció Vegetal & Ciència Forestal, ETSEA, University of Lleida, Spain

Miguel C. Teixeira
IBB - Institute for Biotechnology and Bioengineering, Centre for Biological and Chemical Engineering, Instituto Superior Técnico, Lisboa, Portugal
Department of Bioengineering, Instituto Superior Técnico, Technical University of Lisbon, Lisboa, Portugal

Christel Goudot and Gaëlle Lelandais
Dynamique des Structures et Interactions des Macromolécules Biologiques (DSIMB), France
Univ Paris Diderot, Sorbonne Paris Cité, Paris, France
INTS, 75015, Paris, France

Frédéric Devaux
Laboratoire de Génomique des Microorganismes, Univ Pierre et Marie Curie, Paris, France

Dominika Boguszewska
Plant Breeding and Acclimatization Institute- National Research Institute Division Jadwisin, Poland

Barbara Zagdańska
Warsaw University of Life Sciences, Poland